Fourier Transform
Infrared Spectrometry

CHEMICAL ANALYSIS

A SERIES OF MONOGRAPHS ON
ANALYTICAL CHEMISTRY AND ITS APPLICATIONS

Editors

P. J. ELVING, J. D. WINEFORDNER

Editor Emeritus: **I. M. KOLTHOFF**

VOLUME 83

A WILEY-INTERSCIENCE PUBLICATION

JOHN WILEY & SONS

New York / Chichester / Brisbane / Toronto / Singapore

(*continued on back*)

Fourier Transform Infrared Spectrometry

PETER R. GRIFFITHS
Department of Chemistry
University of California
Riverside, California

JAMES A. de HASETH
Department of Chemistry
University of Georgia
Athens, Georgia

A WILEY-INTERSCIENCE PUBLICATION

JOHN WILEY & SONS

New York / Chichester / Brisbane / Toronto / Singapore

Library of Congress Cataloging in Publication Data:

Griffiths, Peter R., 1942–
 Fourier transform infrared spectrometry.

 (Chemical analysis ; v. 83)
 "A Wiley-Interscience publication."
 Includes bibliographies and index.
 1. Infrared spectroscopy. 2. Fourier transform
spectroscopy. I. de Haseth, James A. II. Title.
III. Series.

QD96.I5G743 1986   543'.08583   85-26301
ISBN 0-471-09902-3

Printed in the United States of America

10 9 8 7 6 5 4 3 2

PREFACE

The past decade has witnessed a virtual revolution in the way that chemists perceive infrared spectrometry. In the early 1970s, infrared spectrometry was commonly believed to be a dying instrumental technique, being superseded by NMR and mass spectrometry for structural determinations and by gas and liquid chromatography for quantitative analysis. Now, in the mid-1980s, the situation has changed dramatically; indeed, several surveys have found infrared spectrometry to be either the first or second most rapidly growing instrumental technique. This change is largely a result of the rapid commercial development of Fourier transform–infrared (FT–IR) spectrometry.

The development of FT–IR spectrometers began with the invention of the two-beam interferometer by Michelson almost a century ago. Shortly after the conception of the Michelson interferometer, both Michelson and Lord Rayleigh recognized that it was theoretically possible to obtain spectra from the interference pattern generated by the interferometer (now called the *interferogram*) through the computation of its Fourier transform. Michelson even built an ingenious analog computer by which an interferogram could be constructed from a given input spectrum. By comparing calculated and measured interferograms, the spectra of line sources could be deduced. At the turn of the century, however, the only reason for measuring spectra using a Michelson interferometer was its capability to resolve the fine structure of atomic lines. Interferometers were not used for the measurement of infrared spectra for another 50 years.

Around 1950, two key discoveries were made. The first was the recognition by Fellgett that information from all spectral elements is measured simultaneously with an interferometer. This *multiplex* (or Fellgett's) advantage is the fundamental theoretical advantage, not only of FT–IR spectrometers, but also of Fourier transform nuclear magnetic resonance and mass spectrometers. A short time after Fellgett's discovery, Jacquinot derived the fact that the maximum allowed solid angle of the "collimated" beam passing through an interferometer is greater than the solid angle of a beam of the same cross-sectional area at the prism or grating of a monochromator measuring at the same resolution. The Fellgett and Jacquinot advantages combine to form the fundamental basis for the improved performance of FT–IR spectrometers over monochromators.

The next step in the slow acceptance of Fourier spectrometry by the chemical community was the discovery (or more accurately rediscovery) of the fast Fourier transform (FFT) algorithm by Cooley and Tukey in 1964. Instead of requiring minutes or even hours to compute a spectrum from an interferogram using the conventional Fourier transform, the FFT permitted low-resolution spectra to be computed in a matter of seconds. Even then, the need to record interferograms on paper tapes or punched cards, transport them to the computer center, and wait until they were processed (often a matter of days rather than hours) still rendered FT–IR spectrometry unattractive to most analytical chemists.

By the late 1960s, however, two technological developments were made that, when applied to the field of FT–IR spectrometry, finally allowed the skepticism of the chemical community to be overcome. These breakthroughs were the fabrication of minicomputers and small gas lasers. The minicomputer allowed spectra to be computed in the laboratory directly after measurement of the interferogram. Helium–neon lasers are used to monitor the travel of the moving mirror of the interferometer, permitting interferograms to be digitized at precisely equal intervals and giving, in effect, an internal wavenumber standard for all measurements.

By 1975, FT–IR spectrometry had become the accepted technique for measuring high-quality infrared spectra. Nevertheless, the cost of the instruments was still sufficiently high that many more grating spectrophotometers were being sold than FT–IR spectrometers. The development of the microcomputer, together with relatively inexpensive peripheral mass memory devices, terminals, and displays, rapidly altered this picture. Several vendors are now able to sell high-quality FT–IR spectrometers for less than $50,000. Indeed, it is possible to purchase certain low-resolution instruments for less than $25,000. At the time of this writing, there are 10 companies in North America (Analect, Beckman, Bomen, Digilab, IBM Instruments, Janos Technology, Mattson, Midac, Nicolet, and Perkin-Elmer) that make FT–IR spectrometers suitable for use by chemists. Other companies in Japan and Europe also manufacture similar instruments. By contrast, only one company in North America (Perkin-Elmer) still makes a grating spectrophotometer. These data are certainly indicative of the revolution that has taken place over the past decade.

In 1974, right at the start of this revolution, one of us (P. R. G.) wrote a book entitled *Chemical Infrared Fourier Transform Spectroscopy*. The present book started its existence 8 years later as the second edition of the earlier volume. By the time we had added 11 chapters and extensively revised most of the other 8, it was obvious that the manuscript represented a new text. A few parts of the original book remain; in particular, much of the theory of the two-beam interferometer described in Chapter 1 is unchanged. In this respect, we crave the reader's indulgence when encountering any reiterated material.

In this book, we attempt to introduce the theory, instrumentation, and applications of FT–IR spectrometry in a way that can be understood and appreciated by chemists at the bachelor's or doctoral level. In introducing the theory of FT–IR spectrometry, we have steered clear of detailed mathematics whenever it seemed appropriate. Occasionally, we have opted for a simple derivation to yield an approximate answer when a more rigorous, but perhaps less comprehensible, approach is needed to give the exact solution. Our goal has been to lay a sufficient foundation to allow actual and potential users of FT–IR spectrometers to plan their experimental procedure correctly.

In an attempt to guard against obsolescence, we have not discussed all commercial spectrometers available in 1985 in detail. Rather we have described selected designs using several of the contemporary instruments as examples. If the number of spectrometers introduced in the past two years is representative, we can expect one or two novel designs to be incorporated in new instruments by the end of 1986. These designs are not, of course, described in this book. Nevertheless, we believe that readers will be able to gain an adequate background to understand the benefits and drawbacks of both the present and future spectrometer designs.

In our coverage of applications of FT–IR spectrometry, we have deliberately not included a detailed bibliography of all papers published on each topic. Several thousand papers and reports have been published over the past 20 years for which FT–IR spectrometry has played a significant role. To describe all this work would, we believe, confuse and overwhelm the reader. Instead, we have selected a few reports of measurements where Fourier spectrometry has been applied in a particularly novel or beneficial manner and have described this work in some detail. Because of this philosophy, it should therefore be recognized that many superb papers have not been referenced in this book. Readers are strongly advised to perform a literature search for each topic of particular interest to them.

Fourier transform infrared spectrometry is a dynamic instrumental technique that can be applied to an enormous variety of samples. We hope that readers of this book will have as much pleasure in working with FT–IR spectrometry as we have.

PETER R. GRIFFITHS
JAMES A. DE HASETH

Riverside, California
Athens, Georgia
January 1986

CONTENTS

CHAPTER

1

THE MICHELSON INTERFEROMETER

I. INTRODUCTION

The design of most interferometers used for infrared spectrometry today is based on that of the two-beam interferometer originally designed by Michelson in 1891 [1,2]. Many other two-beam interferometers have subsequently been designed that may be more useful for certain specific applications than the Michelson interferometer. Nevertheless, the theory behind all scanning two-beam interferometers is similar, and the general theory of interferometry is most readily understood by first acquiring an understanding of the way in which a simple Michelson interferometer can be used for the measurement of infrared spectra.

The Michelson interferometer is a device that can divide a beam of radiation into two paths and then recombine the two beams after a path difference has been introduced. A condition is thereby created under which *interference* between the beams can occur. The intensity variations of the beam emerging from the interferometer can be measured as a function of path difference by a detector. The simplest form of the Michelson interferometer is shown in Fig. 1.1. It consists of two mutually perpendicular plane mirrors, one of which can move along an axis that is perpendicular to its plane. The movable mirror is either moved at a constant velocity or is held at equally spaced points for fixed short time periods and rapidly stepped between these points. Between the fixed mirror and the movable mirror is a *beamsplitter*, where a beam of radiation from an external source can be partially reflected to the fixed mirror (at point F) and partially transmitted to the movable mirror (at point M). After the beams return to the beamsplitter, they interfere and are again partially reflected and partially transmitted. Because of the effect of interference, the intensity of each beam passing to the detector and returning to the source depends on the difference in path of the beams in the two arms of the interferometer. The variation in the intensity of the beams passing to the detector and returning to the source as a function of the path difference ultimately yields the spectral information in a Fourier transform spectrometer.

1

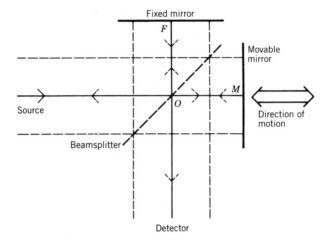

Fig. 1.1. Schematic representation of a Michelson interferometer. The median ray is shown by the solid line, and the extremes of the collimated beam are shown by the broken lines.

The beam that returns to the source is rarely of interest for spectrometry, and usually only the output beam traveling in the direction perpendicular to that of the input beam is measured. Nevertheless, it is important to remember that both of the output beams contain equivalent information. The main reason for measuring only one of the output beams is the difficulty of separating the output beam returning to the source from the input beam. In some measurements, both output beams are measured using two detectors or by focusing both beams onto the same detector. In other measurements, separate input beams can be passed into each arm of the interferometer and the resultant signal measured using one or two detectors. These measurements are generally classified under the heading of dual-beam interferometry and are described in Chapter 8.

II. MONOCHROMATIC LIGHT SOURCES

To understand the processes occurring in a Michelson interferometer better, let us first consider an idealized situation where a source of monochromatic radiation produces an infinitely narrow, perfectly collimated beam. Let the wavelength of the radiation be λ (in centimeters) and its wavenumber be $\bar{\nu}$ (reciprocal centimeters)

$$\bar{\nu} = \frac{1}{\lambda} \qquad (1.1)$$

For this example, we will assume that the beamsplitter is a nonabsorbing film whose reflectance and transmittance are both exactly 50%. We first determine the intensity of the beam at the detector when the movable mirror is held stationary at different positions.

The path difference between the beams traveling to the fixed and movable mirrors is $2(OM - OF)$, see Fig. 1.1. This optical path difference is called the *retardation*, and is usually given the symbol δ. Since δ is the same for all parallel input beams, such as the two broken lines shown in Fig. 1.1, we can relax our criterion for an infinitely narrow input beam, but ideally it should still remain collimated. The effect of having an uncollimated beam is discussed in Section VII of this chapter.

When the fixed and movable mirrors are equidistant from the beamsplitter (zero retardation), the two beams are perfectly in phase on recombination at the beamsplitter (Fig. 1.2a). At this point, the beams interfere *constructively*, and the intensity of the beam passing to the detector is the sum of the intensities of the beams passing to the fixed and movable mirrors. Therefore, all the light from the source reaches the detector at this point and none returns to the source.

If the movable mirror is displaced a distance $\frac{1}{4}\lambda$, the retardation is now $\frac{1}{2}\lambda$. The pathlengths to and from the fixed and movable mirrors are therefore exactly one-half wavelength different. On recombination at the beamsplitter, the beams are out of phase and interfere *destructively* (Fig. 1.2b). At this point, all the light returns to the source and none passes to the detector. A further displacement of the movable mirror by $\frac{1}{4}\lambda$ makes the total retardation λ. The two beams are once more in phase on recombination at the beamsplitter, and a condition of constructive interference again occurs, (Fig. 1.2c). For monochromatic radiation, there is no way to determine whether a particular point at which a signal maximum is measured corresponds to zero retardation or a retardation equal to an integral number of wavelengths.

If the mirror is moved at constant velocity, the signal at the detector will be seen to vary sinusoidally, a maximum being registered each time that the retardation is an integral multiple of λ. The intensity of the beam at the detector measured as a function of retardation is given the symbol $I'(\delta)$. The intensity at any point where $\delta = n\lambda$ (where n is an integer) is equal to the intensity of the source $I(\bar{\nu})$. At other values of δ, the intensity of the beam at the detector, or interference record, is given by

$$I'(\delta) = 0.5I(\bar{\nu}) \left\{ 1 + \cos 2\pi \frac{\delta}{\lambda} \right\} \tag{1.2}$$

$$= 0.5I(\bar{\nu})\{1 + \cos 2\pi\bar{\nu}\delta\} \tag{1.3}$$

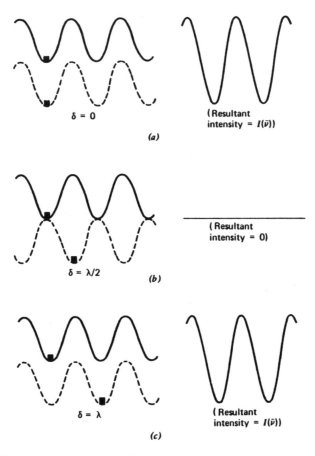

Fig. 1.2. Schematic representation of the phase of the electromagnetic waves from the fixed mirror (solid line) and movable mirror (broken line) at different values of the optical retardation. (*a*) Zero path difference; (*b*) path difference of one-half wavelength; (*c*) path difference of one wavelength. Note that constructive intereference occurs for both (*a*) and (*c*) and all other retardations of integral numbers of wavelengths.

It can be seen that $I'(\delta)$ is composed of a constant (dc) component equal to $0.5I(\bar{\nu})$, and a modulated (ac) component equal to $0.5I(\bar{\nu})$ cos $2\pi\bar{\nu}\delta$. Only the ac component is important in spectrometric measurements, and it is this modulated component that is generally referred to as the *interferogram*, $I(\delta)$. The interferogram from a monochromatic source measured with an ideal interferometer is given by the equation

$$I(\delta) = 0.5I(\bar{\nu}) \cos 2\pi\bar{\nu}\delta \qquad (1.4)$$

In practice, several factors affect the magnitude of the signal measured at the detector. First, it is practically impossible to find a beamsplitter having the ideal characteristics of 50% reflectance and 50% transmittance. The nonideality of the beamsplitter must be allowed for in Eq. 1.4 by multiplying $I(\bar{\nu})$ by a wavenumber-dependent factor less than unity representing the relative beamsplitter efficiency (see Chapter 4, Section II). Second, most infrared detectors do not have a uniform response to all wavenumbers. In addition, the response of many amplifiers is also strongly frequency dependent. It will be seen in Chapter 2 that the amplifier usually contains filter circuits designed to eliminate the signals from radiation outside the spectral range of interest from reaching the detector. It is these filters that cause the amplifier to have a frequency-dependent, and hence a wavenumber-dependent, response (*vide infra*, Chapter 2, Section 1).

In summary, the amplitude of the interferogram as observed after detection and amplification is proportional not only to the intensity of the source but also to the beamsplitter efficiency, detector response, and amplifier characteristics. Of these factors, only $I(\bar{\nu})$ varies from one measurement to the next for a given system configuration while all the other factors remain constant. Therefore, Eq. 1.4 may be modified by a single wavenumber-dependent correction factor, $H(\bar{\nu})$, to give

$$I(\delta) = 0.5H(\bar{\nu})I(\bar{\nu}) \cos 2\pi\bar{\nu}\delta \qquad (1.5)$$

where $0.5H(\bar{\nu})I(\bar{\nu})$ may be set equal to $B(\bar{\nu})$, the single-beam spectral intensity. The simplest equation representing the interferogram is therefore

$$I(\delta) = B(\bar{\nu}) \cos 2\pi\bar{\nu}\delta \qquad (1.6)$$

The parameter $B(\bar{\nu})$ gives the intensity of the source at a wavenumber $\bar{\nu}$ as modified by the instrumental characteristics.

Mathematically, $I(\delta)$ is said to be the cosine *Fourier transform* of $B(\bar{\nu})$. The spectrum is calculated from the interferogram by computing the cosine Fourier transform of $I(\delta)$, which accounts for the name given to this spectrometric technique—*Fourier transform spectrometry*.

In most commercially available Michelson interferometers, the movable mirror is scanned at a constant velocity v (centimeters per second). For these instruments, it is important to understand the way in which the interferogram varies as a function of time, $I(t)$, rather than as a function of retardation, $I(\delta)$. The retardation t seconds after the zero retardation point is given by

$$\delta = 2vt \quad \text{cm} \tag{1.7}$$

Substituting into Eq. 1.6,

$$I(t) = B(\bar{v}) \cos 2\pi\bar{v}\cdot 2vt \tag{1.8}$$

The units of the abscissa of the interferogram (retardation in centimeters or time in seconds) must always be the inverse of the units of the spectrum (wavenumber in reciprocal centimeters or frequency in Hertz, respectively).

For any cosine wave of frequency f the amplitude of the signal after a time t is given by the equation

$$A(t) = A_0 \cos 2\pi f t \tag{1.9}$$

where A_0 is the maximum amplitude of the wave. A comparison of Eqs. 1.8 and 1.9 shows that the frequency $f_{\bar{v}}$ of the interferogram $I(t)$ corresponding to radiation of wavenumber is given by

$$f_{\bar{v}} = 2v\bar{v} \tag{1.10}$$

III. POLYCHROMATIC SOURCES

In the particular case where the spectrum of a source of monochromatic radiation is to be determined, performing the Fourier transform of a measured interferogram is a trivial operation, since the amplitude and wavelength (or frequency) can both be measured directly. However, if the source emits several discrete spectral lines or continuous radiation, the interferogram is more complex and a digital computer is usually required to perform the transform.

When radiation of more than one wavenumber is emitted by the source, the measured interferogram is the resultant of the interferograms corresponding to each wavenumber. For line sources with very simple spectra, interferograms may be found that repeat themselves at regular intervals of retardation. Some simple spectra and their interferograms are shown in Fig. 1.3.

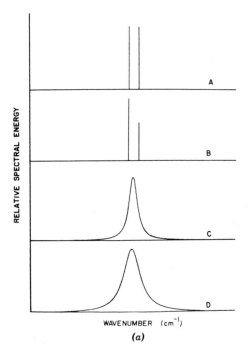

RELATIVE SPECTRAL ENERGY

WAVENUMBER (cm^{-1})

(a)

Fig. 1.3. Simple spectra and interferograms. (A) Two infinitesimally narrow lines of equal intensity. (B) Two infinitesimally narrow lines of unequal intensity; note that the amplitude of the beat signal in the interferogram never goes to zero. (C) A Lorentzian band centered at the mean of the lines in A and C; the frequency of the interferogram is identical to A and B and the envelope decays exponentially. (D) A Lorentzian band at the same wavenumber as C but of twice the width; the exponent of the decay for the interferogram has a value double that of the exponent for C. Interferograms are on the next page.

The upper curves in Fig. 1.3 represent the case when two closely spaced lines are examined. It is interesting to note that this was the situation that occurred when Michelson examined the red Balmer line in the hydrogen spectrum [3]. Although he was not able to resolve the high-frequency wave, Michelson observed the envelope, or visibility curve, of the interferogram. From this visibility curve, he concluded that the Balmer line is actually a doublet. It is now known that this deduction was correct, the separation of the two lines being 0.014 nm.

The lower spectra in Fig. 1.3 both have Lorentzian profiles and yield sinusoidal interferograms with an exponential envelope. The narrower the width of the spectral band, the greater is the width of the envelope

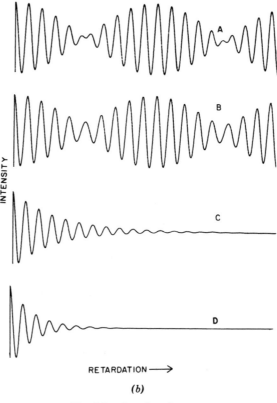

Fig. 1.3. (continued)

of the interferogram. For a monochromatic source, the envelope of the interferogram will have an infinitely large width (i.e., it will be a pure cosine wave). Conversely, for broadband spectral sources, the decay is very rapid.

When the source is a continuum, the interferogram can be represented by the integral:

$$I(\delta) = \int_{-\infty}^{+\infty} B(\bar{\nu}) \cos 2\pi\bar{\nu}\delta \cdot d\bar{\nu} \qquad (1.11)$$

which is one-half of a cosine Fourier transform pair, the other being

$$B(\bar{\nu}) = \int_{-\infty}^{+\infty} I(\delta) \cos 2\pi\bar{\nu}\delta \cdot d\delta \qquad (1.12)$$

It may be noted that $I(\delta)$ is an *even function*, so that Eq. 1.12 may be rewritten as

$$B(\bar{v}) = 2 \int_0^{+\infty} I(\delta) \cos 2\pi\bar{v}\delta \cdot d\delta \qquad (1.13)$$

Equation 1.11 shows that in theory one could measure the complete spectrum from 0 to $+\infty$ (in reciprocal centimeters) at infinitely high resolution. However, Eq. 1.13 shows that in order to achieve this, we would have to scan the moving mirror of the interferometer an infinitely long distance, with δ varying between 0 and $+\infty$ centimeters. Also, if the Fourier transform were to be performed using a digital computer, the interferogram would have to be digitized at infinitesimally small intervals of retardation. In practice, the signal must be digitized at finite sampling intervals. The smaller the sampling interval, the greater is the spectral range that can be measured; this problem will be discussed in more detail in Chapter 2. It is equally apparent that the interferogram cannot be measured to a retardation of $+\infty$ centimeters. The effect of measuring the signal over a limited retardation is to cause the spectrum to have a finite resolution. A more complete description of the effect of measuring the interferogram with a certain limiting retardation is given in the next section.

IV. FINITE RESOLUTION

It is fairly simple to illustrate conceptually how the resolution of a spectrum measured interferometrically depends on the maximum retardation of the scan. As an example, let us consider the case of a spectrum consisting of a doublet, both components of which have equal intensity. Figure 1.4a shows the spectrum, and Fig. 1.4b shows the interferogram from each line. Figure 1.4c shows the resultant of these curves. This case is equivalent to the upper curve in Fig. 1.3.

If the doublet has a separation of $\Delta\bar{v}\ (=\bar{v}_1 - \bar{v}_2)$, the two cosine waves in Fig. 1.4b become out of phase after a retardation of $0.5(\Delta\bar{v})^{-1}$, and are once more back in phase after a retardation of $(\Delta\bar{v})^{-1}$. To go through one complete period of the beat frequency, a retardation of $(\Delta\bar{v})^{-1}$ is therefore required. An interferogram measured only to half this retardation could not readily be distinguished from the interferogram of a source with the profile shown in Fig. 1.3c. The narrower the separation of the doublet, the greater is the retardation before the cosine waves become in phase. It is therefore apparent that the spectral resolution depends on the max-

$$\bar{v} = \frac{1}{\lambda}$$

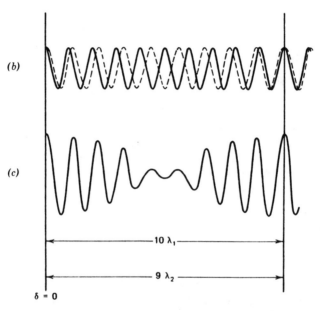

Fig. 1.4. (*a*) Spectrum of two lines of equal intensity at wavenumbers $\bar{\nu}_1$ (solid line) and $\bar{\nu}_2$ (broken line) separated by $0.1\bar{\nu}_1$. (*b*) Interferogram for each spectral line shown individually as solid and broken lines, respectively. (*c*) Resultant interferogram with the first maximum of the beat signal at $10/\bar{\nu}_1$; to resolve these two spectral lines, it is necessary to generate an optical retardation of at least this value.

imum retardation of the interferometer. Intuitively it might be concluded that the two lines could just be resolved if the retardation were increased to the point when the two waves became in phase for the first time after zero retardation. In the above discussion it was shown that the two waves become in phase for the first time after the zero retardation point when

$$\delta = (\Delta\bar{\nu})^{-1}$$

2 cm⁻¹ resolution = 5 mm maximum retardation

Thus, if the maximum retardation of an interferometer is Δ_{max}, the best resolution that could be obtained using this interferometer, $\Delta\bar{\nu}$, is given by

$$\Delta\bar{\nu} = (\Delta_{max})^{-1} \tag{1.14}$$

Although this conclusion was arrived at intuitively, the answer proves to be approximately correct. The next few paragraphs will show a more rigorous mathematical verification of this conclusion.

By restricting the maximum retardation of the interferogram to Δ centimeters, we are effectively multiplying the complete interferogram (between $\delta = -\infty$ and $\delta = +\infty$) by a truncation function, $D(\delta)$, which is unity between $\delta = -\Delta$ and $+\Delta$, and zero at all other points, that is,

$$
\begin{aligned}
D(\delta) &= 1 &&\text{if } -\Delta \leqslant \delta \leqslant +\Delta \\
D(\delta) &= 0 &&\text{if } \delta > |\Delta|
\end{aligned} \tag{1.15}
$$

In view of the shape of this function, $D(\delta)$ is often called a boxcar truncation function. By analogy to Eq. 1.12, the spectrum in this case is given by the equation

$$B(\bar{\nu}) = \int_{-\infty}^{+\infty} I(\delta)D(\delta)\cos 2\pi\bar{\nu}\delta \cdot d\delta \tag{1.16}$$

It can be shown that the Fourier transform (FT) of the product of two functions is the *convolution* of the FT of each function [4]. The effect of multiplying $I(\delta)$ by the boxcar function $D(\delta)$ is to yield a spectrum on Fourier transformation that is the convolution of the FT of $I(\delta)$ measured with an infinitely long retardation and the FT of $D(\delta)$. The FT of $I(\delta)$ is the true spectrum, $B(\bar{\nu})$, while the FT of $D(\delta)$, $f(\bar{\nu})$, is given by

$$
\begin{aligned}
f(\bar{\nu}) &= \frac{2\Delta\,\sin(2\pi\bar{\nu}\Delta)}{2\pi\bar{\nu}\Delta} \\
&\equiv 2\Delta\,\text{sinc}\,2\pi\bar{\nu}\Delta
\end{aligned} \tag{1.17}
$$

This function is shown in Figure 1.5a. It is centered about $\bar{\nu} = 0$ and intersects the $\bar{\nu}$ axis at $\bar{\nu} = n/2\Delta$, where $n = 1, 2, 3, \ldots$, so that the first intersection occurs at a wavenumber of $(2\Delta)^{-1}$.

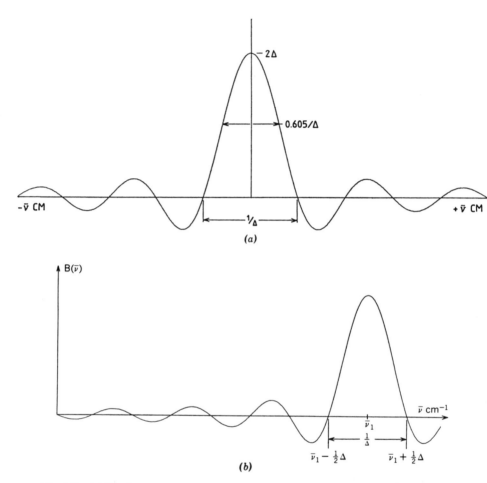

Fig. 1.5. (a) Fourier transform of a boxcar function of unit amplitude extending from $+\Delta$ to $-\Delta$; this function has the shape of a sin x/x, or sinc x, function. (b) Fourier transform of an unweighted sinusoidal interferogram generated by a monochromatic line at wavenumber $\bar{\nu}_1$, the maximum retardation for this interferogram was Δ centimeters.

Mathematically, the convolution of $B(\bar{\nu})$ and $f(\bar{\nu})$ is described by

$$G(\bar{\nu}) = B(\bar{\nu})*f(\bar{\nu})$$

$$= \int_{-\infty}^{+\infty} B(\bar{\nu}')f(\bar{\nu} - \bar{\nu}')\, d\bar{\nu}' \qquad (1.18)$$

To compute the convolution of these two functions, Eq. 1.18 requires that $f(\bar{v})$ be reversed left to right [which is trivial in this case, since $f(\bar{v})$ is an even function], after which the two functions are multiplied point by point along the wavenumber axis. The resultant points are then integrated, and the process is repeated for all possible displacements of $f(\bar{v})$ relative to $B(\bar{v})$. One particular example of convolution is familiar to most infrared spectroscopists. When a spectrum is measured on a monochromator, the true spectrum is convolved with the triangular slit function of the monochromator. The situation with Fourier transform infrared (FT–IR) spectrometry is equivalent, except that the true spectrum is convolved with the sinc function $f(\bar{v})$. Since the FT–IR spectrometer does not have any slits, $f(\bar{v})$ has been variously called the *instrument line shape* (ILS) function, the instrument function, or the apparatus function, of which we prefer the term ILS function.

When the sinc function is convolved with a single spectral line of wavenumber $\bar{v}$, the resultant curve has the formula

$$B(\bar{v}) = 2\Delta B(\bar{v}_1) \text{ sinc } 2\pi(\bar{v}_1 - \bar{v})\Delta \qquad (1.19)$$

The curve is shown in Fig. 1.5b and represents the appearance of a single sharp line measured interferometrically at a resolution considerably broader than the half-width of the line.

Since the curve intersects the wavenumber axis at $(2\Delta)^{-1}$ reciprocal centimeters on either side of $\bar{v}_1$, it can be seen that two lines separated by twice this amount, that is, by Δ^{-1} reciprocal centimeters, would be completely resolved. The result derived earlier by "intuition" is therefore indeed correct. In fact, since two lines separated by Δ^{-1} are *completely* resolved, the practical resolution is somewhat better than this value.

Several criteria have been used to define the resolution of spectrometers. The most popular are the Rayleigh criterion and the full width at half-height (FWHH) criterion. The Rayleigh criterion was used originally to define the resolution obtainable from a diffraction-limited grating spectrometer, the ILS of which may be represented by a function of the form $\text{sinc}^2 x$. Under the Rayleigh criterion, two adjacent spectral lines of equal intensity, each with a $\text{sinc}^2 x$ ILS, are considered to be just resolved when the center of one line is at the same frequency as the first zero value of the ILS of the other. Under this condition, the resultant curve has a dip of approximately 20% of the maximum intensity in the spectrum, as shown in Fig. 1.6a. However, if the same criterion is applied to a line having a sinc x ILS, it is found that the two lines are not resolved, as shown in Fig. 1.6b.

The FWHH criterion is more useful for spectrometers with a triangular slit function. Two triangularly shaped lines of equal intensity and half-

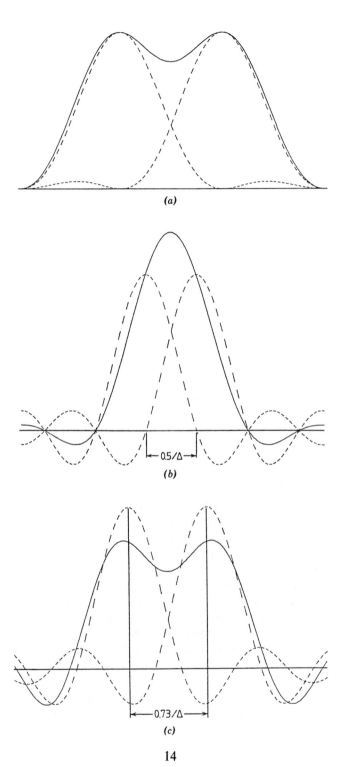

(a)

(b)

(c)

14

width are not resolved until the spacing between the lines is greater than the FWHH of either line. The FWHH of a line whose shape is a sinc function given by Eq. 1.17 is $0.605/\Delta$, but two lines with sinc x line shapes are not resolved when they are separated by this amount. In practice, a dip of approximately 20% is found when the two lines with a sinc x ILS are separated by $0.73/\Delta$, as shown in Fig. 1.6c.

The sinc x function is not a particularly useful line shape for infrared spectrometry in view of its fairly large amplitude at wavenumbers well away from $\bar{v}_1$. The first minimum reaches below zero by an amount that is 22% of the height at $\bar{v}_1$. If a second weak line happened to be present in the spectrum at the frequency of this minimum, it would not be seen in the computed spectrum. One method of circumventing the problem of these secondary minima is through the process known as *apodization*.

V. APODIZATION

From the previous section, we know that when a cosine wave interferogram is unweighted, the shape of the spectral line is the convolution of the true spectrum and a sinc function [i.e., the transform of the boxcar truncation function, $D(\delta)$]. If instead of using the boxcar function, we used a simple weighting function of the form

$$A_1(\delta) = 1 - \left| \frac{\delta}{\Delta} \right| \qquad \text{for } -\Delta \leqslant \delta \leqslant \Delta$$
$$A_1(\delta) = 0 \qquad \text{for } \delta > | -\Delta | \tag{1.20}$$

the true spectrum would be convoived with the Fourier transform of $A_1(\delta)$, and this function would therefore determine the ILS. The Fourier transform of $A_1(\delta)$ has the form

$$f_1(\bar{v}) = \Delta \frac{\sin^2(\pi\bar{v}\Delta)}{(\pi\bar{v}\Delta)^2}$$
$$= \Delta \, \text{sinc}^2(\pi\bar{v}\Delta) \tag{1.21}$$

Fig. 1.6. (*a*) The resultant of two sinc² functions centered $1/\Delta$ reciprocal centimeters apart, with each of the individual sinc² functions shown as a broken line. This condition represented Rayleigh's original definition of resolution for two equally intense lines measured using a diffraction-limited monochromator. (*b*) The resultant of two sinc functions of equal amplitude centered $0.5/\Delta$ apart; note that the features are unresolved in the resultant function. (*c*) The resultant of two sinc functions centered $0.73/\Delta$ apart; the two features are resolved with a dip of approximately 20%, equivalent to the Rayleigh criterion of resolution for a sinc² function.

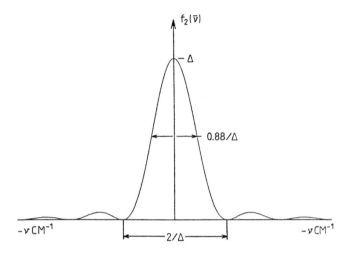

Fig. 1.7 The sinc^2 instrument line shape function computed for triangularly apodized interferograms; note that its full width at half-height is greater than that of the sinc function resulting from boxcar truncation of the same interferogram.

This function is shown in Fig. 1.7. As we saw from the discussion of the Rayleigh criterion of resolution, this is the ILS function of a diffraction-limited grating monochromator. It can be seen that the amplitude of the side lobes has been considerably reduced from that of the side lobes for the sinc function. Suppression of the magnitude of these oscillations is known as *apodization*,† and functions such as $A_1(\delta)$, which weight the interferogram for this purpose, are known as apodization functions.

The function $A_1(\delta)$ is called a *triangular* apodization function and is the most common apodization function used in infrared Fourier transform spectrometry. For lines separated by $1/\Delta$, a 20% dip would be found, as shown in Fig. 1.6a. If the lines were separated by $2/\Delta$, they would be fully resolved. The FWHH for the function $f_1(\bar{v})$ is $0.88/\Delta$, and the lines separated by this amount are just resolved; however, the dip is extremely small, being on the order of 1%.

When an absorbing sample is placed in the beam from a continuous source, the measured interferogram is the sum of the interferogram of the source with no sample present and the interferogram due to the sample. Because energy is being absorbed by the sample, these two interferograms

---

† The word apodization refers to the suppression of the side lobes (or feet) of the ILS; the word is apocryphally derived from Greek, 'ά πόδοs (without feet).

are 180° out of phase. The effect of apodization may be considered separately for the source and sample interferogram. The interferogram due to a broadband continuous source has the appearance of an intense spike at zero retardation about which can be seen modulations that die out rapidly to an unobservably low amplitude. On the other hand, the interferogram caused by a sample with narrow absorption lines will show appreciable modulation even at large retardations. When the measured interferogram is triangularly apodized, the background spectrum due to the source is usually affected imperceptibly since its interferogram has very little information at high retardation. On the other hand, the narrow absorption lines will be apodized and will have a $\mathrm{sinc}^2 x$ line shape.

Any function that has a value of unity at $\delta = 0$ and decreases with increasing retardation will serve as an apodization function. Kauppinen et al. [5] have given explicit forms for a variety of common apodization functions. These are reproduced, together with the equation for the corresponding ILS function, in Figs. 1.8a–h. Also given in these figures is the FWHH, $\Delta \bar{v}_{1/2}$, and the relative magnitude of the strongest side lobe to the central lobe, S. As a general rule, it can be seen that the narrower is $\Delta \bar{v}_{1/2}$, the greater is S. There are exceptions to this "rule". For example, if the "triangular squared" function (Fig. 1.8d) is compared to the Gaussian function (Fig. 1.8h), it can be seen that the $\Delta \bar{v}_{1/2}$ for the ILS computed from the Gaussian function is less than that for the triangular squared function and side lobe suppression is also superior.

Several general studies on apodization functions have been carried out. For example, Filler [6] investigated a variety of trigonometric functions, and Norton and Beer [7] tested over a thousand functions of the general form

$$A(\delta) = \sum_{i=0}^{n} C_i \left[1 - \left(\frac{\delta}{\Delta} \right)^2 \right]^i \qquad (1.22)$$

They found that there is a distinct empirical boundary relation between $\Delta \bar{v}_{1/2}$ and S, see Fig. 1.9. Functions could be found with a wider FWHH and larger side lobe amplitude than the boundary relation; others approached the boundary, but none was found that decreased both the FWHH and the side lobe amplitude beyond the boundary conditions. Three preferred functions, dubbed "weak," "medium," and "strong," were suggested, with all three giving ILS functions with $\Delta \bar{v}_{1/2}$ and S close to the boundary conditions. The three ILS functions are shown in Fig. 1.10, and the coefficients C_i for each are shown in Table 1.1

One other popular apodization function is the Happ–Genzel function,

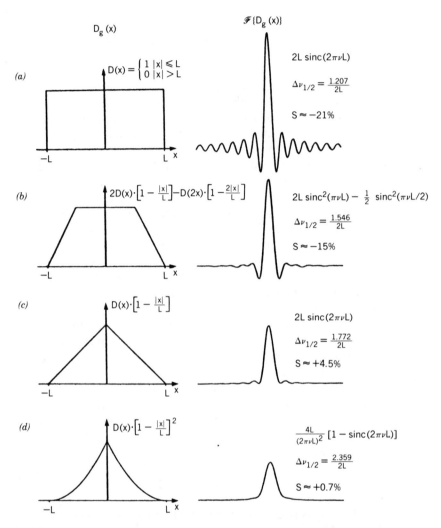

Fig. 1.8 Series of apodization functions and their corresponding instrument line shape functions; in each case the equations representing the shapes of the apodization and ILS functions are given, together with the full width at half-height, $\Delta\nu_{1/2}$, and the amplitude of the largest side lobe, S, as a percentage of the maximum excursion. (*a*) Boxcar truncation; (*b*) trapezoidal; (*c*) triangular; (*d*) triangular squared; (*e*) Bessel; (*f*) cosine; (*g*) sinc²; (*h*) Gaussian apodization. (Reproduced from [5], by permission of the Optical Society of America; copyright © 1981.)

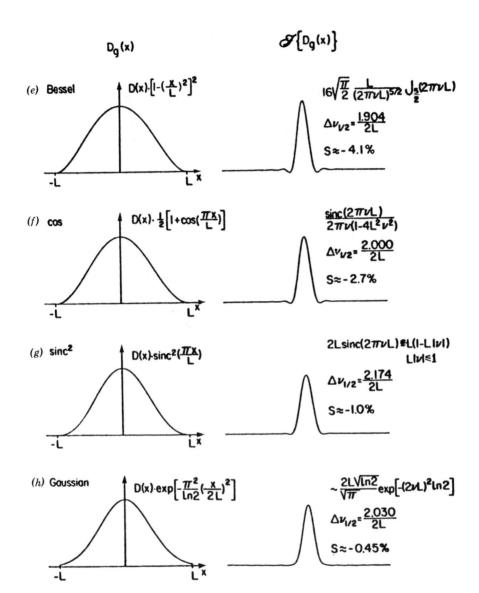

$D_g(x)$ $\qquad$ $\mathscr{F}\{D_g(x)\}$

(e) Bessel $\quad D(x)\cdot\left[1-\left(\frac{x}{L}\right)^2\right]^2$ $\qquad$ $16\sqrt{\frac{\pi}{2}}\,\frac{1}{(2\pi\nu L)^{5/2}}\,J_{\frac{5}{2}}(2\pi\nu L)$

$\Delta\nu_{1/2}=\frac{1.904}{2L}$

$S\approx-4.1\%$

(f) cos $\quad D(x)\cdot\frac{1}{2}\left[1+\cos\left(\frac{\pi x}{L}\right)\right]$ $\qquad$ $\frac{\text{sinc}(2\pi\nu L)}{2\pi\nu(1-4L^2\nu^2)}$

$\Delta\nu_{1/2}=\frac{2.000}{2L}$

$S\approx-2.7\%$

(g) sinc^2 $\quad D(x)\cdot\text{sinc}^2\left(\frac{\pi x}{L}\right)$ $\qquad$ $2L\,\text{sinc}(2\pi\nu L)*L(1-L|\nu|)$
$\qquad\qquad\qquad\qquad\qquad\qquad\qquad L|\nu|\leq 1$

$\Delta\nu_{1/2}=\frac{2.174}{2L}$

$S\approx-1.0\%$

(h) Gaussian $\quad D(x)\cdot\exp\left[-\frac{\pi^2}{\ln 2}\left(\frac{x}{2L}\right)^2\right]$ $\qquad$ $\sim\frac{2L\sqrt{\ln 2}}{\sqrt{\pi}}\exp\left[-(2\nu L)^2\ln 2\right]$

$\Delta\nu_{1/2}=\frac{2.030}{2L}$

$S\approx-0.45\%$

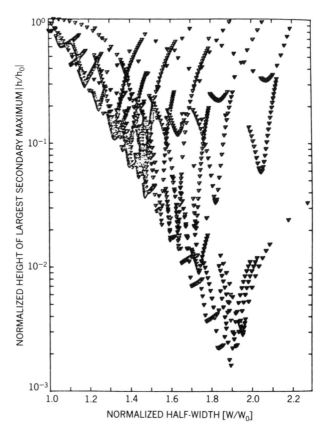

Fig. 1.9. Plot of the normalized height of the largest secondary lobe for a series of apodization functions studied by Norton and Beer (0.01S, from Fig. 1.8) as a function of their half-width relative to the half-width of the corresponding sinc function. (Reproduced from [7], by permission of the Optical Society of America; copyright © 1977.)

given by

$$A(\delta) = 0.54 + 0.46 \cos \pi \frac{\delta}{\Delta} \qquad (1.23)$$

Although the Happ–Genzel function has been recommended by the manufacturers of several commercial FT–IR spectrometers, the FWHH of

→

Fig. 1.10. The weak (F1), medium (F2), and strong (F3) apodization functions proposed by Norton and Beer. The corresponding ILS functions, I_1, I_2 and I_3, are shown, with the sinc function, I_ϕ, for reference. (Reproduced from [7], by permission of the Optical Society of America; copyright © 1977.)

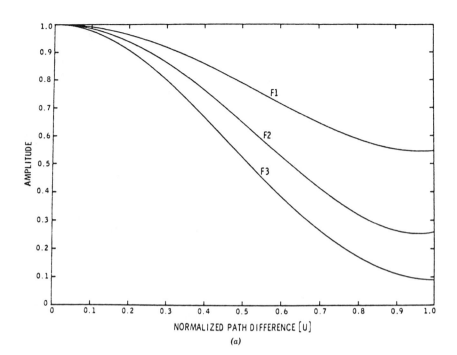

(a)

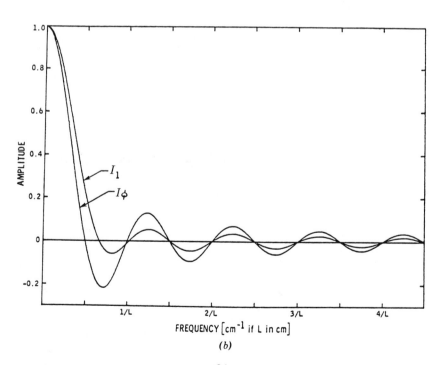

(b)

21

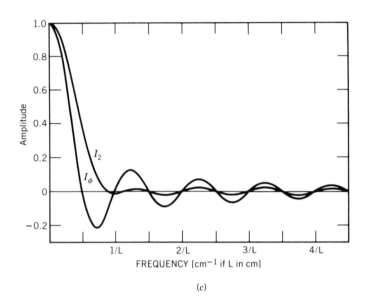

(c)

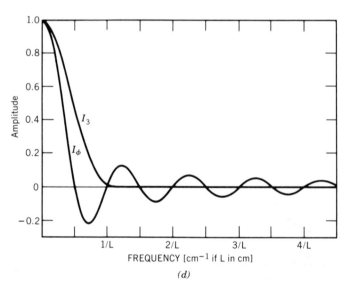

(d)

Fig. 1.10. (continued)

22

Table 1.1. Coefficients C_i for the Norton–Beer Apodization Functions

	C_0	C_1	C_2	C_3
Boxcar	1.000000	0	0	0
Weak	0.348093	−0.087577	0.703484	0
Medium	0.152442	−0.136176	0.983734	0
Strong	0.045335	0	0.554883	0.399782

the ILS functions given by the Happ–Genzel and triangular apodization functions are quite similar. The amplitude of the first side lobe of the sinc^2 ILS given by the triangular apodization function is larger than that of the ILS from the Happ–Genzel function, but this situation is reversed for subsequent lobes. It is noteworthy that the points in Fig. 1.9 that correspond to the triangular and Happ–Genzel apodization functions are both well away from the Norton–Beer boundary conditions. For most measurements, the two apodization functions give rather similar performance.

The measurement of a spectrum with a spectrometer of finite resolution distorts the true spectrum to an extent dependent on the ratio of the FWHH of the instrument line shape function to the FWHH of the band. The distortions arise because the experimentally observed intensity at any wavelength is actually the intensity of the spectrum averaged over the spectral bandpass of the spectrometer. If the intensity of radiation of wavenumber $\bar{\nu}$ incident on the spectrometer is given by $I(\bar{\nu})$, the apparent intensity recorded by the spectrometer at wavenumber $\bar{\nu}_i$ is given by the convolution of $I(\bar{\nu})$ with the ILS function:

$$I^a(\bar{\nu}_i) = \int_0^\infty I(\bar{\nu}) f(\bar{\nu} - \bar{\nu}_i)\, d\bar{\nu} \tag{1.24}$$

For absorption measurements, the intensity of radiation incident on the sample, $I_0(\bar{\nu})$, which is transmitted to the detector is given by

$$I(\bar{\nu}) = I_0(\bar{\nu}) 10^{-A(\bar{\nu})} \tag{1.25}$$

where $A(\bar{\nu})$ is the absorbance. For absorption bands with a Lorentzian shape (which is the most commonly encountered band shape in infrared spectrometry),

$$A(\bar{\nu}) = A^t_{\text{peak}} \frac{\gamma^2}{\gamma^2 + (\bar{\nu} - \bar{\nu}_0)^2} \tag{1.26}$$

where A^t_{peak} is the true peak absorbance, γ is the *half* width at half-height ($=0.5 \times$ FWHH), and $\bar{\nu}_0$ is the wavenumber of maximum absorbance.

Anderson and Griffiths [8] have considered the case of Lorentzian bands convolved with the sinc and $sinc^2$ ILS functions of an FT–IR spectrometer. The apparent (measured) peak absorbance A^a_{peak} was calculated as a function of A^t_{peak} for different values of a *resolution parameter* ρ given by

$$\rho = \frac{\text{resolution}}{\text{FWHH}} = \frac{1}{\Delta_{max} \cdot 2\gamma} \qquad (1.27)$$

These authors showed that $A^t_{peak} - A^a_{peak}$ is greater when weakly absorbing bands ($A^t_{peak} < 0.7$) are measured using triangular apodization than if the interferogram is not apodized, see Figs. 1.11 and 1.12. On the other hand, for strongly absorbing bands ($A^t_{peak} > 2$) measured using boxcar truncation with $\rho > 1.0$, it is possible that bands with an apparent

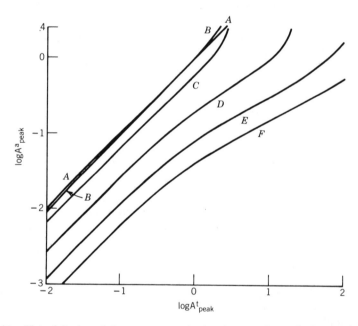

Fig. 1.11. Plot of the log of the apparent peak absorbance, A^a_{peak}, of a Lorentzian band computed with boxcar truncation as a function of the log of the true peak absorbance, A^t_{peak}, for six different values of the resolution parameter ρ: *A*, 0.0; *B*, 1.0; *C*, 3.0; *D*, 10; *E*, 25; *F*, 50. (Reproduced from [8], by permission of the American Chemical Society; copyright © 1975.)

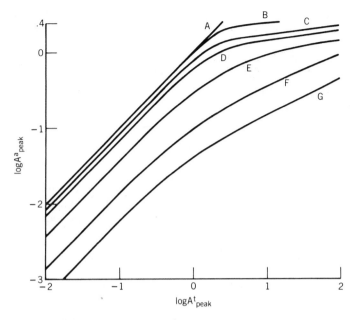

Fig. 1.12. Plot of log A^a_{peak} against log A^t_{peak} for a Lorentzian band computed with triangular apodization for seven different values of the resolution parameter ρ: A, 0.0; B, 0.1; C, 0.5; D, 1.0; E, 3; F, 10; G, 25. (Reproduced from [8], by permission of the American Chemical Society; copyright © 1975.)

peak transmittance below zero (i.e., $A^a_{peak} > \infty$) will result. In this case, the use of triangular apodization is indicated.

In general, it is advisable to use some type of apodization for FT–IR spectrometry. The specific function depends on the experiment being performed. If either high resolution or good quantitative accuracy is required, a function such as the Norton–Beer weak function can be recommended. Stronger apodization can be applied for spectra containing both weak and intense bands, especially when their width is on the same order as the instrumental resolution.

VI. PHASE ERRORS

Up to this point, it has been assumed that Eqs. 1.11 and 1.13 give an accurate representation of the interferogram. In practice, an additional term often has to be added to the *phase angle*, $2\pi\bar{\nu}\delta$, to describe the actual measured interferogram. Corrections to the phase angle may arise due to optical, electronic, or sampling effects; two common examples that lead to a change of the cosine term of these equations may be cited.

1. If we make use of the fact that the interferogram as represented in Eq. 1.11 is symmetrical about $\delta = 0$ but the first data point is actually sampled before the zero retardation point, at $\delta = -\epsilon$, the interferogram takes the form

$$I(\delta) = \int_0^{+\infty} B(\bar{\nu}) \cos 2\pi\bar{\nu}(\delta - \epsilon) \, d\bar{\nu} \qquad (1.28)$$

2. Electronic filters designed to remove high-frequency noise from the interferogram (*vide infra*) have the effect of putting a wavenumber-dependent phase lag, $\theta_{\bar{\nu}}$, on each cosinusoidal component of the interferogram, and the resultant signal is given by

$$I(\delta) = \int_0^{+\infty} B(\bar{\nu}) \cos(2\pi\bar{\nu}\delta - \theta_{\bar{\nu}}) \, d\bar{\nu} \qquad (1.29)$$

Since any cosine wave $\cos(\alpha - \beta)$ can be represented by

$$\cos(\alpha - \beta) = \cos \alpha \cdot \cos \beta + \sin \alpha \cdot \sin \beta \qquad (1.30)$$

the addition of a second term to the phase angle, $2\pi\bar{\nu}\delta$, has the effect of adding sine components to the cosine wave interferogram. Two properties of sine and cosine waves are important in this respect.

(a) $\cos \theta$ and $\sin \theta$ are orthogonal

(b) $\cos \theta - i \sin \theta = \exp(-i\theta)$ $\qquad (1.31)$

where i is the imaginary number $\sqrt{-1}$. In light of this relationship (Euler's equation), Eq. 1.29 may be rewritten in the transcendental exponential notation [9] as

$$I(\delta) = \int_0^{+\infty} B(\bar{\nu}) \exp\{-2\pi i\bar{\nu}\delta\} \, d\delta \qquad (1.32)$$

For this representation, $I(\delta)$ and $B(\bar{\nu})$ are said to be linked through the *complex* Fourier transform. To recover $B(\bar{\nu})$ from $I(\delta)$, the inverse complex transform

$$B(\bar{v}) = \int_{-\infty}^{+\infty} I(\delta) \exp\{2\pi i \bar{v}\delta\} \, d\delta \qquad (1.33)$$

is performed.

The fundamental properties of i, that is,

$$i^2 = -1 \quad \text{and} \quad i^{-1} = -i$$

allow this shorthand notation to accommodate the symmetry of the sine and cosine functions while simultaneously preserving their orthogonality. This notation also helps to explain why the component of the spectrum computed using the cosine Fourier transform is often called the *real* part of the spectrum, $\text{Re}(\bar{v})$, whereas the component computed using a sine transform is called the *imaginary* part of the spectrum, $\text{Im}(\bar{v})$. This will be discussed in greater detail in Chapter 3, Section III.

The cosine Fourier transform of a truncated sine wave has the form shown in Fig. 1.13. In general, the shape of the ILS is intermediate between this function and the sinc function resulting from the cosine transform of a truncated cosine wave. The process of removing these sine components from an interferogram, or removing their effects from a spectrum, is known as *phase correction*.

Connes [10] has discussed the effect on the ILS due to an error in the choice of the zero retardation point. She defined a parameter called phase displacement, which is the ratio of the sampling error, ϵ in Eq. 1.28, to the wavelength of the line being measured, λ. The way in which the ILS changes as the phase displacement varies between 0 (pure cosine wave) and 0.25 (pure sine wave) is reproduced in Fig. 1.14a. For even greater phase displacements, the spectrum can go completely negative (when $\epsilon/\lambda = 0.5$) and return positive when the sampling error is exactly one wavelength. Connes has shown how the ILS varies as the phase displacement varies from -1 to $+1$, and this is shown in Fig. 1.14b. This is equivalent to varying the phase angle $\theta_{\bar{v}}$ from -2π to $+2\pi$. From these diagrams, it can be seen that unless very accurate phase correction is carried out, gross photometric errors can result in the spectrum.

In Connes's treatment, the interferogram was assumed to be symmetrical and the phase angle was only considered to be due to a sampling error, so that the phase angle $\theta_{\bar{v}}$ in Eq. 1.29 was constant. In the more general case, $\theta_{\bar{v}}$ varies with wavenumber. If $\theta_{\bar{v}}$ varies linearly with

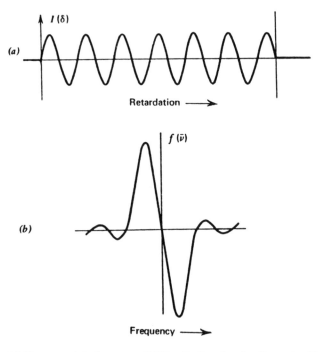

Fig. 1.13. (*a*) Sine wave interferogram. (*b*) Result of performing the cosine Fourier transform on this interferogram.

wavenumber,

$$\theta_{\bar{\nu}} = 2\pi\kappa\bar{\nu} \tag{1.34}$$

then

$$I(\delta) = \int_{0}^{+\infty} B(\bar{\nu}) \cos 2\pi\bar{\nu}(\delta - \kappa) \, d\bar{\nu} \tag{1.35}$$

and the interferogram is symmetrical about the point where $\delta = \kappa$. The position where all frequencies add constructively is called the *point of*

Fig. 1.14. (*a*) Result of performing a cosine Fourier transform on a series of cosine waves with equally spaced values of the phase shift, $\theta_{\bar{\nu}}$, ranging from zero (1) to $\frac{1}{2}\pi$ (7). (*b*) Series of ILS functions computed with phase shifts ranging from $-\pi$ (denoted as -1) to $+\pi$ (denoted as $+1$); note that there is no difference between the ILS functions computed with $\theta_{\bar{\nu}}$ equal to $-\pi$, 0, and $+\pi$. (Reproduced from [10] by permission of the author.)

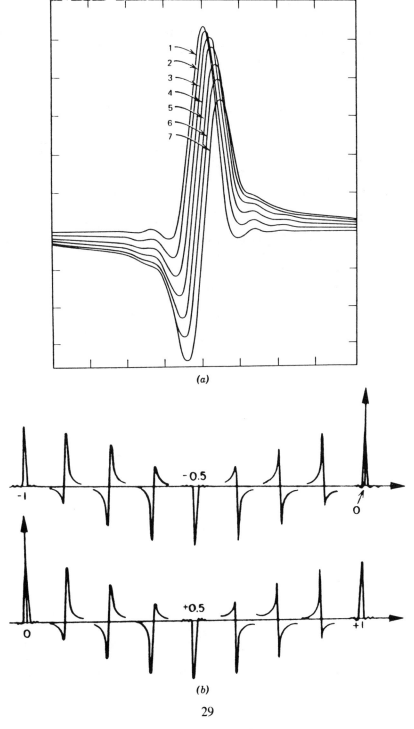

(a)

(b)

29

stationary phase [11]. It can be seen from Eq. 1.35 that this point is not necessarily equal to the zero retardation point.

When higher-order terms are present in the phase spectrum, for example,

$$\theta_{\bar{v}} = A + B\bar{v} + C\bar{v}^2 + \cdots \qquad (1.36)$$

then there is no point of stationary phase and the interferogram is said to be *chirped*. For most interferograms, the amount of chirping is small, but in a few unusual cases, $\theta_{\bar{v}}$ can vary rapidly (see Chapter 4, Section II). Figure 1.15 shows an interferogram exhibiting a small amount of chirping.

A more detailed discussion of phase correction is given in Chapter 3. At this point, it is sufficient to say that if the phase angle $\theta_{\bar{v}}$ in an interferogram measured using a continuous broadband source only varies slowly with wavenumber (as is usually the case), $\theta_{\bar{v}}$ may be computed from a region of the interferogram measured symmetrically on either side of the zero retardation point, or *centerburst*. Because it is only necessary to calculate $\theta_{\bar{v}}$ at very low resolution, data collection is started only a short distance to the left of the centerburst, at $-\Delta_1$, and continued until the desired resolution has been attained, that is, to a point such that $\delta = +\Delta_2$, where $(\Delta_2)^{-1}$ is the resolution desired. The short double-sided region of the interferogram from $-\Delta_1$ to $+\Delta_1$ is used to calculate the phase

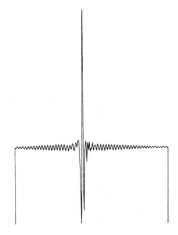

Fig. 1.15. Typical interferogram of a blackbody-type source in the region of the centerburst measured by a rapid-scanning interferometer. The slight asymmetry indicates a very small amount of chirping.

spectrum by computing the sine and cosine transforms and applying the relationship

$$\theta_{\bar{\nu}} = \arctan \frac{\text{Im}(\bar{\nu})}{\text{Re}(\bar{\nu})} \qquad (1.37)$$

In the most commonly used method of phase correction, which was developed by Mertz [12], the *amplitude spectrum* is calculated with reference to the largest data point in the interferogram. The calculated amplitude spectrum is then multiplied at each frequency by the cosine of the difference between the measured phase angle and the reference phase angle to yield the true phase-corrected spectrum.

Prior to the initial calculation of the amplitude spectrum, the original interferogram must be multiplied by a weighting function of the type shown in Fig. 1.16a. If all the points in the interferogram had been weighted equally, the region around the zero retardation point between

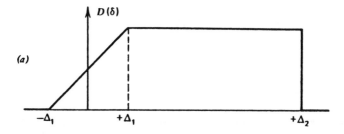

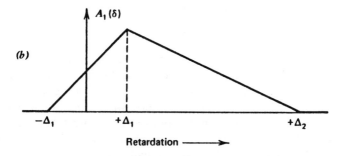

Fig. 1.16. Weighting functions employed for single-sided interferograms measured from $-\Delta_1$ (a short distance to the left of the centerburst) to $+\Delta_2$ (the full distance to the right of the centerburst needed to achieve the desired resolution) (*a*) for boxcar truncation; (*b*) for triangular apodization.

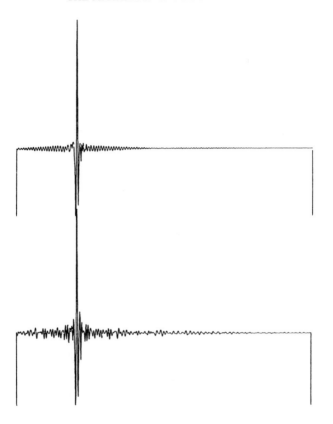

Fig. 1.17. (*Above*) Interferogram measured from a fairly well purged instrument. If an appreciable amount of water vapor had been present in the beam, modulations would be observable in the interferogram at much higher retardation. (*Below*) Interferogram measured from the same instrument with a sheet of 50-μm-thick polystyrene inserted in the beam. Note that the modulations die out at fairly low retardation because the bands of polystyrene are quite wide and do not require a large retardation to effect their resolution.

$-\Delta_1$ and $+\Delta_1$ would have been counted twice with respect to the more distant fringes, giving rise to a source of photometric error. Narrow absorption bands give rise to modulations in the interferogram that do not die out until the retardation is large, whereas the modulations caused by a broadband source occur mainly in the short, double-sided region of the interferogram (Fig. 1.17). Thus, unless the interferogram is weighted in the manner shown in Fig. 1.16a, the intensities of sharp absorption bands would appear to be about half of their true value, as shown in Fig. 1.18. The ILS found using this type of weighting function is a sinc function. If

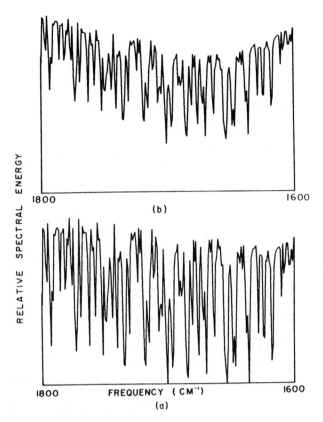

Fig. 1.18. (*Above*) Spectrum of water vapor computed from a single-sided interferogram that had not been weighted in the fashion shown in Fig. 1.16; note that none of the lines has a transmittance much less than 50% even though there is a large amount of water vapor in the beam. (*Below*) Spectrum computed from the same interferogram used for the spectrum above, but weighted with the correct apodization function.

a sinc^2 ILS function is desired, an apodization function of the type shown in Fig. 1.16b must be applied.

VII. EFFECT OF BEAM DIVERGENCE THROUGH THE INTERFEROMETER

In Section II, it was stated that the beam passing through the interferometer must be perfectly collimated. In practice, a perfectly collimated beam could only be generated from an infinitely small source, so that no signal would be measured by the detector. To measure a signal at the detector, a source of finite size must be used. In this section, the limits on source size are discussed, so that the largest possible throughput of radiation may be passed through the interferometer without degrading the spectrum in any way.

If the light from an extended monochromatic source is passed through a Michelson interferometer, a circular fringe pattern is produced in the plane of the image of the source [13] (e.g., the detector position). This effect is shown schematically in Fig. 1.19. The diameter of the rings is a maximum at zero retardation and shrinks as the retardation is increased. As the retardation is increased, the intensity of the signal at any point in this plane changes sinusoidally. Unless the aperture at a plane of the source image is limited to include only the central fringe of the interferogram, there will be no overall change of intensity with retardation and no interferogram would be able to be recorded.

This can be viewed in a simple way so that approximate quantitative

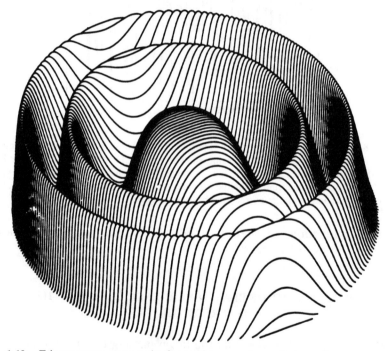

Fig. 1.19. Fringe pattern seen at the focus of a monochromatic beam from a source of finite size. Only the central light fringe must be observed in Fourier transform spectrometry or the resolution of the spectrum will be degraded. (Reproduced from [23], by permission of Plenum Publishing, copyright © 1982.)

relationships can be found. Consider the effect of a noncollimated beam of monochromatic light of wavelength λ passing through the interferometer with a divergence half-angle α. At zero retardation, the path difference between the central ray passing to the fixed and movable mirrors is zero, and there is also no path difference for the extreme rays. Thus, there will be constructive interference for both beams. Let the movable mirror now be moved a distance l (Fig. 1.20). The increase in retardation for the central ray is $2l/\cos \alpha$. Therefore, a path difference x has been generated between the central and the extreme ray, where

$$
\begin{aligned}
x &= \frac{2l}{\cos \alpha} - 2l \\
&= 2l \frac{1 - \cos \alpha}{\cos \alpha}
\end{aligned}
\tag{1.38}
$$

The series expansion for a cosine function is

$$
\cos \alpha = 1 - \frac{\alpha^2}{2!} + \frac{\alpha^4}{4!} - \frac{\alpha^6}{6!} - \cdots
$$

If α is small,

$$
1 - \cos \alpha \sim \tfrac{1}{2}\alpha^2
$$

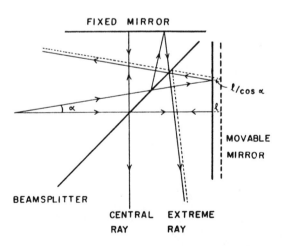

Fig. 1.20. Schematic representation of a diverging beam passing through a Michelson interferometer. The angle between the central ray and the extreme ray is α, and the physical distance moved by the mirror is l, corresponding to a nominal optical retardation of $2l$ for the central ray.

Therefore,

$$x = 2l \cdot \tfrac{1}{2}\alpha^2 = l\alpha^2 \tag{1.39}$$

since cos α is approximately equal to 1.

As l increases, the extreme ray will be out of phase with the central ray for the first time when

$$x = \tfrac{1}{2}\lambda$$

At this point, the fringe contrast at the detector completely disappears, and any further increase in l will add no further information to the interferogram (Fig. 1.21). This retardation ($2l$) gives the highest resolution $\Delta\bar{\nu}$, achievable with this half-angle α for a wavelength λ or a wavenumber $\bar{\nu}$ ($= \lambda^{-1}$). By Eq. 1.14, $(2l)^{-1}$ is equal to the resolution ($\Delta\bar{\nu}$). Fringe contrast would obviously have been lost at a shorter retardation if $\bar{\nu}$ were increased. Therefore, if a resolution of $\Delta\bar{\nu}$ is to be achieved at all wavenumbers in a spectrum whose highest wavenumber is $\bar{\nu}_{max}$, the greatest beam half-angle α_{max} that can be passed through the interferometer is given by

$$l\alpha_{max}^2 = \frac{\alpha^2}{2(\Delta\bar{\nu})} = \frac{1}{2\bar{\nu}_{max}}$$

Therefore,

$$\alpha_{max} = \left(\frac{\Delta\bar{\nu}}{\bar{\nu}_{max}}\right)^{1/2} \tag{1.40}$$

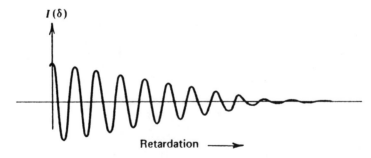

Fig. 1.21. Appearance of the interferogram of a beam of monochromatic radiation diverging rapidly as it passes through the interferometer.

The *maximum solid angle* that can be tolerated is therefore

$$\Omega_{max} = 2\pi\alpha^2 = 2\pi \frac{\Delta\bar{\nu}}{\bar{\nu}_{max}} \quad \text{steradians (sr)} \quad (1.41)$$

The effect of having a large solid angle is similar to the effect of apodization. For this reason, it has been called *self-apodization* [14].

Besides setting a limit on the resolution achievable, beam divergence also has the effect of shifting the wavenumber of a computed spectral line from its true value. Consider the interferograms due to the central and extreme rays from a monochromatic source. The wavelength of the interferogram for the extreme ray is longer than that of the central ray (Fig. 1.20). For a particular divergence half-angle α, the path difference between the central and extreme rays at a retardation Δ is given by Eq. 1.39 as

$$x = \tfrac{1}{2}\alpha^2\Delta \quad (1.42)$$

Between $\delta = 0$ and this retardation there are n maxima in the (cosine) interferogram for the central ray, where

$$\Delta = n\lambda = \frac{n}{\bar{\nu}} \quad (1.43)$$

For the extreme ray, there is an increased retardation $(\Delta + x)$, and the effective wavelength of this ray is therefore changed to a value λ', where

$$\Delta + x = n\lambda' = \frac{n}{\bar{\nu}'} \quad (1.44)$$

Combining Eqs. 1.43 and 1.44 and substituting for x from Eq. 1.42, we obtain

$$\frac{\bar{\nu}'}{\bar{\nu}} = \frac{\Delta}{\Delta + x} = \frac{1}{1 + \alpha^2/2} \quad (1.45)$$

If α is small,

$$\frac{\bar{\nu}'}{\bar{\nu}} = 1 - \tfrac{1}{2}\alpha^2$$

so that

$$\bar{\nu}' = \bar{\nu}\{1 - \tfrac{1}{2}\alpha^2\} \tag{1.46}$$

To a first approximation, the wavenumber of the line computed from this interferogram, $\bar{\nu}''$, is the mean of the wavenumbers of the central and extreme rays, that is,

$$\bar{\nu}'' \sim \frac{\bar{\nu} + \bar{\nu}'}{2} = \bar{\nu}\left\{1 - \frac{\alpha^2}{4}\right\}$$
$$= \bar{\nu}\left\{1 - \frac{\Delta\bar{\nu}}{4\bar{\nu}_{max}}\right\} \tag{1.47}$$

A more complete derivation of this wavelength shift [15,16] leads to a calculated shift of twice the magnitude given in the above equation, since the extreme rays have a larger contribution to the total signal than the central rays.

Two further conclusions may be drawn from this discussion. Most modern interferometers incorporate a He–Ne laser to permit the interferogram to be digitized at equal intervals of retardation (see Chapter 4). If the reference laser is even slightly misaligned (which it usually is), a small wavenumber shift will be introduced. Thus the great wavenumber *accuracy* of an FT–IR spectrometer may be lost, although the abscissa will still be very *precise* (i.e., repeatable). Nevertheless, since the shift always varies linearly with $\bar{\nu}$, it is a simple matter to correct for the shift by changing the value of the laser wavenumber entered into the transform programs by an appropriate amount after calibration against a suitable sample. It should also be noted that the instrument may have to be recalibrated each time the laser is changed or removed from the interferometer.

Similarly, any time the average angle of the beam passing through the interferometer is changed, the wavenumber scale will shift slightly. This effect will occur any time the diameter of an aperture at a focus is changed, for example, through the use of a sample that is smaller than the reference cell [17]. Similar effects can be observed if sampling accessories such as beam condensers are misaligned. The presence of a shift is easily detected in subtraction experiments (see Chapter 10), since absorption bands can never be precisely compensated but rather show up as a "derivative-shaped" residual in the difference spectrum. To avoid this effect, the sample and reference cells must both be larger than the beam or geometrically identical. Errors in the wavelength scale will also be introduced if the beam is deflected by wedged windows or samples [18], and more severe errors will occur if a sample of small diameter is not held at the center of the beam [17].

VIII.   EFFECT OF MIRROR MISALIGNMENT

Two effects dependent on the alignment of the mirrors in a Michelson interferometer can affect the quality of a spectrum. The first depends on the alignment of the fixed mirror relative to the moving mirror, and the second depends on how accurately the plane of the moving mirror is maintained during the scan. These effects will be discussed in this section and in the following one, respectively.

If the moving mirror is held at a different angle than the fixed mirror relative to the plane of the beamsplitter, then the image of the beam to the moving mirror will hit the plane of the detector at a different position than the beam that traveled to the fixed mirror, see Fig. 1.22. Consider the concentric rings forming the image of an extended source at the detector (Fig. 1.19). If the images from the fixed and moving mirrors are not centered at the same point on the detector, fringe contrast can be drastically reduced. Since the diameter of any ring is dependent on the wavelength of the radiation, short-wavelength radiation from a polychromatic source will be affected more than long wavelength radiation.

This effect is illustrated in Fig. 1.23. Spectra of a Globar source are shown for which the two mirrors in the interferometer are (a) well aligned, (b) in fair alignment, and (c) badly aligned. It can be seen that if the interferometer is out of alignment, the wavenumber at which the greatest intensity in the spectrum occurs is reduced. When the interferometer is badly out of alignment, all the information at high wavenumber is lost. To obtain good spectra at short wavelengths, it is vital that the mirrors of the interferometer are maintained in alignment.

For *slow-scanning* interferometers in the far-infrared (see Section X, this chapter), the interferogram is usually perfectly symmetrical when the interferometer is in good alignment. As the alignment deteriorates, not only does the amplitude of the signal at zero retardation decrease but the interferogram also becomes asymmetrical. This behavior demonstrates that the zero phase difference point no longer occurs at the same path difference for all wavelengths. For a perfectly aligned slow-scanning interferometer where all frequencies are in phase at zero retardation, the amplitude of the signal at zero retardation, $I(0)$, should be equal to twice

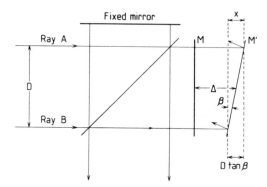

Fig. 1.22. Schematic representation to demonstrate the effect of changing the plane of the moving mirror of an interferometer during a scan. Rays A and B represent the extreme rays of the collimated beam passing through the interferometer, and β is the angle by which the plane of the moving mirror tilts.

the average value of the signal at high retardation, $I(\infty)$. The ratio R where

$$R = \frac{I(0) - I(\infty)}{I(\infty)} \tag{1.48}$$

gives a measure of how well the interferometer is aligned. For well-adjusted interferometers, R should be greater than 0.9.

For rapid-scanning interferometers, the interferogram is always slightly chirped, and so these interferometers should not be aligned using the symmetry of the interferogram as a guide. To align a rapid-scanning interferometer, the moving mirror should be scanned repetitively about the zero retardation point and the fixed mirror is adjusted to give the maximum signal. To check that the alignment is good, it is advisable to have a reference single-beam spectrum at hand taken with the same source, beamsplitter, and detector measured when it is known that the alignment was good, for example, immediately after installation of the spectrometer. Alignment at any other time is tested by ratioing the single-beam spectrum stored after realigning the interferometer against the "good" stored reference spectrum. If the ratio at low wavenumbers is approximately 100% but is substantially lower at higher wavenumbers, there is a high probability that the alignment of the interferometer is still not optimal, see Fig. 1.24. It should be noted that the same effect can also be observed if the source temperature has decreased from the time the original reference spectrum was measured, but usually this is less likely. If the ratio spec-

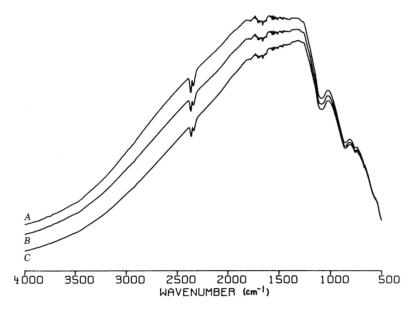

Fig. 1.23. Single-beam spectra measured (*A*) with the fixed mirror of the interferometer in good alignment with the moving mirror, (*B*) with the fixed mirror slightly out of alignment, and (*C*) with the fixed mirror well out of alignment.

trum is flat but at less than 100%, there is a good possibility that the optics (but not the interferometer) have gone out of alignment.

Alignment of the interferometer is almost invariably needed when the beamsplitter is changed. It is also needed at infrequent intervals to peak up performance when the energy at high frequency in the spectrum appears to have decreased. Alignment is generally performed by adjusting the plane of the fixed mirror so that the angle between the plane of the fixed mirror and the plane of the beamsplitter is exactly the same as that for the moving mirror.

IX.   EFFECT OF A POOR MIRROR DRIVE

The quality of the drive mechanism of the moving mirror ultimately determines whether a certain interferometer can be used to measure a spectrum with a resolution corresponding to the maximum retardation of the interferometer. The resolution of a Fourier transform spectrometer is determined by the retardation only if the mirrors maintain good alignment

throughout the entire scan and if the beam passing through the interferometer is sufficiently collimated.

The effect of a drive mechanism that does not allow the plane of the
moving mirror of the interferometer to maintain its angle relative to the
plane of the beamsplitter is somewhat analogous to the effect of beam
divergence discussed in Section VII (this chapter). In the case of the poor
mirror drive, there is an optical path difference generated between the
two extreme rays of the beam passing through the interferometer rather
than a path difference between the extreme rays and the central ray.

Let us consider the effect of a mirror tilt of β radians on a collimated
beam of radiation of wavelength λ and diameter D (in centimeters), see
Fig. 1.22. The increase in retardation, x, for the upper ray (A) over the
lower ray (B) is

$$x = 2D \tan \beta \qquad (1.49)$$

If β is small,

$$x = 2D\beta \qquad (1.50)$$

Loss of fringe modulation will start to become apparent when $x \sim 0.1\lambda$.

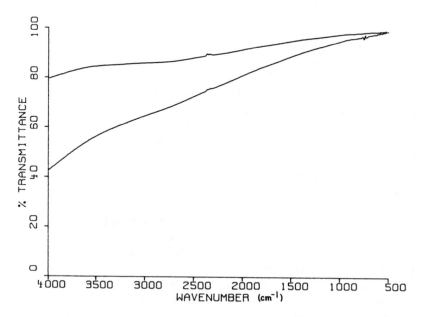

Fig. 1.24. Ratio of spectra (B) (upper trace) and (C) (lower trace) to spectrum (A) in Fig.
1.23.

Thus, to ensure that no degradation of resolution occurs during the scan,

$$x < \frac{\lambda_{min}}{10} = \frac{1}{10\bar{\nu}_{max}}$$

That is,

$$\beta < \frac{1}{20D\bar{\nu}_{max}} \tag{1.51}$$

where $\bar{\nu}_{max}$ is the highest wavenumber in the spectrum. For a mid-infrared spectrometer with $D = 5$ cm and $\bar{\nu}_{max} = 4000$ cm^{-1}, β must be therefore be less than 2.5 μrad throughout the entire scan. Misalignment of the moving mirror with respect to the fixed mirror by this amount will therefore lead to a loss of resolution. The *routine* achievement of such small tolerances over long mirror travels is one very significant factor that has led to the rapid development of FT–IR spectrometry since 1970.

If the resolution is found to have been degraded because of mirror tilt, the effect may be minimized by reducing the diameter of the beam. For example, if the desired resolution is not attained above 2000 cm^{-1} using an interferometer with $D = 5$ cm, the beam may be apertured down to a diameter of 2.5 cm. When the area of the beam is so reduced, however, the amount of energy reaching the detector is reduced in proportion to the area of the beam, that is, by a factor of 4, and the signal-to-noise ratio of the spectrum would be reduced by this amount.

X. SLOW-SCAN AND RAPID-SCAN INTERFEROMETERS

a. Slow-Scanning Interferometers

In practice, interferometers are used in two different ways depending on the scan speed of the moving mirror.

For *slow-scanning interferometers*, the velocity of the moving mirror, v, is sufficiently slow that the modulation frequency $f_{\bar{\nu}}$ of each of spectral wavenumber $\bar{\nu}$ where

$$f_{\bar{\nu}} = 2v\bar{\nu} \tag{1.52}$$

is generally less than 1 Hz. This type of system was commonly used in the late 1960s for far-infrared spectrometry, where the highest wavenumber is generally about 600 cm^{-1}. A typical scan speed is on the order of

4 μm/sec, so that the modulation frequency corresponding to the highest wavenumber in the spectrum is

$$f_{600} = 2 \times 4 \times 10^{-4} \times 600$$

$$= 0.5 \text{ Hz}$$

Far-infrared wavenumbers fall between about 6 and 600 cm^{-1} so that if the mirror velocity used in the above example were used for all measurements, the range of modulation frequencies would be between 5×10^{-3} and 5×10^{-1} Hz. Sometimes, when very far-infrared spectra are measured, scan speeds as low as 0.4 μm/sec can be used, so that a lower limit on the modulation frequency is more like 5×10^{-4} Hz.

Subaudio frequencies in this range are rather difficult to amplify without picking up a large amount of $1/f$ noise. As a result, the usual technique for measuring far-infrared spectra with a slow-scanning interferometer involves modulating the beam with a mechanical chopper. The frequency of the chopper is designed to be considerably higher than the modulation frequency of the shortest-wavelength radiation to be measured and is usually between 10 and 20 Hz. When a chopper is used to modulate the interferogram, a lock-in amplifier is often used to amplify the signal before digitization.

The signal that is measured in this way consists of both the ac and the dc components of the interferogram. Since only the ac portion is needed for the Fourier transform, the average value of the interferogram must be calculated and subtracted from the value of each sample point before the transform is performed.

For accurate sampling of the interferogram, the position of the moving mirror should be monitored by a secondary fringe-referenced device in conjunction with the interferometer drive mechanism. These devices generally involve the use of either a Moiré fringe system or a laser interferometer generating a sinusoidal signal as the mirror travels (see Chapter 4 for further details). Using these devices, the interferogram can be sampled at equal intervals of retardation rather than at equal intervals of time, so that any nonuniformity in the velocity of the moving mirror is compensated.

A lock-in amplifier tuned to the frequency of the chopper is used to effect integration of the signal between sample points. The time constant of the amplifier must be set to match the noise bandwidth to the frequency bandwidth of the signal. If noise of much higher frequency than the maximum modulation frequency $2v\bar{\nu}_{max}$ is not filtered out, the noise level in the computed spectra would increase.

b. Stepped-Scan Interferometers

Several interferometers of the slow-scanning variety use a stepping-motor drive. In stepped-scan interferometers, the movable mirror is held stationary at each sampling and then moved rapidly to the next sampling position, as shown in Fig. 1.25. The interferogram signal is produced by integrating the detector output signal during the time interval that the mirror is held stationary. The time interval during which the mirror is stepped from one sampling position to the next is lost time that does not contribute to the observation and therefore should be kept short relative to the integration periods.

The success of the technique depends on the capability of the stepping motor to achieve a uniform increase in retardation for every step. For far-infrared spectrometry at low resolution, the technique works satisfactorily with a commercial stepping motor using no additional optics or

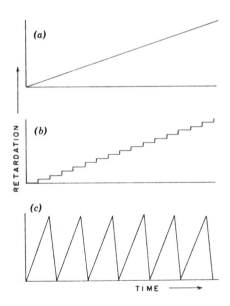

Fig. 1.25. Variation of retardation with time for (*a*) a continuous slow-scanning interferometer, (*b*) a stepped-scan interferometer, (*c*) a rapid-scanning interferometer.

electronics. However, the positioning accuracy required for high-resolution mid- and near-infrared spectrometry is such that the mirror position has to be controlled by a servomechanism actuated after every step by a laser reference interferometer (see Chapter 4).

c. Rapid-Scanning Interferometers

The difference between *rapid-scanning* interferometers and slow-scanning and stepped-scan interferometers is that the mirror velocity of the interferometer is sufficiently high that each wavenumber is modulated in the audio-frequency range. It will be shown in Chapter 2 that rapid scanning is mandatory for the successful measurement for low- or medium-resolution mid-infrared absorption spectra, where the interferogram has an extremely high signal-to-noise ratio (SNR).

A typical mirror velocity used for the measurement of spectra between 4000 and 400 cm^{-1} is 0.158 cm/sec. The modulation frequencies for the upper and lower limit of this range are, therefore,

$$f_{4000} = 2 \times 0.158 \times 4000$$

$$= 1260 \text{ Hz}$$

$$f_{400} = 126 \text{ Hz}$$

Since these modulation frequencies (sometimes called the Fourier frequencies) are in the audio range, they can be easily amplified without the necessity for modulating the beam with a chopper. (If such a chopper were used, it would have to modulate the beam at a considerably higher frequency than the modulation frequency of the shortest wavelength being measured, i.e., at least 10 kHz.)

The signal from the detector of a rapid-scanning Fourier transform spectrometer must be amplified using a bandpass filter. In this way, noise of much higher frequency than the modulation frequency of the shortest wavelength in the spectrum is eliminated, while slow variations in source intensity fall below the lower limit of the bandpass. The measured signal represents the ac component of $I'(\delta)$, so that there is no need to subtract the mean value of the interferogram from each sample point.

To increase the SNR of interferograms measured on a rapid-scanning interferometer, signal-averaging techniques must be used. All corresponding data points in each successive scan must be digitized at exactly the same retardation value. Some commercial instruments use a combined white-light and laser fringe-referenced interferometer to achieve this *coherent addition* (see Chapter 4 for details).

Some of the advantages and disadvantages of rapid-scan and slow-scan interferometry are summarized below.

Rapid Scan	Slow Scan
(i) The SNR to each interferogram is sufficiently low that mid-infrared spectra can be measured using intense sources by signal averaging.	SNR is too high for accurate digitization of the noise level in interferograms of intense sources, so that low-resolution mid-infrared absorption spectra cannot be measured efficiently.
(ii) Rather high scan speeds are necessary to generate Fourier frequencies >10 Hz for very long wavelengths in the far-infrared.	Well suited to far-infrared spectrometry.
(iii) Observation efficiency is reduced by the time taken to retrace or reverse the moving mirror during signal averaging.	Observation efficiency for step-and-integrate systems is reduced by the time taken to increment the retardation between sample points.
(iv) Unaffected by slow variations in source intensity outside the range of Fourier frequencies imposed by the interferometer.	Slow variations in the source intensity are measured, resulting in degradation of spectra, especially at low wavenumber.
(v) No chopper needed; therefore, all the radiation hits the detector all of the time.	A chopper is generally needed (except when phase modulation techniques are used, *vide infra* Section IX, this chapter); therefore, on the average, half the signal is lost.
(vi) Necessitates the use of rapid-response detectors.	Can use detectors with a long response times.
(vii) Only one scan speed need be used, and therefore the relationship between SNR measurement time and resolution (the "trading rules") is easily calculated (see Chapter 7).	The trading rules are more difficult to calculate. When the scan speed is changed, other instrumental parameters (e.g., the filter time constant) must also be changed.

(*continued*)

Rapid Scan	Slow Scan
(viii) Real-time computations (see Chapter 3, Section V) are very difficult to perform at the current state of the art.	Real-time computations can be performed.
(ix) Interferograms are usually asymmetric, and phase correction must be performed.	Interferograms are usually symmetric so that phase correction may be unnecessary.
(x) Not applicable to ultrahigh resolution ($\Delta \nu < 0.01$ cm^{-1}) measurements without dynamic alignment (see Chapter 4).	Have been successfully applied to high-resolution measurements (see Chapter 4).

In summary, the choice of interferometer should be made on the basis of the measurements to be made. For low- or medium-resolution mid-infrared absorption spectrometry, the use of a rapid-scan system is mandatory when the energy reaching the detector is high. On the other hand, if very high resolution measurements are desired, other types of interferometers may also be used. For far-infrared spectrometry, any one of the three types of interferometer can be used.

XI. PHASE MODULATION

One of the advantages of rapid-scanning interferometry over slow-scanning interferometry is that no chopper is used to modulate the radiation. Since the detector sees the beam for the entire measurement of a rapidly scanned interferogram, the SNR of the spectrum is correspondingly greater. Another disadvantage of slow-scanning interferometers is that slow variations in the intensity of the source can result in variations of the baseline of the interferogram which can be of the same frequency as the modulations from the longest wavelength being measured. This effect can give very poor photometric performance at low wavenumber and is often noticed in far-infrared spectrometry when mercury arc sources are used. These sources are rather unstable and their intensity can drift quite significantly. The effect is also prevalent in astronomical spectrometry where the intensity of distant sources may change slowly due to passing clouds, poor following, fluctuations of atmospheric transparency, or fluctuations of refractive index (scintillation). One method of getting around the problem of slowly varying sources used in conjunction with a slow-scanning interferometer, while at the same time picking up approximately a factor of 2 in SNR by eliminating the chopper, is to use a technique known as *phase modulation* (PM).

This technique has been independently developed by Chamberlain's group [19,20] for far-infrared spectrometry and by Connes et al. [21,22] for near-infrared measurements. The method (called internal modulation by Connes) requires the use of a slow-scanning interferometer often employing the step-and-integrate method of traversing the moving mirror. At each step, the beam is modulated not by a chopper [called *amplitude modulation* (AM) by Chamberlain] but by periodically varying the retardation by a low-amplitude oscillation of the fixed mirror.

To illustrate the difference between amplitude and phase modulation, let us consider an idealized boxcar spectrum of the type shown in Fig. 1.26. The interferogram when AM is used for the measurement is a sinc function of the type shown in Fig. 1.27a. However, if PM is used, the measured signal represents the *change* of this AM signal over a small increment of retardation. If the amplitude of the jitter were infinitesimally small, the signal would be the first derivative of the AM interferogram. In practice, even with finite jitter amplitudes, the PM interferogram is to a very good approximation the first derivative of the AM interferogram. Thus, the PM interferogram of the spectrum shown in Fig. 1.26 would have the form shown in Fig. 1.27b.

Consider the case of a monochromatic source. The AM interferogram

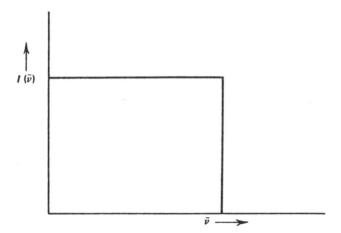

Fig. 1.26. An idealized boxcar or "white" band-limited spectrum.

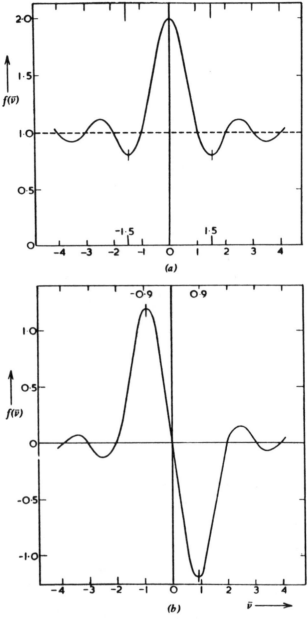

Fig. 1.27. (*a*) Interferogram of a boxcar spectrum calculated using amplitude modulation (in the absence of any phase error). (*b*) Corresponding interferogram calculated using phase modulation. Note the similarity between these interferograms and the *spectra* shown in Figs. 1.13 and 1.14. (Reproduced from [19], by permission of Pergamon Press; copyright © 1971.)

is given by Eq. 1.3:

$$I'(\delta) = 0.5I(\bar{v})\{1 + \cos 2\pi\bar{v}\delta\} \tag{1.3}$$

The derivative of this gives the PM interferogram:

$$\frac{d}{d\delta}[I'(\delta)] = 0.5I(\bar{v}) \sin 2\pi\bar{v}\delta \tag{1.53}$$

Thus, in the absence of phase errors, the spectrum can be computed from an interferogram measured using PM by the sine transform.

In practice, interferograms measured using AM and PM have a shape quite similar to those shown in Fig. 1.27, and typical results for the far-infrared region are shown in Fig. 1.28. To perform these measurements, Chamberlain and Gebbie [20] replaced the moving mirror by a "jittered" mirror, which is a thin plane mirror mounted on the plunger of a small loudspeaker coil. The sinusoidal signal needed to drive the mirror was obtained by a sine wave generator. Experimentally, the zero retardation position is first found using a chopper; then the chopper is stopped and the jitter is started using a frequency chosen to be the same as that of the chopper to allow valid comparisons to be made between the AM and PM measurements. Since the magnitude of the PM interferogram is zero at zero retardation, the position of the movable mirror must be adjusted to the position where the PM interferogram has its maximum value. The amplitude of the jitter is then increased until the greatest signal at the detector is developed. The amplitude should be approximately one-quarter of the central wavelength in the spectral range being measured (which is about one-half of the shortest wavelength in the range). For good performance, it is essential that the equilibrium position of the vibrator remains unchanged during the measurement, that the amplitude of the jitter remains constant, and that the tilt of the vibrating mirror does not alter.

Figure 1.28 shows AM and PM interferograms of the same source measured by Chamberlain and Gebbie [20], and Fig. 1.29 shows the transform of these interferograms shown on the same (arbitrary) scale when the interferograms were recorded with the same gain. It can be seen that the amplitude of the spectrum measured using PM is about twice as great as that of the spectrum measured using AM, which demonstrates the factor of 2 gain obtained on removing the chopper. Although it is rather difficult to see from these spectra, the noise in each spectrum is essentially the same.

The efficiency of PM measurements relative to AM measurements falls

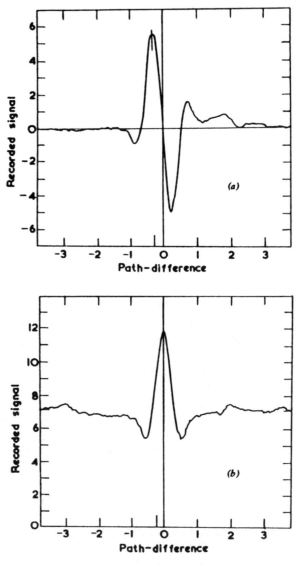

Fig. 1.28. Interferograms of continuous far-infrared sources measured using (*a*) amplitude modulation and (*b*) phase modulation. (Reproduced from [20], by permission of Pergamon Press; copyright © 1971.)

52

off at the extremes of the spectral range being measured. This effect is predominantly due to the fact that modulation of a cosine function (in this case the interferogram) by another cosine function (the modulation of the mirror) generates Bessel coefficients. Since only the lowest-order coefficient is detected, energy is lost in the higher-order coefficients.

It will be seen in Chapter 4, Section II, that the profile of the spectrum in Fig. 1.29 is typical of spectra measured with stretched-film beamsplitters—the type most commonly used for far-infrared spectrometry. These beamsplitters can be used to measure spectra over a wavenumber range 0 to $\bar{\nu}_M$. However, the energy at the low and high wavenumber extremes of this range is always low relative to the central region, thus limiting the useful range to about $0.15\bar{\nu}_M$–$0.85\bar{\nu}_M$. Chamberlain [19] examined the ratio R of the theoretical signal-to-noise ratio in spectra measured using PM and AM when the only source of noise in the interferogram is detector noise; the results are shown in Fig. 1.30. It is seen that R is greater than unity for $0.13\bar{\nu}_M < \bar{\nu} < 0.87\bar{\nu}_M$. Thus, for the total useful range of the far infrared, using stretched-film beamsplitters, PM does indeed give a decided advantage over AM techniques. This result has been demonstrated experimentally by Chamberlain and Gebbie [20].

An experimental comparison between PM techniques with a slow-scanning interferometer and AM techniques with a rapid-scanning interferometer has not yet been made, but the two methods should give rather

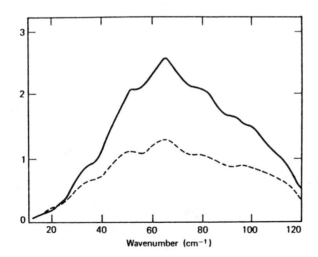

Fig. 1.29. Spectra computed from the interferogram shown in Fig. 1.28 measured with amplitude modulation (broken line) and the interferogram measured with phase modulation (solid line). (Reproduced from [20], by permission of Pergamon Press; copyright © 1971.)

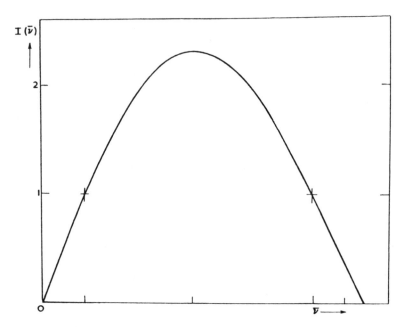

Fig. 1.30. Relative signal-to-noise advantage of phase modulation relative to amplitude modulation as a function of wavenumber for a far-infrared spectrum measured with a Mylar beamsplitter. The characteristics of Mylar beamsplitters are discussed in Chapter 4, but it can be stated here that the curve shown in this figure follows the efficiency curve for Mylar beamsplitters shown in Fig. 4.8. (Reproduced from [19], by permission of Pergamon Press; copyright © 1971.)

similar results in terms of signal-to-noise ratio since neither technique uses a chopper.

Since the mirror jitter has a frequency greater than 1 Hz, the effect of low-frequency source variations is essentially eliminated using PM techniques so that the low-wavenumber performance should be substantially superior to that of the corresponding AM measurement using a slow-scanning interferometer. One other advantage is found that is related to the same cause. Since a PM interferogram has no dc component, it has a more useful dynamic range for digitization so that it is easier to extract all the information required for the spectrum from a given interferogram. Both of these advantages of PM over AM interferometry using a slow-scanning interferometer have also been shown to hold true for a rapid-scan over a slow-scan interferometer, so that the advantage of rapid-scanning interferometry over PM methods (or vice versa) must be decided on a case-by-case basis.

REFERENCES

1. A. A. Michelson, *Phil. Mag.* (5), **31**, 256 (1891).
2. A. A. Michelson, *Light Waves and Their Uses*, University of Chicago Press, Chicago (1902); reissued in the paperback Phoenix Edition (1961).
3. A. A. Michelson, *Studies in Optics*, University of Chicago Press, Chicago (1927); Phoenix Edition (1962).
4. R. Bracewell, *The Fourier Transform and its Applications*, McGraw-Hill, New York (1965).
5. J. K. Kauppinen, D. J. Moffatt, D. G. Cameron, and H. H. Mantsch, *Appl. Opt.*, **20**, 1866 (1981).
6. A. H. Filler, *J. Opt. Soc. Am.*, **54**, 762 (1964).
7. R. H. Norton and R. Beer, *J. Opt. Soc. Am.*, **66**, 259 (1976); *erratum J. Opt. Soc. Am.*, **67**, 419 (1977).
8. R. J. Anderson and P. R. Griffiths, *Anal. Chem.*, **47**, 2339 (1975).
9. C. T. Foskett, in *Transform Techniques in Chemistry* (P. R. Griffiths, ed.), Plenum Publishing, New York (1978).
10. J. Connes, *Rev. Opt.*, **40**, 45, 116, 171, 233 (1961); English translation as Document AD 409869, Clearinghouse for Federal Scientific and Technical Information, Cameron Station, VA.
11. L. Mertz, *Transformations in Optics*, Wiley, New York (1965).
12. L. Mertz, *Infrared Phys.*, **7**, 17 (1967).
13. E. V. Loewenstein, Aspen Int. Conf. on Fourier Spectrosc., 1970 (G. A. Vanasse, A. T. Stair, and D. J. Baker, eds.), AFCRL-71-0019, p. 3 (1971).
14. E. G. Codding and G. Horlick, *Appl. Spectrosc.*, **27**, 85 (1973).
15. J. Chamberlain, *Principles of Interferometric Spectroscopy*, Wiley-Interscience, New York (1979).
16. A. E. Martin, Infrared Interferometric Spectrometers, in *Vibrational Spectra and Structure*, Vol. 8, (J. R. Durig, ed.), Elsevier, Amsterdam (1980).
17. T. Hirschfeld, Chapter 6 in *Fourier Transform Infrared Spectroscopy: Applications to Chemical Systems*, Vol. 2 (J. R. Ferraro and L. J. Basile, eds.), Academic Press, New York (1979).
18. T. Hirschfeld, *Anal. Chem.*, **51**, 495 (1979).
19. J. Chamberlain, *Infrared Phys.*, **11**, 25 (1971).
20. J. Chamberlain and H. A. Gebbie, *Infrared Phys.*, **11**, 57 (1971).
21. J. Connes, P. Connes, and J. P. Maillard, *J. Phys.*, **28**, C2:120 (1967).
22. J. Connes, H. Delouis, P. Connes, G. Guelachvili, J. P. Maillard, and G. Michel, *Nouv. Rev. Opt. Appl.*, **1**, 3 (1970).
23. R. J. Nordstrom, Aspects of Visible/UV Fourier Transform Spectroscopy, in *Fourier, Hadamard, and Hilbert Transforms in Chemistry* (A. G. Marshall, ed.), Plenum Publishing, New York (1982).

CHAPTER

2

SAMPLING THE INTERFEROGRAM

I. SAMPLING FREQUENCY

It was seen from Eqs. 1.11 and 1.12 that in order to compute the complete spectrum from 0 to ∞ reciprocal centimeters, the interferogram would have to be sampled at infinitesimally small increments of retardation. That is, of course, impossible, as an infinite number of data would be collected that would exhaust computer storage space. Even if these data could be collected, the Fourier transform would take forever to be computed. Obviously, interferograms must be sampled discretely. Just how often the interferogram should be sampled is a problem that has been solved mathematically.

Any waveform that is a sinusoidal function of time or distance can be sampled unambiguously using a sampling frequency greater than or equal to twice the bandwidth of the system [1], known as the *Nyquist Criterion.* The signal may then be effectively recorded *without any loss of information.* Consider the case of a spectrum in which the highest frequency reaching the detector is $\bar{\nu}_{max}$. The frequency of the cosine wave in the interferogram corresponding to $\bar{\nu}_{max}$ is $2v\bar{\nu}_{max}$ (from Eq. 1.10), and therefore the interferogram must be digitized at a frequency of $4v\bar{\nu}_{max}$ Hertz or once every $(4v\bar{\nu}_{max})^{-1}$ seconds. This is equivalent to digitizing the signal at retardation intervals of $(2\bar{\nu}_{max})^{-1}$ centimeters. It is better to design a system where the signal is sampled at equal intervals of retardation rather than at equal intervals of time because if the mirror velocity varies slightly during the scan, the signal is still sampled at the correct interval. Using this type of sampling, only gross deviations in mirror velocity can cause deleterious effects in the computed spectrum [2].

It is a property of Fourier transform mathematics that multiplication in one domain is equivalent to convolution in the other. (Convolution has already been introduced with regard to apodization, Chapter 1, Section V.) If we sample an analog interferogram at regular discrete retardations, we have in effect multiplied the interferogram by a repetitive impulse function. The repetitive impulse function is in actuality an infinite series

56

of Dirac delta functions spaced at an interval $1/x$. That is,

$$\text{Ш}_{1/x}(\delta) = \frac{1}{x} \sum_{n=-\infty}^{\infty} \Delta \left(\delta - \frac{n}{x} \right) \qquad (2.1)$$

where Δ is the Dirac delta function and $\text{Ш}_{1/x}(\delta)$† is the Dirac delta comb as a function of retardation δ. Figure 2.1a shows a Dirac delta comb, where the up arrows indicate the impulses. The Fourier transform of the Dirac delta comb is another Dirac delta comb of period x as opposed to $1/x$, that is,

$$\text{Ш}_x(\bar{\nu}) = x \sum_{-\infty}^{\infty} \Delta(\bar{\nu} - nx) \qquad (2.2)$$

Figure 2.1b illustrates the Dirac delta comb as a function of wavenumber, $\bar{\nu}$. The function in Eq. 2.2 is an infinite series. We multiply the analog interferogram with the Dirac delta comb of Eq. 2.1, and consequently, the Fourier transform of the interferogram (i.e., the spectrum) is convolved with the transformed comb. The effect of this convolution is to repeat the spectrum ad infinitum. If the spectrum covers the bandwidth 0 to $\bar{\nu}_{max}$, the transformed Dirac delta comb must have a period of at least $2\bar{\nu}_{max}$; otherwise the spectra will overlap as a result of the convolution. In other words,

$$x > 2\bar{\nu}_{max} \qquad (2.3)$$

and

$$\frac{1}{x} \le (2\bar{\nu}_{max})^{-1} \qquad (2.4)$$

This process is shown in Fig. 2.2. Figure 2.2a shows a discretely sampled sinc^2 function sampled at a retardation of $(2\bar{\nu}_{max})^{-1}$, and Fig. 2.2b shows its Fourier transform. As can be seen clearly, there is no overlap of the resulting spectra (triangular functions, in this case). If the sampling frequency were decreased so that the signal was sampled at $(\bar{\nu}_{max})^{-1}$, the transformed spectra, which are necessarily convolved with the spectral domain Dirac delta comb, would overlap and spectral features would

† Ш is the cyrillic character *shah*.

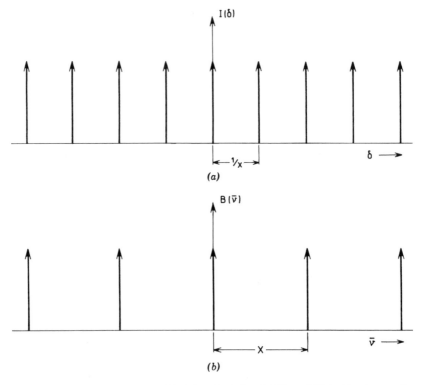

Fig. 2.1. (*a*) Dirac delta comb (or Shah function) of period $1/x$ and (*b*) its Fourier transform, another Dirac delta comb of period x.

appear at incorrect wavenumbers. This phenomenon is known as *aliasing* or *folding*.

An alternate description of folding is as follows: If the sampling frequency is equivalent to $2\bar{\nu}_{max}$ and there is a frequency in the signal equivalent to $\bar{\nu}_{max} + \bar{\nu}_1$, the feature at this frequency is computed to occur at $\bar{\nu}_{max} - \bar{\nu}_1$ in addition to its true frequency. Figure 2.3*a* shows a signal $\bar{\nu}_{max}$, sampled at its appropriate frequency $2\bar{\nu}_{max}$. For clarity, the sampling is performed at each zero value of the sinusoidal wave, and the sampling is thus exactly twice the frequency, as required. However, if a second signal at $\bar{\nu}_{max} + \bar{\nu}_1$ is sampled as shown in Fig. 2.3*b*, data are collected less than once every half wavelength. If another sinusoid is drawn at a frequency $\bar{\nu}_{max} - \bar{\nu}_1$, as shown in Fig. 2.3*c*, this curve fits the sampled data from Fig. 2.3*b* perfectly. Therefore, the sampled waveform will pro-

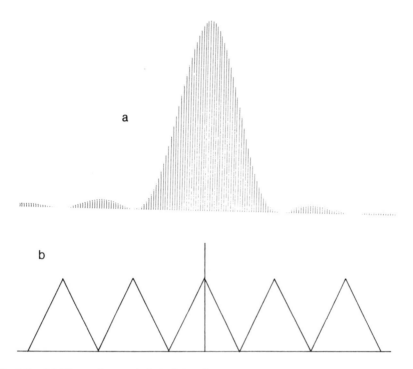

Fig. 2.2. (*a*) Discretely sampled sinc^2 function and (*b*) its Fourier transform, a series of triangle functions.

duce a feature at $\bar{\nu}_{max} - \bar{\nu}_1$ after computation, although there may never have been a feature at that frequency originally.

It can also be seen that a higher-frequency wave (in fact, many higher-frequency waves) can be drawn through the data points in Fig. 2.3*b* (see Fig. 2.4). It is a property of discrete Fourier transform mathematics that the true wavelength of the signal cannot be determined unequivocally unless precautions are taken to ensure that all frequencies are less than $\bar{\nu}_{max}$. It is therefore most important to limit either the range of frequencies reaching the detector by means of optical filters or the bandwidth for which the detector or amplifier has any response.

If the sampling frequency is $2F$, all the frequencies below F will be transferred through the sampling process unambiguously. On the other hand, a feature whose frequency is higher than F will appear at a frequency below F. Figure 2.5 shows a spectrum in schematic form where $2F$ is the sampling frequency [3]. Two features, L' and M', show where

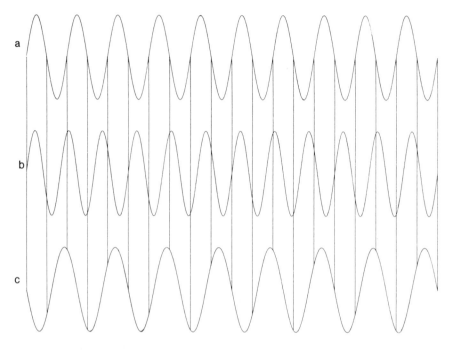

Fig. 2.3. (*a*) A sine wave sampled at exactly twice its frequency. (*b*) A higher-frequency wave sampled at the same frequency as in (*a*). (*c*) A sine wave of lower frequency than the wave in (*a*) and sampled at the same frequency. Note that the sampling amplitudes in (*b*) and (*c*) are identical.

these features are also computed to occur because of folding. It can be seen that there is no way to distinguish which of the features is real and which arises because of folding. Thus, if the sampling frequency is $2F$, any frequency region in bandwidth F can be examined.

This statement has great importance for high-resolution spectrometry. By limiting the bandwidth of the spectrum with an optical filter, a lower sampling frequency can be used than if the complete spectrum from zero to $\bar{\nu}_{max}$ were measured. This cuts down the number of data points that has to be collected for a given retardation and the computer time required for the Fourier transform. However, the collection time is not reduced as the full retardation of the interferometer at that resolution must still be used. When a small computer is being used for data collection and computation, restriction of the spectral range and the use of a longer sampling interval than $(2\bar{\nu}_{max})^{-1}$ often allows measurements to be taken at a high resolution and a relatively short computation time.

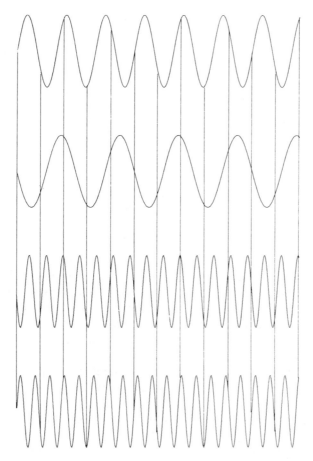

Fig. 2.4. Multiple sine waves, all with identical amplitudes at the sampling frequency shown in Fig. 2.3.

In summary, when all the spectrum from zero to $\bar{\nu}_{max}$ is required at a resolution of $\Delta\bar{\nu}$, the number of points to be sampled, N_s, is given by

$$N_s = \frac{2\bar{\nu}_{max}}{\Delta\bar{\nu}} \tag{2.5}$$

If the spectral range is restricted to fall between a minimum wavenumber $\bar{\nu}_{min}$ and a maximum wavenumber $\bar{\nu}_{max}$, the number of points required is

$$N_s = \frac{2(\bar{\nu}_{max} - \bar{\nu}_{min})}{\Delta\bar{\nu}} \tag{2.6}$$

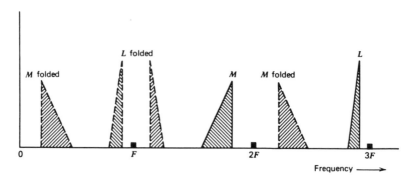

Fig. 2.5. Regions of folding: Two spectral features L and M having frequencies greater than F, the maximum allowable frequency under the sampling conditions, are computed to occur at frequencies less than F. (Reproduced from [3], by permission of Marcel Dekker, Inc.; Copyright © 1972.)

One very important fact mut be remembered when sampling considerations are being discussed. A data system cannot distinguish between signal and noise frequencies, so that any noise whose frequency falls outside the range $2v\bar{\nu}_{min}-2v\bar{\nu}_{max}$ will be treated in the same fashion as any real signal outside the range and will be folded back into the spectrum. Thus, the noise between $\bar{\nu}_{min}$ and $\bar{\nu}_{max}$ will increase. It is therefore most important to restrict the bandwidth of the amplifier so that only the wavenumber range of interest is studied and all the other electrical frequencies (whether they correspond to signal or noise) are filtered out. In practice, it can be quite difficult to design a filter with a very sharp cutoff, and so generally a slightly higher sampling frequency than required by $\bar{\nu}_{max}$ and $\bar{\nu}_{min}$ is used. In this way, any noise that is folded back into the spectrum is folded back into a region where there is known to be no real spectral information.

An example of a real system where these factors have been brought into consideration may be found in many commercial Fourier transform infrared spectrometers that collect high-resolution spectra. As discussed above, a lower sampling frequency than necessary for a bandwidth of 0 to $\bar{\nu}_{max}$ can be employed for a bandwidth $\bar{\nu}_{min}$ to $\bar{\nu}_{max}$. In this case, optical and electronic filters are used to restrict the bandwidth of the detector output. Most mid-infrared FT–IR spectrometers use a helium–neon laser to measure the optical retardation of the interferometer; the normal sam-

Fig. 2.6. (a) Optical filter used for undersampling in the region 1317–2634 cm^{-1}. (b) Electronic filter used in the same region as (a). (c) Combination of electronic and optical filters. Note the radiation outside the desired bandwidth is near zero.

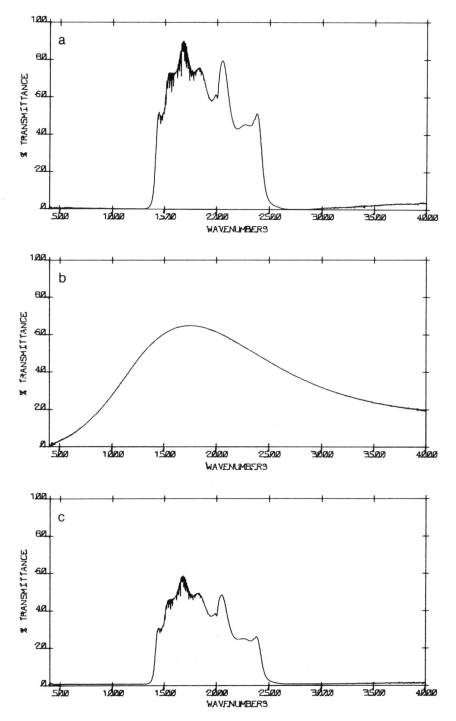

63

pling interval is 632.8 nm (see Chapter 4). Multiples of this He–Ne wavelength are used to keep the number of sampling points to a minimum. Thus, in mid-infrared spectrometry, the sampling interval is 632.8 nm, which corresponds to a maximum wavelength of 2×632.8 nm, which in turn is equivalent to 7902 cm^{-1}. If a sampling interval of 2×632.8 nm is used, the bandwidth must not exceed 3951 cm^{-1}. For a high-resolution spectrum, the number of data points can be greatly reduced by using a large sampling interval, or as it is often called, a high *undersampling ratio*.

An example is where the relatively narrow bandwidth of 1317 cm^{-1} is desired in which a large sampling interval of 6×632.8 cm^{-1} can be used. To prevent spectral features from outside the bandwidth, 1317–2634 cm^{-1} in this case, both optical and electronic filters are employed. Figure 2.6a shows the actual bandpass of an optical filter used for this bandwidth. As can be seen, there is an additional bandpass at higher wavenumbers than the desired bandwidth, and these are removed using the electronic filter shown in Fig. 2.6b. The combined effort of both filters is shown in Fig. 2.6c. The regions outside the desired bandwidth are effectively nulled so there is little folding of signal or noise into the desired bandwidth. The combined optical and electronic filters do not produce a simple boxcar function (i.e., a constant signal versus wavenumber across the bandwidth), as might be expected. The bandpass function of the filters becomes part of the *instrument function* and is removed upon ratioing the sample and reference spectra. An alternate method of bandwidth reduction, numerical filtering, is explained in Chapter 3, Section III.

Another important aspect to be considered is any possible nonlinearity of the detection system. Specifically, a detection system may not respond linearly to signal intensity to the extent that the signals of high intensity may be clipped. A normal signal is a sinusoid, but a clipped signal begins to resemble a square wave. It was stated that any waveform must be digitized at twice the bandwidth of the system so that the signal may be completely recovered. If harmonics of the signal are generated in the detection system, deleterious effects may occur. Such is the case of a square wave, which can be expressed as a series:

$$\sin x + \tfrac{1}{3} \sin 3x + \tfrac{1}{5} \sin 5x + \cdots$$

If the harmonics are outside the bandwidth allowed by the sampling frequency, they will naturally be folded into the real spectrum, thereby causing either spurious lines to appear (in the case of a line spectrum) or photometric inaccuracy (when the source is continuous). An example of a nonlinear spectrum is given in Fig. 2.7.

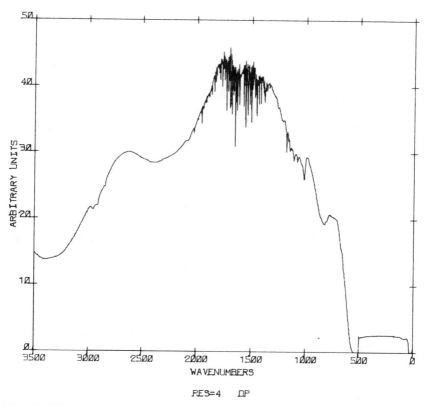

Fig. 2.7. Single-beam spectrum of a mercury cadmium telluride (MCT) detector operating in the nonlinear range. Spectral artifacts can be seen below 500 cm^{-1}, where the response should be zero.

II. DYNAMIC RANGE

In the previous section, we discussed how often the interferogram should be digitized; in this section, we discuss to what accuracy the amplitude of the signal should be sampled. In view of the fact that the intensity of the interferogram at zero retardation represents the summation of the amplitudes of all the waves in the interferogram, the signal-to-noise ratio at this point can be extremely high when an incandescent broadband source is being measured. On the other hand, the signal-to-noise ratio in the regions of the interferogram well displaced from zero retardation can be rather low.

To give an example of the range of the signals found in interferometry, consider the case of the interferogram of an incandescent blackbody source generated by a rapid-scanning interferometer and detected by a mid-infrared bolometer. The ratio of the intensity of the signal to zero retardation to the root-mean-square (rms) noise level (often called the *dynamic range*) can often be higher than $10^4:1$. State-of-the-art analog-to-digital converters (ADCs) have a resolution of approximately 16 bits, which means that the signal can be divided up into a maximum of 2^{16} (65,536) levels. Thus, if the dynamic range of the interferogram is 20,000, only the two least significant bits of the ADC would be used to digitize the noise level. If the dynamic range of the interferogram were, say, an order of magnitude higher, the noise level would fall below the least significant bit of the ADC and real information would be lost from the interferogram.

This example illustrates an important point for the sampling of interferograms in that *at least* one or two bits of the ADC should be sampling detector noise. If no detector noise is seen, the inaccurate sampling of the interferogram can lead to the generation of another type of noise in the spectrum called quantization, or *digitization noise*. The source of digitization noise is illustrated in Fig. 2.8, where it can be seen that there is a difference between the sampled values of the waveform and the true values. This is equivalent to noise in the signal and will obviously transform into spectral noise.

For a spectrum measured between $\bar{\nu}_{max}$ and $\bar{\nu}_{min}$ at a resolution of $\Delta\bar{\nu}$, there are M resolution elements, where

$$M = \frac{(\bar{\nu}_{max} - \bar{\nu}_{min})}{\Delta\bar{\nu}} \qquad (2.7)$$

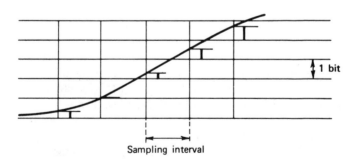

Sampling interval

Fig. 2.8. The effect of poor digitization accuracy: The difference between the actual and sampled signals is shown for each data point. This error can be considered to be equivalent to a peak-to-peak detector noise equal to approximately half a bit.

For a broadband source, the dynamic range of the interferogram is given to a good approximation by the dynamic range of the spectrum multiplied by $\sqrt{M}$. For an average spectral dynamic range of, say, 300:1 at a resolution of 2 cm^{-1} for a bandwidth of 4000 cm^{-1}, the dynamic range of the interferogram is $(2000)^{1/2} \times 300$ or 13,500:1. Thus, at least a 14-bit ADC would be needed to sample the signal.

The effect of grossly insufficient resolution of an ADC can be seen in Fig. 2.9. It is seen that as the dynamic range of the ADC is decreased, wavenumber resolution and band intensities are decreased in the spectrum. The reason for this effect is fairly easily understood by comparing the same interferogram measured with a 16-bit ADC and a 6-bit ADC. These interferograms are shown in Fig. 2.10a,b. It can be easily seen that most of the later points in the interferogram have a value of zero. The only data points that influence the spectrum are those at low retardation, thus explaining the low resolution of spectra measured with an ADC of low dynamic range. Because the spectrum is measured at such low resolution, the band intensity is also reduced.

Perry et al. [4] have discussed digitization needs more quantitatively. They have approximated the shape of an emission background to a Gaussian curve and have expressed the spectral intensity $S_b(\bar{\nu})$ at any frequency $\bar{\nu}$ as

$$S_b(\bar{\nu}) = S_b \exp\left\{ -\frac{(\bar{\nu} - \bar{\nu}_b)^2}{2\sigma_b^2} \right\} \qquad (2.8)$$

where S_b = peak amplitude
σ_b = bandwidth (standard deviation)
$\bar{\nu}_b$ = center frequency

The interferogram, $I_b(\delta)$, can be shown to be

$$I_b(\delta) = \left(\frac{\pi}{2}\right)^{1/2} S_b\sigma_b\left\{ \exp\left[-\frac{(\pi\sigma_b\delta)^2}{2} \right] \right\} \cos(2\pi\bar{\nu}_b\delta) \qquad (2.9)$$

For a Gaussian-type absorption band superimposed on the background with a center frequency $\bar{\nu}_s$, peak amplitude S_s, and width σ_s (Fig. 2.11), the corresponding interferogram can be written

$$I_s(\delta) = \left(\frac{\pi}{2}\right)^{1/2} S_s\sigma_s\left\{ \exp\left[-\frac{(\pi\sigma_s\delta)^2}{2} \right] \right\} \cos(2\pi\bar{\nu}_s\delta) \qquad (2.10)$$

In order to observe this interferogram, $I_s(\delta)$, superimposed on the back-

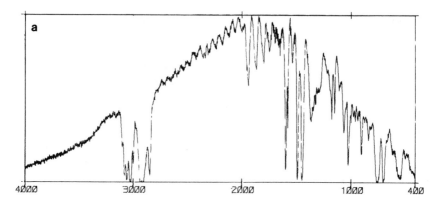

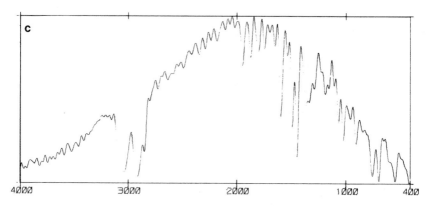

Fig. 2.9. The effect of insufficient resolution of the ADC. Single-beam spectra of polystyrene were collected using (*a*) 16-bit, (*b*) 12-bit, (*c*) 8-bit and (*d*) 6-bit ADCs.

ground interferogram $I_b(\delta)$, the dynamic range of the digital output must be greater than $I_b(\delta)/I_s(\delta)$ at the zero retardation point. Consider the special case that the small band absorbs at the center wavenumber of the background, that is, $\bar{\nu}_b = \bar{\nu}_s$. Let the amplitude of the small spectral band be a factor α less than the amplitude of the background. For this special case, the dynamic range of the ADC must be greater than $\sigma_b/\alpha\sigma_s$. In the general case, the dynamic range must be greater than the ratio of the area of the background spectrum to the area of the absorption band. If quantitative intensity information is needed on the small band, the dynamic range should be at least 10 times larger than the ratio of the two areas.

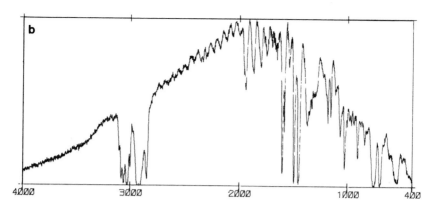

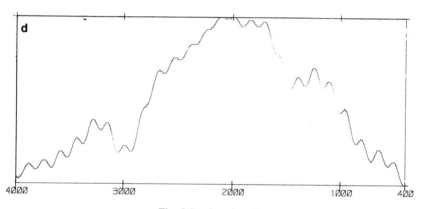

Fig. 2.9. (continued)

As the observation time per point increases, so does the signal-to-noise ratio in the sampled interferogram. Thus, in order to adjust the dynamic range of any interferogram to the level that it can be sampled correctly, the observation time must be changed until the least significant bit of the ADC is sampling noise. The only way that the dynamic range can be reduced while maintaining a high observation efficiency is to increase the scan speed of the moving mirror (see Chapter 7). This is the reason only rapid-scanning interferometers have been used for mid-infrared Fourier transform spectrometry when a large incandescent source is being measured. It may also be noted that if only the last bits of the ADC are being

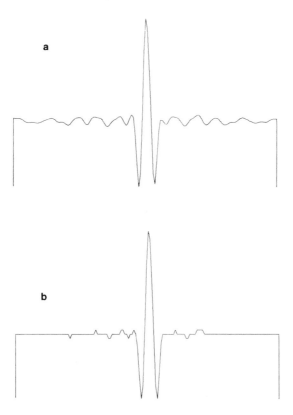

Fig. 2.10. Interferogram of polystyrene using a 16-bit ADC and (*b*) an interferogram of the same sample using a 6-bit ADC.

used to digitize the noise level, a change to a more sensitive detector gives little gain in spectral signal-to-noise ratio under these conditions. On the other hand, slow-scanning interferometers can be used effectively for the measurement of weak sources, often in conjunction with very sensitive detection systems.

There have been several attempts made to increase the effective dynamic range of digitizing systems. One method involves the use of dual-beam Fourier transform spectrometry. Since this technique is discussed in greater detail in Chapter 8, it will be only briefly described here. Basically, the technique involves mixing the signals from both arms of the interferometer. Since these beams are exactly 180° out of phase, if the paths were identical, they would destructively interfere to give no net signal. If an absorbing material were placed in one path between the beamsplitter and detector, the radiation absorbed by this sample would not be totally compensated and an interferogram caused only by the sample absorption bands would be measured. Thus, the dynamic range of the interferogram is substantially less than if all the frequencies from the source were being measured. This is equivalent to being able to measure $I_s(\delta)$ of Eq. 2.10 without having it superimposed on $I_b(\delta)$.

As mentioned above, the dynamic range of any interferogram can be reduced by scanning the moving mirror faster, and any signal can be sampled accurately (i.e., so that the least two or three significant bits of the ADC digitize noise) provided the scan speed of the interferometer can be increased in order that the signal is reduced. The limiting factor for this method therefore becomes the response time of the detector, which can sometimes become a problem. For example, mid-infrared radiation is usually detected by thermal detectors (thermocouples, thermistor bolometers, etc.). However, in order to reduce the dynamic range of the interferogram measured from an incandescent source, the scan speed has to be increased to the point that the modulation frequencies of the shortest wavelengths in the spectrum are so high (>1 kHz) that their period is less than the response time of the detector (>1 msec). This presented a real problem in the measurement of mid-infrared spectra over a wide range until the advent of the pyroelectric bolometer. These detectors have a broad spectral response with a short response time (<1 msec) even though

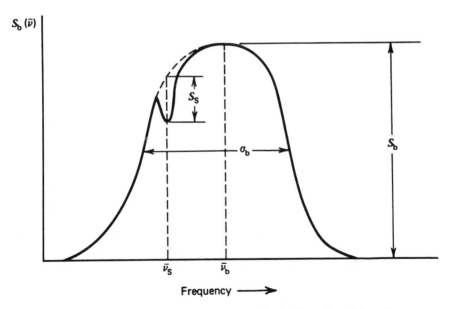

Fig. 2.11. Spectrum showing the parameters used in the discussion of dynamic range.

their noise-equivalent power is approximately equal to that of a thermistor bolometer.

Even when a pyroelectric bolometer (such as triglycine sulfate, TGS) is used for these measurements, the dynamic range of an interferogram from an incandescent source measured using a single scan of a rapid-scanning interferometer sometimes does not allow weak spectral features to be seen. For instance, for a spectrum with $\Delta \bar{\nu} = 2 \text{ cm}^{-1}$ and $\bar{\nu}_{max} = 4000 \text{ cm}^{-1}$ measured with a 15-bit ADC, any feature in this spectrum absorbing less than 0.25% of the radiation would be missed. To improve the signal-to-noise ratio achievable in the interferogram, signal-averaging techniques must be used. In this case, it is possible that the final spectral signal-to-noise ratio could be decided not by the ADC but by the word length of the signal averager. Thus, if a data system with a 16-bit word length were used in single precision for the signal-averaging process and computation, the ultimate dynamic range of the interferogram could be no larger than 16 bits, and hence the spectrum would show bands absorbing 0.1% (but no less). To obtain a greater dynamic range in the spectrum, either double-precision techniques (i.e., using two words per data point) or a computer with a longer word length would have to be used. Foskett [5] has addressed the question of computer word length requirements for FT–IR spectrometry in some detail. He showed that it is rarely necessary to use greater than 24 or 25 bits per data point, and that a 27-bit word probably represents the upper limit needed in any practical experiment. (This case would require data to be signal averaged for more than a day.) Even when data are acquired in "single precision" on computers with shorter word lengths, it is common to collect the region of the interferogram around the centerburst in double precision to ensure that the dynamic range of the computer word is not exceeded.

If the noise level in any waveform is much smaller than the resolution of the ADC, signal averaging will not improve the signal-to-noise ratio of the signal-averaged waveform over the single-scan case, since the same value of any point will be measured for each scan. This leads to the odd situation that it is conceivably possible to obtain a greater spectral signal-to-noise ratio ultimately from an interferogram of small dynamic range than from one of high dynamic range. It has been suggested that a random noise signal of limited bandwidth or a single coherent sinusoidal wave outside the frequency range of interest but below the folding frequency can be added to the interferogram. This raises the instantaneous signal so that it will be digitized accurately, and signal averaging will allow the true value to be approached (Fig. 2.12).

The technique of noise injection was first proposed by Mertz [6]. The band of noise frequencies must be *outside* the range of modulation frequencies imposed by the interferometer. Remembering that the highest Fourier frequency in the interferogram is $2v\bar{\nu}_{max}$, it is apparent that the *lowest* frequency in the band of injected noise must be greater than this frequency. The highest frequency in the noise band should be less than the Nyquist frequency (i.e., one-half of the sampling frequency). Alternatively, noise could be injected between 0 and $2v\bar{\nu}_{min}$ if there is a large enough frequency band to permit this to be achieved. One way that has been proposed of injecting "noise" into the interferogram is through the leakage of 60 Hz line frequency into the signal. For mid-infrared measurements using KBr optics ($\bar{\nu}_{min} = 400$ cm^{-1}) with a rapid-scanning interferometer having a mirror velocity of 1.6 mm/sec, $2v\bar{\nu}_{min} = 128$ Hz; therefore, the 60 Hz line frequency would fall in an ideal region provided higher harmonics are not present (see Chapter 7). When successfully achieved, noise injection has the effect of randomly changing the least significant bit of the ADC without reducing the SNR of the spectrum between $\bar{\nu}_{min}$ and $\bar{\nu}_{max}$. This technique is incorporated in at least one commercial FT–IR spectrometer.

A technique that is now almost universally used to increase the dynamic range of the interferogram is the use of a *gain-ranging amplifier*. In this method, an amplifier is used whose gain can be rapidly changed by a

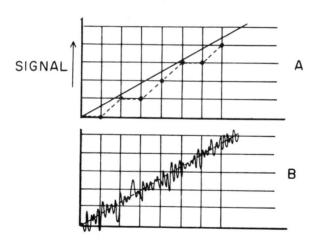

Fig. 2.12. The effect of digitizing a signal with an ADC of limited dynamic range for signal-averaging systems. For a noise-free signal [solid line in (*a*)], the same value of the signal is measured each scan, and the apparent signal after one scan or many signal-averaged scans would be the same (broken line). However, if the peak-to-peak noise level is greater than the least significant bit [solid line in (*b*)], signal averaging takes place and the final signal (broken line) is a much closer approximation to a straight line than if no noise were present.

factor of 2, 4, 8, and so forth, a certain distance either side of zero retardation (Fig. 2.13). In this way, the region of less dynamic range away from zero retardation can be amplified and therefore sampled more accurately than if such an amplifier were not used [7]. At the end of the measurement, the least significant bits of the signal-averaged interferogram in the region where the greater gain had been used were dropped from the interferogram. If the increase in gain is a factor of 2^N in the region of small dynamic range, then N bits would be dropped before the transform is performed. The effect of gain ranging is discussed in more detail in Chapter 7.

Mark and Low [8] have suggested a technique for decreasing the dynamic range that involves "clipping" or "blanking" the interferogram around zero retardation. Clipping involves increasing the amplifier gain so that the region or largest amplitude has a greater intensity than the full dynamic range of the ADC. Blanking involves bringing the region around zero retardation to zero. These effects are illustrated in Fig. 2.14. In either case, a higher gain can be used so that the regions away from zero retardation (the "wings" of the interferogram) can be more accurately sampled. These techniques have the great disadvantage of putting a discontinuity in the interferogram, thereby adding a great deal of noise to all frequencies in the spectrum. Any benefit gained by sampling the wings of the interferogram is more than compensated by this increase in noise, and therefore the technique shows no practical purpose. On the other hand, if the phase spectrum is very accurately known, it may become

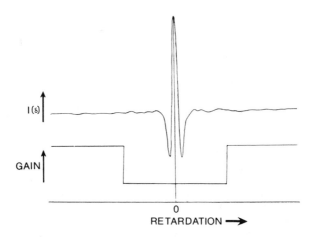

Fig. 2.13. Action of a gain-ranging amplifier.

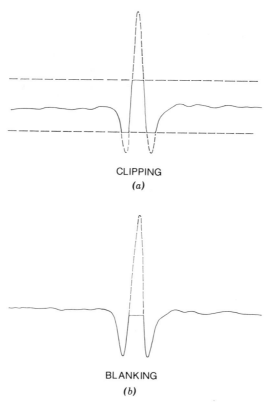

CLIPPING

(a)

BLANKING

(b)

Fig. 2.14. (a) "Clipping" the interferogram and (b) "blanking" the interferogram around zero retardation.

possible to sample the interferogram entirely to one side of the zero retardation region, thereby eliminating the centerburst completely. The results of such an experiment have not yet been reported.

Another method of circumventing the problems of dynamic range has been used in the Infrared Interferometer Spectrometer (IRIS) experiment aboard the Nimbus B satellite and described by Forman [9]. All points that are less than 10% of the maximum are quantized to 256 levels (8-bit ADC). The remaining points are then divided by 10 and quantized to the same number of levels, and the operation is recorded by 1 bit of the 10-bit telemetry word. This system was designed because of the telemetry unit rather than because of the dynamic range of the ADC. Forman's data show not only the usefulness of this technique but also the effect of insufficient resolution of the ADC on computed spectra. His spectra are shown in Fig. 2.15.

In all the interferograms discussed to date, there is a very large signal at one point (zero retardation) since each wavelength is in phase (or very close to it) at this point. It is possible to modify the interferogram such

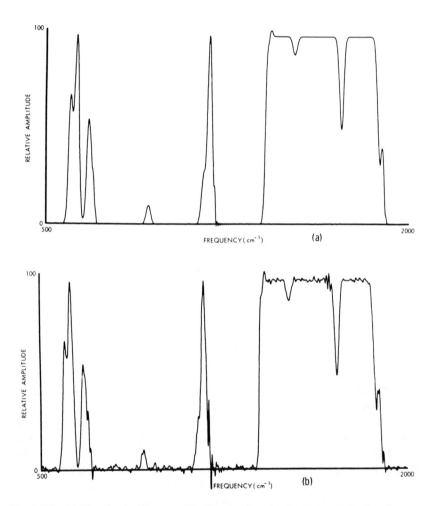

Fig. 2.15. (a) Simulated spectrum; (b) effect on the simulated data if the interferogram were sampled with an 8-bit ADC; (c) effect on the simulated data if the interferogram were sampled with a 10-bit ADC; and (d) effect on the simulated data if the interferogram were sampled using Forman's divide-by-ten method using an 8-bit ADC. (Reproduced from [9], by permission of the author.)

that each wavelength has its own origin, thus avoiding the large central fringes that normally occur. This distribution of wavelength centers is accomplished by placing a flat optical element into each beam before the point of beam recombination. If the refractive index of one of the plates is frequency dependent or if the plates are of different thickness, then

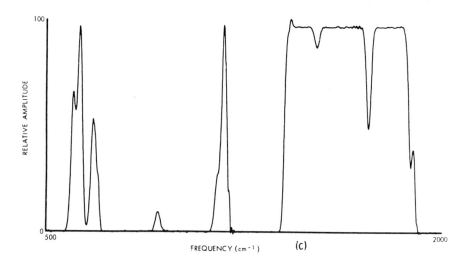

FREQUENCY (cm⁻¹)   (c)

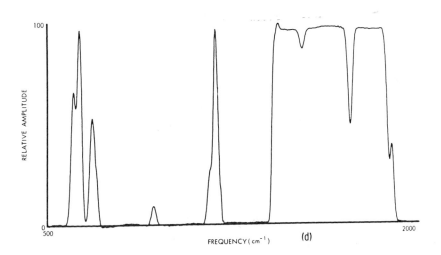

FREQUENCY (cm⁻¹)   (d)

Fig. 2.15. (continued)

each wavelength will have its center shifted by a different amount, and the interferogram becomes strongly chirped. Therefore, there is no single path difference at which all wavelengths are in phase to produce a central fringe, so that the dynamic range is reduced while Fellgett's advantage is retained.

Two materials that have been used for this purpose are Irtran 4 and Irtran 5. The displacement in the zero phase difference point for different frequencies (with CaF_2 in the other arm) is shown in Fig. 2.16. It can be seen that Irtran 5 spreads the zero *phase* difference fringe for the 2000–4000-cm^{-1} region over 500 μm of *path* difference. A PbSe detector has its peak response over this region and has been used in a "chirped" interferometer by Hohnstreiter et al. [10]. A typical interferogram is shown in Fig. 2.17. The interferogram shows that the low frequencies are spread over a wide range and the high frequencies converge to a single point. This is expected in the light of Fig. 2.17 since chirping is only expected to occur strongly below 4000 cm^{-1} for an Irtran 5–CaF_2 combination.

In the measurements described by Hohnstreiter et al. [10], the mag-

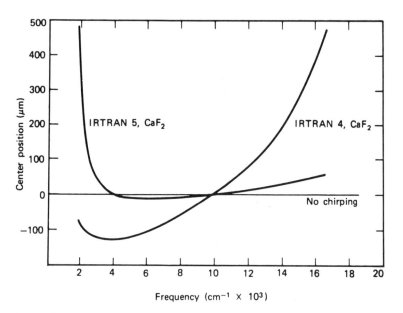

Fig. 2.16. Dispersion of the white-light fringe location for two possible chirping configurations. (Reproduced from [10], by permission of the author.)

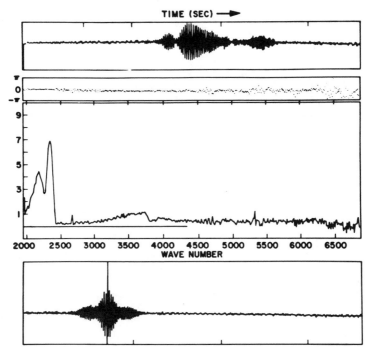

Fig. 2.17. (*Above*) Chirped interferogram and (center) the phase-corrected spectrum of the source calculated from this interferogram. (*Below*) An almost completely unchirped interferogram reconstructed from this spectrum. (Reproduced from [12], by permission of the Society for Applied Spectroscopy; copyright © 1974.)

nitude of the signal could vary by quite large amounts, and eight levels of automatic gain changing were required to keep the signal level within the telemetry range. To obtain the maximum information from the interferogram under these conditions, it was found necessary to set the gain of the amplifier so that the maximum signal had a value small enough that it can be modulated for telemetry without clipping the interferogram yet large enough that the principal source of noise was not the telemetry circuits. To achieve this objective, successive scans were made, and the gain was changed by factors of approximately 4 between scans if the maximum voltage from the previous interferogram exceeded a maximum, or was less than a minimum, threshold value. In this way, the telemetered interferogram could be transmitted at the optimum gain. The complex data-handling equipment has been described by Sheahen et al. [11].

REFERENCES

1. M. Woodward, *Probability and Information Theory*, Pergamon Press, New York (1955).
2. A. S. Zachor, I. Coleman, and W. G. Mankin, Effects of Drive Nonlinearities in Fourier Spectroscopy, in *Spectrometric Methods*, Vol. II (G. A. Vanasse, ed.), Academic Press, New York (1981).
3. P. R. Griffiths, C. T. Foskett, and R. Curbelo, *Appl. Spectrosc. Revs.*, **6**, 31 (1972).
4. C. H. Perry, R. Geick, and E. F. Young, *Appl. Optics*, **5**, 1171 (1966).
5. C. T. Foskett, *Appl. Spectrosc.*, **30**, 531 (1976).
6. L. Mertz, *Transformations in Optics*, Wiley, New York (1965).
7. T. Hirschfeld, *Appl. Spectrosc.*, **33**, 525 (1979).
8. H. Mark and M. J. D. Low, *Appl. Spectrosc.* **25**, 605 (1971).
9. M. L. Forman, Aspen Int. Conf. on Fourier Spectrosc., 1970 (G. A. Vanasse, A. T. Stair, and D. J. Baker eds.), AFCRL-71-0019, p. 305 (1971).
10. G. F. Hohnstreiter, W. Howell, and T. P. Sheahen, Aspen Int. Conf. on Fourier Spectrosc., 1970 (G. A. Vanasse, A. T. Stair, and D. J. Baker, eds.), AFCRL-71-0019, p. 243 (1971).
11. T. P. Sheahen, G. F. Hohnstreiter, W. R. Howell, and I. Coleman, Aspen Int. Conf. on Fourier Spectrosc., 1970 (G. A. Vanasse, A. T. Stair, and D. J. Baker, eds.), AFCRL-71-0019, p. 255 (1971).
12. T. P. Sheahen, *Appl. Spectrosc.*, **28**, 283 (1974).

CHAPTER

3

COMPUTING TECHNIQUES

I. THE CONVENTIONAL FOURIER TRANSFORM

a. Elementary Concepts

In the two previous chapters, the various factors for the generation of an interferogram were discussed. In this chapter, the techniques for computing the spectrum from this digitized interferogram are described.

Prior to 1966, spectroscopists who measured spectra interferometrically used the same basic algorithm for their computations. This involved the use of what is now known as the *conventional*, or *classical*, or *discrete* Fourier transform. Although it is true that few people today use this algorithm in view of the substantial time advantages to be gained by the use of the fast Fourier transform technique (which will be described in the next section), an understanding of the conventional Fourier transform leads to a better comprehension of more advanced techniques.

The concept of the classical Fourier transform is fairly simple to understand. As an example, let us consider the case of a long cosine wave interferogram digitized at equal intervals. To perform the classical Fourier transform, we simply multiply each point by the corresponding point of an analyzing cosine wave of unit amplitude and add the resultant values. For each analyzing wave whose frequency is *different* from that of the cosinusoidal interferogram, the sum will be zero (i.e., the two waves are orthogonal). If the frequencies of the interferogram and the analyzing wave are the same, the resultant will be a large positive number, since for each point the interferogram takes a positive value, the analyzing wave will also be positive and vice versa. The magnitude of the sum will be proportional to the amplitude of the cosine wave interferogram.

For an analyzing wave whose frequency is very close to that of the monochromatic interferogram, it requires a greater retardation for the resultant to take a zero value than for analyzing waves whose frequencies are well separated from that of the interferogram. In fact, if the frequency of the analyzing wave is gradually increased from that of the interferogram, it is not too difficult to visualize that the resultant will decrease from the large positive magnitude discussed in the previous paragraph, pass through zero to a negative value, and then oscillate between negative and positive values whose amplitude decreases as the frequency difference between the interferogram and analyzing wave increases. This oscillatory pattern is described mathematically by the sinc function shown in Fig. 1.5b. The greater the retardation over which the interferogram is sampled, the higher is the frequency of this oscillation, that is, the narrower is the FWHH of the sinc function and the higher is the spectral resolution.

The way in which *phase errors* are introduced during the transform may also be understood conceptually from the above example. For example, if the sinusoidal interferogram was 180° out of phase with the analyzing cosine wave ($\theta_{\bar{\nu}} = \pi$ in Eq. 1.29), then at each retardation the interferogram has a positive value, the analyzing wave will be negative, and vice versa. Therefore, the product of the interferogram and analyzing wave will be negative for each data point, and the resultant will be equal in magnitude but opposite in sign to the case when $\theta_{\bar{\nu}} = 0$. Obviously, correction for phase errors is important if photometrically accurate spectra are to be calculated.

b.   Mathematical Basis

Let us now consider the classical Fourier transform from a more mathematical basis by considering the case of a symmetrical interferogram that has been measured from the zero retardation point [1]. The integral

$$B(\bar{\nu}) = \int_{-\infty}^{+\infty} I(\delta)A(\delta) \cos 2\pi\bar{\nu}\delta \cdot d\delta \qquad (3.1)$$

where $A(\delta)$ represents any truncation or apodization function previously referred to as $D(\delta)$ and $A_1(\delta)$, $A_2(\delta)$, and so forth, can be written as

$$B(\bar{\nu}) = 2 \int_0^{+\infty} I(\delta)A(\delta) \cos 2\pi\bar{\nu}\delta \cdot d\delta \qquad (3.2)$$

because of the symmetry of the interferogram about the point of zero retardation.

It was stated in Chapter 2 that if the signal was digitized with a small enough interval between data points, no information from the analog signal would be lost. The minimum sampling interval h is given by

$$2h = (\bar{\nu}_{max} - \bar{\nu}_{min})^{-1} \qquad (3.3)$$

If the interferogram has been sampled correctly, we can replace the integral of Eq. 3.2 by a summation and calculate the intensity $B'(\bar{\nu}_1)$ at any wavenumber $\bar{\nu}_1$. Let the retardation at any sampling point be given by nh, where n is an integer, and let $I_a(n)$ be the value of the interferogram at this point:

$$B'(\bar{\nu}_1) = I_a(0) + 2I_a(1)\cos(2\pi\bar{\nu}_1 \cdot h) + 2I_a(2)\cos(2\pi\bar{\nu}_1 \cdot 2h)$$
$$+ \cdots + 2I_a(N-1)\cos[2\pi\bar{\nu}_1 \cdot (N-1)h] \quad (3.4)$$

where N is the total number of points sampled.

$$B'(\bar{\nu}_1) = I_a(0) + 2\sum_{k=1}^{N-1} I_a(k)\cos(2\pi\bar{\nu}_1 \cdot kh) \quad (3.5)$$

The summation in Eq. 3.5 is performed for all wavenumbers of interest in the spectrum. It need not, however, be carried out for all wavenumbers from $\bar{\nu}_{max}$ to $\bar{\nu}_{min}$.

Even for fairly short spectral ranges, the computing time involved for this operation can be quite high, especially if each cosine term required is calculated from a series. Cosine tables could be stored in the computer memory, but for large interferograms this would take a large amount of space. It has been found that a successful method of circumventing this problem is by the use of recursion relationships, whereby a value for $\cos(2\pi\bar{\nu}_1 \cdot kh)$ can be calculated from the value of $\cos(2\pi\bar{\nu}_1 h)$. In this way, only one cosine value has to be calculated for each spectral wavenumber. The Chebyshev recursion formula

$$\cos(p + 1)x = 2\cos x \cos px - \cos(p - 1)x \quad (3.6)$$

is the one used most frequently. It has been shown that if the cosines are calculated to six significant figures, the difference between the value computed by a series expansion and the values obtained from 12,000 iterations is less than 1 in 50,000. The first cosine, $\cos 2\pi\bar{\nu}_1 h$, is computed using a series formula and all the cosines that follow are computed using a recursion formula for each new value $\bar{\nu}_1$. Each of the values of $I_a(k)\cos(2\pi\bar{\nu}_1 \cdot kh)$ is added to the sum, and the resultant number is proportional to the value of $B'(\bar{\nu}_1)$.

If a double-sided interferogram has been collected and the centerburst is near the middle of the interferogram, the summation in Eq. 3.5 may be replaced by the summation

$$| B'(\bar{v}_1) | = \left\{ \left[\sum_{k=-N/2}^{(N/2)-1} I_a(k) \cos(2\pi\bar{v}_1 \cdot kh) \right]^2 \right.$$
$$\left. + \left[\sum_{k=-N/2}^{(N/2)-1} I_a(k) \sin(2\pi\bar{v}_1 \cdot kh) \right]^2 \right\}^{1/2} \quad (3.7)$$

The sine values are calculated using another recursion formula also attributed to Chebyshev:

$$\sin(p + 1)x = 2 \sin px \cos x - \sin(p - 1)x \quad (3.8)$$

The summation of Eq. 3.7 produces the *magnitude* spectrum of $B'(\bar{v}_1)$, denoted $B(\bar{v}_1)$. The magnitude spectrum will exhibit no phase error, but the effect of noise on the spectrum is nonlinear, as all noise will be positive and increase as the signal decreases.

If there are N points in the interferogram, let us consider how many operations are required during the computation of a single-sided cosine transform. For each spectral wavenumber $\bar{v}_j$, we perform N multiplications for $I_a(k) \cos(2\pi\bar{v}_j \cdot kh)$. To add the values, we must do $N - 1$ additions, so that a total of $2N$ operations are needed for each wavenumber $\bar{v}_j$. To examine the complete spectrum, we must compute the spectrum at $\frac{1}{2}N$ wavenumber values at least. Thus, a total of approximately N^2 operations are needed for the conventional Fourier transform, not counting the time required to compute the cosine value.

If the double-sided complex transform is performed, the number of operations increases. For each value of $\bar{v}_j$, we now perform N multiplications for $I_a(k) \cos(2\pi\bar{v}_j \cdot kh)$, and N multiplications are required for $I_a(k) \sin(2\pi\bar{v}_j \cdot kh)$. Then $N - 1$ additions must be made for the cosine transform and $N - 1$ for the sine transform, so a total of approximately $4N$ operations must be made for each wavenumber. If the full bandwidth of the spectrum is to be examined at the same resolution as the single-sided spectrum, twice as many data points must be taken to achieve the same resolution, so that the number of operations is increased to $8N$. The number of output points is the same as for the single-sided case, so that $4N^2$ operations are required. This number does not include the time required for the cosine, sine, square, and square root calculations required for each value of $\bar{v}_j$.

As the number of points to be calculated increases beyond about 10,000, the calculation time for a spectrum can become prohibitive. In view of the fact that interferometry shows its greatest advantage over dispersive spectrometry at high resolution, that is, when large numbers of points have to be calculated, this problem was especially annoying and did not appear to be resolvable until about 1966. At this time, Forman [2] published a paper on the application of the fast Fourier transform technique to Fourier spectrometry. This technique had been described in the literature by Cooley and Tukey [3] one year earlier. This algorithm extended the use of Fourier transform spectrometry to encompass high-resolution data in all regions of the infrared spectrum. It is described in the next section.

II. THE FAST FOURIER TRANSFORM

The fast Fourier transform (FFT) is an algorithm that was described by Cooley and Tukey [3] in which the number of necessary computations is drastically reduced when compared to the classical Fourier transform. To understand the algorithm, the interferogram and the spectrum should be regarded as the complex pair [4]

$$I(\delta) = \int_{-\infty}^{+\infty} B(\bar{\nu})e^{i2\pi\bar{\nu}} \, d\bar{\nu} \tag{3.9}$$

$$B(\bar{\nu}) = \int_{-\infty}^{+\infty} I(\delta)e^{-i2\pi\bar{\nu}\delta} \, d\delta \tag{3.10}$$

where Eq. 3.9 is the complex inverse Fourier transform of the spectrum to produce the interferogram. This is an alternative form of Eq. 1.29.

The point of departure for the FFT from the continuous Fourier transform is the discrete Fourier transform (DFT). The DFT of an interferogram of N points to produce a spectrum of N points may be written as

$$B(r) = \sum_{k=0}^{N-1} I_0(k)e^{-i2\pi rk/N} \qquad r = 1, 2, \ldots, N - 1\dagger \tag{3.11}$$

where $B(r)$ is the spectrum expressed at discrete wavenumbers r. Let us define a parameter W such that

$$W = e^{-i2\pi/N} \tag{3.12}$$

† This equation is identical to Eq. 3.5 if $h = 1/N$.

Substituting this value of W in Eq. 3.11, we obtain

$$B(r) = \sum_{k=0}^{N-1} I_0(k) W^{rk} \qquad (3.13)$$

From this equation, it can be seen that each discrete value of $B(r)$ requires N complex multiplications and $N - 1$ complex additions. Since there are N terms of $B(r)$, N^2 complex multiplications are required, as stated in the last section. The FFT is based on the idea that Eq. 3.13 can be expressed in general matrix form, and that matrix can be factored in a manner that will reduce the overall number of computations.

The Cooley–Tukey algorithm is general and can be applied to any Fourier transform, but the computation is greatly simplfied if N is a base 2 number, that is, $N = 2^\alpha$, where α is a positive integer. The factorization procedure may be demonstrated easily for the case when $N = 4 = 2^2$. In this example, Eq. 3.13 becomes

$$B(r) = \sum_{k=0}^{3} I_0(k) W^{rk} \qquad r = 0, 1, 2, 3 \qquad (3.14)$$

which may be expressed more fully as

$$
\begin{aligned}
B(0) &= I_0(0)W^0 + I_0(1)W^0 + I_0(2)W^0 + I_0(3)W^0 \\
B(1) &= I_0(0)W^0 + I_0(1)W^1 + I_0(2)W^2 + I_0(3)W^3 \\
B(2) &= I_0(0)W^0 + I_0(1)W^2 + I_0(2)W^4 + I_0(3)W^6 \\
B(3) &= I_0(0)W^0 + I_0(1)W^3 + I_0(2)W^6 + I_0(3)W^9
\end{aligned}
\qquad (3.15)
$$

The set of equations in 3.15 may be expressed as matrices:

$$
\begin{vmatrix} B(0) \\ B(1) \\ B(2) \\ B(3) \end{vmatrix}
=
\begin{vmatrix} W^0 & W^0 & W^0 & W^0 \\ W^0 & W^1 & W^2 & W^3 \\ W^0 & W^2 & W^3 & W^6 \\ W^0 & W^3 & W^6 & W^9 \end{vmatrix}
\begin{vmatrix} I_0(0) \\ I_0(1) \\ I_0(2) \\ I_0(3) \end{vmatrix}
\qquad (3.16)
$$

or, more compactly,

$$\mathbf{B}(r) = \mathbf{W}^{rk} \mathbf{I}_0(k) \qquad (3.17)$$

Several simplifications can be immediately applied to reduce the number of computations. Most obviously, knowing that $W^0 = 1$ saves computation of seven exponentials. Two other simplifications, which are less obvious, are based on the cyclic nature of the exponential term W (W is an exponential function of 2π). First,

$$W^M = W^{\mathrm{mod}(M,N)} \tag{3.18}$$

where $\mathrm{mod}(M, N)$ is the integer function equal to $M - (M/N)*N$, where the operator $*$ is multiplication. Recall that in integer arithmetic M/N produces a truncated quotient, so the mod function calculates the remainder after division. The proof of Eq. 3.18 is straightforward. If $M = nN + k$, where M, n, N, and k are all integers, then

$$W^M = e^{-i2\pi M/N} = e^{-i2\pi(nN+k)/N}$$

$$= e^{-i2\pi k/N} \cdot e^{-i2\pi n} = e^{-i2\pi k/N}$$

$$= W^{\mathrm{mod}(M,N)}$$

Therefore, Eq. 3.16 may be further reduced by substituting $W^{\mathrm{mod}(rk,N)}$ for W^{rk}; thus,

$$\begin{vmatrix} B(0) \\ B(1) \\ B(2) \\ B(3) \end{vmatrix} = \begin{vmatrix} 1 & 1 & 1 & 1 \\ 1 & W^1 & W^2 & W^3 \\ 1 & W^2 & W^0 & W^2 \\ 1 & W^3 & W^2 & W^1 \end{vmatrix} \begin{vmatrix} I_0(0) \\ I_0(1) \\ I_0(2) \\ I_0(3) \end{vmatrix} \tag{3.19}$$

The second cyclic property of W is

$$W^{k+N/2} = e^{-i2\pi(k+N/2)/N}$$

$$= e^{-i2\pi k/N} \cdot e^{-i\pi} = -e^{-i2\pi k/N} \tag{3.20}$$

$$= -W^k$$

Thus, Eq. 3.19 may be further reduced by subsituting $-W^1$ for W^3 and $-W^0$ $(= -1)$ for W^2. Although the DFT can be greatly simplified by these substitutions and only $\frac{1}{2}N$ complex terms need be calculated, the DFT still requires N^2 multiplications for the complete calculation of $B(r)$.

If we consider Eq. 3.14, we see that both k and r can be represented as binary numbers, specifically 2-bit binary numbers. For $N = 4$,

$k = 0, 1, 2, 3$ (decimal)   or $k = (k_1, k_0) = 00,01,10,11$ (binary)

$r = 0, 1, 2, 3$ (decimal)   or $r = (r_1, r_0) = 00,01,10,11$ (binary)

Thus,

$$k = 2k_1 + k_0 \quad \text{and} \quad r = 2r_1 + r_0$$

where k and r are the base 10 representations of the two numbers. With k and r expressed as binary numbers, Eq. 3.14 can be rewritten

$$B(r_1, r_0) = \sum_{k_0=0}^{1} \sum_{k_1=0}^{1} I_0(k_1, k_0) W^{(2r_1 + r_0)(2k_1 + k_0)} \tag{3.21}$$

When Eq. 3.14 is rewritten, it must contain a summation sign for each bit in the binary number. The FFT is based on factorization of the DFT, but the exponential term W is not as yet in the appropriate form; thus,

$$\begin{aligned}
W^{(2r_1 + r_0)(2k_1 + k_0)} &= W^{(2r_1 + r_0)2k_1} \cdot W^{(2r_1 + r_0)k_0} \\
&= W^{4r_1 k_1} \cdot W^{2r_0 k_1} \cdot W^{(2r_1 + r_0)k_0} \\
&= W^{2r_0 k_1} \cdot W^{(2r_1 + r_0)k_0}
\end{aligned} \tag{3.22}$$

Eq. 3.21 can be written

$$B(r_1, r_0) = \sum_{k_0=0}^{1} \sum_{k_1=0}^{1} I_0(k_1, k_0) W^{2r_0 k_1} \cdot W^{(2r_1 + r_0)k_0} \tag{3.23}$$

At this point, the DFT has been factorized into the FFT form. To effect the FFT, it is necessary only to calculate each summation individually. The inside summation is evaluated first, and the result is used to calculate the outside summation. The inside summation is

$$I_1(r_0, k_0) = \sum_{k_1=0}^{1} I_0(k_1, k_0) W^{2r_0 k_1} \tag{3.24}$$

which may be expanded to

$$\begin{aligned}
I_1(0, 0) &= I_0(0, 0) + I_0(1, 0)W^0 \\
I_1(0, 1) &= I_0(0, 1) + I_0(1, 1)W^0 \\
I_1(1, 0) &= I_0(0, 0) + I_0(1, 0)W^2 \\
I_1(1, 1) &= I_0(0, 1) + I_0(1, 1)W^2
\end{aligned} \tag{3.24}$$

Based on the cyclic properties of W as explained above (cf. Eq. 3.20),

$I_1(0, 0)$ and $I_1(1, 0)$ may be calculated simultaneously as

$$
\begin{aligned}
I_1(1, 0) &= I_0(0, 0) + I_0(1, 0)W^2 \\
&= I_0(0, 0) - I_0(1, 0)W^0
\end{aligned}
\tag{3.25}
$$

Equation 3.25 shows that $I_1(1, 0)$ differs from $I_1(0, 0)$ only by the sign of the second term. Thus, for $I_1(0, 0)$ an addition is performed, and, for $I_1(1, 0)$ a subtraction is performed. Figure 3.1a is a schematic shorthand notation for the summation in Eq. 3.24. The schematic is read by interpreting the symbols near the head and tail of the arrows. For example, $I_1(0, 1) = I_1(1_{10}) = I_0(1_{10}) + I_0(3_{10})W^0$, where the subscript 10 indicates the numbers are to the base 10. The $I_1(1_{10})$ and $I_1(3_{10})$ are linked by a dual-node pair arrow, which indicates that they are calculated simultaneously as all their terms are identical. This summation has required only one complex term to be calculated (W^0) and two multiplications.

The outside summation uses the results of the inside summation, or $I_1(r_0, k_0)$; then

$$
I_2(r_0, r_1) = \sum_{k_0 = 0}^{1} I_1(r_0, k_0)W^{(2r_1 + r_0)k_0}
\tag{3.26}
$$

Similarly to Eq. 3.24, this summation can be enumerated:

$$
\begin{aligned}
I_2(0, 0) &= I_1(0, 0) + I_1(0, 1)W^0 \\
I_2(0, 1) &= I_1(0, 0) + I_1(0, 1)W^2 \\
I_2(1, 0) &= I_1(1, 0) + I_1(1, 1)W^1 \\
I_2(1, 1) &= I_1(1, 0) + I_1(1, 1)W^3
\end{aligned}
\tag{3.27}
$$

Figure 3.1b shows the schematic form Fig. 3.1a with the computations from Eq. 3.26 added. Again only 2 complex multiplications in the second summation have been performed, giving a total of 4 from the FFT algorithm as opposed to the 16 (N^2) for the DFT. In general, the FFT requires only $\frac{1}{2}N$ complex multiplications and $N(N - 1)$ complex additions. The ratio of FFT operations to DFT operations can be attributed to the complex multiplications; hence

$$
\frac{N^2}{N\alpha/2} = \frac{2N}{\alpha}
\tag{3.28}
$$

Clearly, as N increases, the efficiency of the FFT increases dramatically.

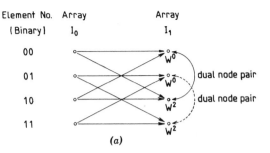

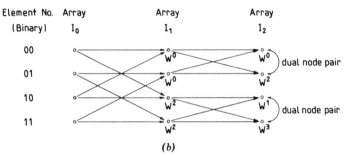

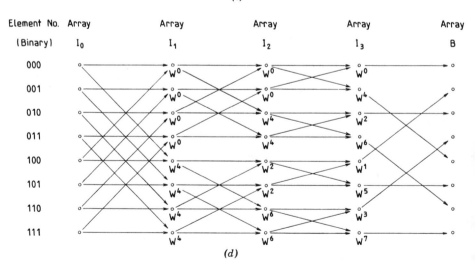

90

The FFT does not end with the second summation in Eq. 3.26 for the case $N = 4$. The spectral datum $B(r) = B(r_1, r_0)$ is not equal to $I_2(r_0, r_1)$. The only difference between the two data vectors is that the bits in the subscripted variables have been reversed. That is, the elements of B have the bits in the order r_1, r_0 whereas I_2 has the bits in the order r_0, r_1. It is a simple matter to unscramble the bits to produce the desired result. Figure 3.1c shows the full schematic for the FFT of Eq. 3.14.

The general case, where $N = 2^\alpha$, requires that the DFT be factored into summations. The flow graph schematic for the FFT of an interferogram where $N = 2^3 = 8$ is shown in Fig. 3.1d. This flow graph illustrates that the FFT has more steps, but in the evaluation of each element of each array, only one complex multiplication and one complex addition is required.

It is interesting to note that there are several canonic forms of the FFT algorithm. If the original input data are scrambled according to the bit reversal algorithm, the spectrum can be computed in place. In a real sense, the procedure of the FFT is reversed, and the flow graph of this canonic form would begin with the dual-node pairs adjacent and end with the pairs $\frac{1}{2}N$ elements apart. The results are identical, but the procedure varies. The latter canonic form has been used by Connes [5]. These two canonic forms of the Cooley–Tukey algorithm are termed "decimation in time" because some alternative derivations imply that the number of samples is reduced.

The other two canonic forms of the FFT algorithm are based on the Sande–Tukey algorithm. This algorithm factors the DFT in r as opposed to k, leading to a "decimation in frequency" procedure. All four forms produce identical results, although two of the forms are computationally more efficient. In general, the FFT can be performed on any size array as long as the number of data points in the interferogram is an integer power of 2. Consequently, all commercial FT–IR instruments collect interferograms that are 2^α data points in length.

Advantage can be taken of the fact that interferograms are real functions and thus there is no imaginary input as the imaginary portion is zero. Computer space can be conserved by exploiting this fact. Algorithms exist whereby a real input array of N points is Fourier transformed to give real

Fig. 3.1. Schematic shorthand notation for the FFT procedure. (a) Schematic representation for Eq. 3.24 that is interpreted by the convention that the datum at the head of the arrows is the sum of the data at the tails of the arrows. A term W may be near one of the arrowheads and the datum at the tail of that arrow is multiplied by the W term prior to addition. (b) Schematic representation for Eqs. 3.24 and 3.26. (c) Full schematic for four data FFT, including bit unscrambling operation. (d) Eight-point FFT schematic representation.

and imaginary spectral domain arrays of $\frac{1}{2}N$ points each. Because the spectral domain arrays are symmetric, all the information is present in one-half (or $\frac{1}{2}N$ points) of the array. A complex FFT requires $2N$ data locations in computer memory, whereas a real function FFT requires only N. Besides the space saving, the number of operations in the FFT computation is reduced by approximately 2.

If the interferogram is not only real but is also even (i.e., symmetric about the point of zero retardation), further savings may be realized. By applying certain properties of real and even functions, Connes [5] has shown that computer space can be reduced by a factor of 4 over the case of a complex function and by a factor of 2 over a real function. When extremely large transforms are being performed, these savings in computer space can be very important since computer memory size is usually limited. When mass storage must be used to contain part of the transform array because insufficient memory is available, the mass storage to memory transfer time drastically increases the transform time.

In view of additional operations besides those directly attributable to the FFT, the time savings in these computations are not quite as great as the space savings. Connes has reduced the computing time by a factor of 1.5 by considering the interferogram to be an arbitrary real function and by 2.8 by considering it as a real, even function.

If phase modulation (see Chapter 1) is used in the measurement of the interferogram, the interferogram is no longer a real, even function but a real, odd function. Although the mathematics of the transform are formally different from the usual case, equal time and space savings can result in the computation of the spectrum from a phase-modulated interferogram using the FFT algorithm.

The fast Fourier transform provides tremendous time savings for the chemical spectroscopist. For example, the Coblentz Society [6] has suggested that reference spectra be collected at a resolution of 2 cm^{-1} over a minimal bandwidth of 3800 to 450 cm^{-1}. If the measurement is performed interferometrically, 8192 data points must be collected for each of the sample and reference interferograms. The time savings for an 8192-point interferogram transformed using an FFT as opposed to a DFT is approximately a factor of 1250, according to Eq. 3.28. Most commercial FT–IR systems that have a dedicated minicomputer can perform an 8192-point FFT in under 2 min, and some systems can perform it in under 5 sec (see Chapter 6). An 8192-data-point transform using a DFT that includes the Chebyshev recursion algorithm takes approximately 15 min on most mainframe computers. A reference spectrum that meets the Coblentz Society specifications can be measured in a little over a half an hour using a good grating spectrophotometer. Thus there is no discernible time advantage when an FT–IR spectrometer is used and the computations are performed using the DFT, even though the interferograms can be measured in a matter of seconds.

A thorough discussion of the FFT is available [4]. An explanation for the application of commercial FFT software routines to FT–IR interferograms may be found elsewhere [7]. Appropriate transformation of interferograms to produce accurate spectra is not restricted to Fourier transform algorithms, and some of the additional procedures, such as phase correction and zero filling, are presented in the following two sections.

III. PHASE CORRECTION

The sources of phase error have been presented in Chapter 1, Section VI, where it was stated that the phase angle $\theta_{\bar{v}}$ is usually a function of frequency. It should also be noted that phase is a complex quantity; that is, it has real and imaginary parts.

When a recorded interferogram $I(\delta)$ is transformed to produce a spectrum $B'(\bar{v})$, a complex Fourier transform must be used unless the interferogram is symmetric. Hence, from Eq. 3.10,

$$B'(\bar{v}) = \int_{-\infty}^{+\infty} I(\delta)e^{-i2\pi\bar{v}\delta}\, d\delta \tag{3.29}$$

After transformation, $B'(\bar{v})$ is calculated by the complex addition

$$B'(\bar{v}) = \mathrm{Re}(\bar{v}) + i\,\mathrm{Im}(\bar{v}) \tag{3.30}$$

where $\mathrm{Re}(\bar{v})$ and $\mathrm{Im}(\bar{v})$ are the real and imaginary parts of $B'(\bar{v})$, respectively. If $I(\delta)$ is symmetrically sampled (i.e., there is an equal number of points on both sides of the centerburst), we may calculate the magnitude spectrum of $B'(\bar{v})$, which is denoted $|B(\bar{v})|$:

$$|B(\bar{v})| = \{[\mathrm{Re}(\bar{v})]^2 + [\mathrm{Im}(\bar{v})]^2\}^{1/2} \tag{3.31}$$

The magnitude spectrum exhibits zero phase error but has noise nonlinearities. The magnitude (Eq. 3.31) and complex spectra (Eq. 3.30) are related by the phase angle $\theta_{\bar{v}}$:

$$B'(\bar{v}) = |B(\bar{v})|\, e^{i\theta_{\bar{v}}} \tag{3.32}$$

The complex spectrum $B'(\bar{v})$ contains all the spectral information, but it is dispersed into two complex planes by the phase. The magnitude spectrum $|B(\bar{v})|$ is a real (in the complex sense) representation of the spectrum, but it is only the absolute value of that representation. The true spectrum $B(\bar{v})$, which is also real, lacks the noise nonlinearities of the magnitude spectrum. The object of phase correction is to produce the true spectrum $B(\bar{v})$. Since $\theta_{\bar{v}}$ usually varies slowly with wavenumber, it is possible to factor $e^{i\theta_{\bar{v}}}$ from Eq. 3.32. In this case,

$$
\begin{aligned}
B(\bar{v}) &= B'(\bar{v})e^{-i\theta_{\bar{v}}} \\
&= \text{Re}(\bar{v}) \cos \theta_{\bar{v}} + \text{Im}(\bar{v}) \sin \theta_{\bar{v}}
\end{aligned}
\tag{3.33}
$$

When $e^{i\theta_{\bar{v}}}$ is transposed from one side to the other of Eq. 3.32, only the real terms are retained from the trigonometric expansion because the true and amplitude spectra are real functions. Equation 3.33 represents a phase correction algorithm for double-sided interferograms.

As was stated above, the phase angle is, in practice, a slowly varying function with wavenumber; therefore, $\theta_{\bar{v}}$ does not need to be measured to a high resolution. The phase angle can be adequately calculated from a short, symmetrically sampled interferogram and subsequently applied to a much higher resolution spectrum. Of course, when this is done, the phase angle spectrum must be interpolated to the same resolution as the amplitude spectrum from the whole interferogram. Equation 3.33 holds in this case. Therefore, an asymmetrically sampled interferogram can be collected as long as there is a short symmetrically sampled portion that will provide the phase angle. This method of phase correction was developed by Mertz [8,9].

The Mertz method entails extracting a short double-sided interferogram, usually of 128, 256, or 512 total data points, from the recorded signal. This interferogram may be zero filled (see Section IV of this chapter.) The short interferogram is Fourier transformed and the phase angle calculated as shown:

$$
\theta_{\bar{v}} = \arctan \frac{\text{Im}(\bar{v})}{\text{Re}(\bar{v})}
\tag{3.34}
$$

where $\text{Re}(\bar{v})$ and $\text{Im}(\bar{v})$ are the real and imaginary parts of the complex transform, respectively. The complex Fourier transform (DFT or FFT) is performed on the entire interferogram after apodization. Before the phase correction can be performed (Eq. 3.33), the phase angle spectrum $\theta_{\bar{v}}$ is interpolated to bring it to the same resolution as the resolution of the entire interferogram.

Because a single-sided (or a symmetrically sampled) interferogram has a short signal on one side of the centerburst, the area around the centerburst is counted twice during transformation. To prevent photometric errors, the signal is weighted so that the symmetric area around the centerburst is evaluated correctly, as explained in Chapter 1, Section VI. The double-sided portion of the interferogram is multiplied by a ramp function that has a value of 0 at the beginning of the interferogram ($\delta = -\Delta_1$) and ends at a value of 1 at the end of the double-sided section ($\delta = +\Delta_1$). The interferogram, which extends to $\delta = \Delta_2$ ($\Delta_1 < \Delta_2$), is multiplied by the apodization function (boxcar, triangular, Norton–Beer, etc.) from $\delta = +\Delta_1$ to $\delta = \Delta_2$. An illustration of two such apodization functions is presented in Fig. 1.16.

The second method of phase correction, which was developed by Forman, Steele, and Vanasse [10], is mathematically equivalent to the Mertz method but is performed in the interferogram domain. Recalling that a multiplication in one domain is a convolution in the other, phase correction may be accomplished by taking the inverse Fourier transforms of the two right-hand terms of Eq. 3.33 and convolving them to produce a phase-corrected interferogram. The inverse Fourier transform of $B'(\bar{v})$,

$$J'(\delta) = \int_{-\infty}^{\infty} B'(\bar{v})e^{i2\pi\bar{v}\delta}\, d\bar{v} \qquad (3.35)$$

yields $J'(\delta)$, which is the recorded signal. Similarly,

$$\theta_\delta = \int_{-\infty}^{\infty} e^{-i\theta_{\bar{v}}}\, e^{i2\pi\bar{v}\delta}\, d\bar{v} \qquad (3.36)$$

where θ_δ is the interferogram of the phase angle spectrum. The two functions to be convolved may be of different dimensions; therefore, no interpolation of θ_δ is necessary. The phase-corrected interferogram, $J(\delta)$, is then

$$J(\delta) = \int_{-\infty}^{\infty} J'(\delta)\theta(\delta - \delta')\, d\delta$$
$$= J'(\delta)*\theta_\delta \qquad (3.37)$$

where $*$ denotes convolution.

The phase-corrected interferogram is symmetric and is thus an even function. A complex Fourier transform is not necessary and only a cosine Fourier transform need be performed if the short side of the interferogram is discarded. Apodization is accomplished by multiplying the interferogram from $\delta = 0$ to $\delta = \Delta$ (where Δ is maximum retardation) with the apodizing function. Because the short side of the interferogram has been discarded, no ramp function as used in the Mertz method is needed. Mathematically, both phase correction methods are equivalent. However, Chase [11] has found that the Mertz method is not as accurate as the Forman method, not due to the phase correction itself, but by the inclusion of the ramp function in the apodization. The error in the Mertz method is very small and can be discounted in most applications. The Mertz method is by far the more computationally efficient of the two methods due to the high computational requirements of a convolution.

An interesting aspect of the Forman method is the opportunity to include numerical filtering. Because $\theta_{\tilde{\nu}}$ is a quotient, it is not defined where there is no signal and only noise is present. This is because $\theta_{\tilde{\nu}}$ varies wildly and provides no useful information. Thus, by necessity the phase angle spectrum must be zeroed outside the signal bandwidth. This amounts to truncating the bandwidth of both the phase angle spectrum and the spectrum itself after convolution because the phase angle bandwidth is carried through the convolution. This property may be used to some advantage.

If a high-resolution spectrum is collected, the number of data points in the interferogram can be large (10^6 is not uncommon). Undersampling may be used to examine a restricted spectral region, but appropriate optical and electronic filters are required. An effective sampling can be achieved by numerical filtering. The entire interferogram is collected at the normal sampling interval and the phase angle spectrum calculated. Rather than compute the inverse Fourier transform of the entire bandwidth, the bandwidth of the phase angle spectrum is restricted to the region of interest. The only stipulation is that the restricted bandwidth is an integer fraction of the total bandwidth. Convolution produces a symmetric interferogram with an effective oversampling. If the restricted bandwidth is one-sixtieth the total bandwidth, the effective oversampling is a factor of 60 and only every sixtieth data point need be retained for cosine Fourier transformation. A huge reduction in storage space and transform computation time is achieved and no electronic or optical filter is required. This method of data reduction has been described by Bell [12], and at least one commercial FT–IR spectrometer vendor has implemented this technique [13].

Phase correction may not always be possible. As stated above, the phase angle spectrum is defined only where a signal exceeds the noise, which is not the case for discrete emission spectra. In this instance, a double-sided interferogram must be collected and the magnitude spectrum computed. Discrete emission spectra are not commonly encountered in infrared spectrometry but occur in the ultraviolet–visible region for atomic emission spectrometry (see Chapter 16).

One final aspect of interferogram transformation is concerned with the choice of the origin of the interferogram. It is common practice to shift the interferogram so that the first data point corresponds to the center-burst and the points that preceeded the centerburst are shifted to the end of the interferogram signal. Hence, the interferogram has maxima at the extrema and approaches zero in the center. This shift is performed after apodization, just before tranformation. The reason for this shift is to reference the phase angle to the zero phase, that is, to the centerburst. A very thorough treatment of apodization and phase correction may be found elsewhere [14].

IV. INTERFEROGRAM DOMAIN PROCESSING

We saw in Chapter 1, Section V, how the instrument line shape (ILS) function may be changed by multiplying the interferogram by an *apodization* function. Apodization functions not only control the ILS but also act as smoothing functions since the amplitude of the high spatial frequencies in the interferogram is reduced. When the absorption bands in the spectrum are broad, their spatial frequencies are predominantly low and are less attenuated by an apodization function. Thus, the overall effect of the apodization function in this case is to increase the SNR of the spectrum.

Similarly, *truncation* of the interferogram, that is, measuring at lower resolution, will also increase the SNR of a spectrum with broad spectral features since the high spatial frequencies are not measured at all.

These examples illustrate the fact that it is often very convenient to manipulate spectra in the interferogram, or time domain. It does not actually matter whether the spectrum is measured by an FT–IR or a dispersive spectrometer, except that for certain operations such as derivation or deconvolution the high SNR of FT–IR spectra is often necessary to obtain interpretable results. The measured absorbance spectrum is first transformed into an absorbance interferogram by calculating the inverse FT; this data array is then operated on in some way and then the forward FT is computed to recover the spectrum. Several examples will be presented to illustrate the use of interferogram domain processing.

a. Smoothing

If noise has been introduced into a spectrum by measuring it at an unnecessarily high resolution, the spectrum may be easily smoothed by computing its Fourier transform after first zeroing out the last part of the array and computing the inverse transform. If the last half of the array is zeroed out, the noise is reduced by a factor of approximately $2^{1/2}$. It should be noted, however, that this is a rather inefficient method for spectral smoothing, since the same effect could have been achieved in half the time if the measurement had been made *originally* at the lower resolution. In addition, the possibility of introducing side lobes on sharp spectral features exists when smoothing is effected by this simple truncation process. This may be avoided through the use of a smoothly decaying apodization function to reduce the amplitude of the high spatial frequencies without introducing side lobes on narrow bands.

b.   Interpolation by Zero Filling

As described above, the complex FFT produces one real (spectrum) point per resolution element. Although it is true that these points define the spectrum, a linear interpolation between each point results in an obvious stepping. A direct result of this stepping effect is a loss in the photometric accuracy of Fourier transform spectrometry since it is rare that a computed datum corresponds to the wavenumber of maximum absorption (or emission) of any spectral feature.

When the complex FFT of N interferogram points is performed, the real output array contains $\frac{1}{2}N$ points and the imaginary array also contains $\frac{1}{2}N$ points across the spectral bandwidth (see Section VI, this chapter). If N zeros are added to the input array, the real and imaginary output bandwidth then each contain N points, of which $\frac{1}{2}N$ are linearly independent and the rest represent interpolations between these points. The interpolation function is determined by the apodization line shape. Although a smoother plot results from calculating more spectrum points in this fashion, distinct breaks are still seen between each data point, as evidenced by the portion of the spectrum of carbon monoxide shown in Fig. 3.2a, which was calculated after adding 8-k zeros to an 8-k-point interferogram (1-k = 1024).

In the general case, if $(2^m - 1)N$ zeros are added to an N-point interferogram, where m is an integer greater than 1, the bandwidth after the complex FFT will contain $2^{(m-1)}N$ points, of which $\frac{1}{2}N$ are linearly independent and the rest are interpolated. This procedure, which is known as *zero filling* [15], results in a far smoother spectrum, as demonstrated

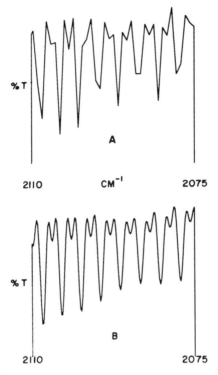

Fig. 3.2. Spectra of carbon monoxide from 2110 to 2075 cm^{-1} showing the effect of zero filling; spectra were measured at an effective resolution of 2 cm^{-1} with boxcar truncation. (*a*) $m = 1$; one independent and one interpolated point per resolution element; (*b*) $m = 3$; one independent and seven interplated points per resolution element. (Reproduced from [15], by permission of the Society for Applied Spectroscopy; copyright © 1975.)

in Fig. 3.2. Figure 3.2*a* shows a spectrum of carbon monoxide computed with $m = 1$, whereas Fig. 3.2*b* shows the same interferogram computed with $m = 3$. Here $m = 1$ represents the minimum amount of zero filling that should be carried out on any interferogram that is not symmetrical about the zero retardation point and for which a complex FFT must be performed. However, for really good photometric precision, at least *eight* output points per resolution element are necessary, for which $m = 3$. Zero filling obviously increases the computation time considerably, so that there is a tradeoff between computing time and photometric precision. In addition, when data systems with limited storage are being used for the computations, it may not even be possible to perform zero filling to any great extent.

To compute a spectrum from a zero-filled interferogram so that it has

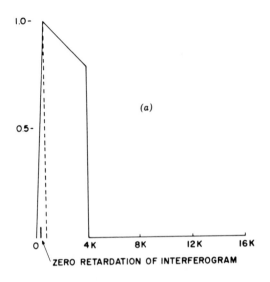

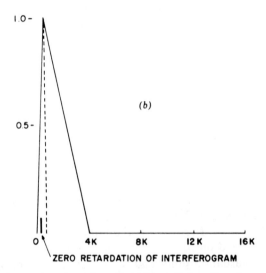

Fig. 3.3. Apodization function, triangular in this case, for an interferogram with zero filling. Note only the interferogram data are apodized, not the zeros. (Reproduced from [15], by permission of the Society for Applied Spectroscopy; copyright © 1975.)

the same resolution and ILS as the spectrum computed from a triangularly apodized, non-zero-filled interferogram, only the measured data should be multiplied by the apodization function. This is the function of the type shown in Fig. 3.3. This particular function is illustrated for a 4-k interferogram zero filled with 12-k zeros ($m = 2$).

It should be noted at this point that whereas zero filling is the fastest method of computing interpolated data points over the *complete* spectrum, other methods of adding interpolated data points can also be used. In particular, a polynomial of two or three terms can be fitted to three or four data points very rapidly, and the interpolations between small numbers of data points can be made more quickly using this method than the zero-filling technique.

Forman [16] has discussed the tradeoffs between the use of zero filling and other interpolation techniques, concluding that zero filling is the most computationally efficient means of interpolating the complete spectrum. If only a short region of the spectrum has to be interpolated, however, other techniques can be more efficient.

c. Derivative Spectrometry

One of the properties of Fourier transforms is that if a single-sided absorbance interferogram is multiplied by a function of the form ax^n,

$$F(x) = ax^n$$

Computation of the forward transform gives the nth derivative of the spectrum. In this discussion, the abscissa will be denoted x to indicate that the array being used is not the original interferogram but is rather the inverse FT of the phase-corrected (and usually ratioed) spectrum. In practice, even-order derivative spectra are more easily interpreted than odd-order derivatives, and more applications of second-derivative FT–IR spectra have been described than for any other order. The computation of a second-derivative spectrum is illustrated in Fig. 3.4. Since complex FTs are usually used for the computation of derivative spectra, it may be noted that even-order derivatives are located in the real array whereas odd-order derivatives are found in the imaginary array.

The effect of the parabolic weighting function used in the calculation of second-derivative spectra is to amplify the noise at high retardations. This is because the signal level tends to be low in these regions and the noise is constant across the interferogram. Spectra should therefore not be measured at a resolution that is unnecessarily high. On the other hand, if the resolution is too low, the amplitude of the signal in the interferogram

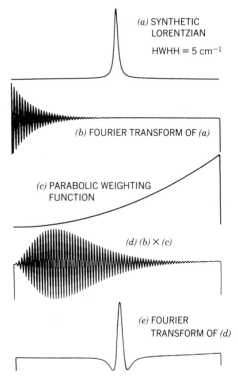

Fig. 3.4. Computation of the second derivative of a Lorentzian band. (*a*) Synthetic Lorentzian with $\gamma = 5$ cm^{-1}; (*b*) Fourier transform of (*a*); (*c*) Parabolic weighting function; (*d*) Product of (*b*) and (*c*); (*e*) Fourier transform of (*d*), yielding the second derivative of (*a*).

due to the band will not have decayed sufficiently before it is truncated so that side lobes will be introduced into the spectrum (see Fig. 3.5). Two rules of thumb may be cited for the measurement of derivative spectra:

1. Each time the order of the derivative is increased by 2, the SNR of the spectrum is degraded by approximately an order of magnitude.
2. For the measurement of second-derivative spectra, the resolution should be about 2.5 times less (numerically) than the FWHH of the bands in the spectrum.

d. Fourier Self-Deconvolution

Most bands in the infrared absorbance spectra of condensed phase samples and gases whose line shape is determined by pressure broadening

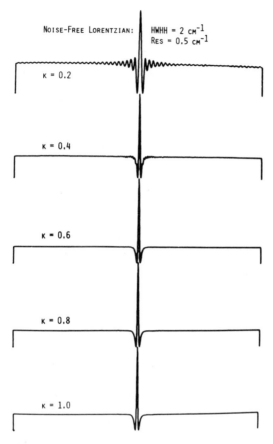

Fig. 3.5. Second derivative spectra of synthetic Lorentzian with HWHH = 2 cm^{-1} computed at retardations of 2 cm (lowest curve, K = 1.0); 1.6 cm (K = 0.8); 1.2 cm (K = 0.6); 0.8 cm (K = 0.4); 0.4 cm (K = 0.2). For the upper curve the nominal resolution is of the same order as the original bandwidth and intense side lobes are observed.

have a Lorentzian profile, given by Eq. 1.26 as

$$A(\bar{\nu}) = A_{\text{peak}} \frac{\gamma^2}{\gamma^2 + (\bar{\nu} - \bar{\nu}_0)^2} \tag{3.38}$$

The Fourier transform of $A(\bar{\nu})$ is given by

$$y(x) = \int_{-\infty}^{+\infty} A(\bar{\nu}) \exp(+2\pi i x \bar{\nu}) \, d\bar{\nu} \tag{3.39}$$

$$= \tfrac{1}{2}(\gamma A_{\text{peak}}) \exp(-2\pi i \bar{\nu}_0 \mid x \mid) \exp(-2\pi \gamma \mid x \mid) \tag{3.40}$$

The effect of the Lorentzian broadening function is to multiply the sinusoid that would be given by an infinitely sharp band at $\bar{v}_0$ by an exponential decay function $\exp(-2\pi\gamma \mid x \mid)$. If $y(x)$ is multiplied by a function of the form

$$f(x) = \exp(+2\pi\gamma' \mid x \mid) \tag{3.41}$$

the rate of decay of $y(x)$ will be decreased. If $\gamma' = \gamma$, the effect of the Lorentzian broadening function will be completely removed. When the Fourier domain array is transformed, the band will have the characteristic shape of a sinc function. The magnitude of the side lobes can, of course, be reduced by multiplying the array by a suitable apodization function prior to performing the transform. The entire process is illustrated in Fig. 3.6.

If γ' is less than γ, the sinusoidal signal is left with a residual decay, which is equivalent to an apodization function of the form

$$A(x) = \exp[-2\pi(\gamma - \gamma') \mid x \mid] \tag{3.42}$$

Nevertheless, the width of the band resulting in performing the transform is less than the original width γ so that the effective resolution is enhanced.

Recalling that convolution in the spectrum is equivalent to multiplication in the interferogram, it is apparent that deconvolution may be achieved by division in the interferogram. In the above example, multiplication of the interferogram by $\exp(2\pi\gamma' \mid x \mid)$ is the same as division by $\exp(-2\pi\gamma' \mid x \mid)$. Since the Lorentzian broadening function is being removed, this process has been called *Fourier self-deconvolution*. The technique was first described in 1962 by Stone [17] and more recently has been discussed in some detail in a series of three papers by Kauppinen et al. [18–20].

The type of weighting functions used in Fourier self-deconvolution are similar to those used in the generation of derivative spectra in that their magnitude is greatest for large retardations. It is again necessary for the noise level of the spectrum to be very low if self-deconvolution is to be successfully achieved. It is also important to realize that the amount by which the FWHH of bands can be reduced is determined by the resolution at which the spectrum was originally measured. It is theoretically impossible, even for noise-free spectra, to create a linewidth less than the instrumental resolution unless severe side lobes can be tolerated. (This would be the case if a value of γ' greater than γ were selected.) Since the selection of the parameters used in Fourier self-deconvolution is usually done empirically, it is always necessary to watch for the appearance

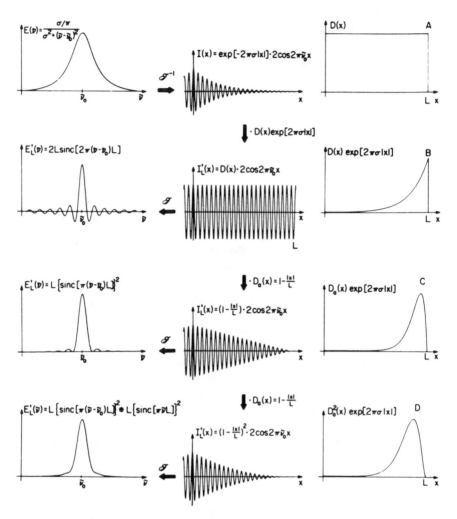

Fig. 3.6. Illustration of the Fourier self-deconvolution process starting with a Lorentzian line (top, left corner) and applying different weighting functions. The middle column illustrates the effect of various functions (right column) on the Fourier transform of the Lorentzian line (top, middle column). The resulting line shapes are shown in the left column. (Reproduced from [18], by permission of the Society for Applied Spectroscopy; copyright © 1981.)

of side lobes, which indicates that the spectrum has been "overdeconvolved." The best way of checking that all the features in deconvolved spectra are real is to verify their existence by remeasuring the spectrum at higher resolution and SNR, if possible. If a spectral feature disappears under these circumstances, it was almost certainly an artifact in the lower-resolution spectrum.

V. REAL-TIME FOURIER TRANSFORMS

The techniques that have been described in the first two sections can only be used to compute the spectrum after the end of the measurement of the interferogram. It is occasionally useful to compute the spectrum during the measurement of the interferogram. This may be useful, for example, to determine whether the noise level in the spectrum is too great to justify any further increase in the retardation or to check if the spectral information is increasing with increasing retardation. The techniques to be described in this section can only be used if the data rate is sufficiently low that the computations for one data point are finished before the next data point is measured. Thus, spectra can rarely be computed from rapid-scanning interferometers using this techniques, especially in view of the fact that signal averaging is commonly used to improve the signal-to-noise ratio in the interferogram before the transform.

The techniques that have been developed for the purpose of computing the spectrum during the measurement of the interferogram are generally known as real-time Fourier transforms. It is generally only possible to examine a small spectral range (1-k output points is a typical number), but this can make it possible to examine a complete far-infrared spectrum at medium resolution or part of a high-resolution spectrum at any wavenumber. These techniques are not employed in any commercial system, but they are presented here because of their use in "home-made" systems.

The conventional Fourier transform involves several sums for each wavenumber value:

$$B(\bar{\nu}_1) = \tfrac{1}{2}I_a(0) + I_a(1) \cos(2\pi\bar{\nu}_1 \cdot h) + I_a(2) \cos(2\pi\bar{\nu}_1 \cdot 2h) + \cdots$$

$$B(\bar{\nu}_2) = \tfrac{1}{2}I_a(0) + I_a(1) \cos(2\pi\bar{\nu}_2 \cdot h) + I_a(2) \cos(2\pi\bar{\nu}_2 \cdot 2h) + \cdots \quad (3.43)$$

$$\vdots$$

$$B(\bar{\nu}_i) = \tfrac{1}{2}I_a(0) + I_a(1) \cos(2\pi\bar{\nu}_i \cdot h) + I_a(2) \cos(2\pi\bar{\nu}_i \cdot 2h) + \cdots$$

The usual calculation is made in each row after scanning of the interferogram. For real-time transforms [21], the calculation is made in each column, that is, in parallel for each wavenumber $\bar{v}_i$. The cosines cos $2\pi\bar{v}\cdot kh$ must therefore be calculated at each sampling point or taken from a stored cosine table, if necessary using interpolation routines. Recursion relationships cannot be used in this case.

Let the number of wavenumber points in the output be M, so that the spectral intensity is being computed at wavenumbers $\bar{v}_1, \bar{v}_2, \bar{v}_2, \ldots, \bar{v}_i,$ $\ldots, \bar{v}_M$. After k input points have been measured, M computer locations corresponding to the wavenumbers 1 through M contain the sum $B_{(k-1)}(\bar{v}_i)$, where

$$B_{(k-1)}(\bar{v}_i) = \tfrac{1}{2}I_a(0) + \sum_{j=1}^{k-1} I_a(j)\cos(2\pi\bar{v}_i\cdot jh) \qquad (3.44)$$

When the $(k+1)$st point is taken, the product $I_a(k)\cos(2\pi\bar{v}_i kh)$ is calculated and added to each value of $B_{(k-1)}(\bar{v}_1)$ replacing the value of $B_{(k-1)}(\bar{v}_i)$ in the computer memory. The current values of the M locations can be viewed on an oscilloscope to check visually how the spectrum is changing with retardation, and the interferogram can be simultaneously recorded for computation of the complete measurement at the end of the measurement.

In practice, real-time computing techniques have fairly limited application to chemical Fourier transform spectrometry. One reason for this is the high data rate required for reasons of dynamic range when an intense source is being measured. This type of measurement necessitates the use of signal averaging for the improvement of signal-to-noise ratio so that the use of real-time computing of the spectrum presents no positive benefits. Another reason is that even if there is as much as a 1-sec interval between data points, the number of output points is still limited since the equivalent of a Cooley–Tukey FFT usually cannot be performed in the real-time mode. Thus, for mid-infrared chemical spectrometry where relatively large numbers of data points are collected per interferogram at a high data rate using signal-averaging techniques, the optimum computing technique is usually an FFT at the completion of data collection.

For low-resolution far-infrared spectrometry using a slow- or stepped-scan interferometer or very high resolution spectrometry using a stepped-scan interferometer, the complete spectrum can be computed between input data points. In these cases, however, there is the possibility of two sources of error. First, it was noted in Chapter 1, Section II, that an interferogram measured using a slow-scan interferometer consists of an ac and a dc component. The first step in the Fourier transform calculation involves determining the mean value of the interferogram and subtracting this value from each measured data point. This is most important for computing spectra using real-time techniques, and therefore the mean value of the interferogram must be determined before the measurement is started. Levy et al [22] have performed this subtraction electronically, backing off the signal at the infrared detector by a stable constant voltage so that the average value of the interferogram well away from zero retardation is set as precisely as possible at zero volts. The effect of a drift of the total signal measured at the detector during the measurement would be the introduction of a large amount of false energy at low wavenumber. This effect is not an effect of the real-time technique and would be equally noted if the transform were carried out at the end of the measurement. The only way to compensate for such a drift is by the use of a rapid-scanning interferometer whose bandpass is outside the frequency of the source drift or by phase modulation.

The second source of error involves the determination of the zero retardation point. Any slight sampling error in the position at which the first data point is taken means that the instrumental line shape will be distorted, as discussed in Chapter 1, Section VI. To get the best results with real-time computing techniques, the sampling error ϵ can then be computed and applied to the summation shown in Eq. 3.43 to give

$$B_{(k-1)}(\bar{\nu}_i) = \tfrac{1}{2}I_a(\epsilon) \sum_{j=1}^{k-1} I_a(j) \cos[2\pi\bar{\nu}_i(jh + \epsilon)] \qquad (3.45)$$

If the interferogram is not symmetric about the zero retardation point, it may be deduced that the phase angle $\theta_{\bar{\nu}}$ is finite and that it probably varies with wavenumber. In this case, the more sophisticated phase correction routines described earlier in this chapter must be applied. These routines involve calculations on a relatively large (64–512 points) region of the interferogram on both sides of zero retardation; real-time computations could not be started until after the first points in the interferogram had been sampled, which would be an extremely difficult computing problem unless the mirror drive were stopped at this point.

VI. FOURIER TRANSFORM: A PICTORIAL ESSAY

The nuances of a practical computational technique are often not clear from a discussion of the theory. With this in mind an example of the Fourier transformation of an interferogram into a spectrum is illustrated for each step of the procedure. Both methods of phase correction are shown, that is, the methods devised by Mertz [8,9] and Forman [10].

a. Mertz Method

The Fourier transformation of an interferogram by the Mertz method is illustrated in Fig. 3.7. Figure 3.7a is an interferogram of acetone in the vapor phase. This interferogram is somewhat chirped so that some phase error can be expected upon direct transformation. The first step in the Mertz method is to extract a small double-sided interferogram from around the centerburst (in this case 256 data points), which is shown in Fig. 3.7b. Figure 3.7c illustrates the triangular apodization function used to apodize the double-sided interferogram. After the arrays in Figs. 3.7b,c are multiplied together, the data are shifted about the maximum value to reference the function as close to zero phase as possible, see Fig. 3.7d. This short double-sided interferogram is used to calculate the phase curve. Upon complex Fourier transformation of the function in Fig. 3.7d, both real and imaginary portions of the phase curve are produced, as the original interferogram has both cosine and sine components. The real and imaginary parts of the phase curve, shown in Fig. 3.7e, are even functions by virtue of the fact that the interferogram is even (or approximately so). In these curves, the central point is at the highest wavenumber of the bandpass (7900 cm^{-1} for this spectrum) whereas the two end points correspond to 0 cm^{-1}. The actual phase curve (Fig. 3.7f) is calculated using Eq. 3.34. The final step in the preparation of the phase correction curve is the interpolation of the curve to the same resolution as the final spectrum.

Once interpolated, the cosine and sine functions are calculated from the phase curve and are shown in Fig. 3.7g. Only one-half of each curve of Fig. 3.7g has been retained since each half is a mirror image of the other and no information is lost when one half is discarded. It should also be noted that the phase curve has a value of zero where the phase is not defined. These regions where the phase is not defined are those regions outside the spectral bandwidth, that is, at wavenumbers greater than 4400 cm^{-1} and less than 550 cm^{-1}. The cosine and sine phase curves are not calculated in the undefined regions.

The interferogram in Fig. 3.7a is apodized with a triangular apodization function, such as the one shown in Fig. 3.3b, zero-filled by a factor of 2, and shifted so that the maximum valued datum is the first point in the array. The result of these operations is shown in Fig. 3.7h. Upon Fourier transformation, the real and imaginary parts of the spectrum are produced in Fig. 3.7i (only one-half of each curve is shown). The data in Figs. 3.7g,

a

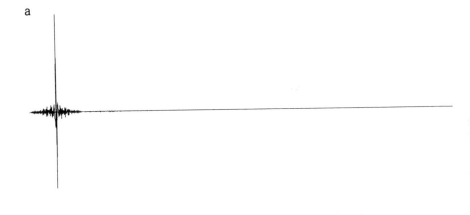

b

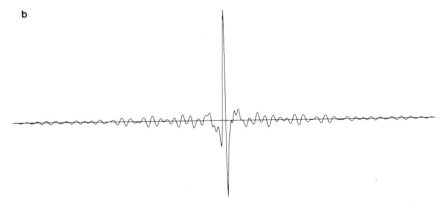

Fig. 3.7. Pictorial essay of transformation and phase correction using the Mertz method. (*a*) Interferogram of acetone in the vapor phase. (*b*) Short double-sided interferogram, 256 points total. (*c*) Apodization function for double-sided interferogram. (*d*) Product of interferogram in (*b*) and apodization function in (*c*). The interferogram is rotated for zero phase. (*e*) Real and imaginary portions of complex FFT of interferogram in (*d*). (*f*) Phase curve from data in (*e*). (*g*) Cosine and sine values of phase curve. These have been interpolated to the full resolution of the spectrum and only half the data are shown because these functions are even. (*l*) Interferogram in (*a*) after apodization and shift. (*i*) Real and imaginary parts of complex FFT of data in (*l*). (*j*) Products of curves in (*g*) and (*i*). The real spectrum is multiplied by the cosine phase curve and the imaginary spectrum by the sine phase curve. (*k*) Sum of spectra in (*j*). (*l*) Transmittance spectrum after ratioing single-beam spectrum in (*k*) with a reference single-beam spectrum.

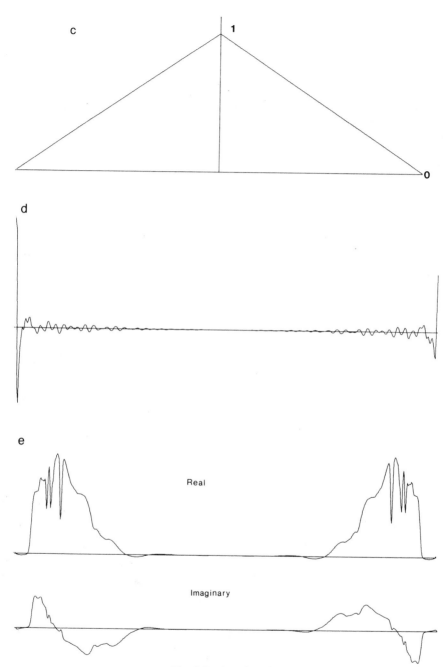

Fig. 3.7. (continued)

111

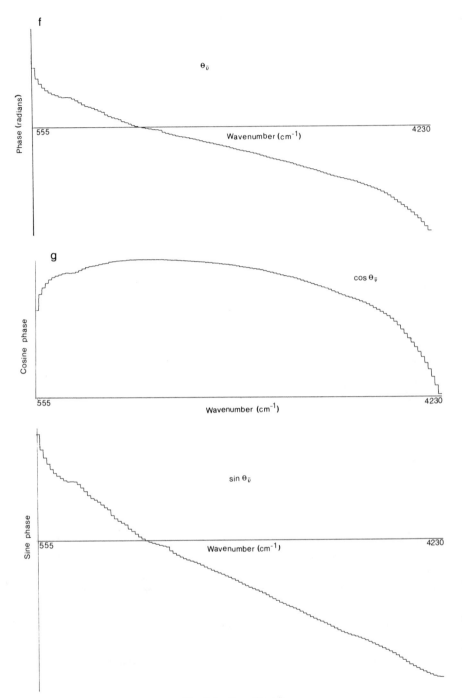

Fig. 3.7. (continued)

112

h

i

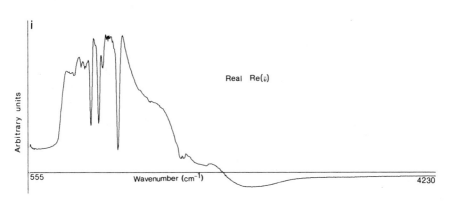

Real Re($\tilde{v}$)

Arbitrary units

555   Wavenumber (cm^{-1})   4230

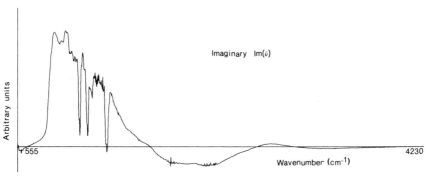

Imaginary Im($\tilde{v}$)

Arbitrary units

555   4230

Wavenumber (cm^{-1})

Fig. 3.7. (continued)

113

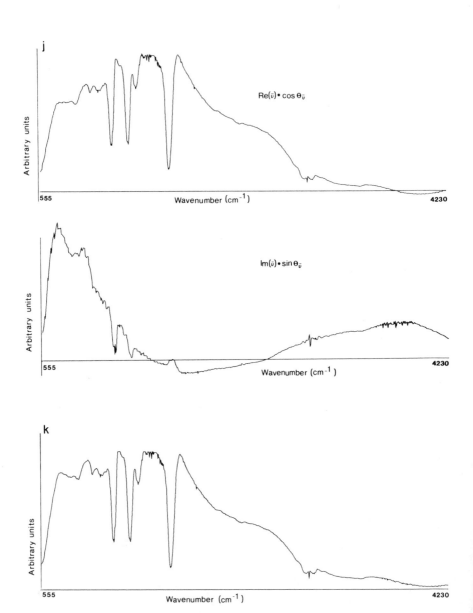

Fig. 3.7. (continued)

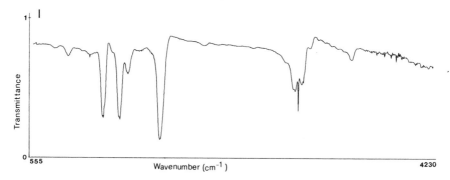

Fig. 3.7. (continued)

i are multiplied together to produce the curves in Fig. 3.7*j*. Addition of the real or imaginary parts (Fig. 3.7*k*) produces the single-beam phase-corrected spectrum. The results of Figs. 3.7*j,k* come from the application of Eq. 3.33. A similar procedure is carried out with a reference signal (no sample in the infrared path) and the two single-beam spectra are numerically ratioed to produce the phase-corrected transmittance spectrum, Fig. 3.7*l*.

b. Forman Method

The identical interferogram to the one in Fig. 3.7*a* is shown in Fig. 3.8*a*. Because the Mertz and Forman methods both use a short double-sided interferogram to calculate the phase of the spectrum, the early steps of both methods are identical, and the procedure described for Figs. 3.7*b*–*g* is carried out in the Forman method, with two exceptions. First, no apodization is applied to a double-sided interferogram (Fig. 3.7*c*), and second, no interpolation of the phase curve (Fig. 3.7*f*) is performed. The complete cosine and sine phase curves are calculated from the phase curve, and the regions where the phase is not defined are left as zero. An inverse complex FFT is taken of the cosine (real) and sine (imaginary) parts of the phase curve to produce an interferometric signal, Fig. 3.8*b*. The interferogram in Fig. 3.8*b* has maxima at both ends, reflecting the fact that the phase is referenced to zero. The data are shifted so the centerburst is in the middle of the array, Fig. 3.8*c*. The interferogram of the phase is apodized prior to convolution with an apodization function $A(\delta)$, devised by Forman [10]:

$$A(\delta) = \left[1 - \left(\frac{\delta}{\Delta} \right)^2 \right]^2 \qquad (3.46)$$

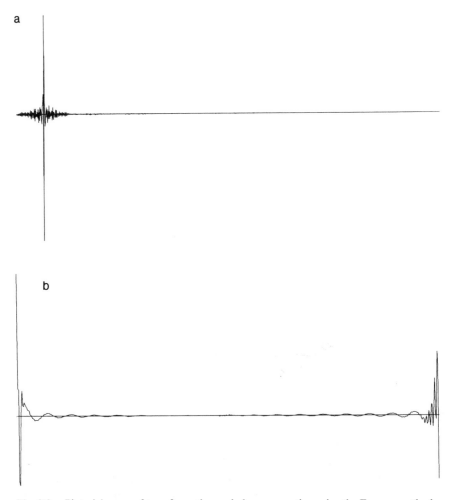

Fig. 3.8. Pictorial essay of transformation and phase correction using the Forman method. (*a*) Interferogram from Fig. 3.7*a*. (*b*) Inverse complex FFT of phase curve from short double-sided interferogram. (*c*) Shift of data to put centerburst in the middle of the data. (*d*) Forman apodization function for interferogram in (*c*). (*e*) Apodized interferogram from (*d*). (*f*) Result of convolution of interferograms in (*a*) and (*e*). (*g*) Interferogram from (*f*) after zero filling and shift prior to cosine FFT. (*h*) Single-beam spectrum. (*i*) Transmittance spectrum after ratioing single-beam spectrum in *h* with a single-beam reference spectrum.

116

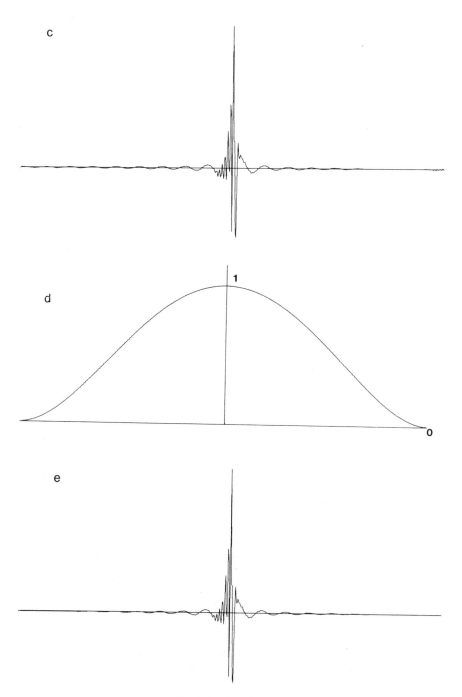

c

d

1

0

e

Fig. 3.8. (continued)

117

f

g

h

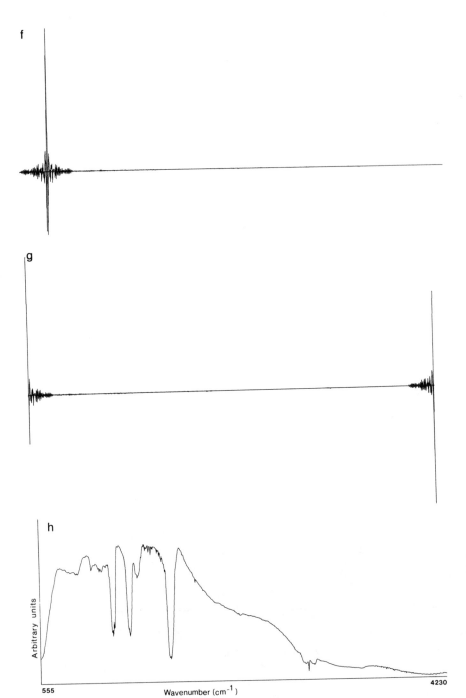

Arbitrary units

555   Wavenumber (cm^{-1})   4230

Fig. 3.8. (continued)

118

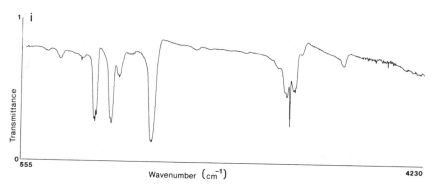

Fig. 3.8. (continued)

which is illustrated in Fig. 3.8*d*. The Forman apodization function is used to reduce the effects of the numerical filtering operation (Section III, above), which were introduced by default, by nulling the phase in the regions in which it is not defined. The apodized phase interferogram is shown in Fig. 3.8*e*, and this is convolved with the interferogram in Fig. 3.8*a*, yielding the phase-corrected interferogram in Fig. 3.8*f*.

The interferogram that has been phase corrected exhibits no chirping and is by definition a real and even function. The remainder of the data is zero filled and the short side moved to the end prior to Fourier transformation, see Fig. 3.8*g*. The single-beam spectrum is calculated from the data in Fig. 3.8*g* by a cosine FFT, the result of which is shown in Fig. 3.8*h*. A reference spectrum is calculated in an identical manner and the two single-beam spectra are ratioed, as in the Mertz method, to produce a ratioed spectrum, Fig. 3.8*i*.

REFERENCES

1. J. Connes, *Rev. Opt.*, **40**, 45, 116, 171, 233 (1961); English translation as Document AD 409869, Clearinghouse for Federal Scientific and Technical Information, Cameron Station, VA.
2. M. L. Forman, *J. Opt. Soc. Am.*, **56**, 978 (1966).
3. J. W. Cooley and J. W. Tukey, *Math. Comput.*, **19**, 297 (1965).
4. E. O. Brigham, *The Fast Fourier Transform*, Prentice-Hall, Englewood Cliffs, N.J. (1974).
5. J. Connes, Aspen Int. Conf. on Fourier Spectrosc., 1970 (G. A. Vanasse, A. T. Stair, and D. J. Baker, eds.), AFCRL-71-0019, p. 83 (1971).
6. Coblentz Society Board of Management, *Anal. Chem.*, **38**, 27A (1966).

7. R. B. Lam, R. C. Wieboldt, and T. L. Isenhour, *Anal. Chem.*, **53**, 889A (1981).

8. L. Mertz, *Transformations in Optics*, Wiley, New York (1965).

9. L. Mertz, *Infrared Phys.*, **7**, 17 (1967).

10. M. L. Forman, W. H. Steele, and G. A. Vanasse, *J. Opt. Soc. Am.*, **56**, 59 (1966).

11. D. B. Chase, *Appl. Spectrosc.*, **36**, 240 (1982).

12. R. J. Bell, *Introductory Fourier Transform Spectroscopy*, Academic Press, New York, p. 59 (1972).

13. Bomem, Inc., 625 Marais Street, Vanier, Quebec GlM 2Y2, Canada.

14. J. E. Bertie, Apodization and Phase Correction, in *Analytical Applications of FT-IR to Molecular and Biological Systems*, (J. R. Durig, ed.), D. Reidel, Dordrecht, Holland (1980).

15. P. R. Griffiths, *Appl. Spectrosc.*, **29**, 11 (1975).

16. M. L. Forman, *Appl. Opt.*, **16**, 2081 (1977).

17. H. Stone, *J. Opt. Soc. Am.*, **52**, 998 (1962).

18. J. K. Kauppinen, D. J. Moffatt, H. H. Mantsch, and D. G. Cameron, *Appl. Spectrosc.*, **35**, 271 (1981).

19. J. K. Kauppinen, D. J. Moffatt, H. H. Mantsch, and D. G. Cameron, *Anal. Chem.*, **53**, 1454 (1981).

20. J. K. Kauppinen, D. J. Moffatt, D. G. Cameron, and H. H. Mantsch, *Appl. Opt.*, **20**, 1866 (1981).

21. H. Yoshinaga et al., *Appl. Opt.*, **5**, 11 59 (1966).

22. F. Levy, R. C. Milward, S. Bras, and R. Le Toullec, Aspen Int. Conf. on Fourier Spectrosc., 1970 (G. A. Vanasse, A. T. Stair, and D. B. Baker, eds.), AFCRL-71-0019, p. 331 (1971).

TWO-BEAM INTERFEROMETERS

I. DRIVE SYSTEMS AND DESIGNS FOR COMMON INTERFEROMETERS

a. Introduction

The simplest version of the Michelson interferometer consists of two plane mirrors, one of which can move in a direction perpendicular to the plane of the other, with a beamsplitter between them. The interferometer may be considered as consisting of two separate parts, the drive mechanism for the moving mirror (including the sampling triggers), which will be described in the first section of this chapter, and the beamsplitter, which will be described in the second section.

The higher the resolution and maximum wavenumber in the spectrum, the higher must be the quality of the drive mechanism. The crudest drive mechanisms have been made for far-infrared spectrometry at low and medium resolution. Many of these are "home-built" instruments and are relatively easy to construct. Although the tolerances of the mirror drive for medium-resolution ($\Delta \bar{\nu} \geq 0.05$ cm^{-1}) mid-infrared spectrometry must be higher than for the far-infrared systems, many interferometers falling into this category are now commercially available at moderate cost. When ultrahigh resolution ($\Delta \bar{\nu} \leq 0.01$ cm^{-1}) is required in the mid or near infrared, great care must be taken in the design of the interferometer. Although most of these instruments have been constructed by the spectroscopists who wished to use them, at least one very high resolution interferometer is now commercially available. Interferometers in each of these categories will be described, with primary emphasis being placed on medium-resolution mid-infrared interferometers of the type most likely to be used by chemical spectroscopists.

b. Medium-Resolution Far-Infrared Interferometers

The earliest interferometers designed specifically for far-infrared spectrometry at low or medium resolution were usually of the slow-scanning or stepped-scan variety. The first commercially available slow-scanning interferometers to be built for far-infrared spectrometry incorporated an associated Moiré fringe reference device so that the interferogram could be sampled at equal intervals of retardation. Conversely, for far-infrared step-and-integrate interferometers, fringe referencing is unnecessary, since the accuracy of stepping motors is sufficiently high that the retardation can be increased by equal increments without adding any noise to the signal.

The scan speed of continuous scanning interferometers for far-infrared spectrometry is variable, typically between about 10 and 0.1 μm sec^{-1}. (The effect of scan speed on signal-to-noise ratio is discussed in Chapter 7.) For step-and-integrate systems, the time during which the movable mirror is held at each position can be varied between about 0.5 and 5 sec and occasionally longer. All the early far-infrared instruments incorporated an external chopper to modulate the beam from a water-cooled mercury arc, most commonly at a frequency between 10 and 15 Hz. In this case, both the ac and dc components of the interference record are measured; the dc component is then subtracted from $I'(\delta)$ to give the interferogram prior to the actual Fourier transform. The dc component is usually estimated from the average of the last 100 points in the interferogram. If the source intensity drifts slowly, however, the intensity of the first half of the interferogram may differ from that of the final 100 points. In this case, a large spike near 0 cm^{-1} will be observed in the spectrum.

None of the original commercial far-infrared Fourier spectrometers is now available. Indeed, the companies involved—Research and Industrial Instruments (later Beckman–RIIC) and Grubb-Parsons in England, Coderg in France, and Polytec in Germany—stopped making Fourier spectrometers of any sort several years ago. It is, however, interesting to note that most of these instruments were descended from the interferometers originally developed by Gebbie's group at the National Physical Laboratory in England. The NPL Cube is still being used in many laboratories around the world, and a variant of this instrument was being marketed, at the time of this writing, by Imperial College Instruments in England.

c. Medium-Resolution Mid-Infrared Interferometers

(i) Michelson Interferometers

The mirror drive for an interferometer being used for mid-infrared spectrometry (with $\bar{\nu}_{max} = 4000$ cm^{-1}) must have at least an order of magnitude higher precision than the drive of a far-infrared interferometer ($\bar{\nu}_{max} = 400$ cm^{-1}) operating at the same maximum retardation. In addition, the intensity of mid-infrared sources used for laboratory measurements is much higher than that of far-infrared sources. Thus, if a slow-scanning interferometer were used for mid-infrared Fourier spectrometry, the SNR of the interferogram near the centerburst would generally exceed the dynamic range of the analog-to-digital converter (ADC). It is for this reason that most interferometers designed for mid-infrared spectrometry permit high mirror velocities to be attained. On many instruments, the mirror velocity may be varied, often between about 0.5 and 60 mm sec^{-1}. For the few low-cost instruments with a single-scan speed, a velocity of 1.6 mm sec^{-1} is typical if a TGS detector is incorporated.

Rapid-scanning interferometers usually operate by signal averaging repetitive scans. For exact signal averaging each interferogram must be sampled at exactly the same retardation interval for every scan. Mid-infrared interferometers must therefore be free of any short- or long-term drifts in alignment so they are sometimes thermostated. It is equally important that great care is taken to ensure that the first data point of each successive interferogram is sampled at exactly the same retardation for each scan.

All the current rapid-scanning interferometers are direct descendants of the first interferometers designed by Mertz [1]. These early instruments were small and simple but of rather too low resolution to be really useful for chemical spectrometry. They derived much of their sensitivity from the large acceptance angle of the interferometer. The large divergence angle of the beam together with the low precision drive and low resolution meant than interferometers of this type (such as the Block Engineering Model 200) were used successfully for measurements of extended remote sources. They were not used for many laboratory measurements where either the source or the sample was of small size.

The moving mirror of these interferometers was mounted on a spring and displaced by an electromagnetic transducer similar to the voice coil of a loudspeaker. A slowly increasing current was applied to the coil to drive the mirror at a constant velocity. At the end of the scan, the direction of the current was altered rapidly, thereby retracing the mirror to its rest position. The drive current was again reversed and gradually increased so that a second drive cycle began. The trigger for data collection was actuated by a constant-frequency clock, so that sampling was strictly linear in time rather than linear in retardation. Nevertheless, because of the short scan, this method allowed signal-averaging techniques to be used successfully to increase the signal-to-noise ratio of interferograms from weak sources.

The constant-frequency time base that was used for these early rapid-scanning interferometers gave good results primarily because the retardation was so short (<1 mm). There was not enough time or long enough retardation for the interferograms to get significantly out of phase. Had the retardation been much longer, the chances of digitizing the interfer-ogram at corresponding points along the scan (sometimes called *coherent addition* or "coadding") would have been reduced considerably. For rapid-scanning interferometers designed for higher-resolution operation in the mid and near infrared, a method of coherently signal averaging successive interferograms was developed by Mertz and Curbelo at Block Engineering [2].

This method involved the use of a second (reference) interferometer, the moving mirror of which was attached to the moving mirror of the main, or signal, interferometer. When monochromatic light is passed through this reference interferometer to a detector, a sinusoidal signal is measured. Since each zero crossing of the laser interferogram is found at equal intervals of retardation, they can be used to trigger the ADC sampling of the signal from the main interferometer. For a 632.8-nm he-lium–neon laser, each zero crossing occurs at retardation intervals of 0.3164 μm. If sampling was triggered by each zero crossing, the shortest wavelength in the spectrum allowed by the Nyquist sampling criterion (Chapter 2) would be 0.6328 μm, that is, $\bar{\nu}_{max} = 15804$ cm^{-1}; this interval would be used for *near*-infrared spectrometry. For mid-infrared spectro-metry, samples would usually be taken every second zero crossing, so that $\bar{\nu}_{max} = 7902$ cm^{-1}. When computer memory is limited, every fourth zero crossing may be used, in which case $\bar{\nu}_{max}$ would be equal to 3951 cm^{-1}. If the same laser were to be used for visible spectrometry, a higher sampling frequency could be achieved by electronically frequency dou-bling the laser interferogram. It should be noted that care must be taken to filter out all high-frequency spectral information *and noise* correctly when long sampling intervals are used so that the computed spectrum does not show additional noise or spectral features due to folding.

The laser fringe-referencing system is not the only additional feature required by signal averaging types of interferometers, since the sinusoidal signal at the detector of the reference interferometer gives no indication as to where either the zero retardation position of the main interferometer is situated or, more important still, where the first sample point should be recorded. Without this knowledge, the first data point could be recorded at different points on successive scans, and rather than the sig-nal-to-noise ratio of the interferogram being improved on signal averaging, it could actually be degraded.

The technique that was used in the Mertz–Curbelo interferometer to ensure that the first data point is always sampled at the same retardation involved passing the radiation from a source of white visible light through the reference interferometer. The white light interferogram was measured with yet another detector. Since visible wavelengths are much shorter than infrared wavelengths, this interferogram is sharp relative to the infrared interferogram. The zero retardation position as measured by the white-light detector always occurs in the same place relative to the centerburst of the signal interferogram. Therefore, this signal can be used as a "fiducial mark" to initiate data collection at the next or some subsequent zero crossing of the laser interferogram. The position of the fixed mirror of the reference interferometer can usually be adjusted so that the zero retardation point of the white-light interferogram occurs at any desired point along the infrared interferogram.

If a single-sided interferogram is being measured, this white-light fringe is usually set to occur between 100 and 5000 μm before the zero retardation point of the signal interferogram, the exact value being selected by the phase correction procedure used (Chapter 3). Most of the data points are sampled on the other side of zero retardation so that the desired resolution can be achieved with the minimum scan length and hence the smallest demand on computer memory.

These techniques were first successfully incorporated into a series of interferometers by Block Engineering and its subsidiary Digilab. The first of their interferometers that used this principle had the moving mirrors of the main and reference interferometers back to back, using two interferometers with a common drive, similar to the Block Model 200. Radiation from the infrared source was passed into one interferometer and measured with the appropriate detector. Into the other interferometer was passed monochromatic light from a He–Ne laser (measured with a germanium photodiode) and white light from a small lightbulb (measured with a silicon photodiode), see Fig. 4.1. The first time the white-light interferogram reached a certain voltage, a counter was initiated and zero crossings from the laser signal were counted. Data collection was based on these zero crossings.

After this system had been used to verify the validity of the technique, Block Engineering developed other interferometers of higher maximum retardation. The first of these systems was the Model 296, in which the reference interferometer is a small "cube" that is not back to back with the signal interferometer but is still in parallel with it. This arrangement formed the basis of many of the Digilab interferometers (Fig. 4.2). The largest of these laser fringe-referenced interferometers (the Model 496) will allow a resolution slightly better than 0.1 cm^{-1} to be attained. The drive in most of the current Digilab interferometers incorporates a single air bearing to ensure a smooth travel, whereas the lower-resolution systems used an oil bearing similar to that of the earliest Block interferometers.

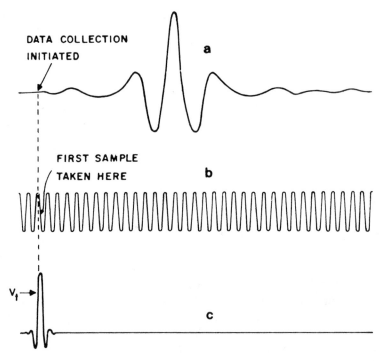

Fig. 4.1. Interferograms measured using a rapid-scanning interferometer equipped with a laser and white-light reference interferometer: (*a*) signal at infrared detector; (*b*) signal measured by laser detector; (*c*) signal measured by white-light detector. Note that the reference interferometer is positioned so that the centerburst of the white-light interferogram occurs a short distance before the infrared centerburst. After the amplitude of the white-light interferogram exceeds the threshold voltage V_t, data acquisition is initiated at the next zero crossing of the laser interferogram.

The mirror velocity is set so that the frequency of the laser reference signal is constant. Any variations from this frequency cause a servoloop to increase or decrease the drive voltage to reduce the velocity error and maintain a constant mirror velocity. A uniform speed is necessary because major variations in the mirror velocity cause an increase in the noise level of the spectrum and a distortion of the instrument line shape function [3].

Many Michelson interferometers designed to achieve a retardation greater than 1 cm incorporate at least one air bearing in their drive systems. Besides single air-bearing instruments, such as the Digilab interferometers and the IBM IR/85, several of the Nicolet interferometers and the Midac interferometer incorporate a dual-air-bearing drive. In these instruments, the moving mirror is driven on two air bearings mounted at the center of mass of the scanning mirror carriage. Much has been spoken (and very little written) about the relative strengths and weaknesses of single- and dual-bearing designs and so a brief discussion of their merits will be given here.

Ideally the cylinder of the bearing should cover the shaft over the full mirror travel. For measurements made at a resolution of 0.1 cm^{-1}, where the moving mirror of a standard Michelson interferometer travels 5 cm, it is therefore preferable that a 10-cm-long bearing be used. Since the spacing between the shaft and cylinder of most air bearings is about 0.2 μm, the *maximum* tilt along the entire drive is 2×10^{-6} rad. According to the discussion in Chapter 1, Section IX, this is adequate for measurements made at 0.1 cm^{-1} resolution. The dual-bearing design usually involves shorter bearings (typically 5 cm long), so that the maximum tilt would be approximately 4×10^{-6} rad, barely sufficient to avoid loss of fringe modulation. It may also be noted that the rods on which the bearings travel must be mounted at each end to a precision of better than 0.1 μm, and the rods themselves must be exceptionally straight. Midac [4] claims that the total bow of the shaft used on their interferometers does not exceed a tenth wavelength of visible light.

The biggest potential drawback to a single-bearing design is the possibility of motion about the axis of motion (the *roll* axis). In several of the early low-resolution, rapid-scanning interferometers, the roll motion

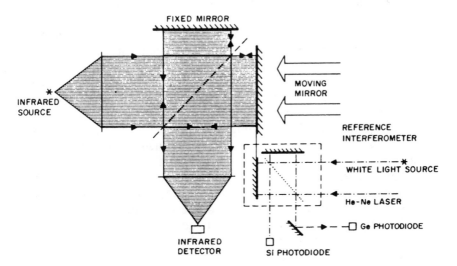

Fig. 4.2. Schematic representation of an interferometer in which there is a separate reference interferometer for the laser and white-light channels.

was restrained through the use of springs mounted on the rear end of the mirror drive. In the Digilab 496 in which the maximum retardation is about 10 cm, the roll motion is prevented by a stainless steel extension attached to this moving mirror assembly, which slides over a Teflon-coated bar. The dual-bearing design eliminates any roll, but it gives a larger possibility for rigid-body rotational vibrations about the vertical axis (the *yaw* mode). A description of this problem for interferometers with large-diameter mirrors given in a commercial catalog [4] attests to its seriousness. Additional motions in a direction perpendicular to the direction of the drive can be set up in dual-bearing interferometers if a small lateral force is applied (e.g., because of the bearings binding).

Roll and yaw have been reduced on other interferometers by using rectangular bearings. The Nicolet 60-SX incorporates a double in-line cylindrical bearing for the moving mirror. This is effectively one long bearing, the mass of which is quite low to reduce the effects of inertia and momentum. A rectangular parallel bearing is used to prevent roll and yaw. The Perkin-Elmer Model 1800 interferometer incorporates a single rectangular air bearing. The tilt tolerances are as good as the cylindrical bearings, and the shape of the bearing prohibits roll and yaw. One interesting aspect of this design is that the central bearing shaft is stationary and the bearing sleeve moves. This permits the mirror to be cradled on a long bearing, which further reduces the tendency to tilt. We believe that the Jasco Model FT/IR-3 interferometer incorporates a trapped gas rectangular bearing; however, extensive details of the design are not known to these authors.

On a practical basis, a detailed understanding of the factors affecting the quality of mirror drives is somewhat academic for most *users* of FT–IR spectrometers, since almost all commercial Fourier spectrometers perform to specification on delivery and for a long time thereafter. Problems of the type described above may show occasionally during high-resolution measurements after the interferometer has been in use for several years.

Another difference between interferometers from various vendors is the path taken by the He–Ne laser and white light beams. In the early Digilab design, shown schematically in Fig. 4.2, these beams pass through a reference interferometer. The physical separation of the laser and white light beams from the infrared beam could lead to a reduction in the short-term stability of the measured infrared interferograms, so that systems with this design must be thermostated to attain adequate stability. A superior approach for mid- and near-infrared interferometers is to pass the laser and white-light beams through different regions of the main beamsplitter. If the laser beam is passed through the center of the beamsplitter, small mirror tilts lead to reduced sampling errors, and the need to thermostat the interferometer is reduced (but not entirely eliminated).

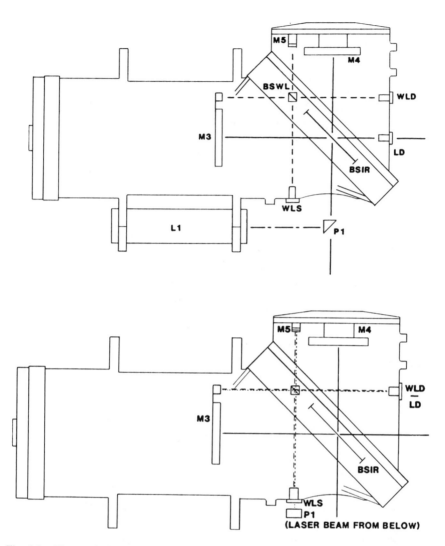

Fig. 4.3. Two optical designs used in Nicolet interferometers in which the reference laser beam is either passed through the center of the infrared beamsplitter (above) or through the side (below). In each case the beam from the white light source (WLS) is passed through a small visible interferometer mounted at the side of the infrared interferometer. The latter case would be more suitable for far-infrared measurements using a flexible Mylar beamsplitter. The laser (L1) is usually mounted on the optical bench, but for evacuated systems the laser is mounted below the instrument under ambient conditions and the beam is deflected into the interferometer by the prism P1. M3 and M4 are the moving and fixed mirrors of the infrared interferometer, respectively, and M5 is the fixed mirror of the white-light interferometer, BSIR and BSWL are the infrared and visible beamsplitters, respectively, and WLD and LD are the white-light and laser detectors, respectively. (Reproduced by permission of Nicolet Instrument Corporation.)

This type of design works well for interferometers equipped with a rigid beamsplitter. In the next section we will see that far-infrared interferometers incorporate a stretched film of Mylar as the beamsplitter. These films can be quite microphonic and often pick up vibrations from a variety of sources. When the reference laser beam is passes through the center of a Mylar beamsplitter (where the amplitude of the vibrations will be largest), the spacing of the sampling points can become quite nonuniform, and noise will be generated in the spectrum (see Chapter 7). This fact explains why the interferometers used in the Nicolet 170-SX and 200-SX FT–IR spectrometers allow the operator to select whether the He–Ne reference laser beam is passed through the center of the infrared beamsplitter or through a separate interferometer mounted beside the main beamsplitter, see Fig. 4.3. When a rigid beamsplitter is mounted in the interferometer, it is preferable to pass the reference laser beam through its center. On the other hand, the alternative path is far preferable when a Mylar beamsplitter is installed. It is noteworthy that the 200-SX is a vacuum instrument, and the laser is mounted outside the evacuated area to prevent overheating or arcing. In addition, special sealed air bearings are incorporated in this instrument so that a high vacuum can be maintained. Several other "vacuum" instruments with air-bearing drives cannot achieve a high vacuum because of the constant air bleed from the bearings. The reduced vacuum in these instruments reduces the possibility of arcing, but very dry air or nitrogen should be fed to the bearings so that the effect of water vapor is not observed in the spectrum.

Some of the more recently introduced interferometers made by Beckman, Digilab, Mattson, Nicolet. and Perkin Elmer incorporate a microcomputer-controlled drive in which the position of the moving mirror is monitored throughout the measurement by continuously keeping a record of the fringes of the laser reference interferogram. In earlier instruments the mirror position was only monitored between the white-light centerburst and the end of data acquisition. With "fringe-counting" electronics, there is no longer any need to generate a white-light interferogram, and interferograms can be measured during both the "forward" and the "reverse" scans.

The ways in which fringe-counting has been achieved are very interesting from an instrumental viewpoint. In the Nicolet 60-SX, for example, the beam from the He–Ne laser is passed through a quarter-wave retarder so that a pair of monochromatic signals of the same frequency, but in phase quadrature, is generated. Separate detection of each signal allows the retardation to be monitored accurately as the direction of motion is reversed. Interferograms measured during "forward" and "reverse" scans may be co-added into separate files. At the end of data collection, each interferogram is separately phase-corrected and transformed. When spectra computed from the two arrays are ratioed, the deviation from the 100% line is less than ± 1%. This data acquisition scheme can substantially improve the duty cycle efficiency over unidirectional data acquisition, especially at high mirror velocities where it is difficult to attain a retrace velocity that is much greater than the velocity during the active part of the cycle. Nevertheless, it should still be noted that the duty cycle efficiency of these interferometers at a high scan speed and low resolution may still be only on the order of 50% since the mirror usually travels well past the last point at which the interferogram is digitized as it decelerates and reaccelerates in the opposite direction.

The method of controlling mirror motion incorporated in the Beckman interferometers is analogous but more sophisticated. These instruments make use of heterodyne laser control using the signal from a Zeeman-split He–Ne laser. Two signals with a beat frequency of 250 kHz are generated by this laser. At the same time, a frequency synthesizer generates a reference signal of the same frequency. Both signals are passed into a phase-lock loop. The difference between the two signals is used to control the velocity of the moving mirror. The phase relationship between the two interferograms is applied for data acquisition in an analogous fashion to the approach described in the previous paragraph. The fact that the beat frequency is about 50 times greater than the typical (5 kHz) frequency of the laser reference interferogram means that the position of the moving mirror is known to one-fiftieth of the wavelength of the He–Ne laser and, again, the need for a white-light reference interferogram is obviated. However, it should be noted that as the interferometer scan speed is increased, the difference between the beat frequency and the laser reference frequency decreases. Hence some of the precision is lost.

An alternative approach designed to increase the duty cycle efficiency has been developed by Digilab. At each zero crossing of the laser interferogram, a pulse is generated that initiates the count of a periodic pulse train having a frequency that is a large multiple of the frequency of the He–Ne laser reference sinusoid. A resettable counter, initially set for a predetermined number of counts of the high pulse rate signal, provides a time interval that is compared with the interval between successive zero crossings of the He–Ne laser signal. The duration of this pulse train is one-half of the desired interval between the zero crossings. The time-averaged value of this train provides an error signal that is used to control the velocity of the moving mirror.

The zero crossings of the laser reference interferogram are used to index a reversible up–down counter that may be zeroed at the infrared interferogram centerburst. When the pulse count from this point reaches a value that is determined by the desired resolution, a control sequence is initiated to change the length of the pulses in the pulse train being compared with the zero crossings of the laser reference signal. As a result, the error signal providing the scan control is changed, accelerating the moving mirror accordingly. The predetermined sequence is selected so as to decelerate and reverse the direction of the mirror controllably within one-quarter of a wavelength of the He–Ne laser after it passes the desired maximum retardation. At the same time, the direction of counting in the up–down counter is reversed. This system has the advantage of not only eliminating the need for a white-light reference but also of allowing data acquisition during the forward and reverse scans at a very high efficiency of the duty cycle.

In recent years there has been a movement toward the development of less expensive Fourier spectrometers. Some of the low-priced instruments incorporate scaled-down versions of the larger interferometers. The IBM Instruments IR/32 uses exactly the same interferometer found in their IR/85 with a scaled-down data system. The IR/85 and IR/32 incorporate a standard Michelson interferometer with a single air bearing, a laser reference, and a white-light interferometer.

The interferometer used in the Digilab Qualimatic and FTS-40, FTS-50 and FTS-60 spectrometers also incorporates a single air bearing. As discussed earlier in this chapter, this interferometer uses laser fringe counting and does not incorporate a white-light reference. The angle between the fixed and moving mirrors is 60° in the Michelson interferometer rather than the more conventional 90°. A 60° interferometer allows slightly greater throughput for a given solid angle and beamsplitter area. Since the beamsplitter is the most expensive component in a mid-infrared interferometer, the advantage of this feature is obvious. The Bomen and Perkin-Elmer FT–IR spectrometers also incorporate a 60° interferometer.

A few interferometers designed for low-resolution spectrometry do not incorporate an air-bearing drive. The first system for medium-resolution measurements that did not incorporate an air bearing was introduced by Nicolet as the MX-S, which is known now as either the 5-MX or the 5-DX (depending on the nature of the data system). The drive for the interferometer in these spectrometers is based on the "flex pivot" or "porch swing" design first described by Walker and Rex [5]. Their design is shown schematically in Fig. 4.4. The mirror is attached to a hollow aluminum beam suspended on a pair of aluminum box-beam arms. The rear arm has a heavy lead weight attached to its upper cross-member to damp out linear vibrations. The pivot points (where the arms are attached) are equipped with flexure pivots that do not depend on sliding or rolling sur-

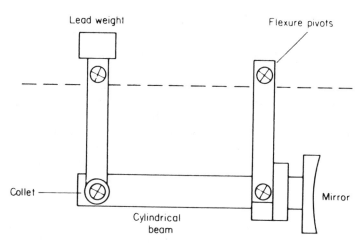

Fig. 4.4.   Schematic representation of a "porch-swing" or "flex-pivot" interferometer; see text for details.

faces for their operation. They do not rotate precisely about their exact centers, and the spacings between the mounting holes in the block and between the holes in the mirror beam are never exactly the same. Unless the effect of this nonuniform spacing were corrected, the translation of the mirror would not permit the plane of its surface to remain parallel throughout the scan.

The error is corrected by mounting the rear mirror pivot in split bushings, or collets, whose centerholes have been bored very slightly off center. The amount of eccentricity corrects for the small spacing errors between the pairs of pivot holes in the block and mirror beam. The slightly eccentric pivot mount can be rotated to an appropriate position to establish parallel translation of the mirror beam. Although such a drive is not suitable for high-resolution measurements, planarity of as little as 5×10^{-6} rad can be achieved for retardations of up to 0.25 cm.

Another interferometer that does not incorporate an air-bearing drive was designed by Beckman Instruments. This instrument incorporates a glass bearing that does not give a sufficiently precise mirror motion for operation in the same fashion as the other interferometers described in this section. Instead, the plane of the fixed mirror must be continuously adjusted throughout the scan so that the resulting spectrum is not distorted. The technique of *dynamic alignment* is described in more detail later in this chapter.

(ii)　Genzel Interferometers

A novel type of interferometer first developed by Genzel in Germany [6–8] forms the basis of the first FT–IR systems built by Bruker. These systems are sold in the United States by IBM Instruments as the IR/90 series. The difference between these instruments and the standard Michelson interferometer is the fact that the beamsplitter is located at a focal plane and can therefore be much smaller than the beamsplitter of a standard Michelson interferometer, see Fig. 4.5. The beams which are transmitted and reflected from the beamsplitter are collimated by mirrors located at a position such that the two beams travel collinearly toward each other.

A moving double-sided mirror carriage is located midway between the two collimating mirrors so that the beams return to the beamsplitter after a path difference has been introduced. It may be noted that a mirror carriage displacement x results in an optical retardation of $4x$, which should be compared to the value of $2x$ generated in a standard Michelson interferometer. This possible advantage is offset by the fact that a small

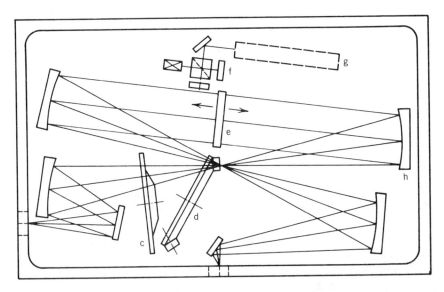

Fig. 4.5. Genzel-type interferometer incorporated in the IBM Instruments IR/90 series of FT-IR spectrometers and the Bruker IFS-113 and IFS-114 spectrometers. c, filter wheel; d, carousel on which up to six beamsplitters may be mounted; e, moving two-sided mirror; f, reference interferometer; g, He–Ne laser; h, spherical collimating mirror. (Reproduced by permission of IBM Instruments Corporation.)

tilt in the moving mirror assembly results in twice the angular displacement of the beam.

The real advantage of the Genzel interferometer is derived from the small size of the beamsplitter. It is possible to mount up to six beamsplitters on a carousel and switch from one spectral region to another without breaking the vacuum or purge in the instrument. This feature is particularly useful for far-infrared measurements where several Mylar beamsplitters are required to cover the entire region from 500 to 10 cm^{-1}. The optics on this instrument allow the user to select between either of two sources or detectors so that the entire mid and far infrared can be covered without breaking the vacuum or purge. (It should be noted that Nicolet has developed a beamsplitter interchange for their 60-SX spectrometer, but this is a much less compact device due to the large beamsplitters required for a 90° Michelson interferometer.)

(iii)   Refractively Scanned Interferometers

Although most spectrometric measurements are made with small sources or samples, there are certain types of emission measurements where the source can be quite large. In particular, for measurements of twilight and night airglow, the source is very extensive and extremely weak. Thus, the signal-to-noise ratio of the interferogram could be improved if the solid angle of the radiation passing through the interferometer could be increased without sacrificing resolution.

The reason the acceptance angle for a given resolution has to be kept below a certain limit was discussed in Chapter 1, Section VII. If some means of making the path difference invariant with the entrance angle were found, then a greater throughput from an extended source could be achieved and hence a greater signal-to-noise ratio attained.

A technique suggested by Bouchareine and Connes [9] accomplishes such *field compensation* to the first order by using prisms in place of the usual mirrors of the Michelson interferometer as shown in Fig. 4.6. The back sides of the prisms are silvered and the optical system is aligned as in a typical interferometer. Instead of obtaining differential pathlengths by moving one mirror parallel with the rays of light that impinge on it, in the field-widening technique a prism is moved parallel with its "apparent mirror position" inserting more material in one beam to effect an increase in retardation. "Apparent mirror position" refers to the apparent position of the coated back surface of the prism, looking through the optical material. Thus the direction of the drive is perpendicular to the light rays. It can be shown that retardation is independent of the incident angle, so that angles larger than α_{max} (Eq. 1.40) can be tolerated. To

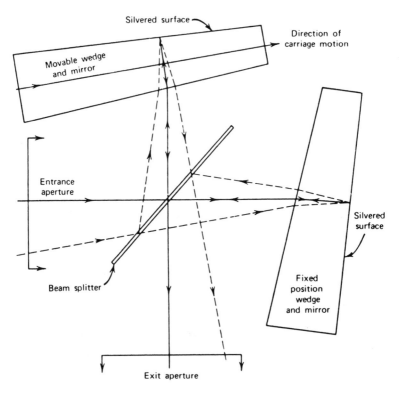

Fig. 4.6. One method used for widening the field of an interferometer. The movable rear-silvered prism is translated in a direction perpendicular to that of the movable mirror in a conventional Michelson interferometer so that the retardation is increased by increasing the distance traveled through the prism. (Reproduced from [9], by permission of the author.)

achieve the field-widening increase in solid angle, large prisms must be constructed and displaced during operation, so that the wavefront distortions are less than $\frac{1}{10}\lambda$ across the full travel of the prism, see Fig. 4.6. An interferometer using this principle for operation in the near-infrared and visible regions of the spectrum has been described [10–12]. This interferometer consisted of a prism with a wedge angle of 8° constructed from quartz because of its low dispersion and expansion coefficient. With this particular system, it has been shown that the retardation δ is equal to 0.21 times the distance x over which the wedge is moved. Thus, to achieve a resolution of 1 cm^{-1}, a drive length of at least 4.7 cm is needed. Maintaining accurate positioning of the prism over the full length of this scan presented a real problem in stability. The base of the interferometer had to be constructed of granite, and the optical carrier was made of invar stainless steel, whose very low expansion coefficient is close to that of granite and quartz.

This instrument can be used with either a photomultiplier or a photoconductive detector. When a photomultiplier is used, the ratio of the times required to measure spectra between 0.4 and 0.7 μm at the same signal-to-noise ratio using a standard Michelson interferometer and the field-widened interferometer is about 19:1. If a photoconductive detector is used to measure spectra at 2.5 μm with the same signal-to-noise ratio using standard Michelson and field-widened interferometers, the ratio of the measurement times is increased to about 360:1.

The principle of moving a refractive optical element instead of a mirror also lies behind the design of the interferometer used in the Analect fX-6200 and fX-6250 spectrometers. The design of this interferometer is shown in Fig. 4.7. The use of cube-corner retroreflectors reduces the criticality of motion control and position measurement of the moving element and reduces the amount of angular alignment needed for the scanning mechanism.

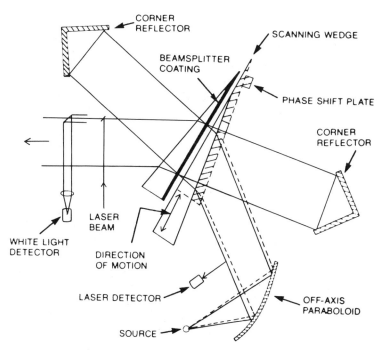

Fig. 4.7. Schematic representation of the refractively scanned interferometer designed by Analect Instruments (sometimes known as the Transept® or Doyle interferometer). (Reproduced by permission of Analect Instruments Division of Laser Precision Corporation.)

Because of its design, this instrument achieves a modest amount of field widening, with an improvement over a conventional Michelson interferometer of the same aperture given approximately by the square of the refractive index of the moving wedge [13]. For KBr wedges with n = 1.53 and a wedge apex angle of 10°, this advantage amounts to a factor of 2.3. Therefore, even though the aperture of the Analect interferometer is only 2.5 cm, its throughput is close to that of a 5-cm Michelson interferometer.

In theory, the refractive scanning technique leads to about an order of magnitude reduction in the accuracy to which the sampling position must be known when compared with a standard Michelson interferometer. Thus, in the original design of this instrument, a simple Moiré scale was used to generate the sampling trigger. Since a reduction in alignment stability of the moving element by two orders of magnitude should have been able to be tolerated with this design, it was believed that there would be no need for an air-bearing drive. Nevertheless, it was found that the use of laser fringe referencing and the installation of an air-bearing drive in this instrument led to improved performance. The current refractively scanned interferometers made by Analect incorporate both these components and the SNR of spectra measured by the fX-6250 (which has a 3-cm aperture) closely matches that of Michelson interferometers with 5-cm apertures, with the same detector being used for each measurement.

One potential source of performance degradation for these instruments is dispersion of the refractive index of the moving wedge. This dispersion should be constant at any given wavelength, so that the nonlinear wavenumber scale that results is correctable by software. Surprisingly, perhaps, no correction for the change in dispersion with temperature is needed. To verify this, Doyle and McIntosh [14] performed two measurements of the spectrum of indene, one with the interferometer at 27°C and the other at 39°C. No residual features were observed in the difference spectrum (see Fig. 4.8, indicating a wavenumber scale temperature coefficient of less than 5×10^{-3} cm^{-1}/°C. It is obvious that this instrument represents a real competition to Michelson interferometers at the low cost end of the commercial market, especially for process control or operation in noisy environments, in spite of the high cost of the refractive element.

A different type of refractively scanned interferometer was recently introduced by Janos Technology in which the path difference is introduced by a rotating refractor plate [15]. A schematic diagram of the interferometer is shown in Fig. 4.9. Collimated radiation is passed onto a germanium beamsplitter. A compensator plate of the same thickness as the substrate for the germanium layer is spaced about a millimeter from the beamsplitter and slightly out of parallelism to allow for escape of spurious

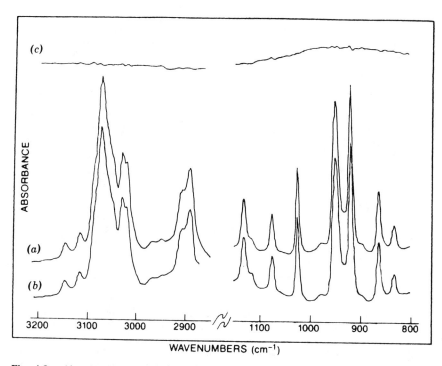

Fig. 4.8. Absorbance spectra of indene measured using the Transept interferometer at temperatures of (*a*) 27°C and (*b*) 39°C. (*c*) Difference spectrum of (*a*) and (*b*). (Reproduced by permission of Analect Instruments Division of Laser Precision Corporation.)

reflections. The two beams from the beamsplitter are directed to either side of a rotating refractive plate that is flat to better than one-tenth wave with faces parallel to better than 1 second of arc. Rotation of the plate creates the optical path difference between the beams.

The prototype instrument operated with a constant angular velocity of the rotating plate. In this case, the retardation does not vary linearly with time. Ideally, the speed of the motor should be varied so that the variation of retardation with time is constant, possibly by locking the motor drive to the laser frequency. The biggest advantage of an interferometer of this design is the fact that it is self-compensating and therefore not particularly susceptible to shock or vibrations. The manufacturer believes that this instrument will be most suitable for process monitoring since it can be made almost as cheaply as a monitor incorporating a circularly variable filter.

Another interferometer involving a rotating element design has been employed by Perkin-Elmer for their Model 1700. This interferometer pro-

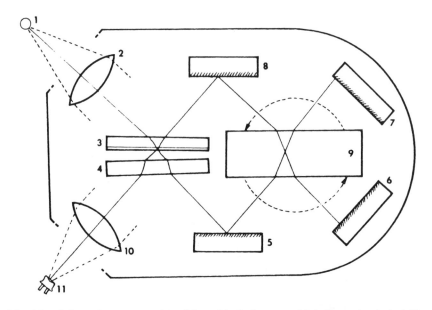

Fig. 4.9. Schematic representation of the refractively scanned interferometer designed by Janos Technology, Inc. Radiation from the source (1) is collimated (2) and passed onto the beamsplitter, which consists of a substrate plate coated with a layer of germanium and a compensator plate of the same material as the substrate. The transmitted and reflected beams are reflected from plane mirrors 5 and 8 onto a rotating refractor and hence to two more plane mirrors, which return the beams to the refractor plate and thence to the beamsplitter. One beam from the beamsplitter is collected by the optical element (10) and focused onto the detector (11). (Reproduced from [15], by permission of International Scientific Communications, Inc.; copyright © 1984.)

duces optical retardation by rotation of an assembly containing the beamsplitter and two plane mirrors [16]. An optical diagram is shown in Fig. 4.10. Radiation from the infrared source travels to paraboloid P1 where it is collimated before striking the beamsplitter. The beamsplitter and two mirrors R1 and R2 are mounted on a rotating stage that has an axis at x. The transmitted beam passes to the fixed mirror F2 and the reflected beam goes to another fixed mirror F3. Because the beamsplitter stage rotates, the pathlength for the reflected beam changes at a greater rate than the pathlength for the transmitted beam. The rotation produces the path difference necessary to produce a maximum resolution of 2 cm^{-1}. The second mirror on the rotating stage, R2, compensates for the displacement of the output beam and ensures that the beam will not shift.

It is interesting to note that there is no roll, yaw, or tilt in this design. Any such misalignment is compensated by the design of the rotating stage.

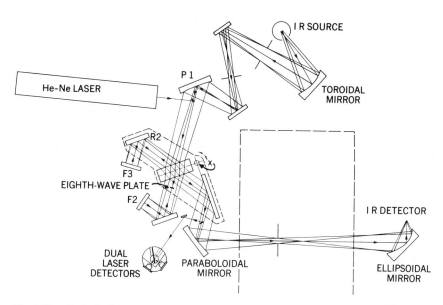

Fig. 4.10. Optical schematic of the Perkin-Elmer Model 1700 FT–IR spectrometer showing the rather unusual rotating interferometer design used in this instrument. Radiation from the IR source is collimated by the paraboloid, P1. F2 and F3 are fixed plane mirrors, R2 is a plane mirror mounted on the rotating stage, and x is the axis of rotation of the stage. (Reproduced by permission of the Perkin-Elmer Corporation.)

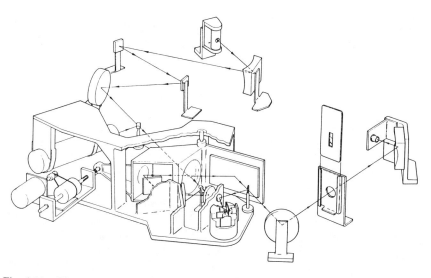

Fig. 4.11. Three-dimensional drawing of the Perkin-Elmer Model 1700 spectrometer with an exploded view of the interferometer. (Reproduced by permission of the Perkin-Elmer Corporation.)

Consequently, even if the bearing at x becomes worn, no degradation in performance is seen. As with several other systems described above, laser fringes are counted (which eliminates the need for a white-light reference), and data can be collected in both directions. The laser fringe-counting system (also used in the Perkin-Elmer Model 1800 spectrophotometer) operates on a somewhat similar principle to that of the Nicolet 60-SX but is different in several respects including the use of an eighth-wave plate. A three-dimensional drawing of the optics is provided in Fig. 4.11 to help the reader confirm that all movement (other than the pathlength) is compensated.

d. Tilt Compensation

To achieve a very high resolution at wavenumbers as high as $10,000 \, \text{cm}^{-1}$, great care has to be taken in the design of the interferometer drive to avoid the effects described in Chapter 1, Section IX. For interferometers designed for high resolution, or even interferometers designed to circumvent the limitations of standard Michelson interferometers, special precautions have to be taken to guard against the effect of mirror tilt. Two approaches have been taken. The first is to use a standard Michelson interferometer, in which the plane of the mirror is monitored and connected continuously throughout the scan. This technique has been called *dynamic alignment*, and it is the basis of the Fourier spectrometers made by Bomem in Canada and by Beckman in the United States. The second involves the installation of *cats-eye* or *cube-corner retroreflectors* in the interferometer to compensate optically for the effect of tilt.

(i) Dynamic Alignment

In the dynamic alignment technique, three separate laser interferograms are usually generated. Transducers to tilt an optical element are actuated when the phase of each of these sinusoidal interferograms varies. The servocontrol for the transducers is used to maintain the angle between the moving mirror and the beamsplitter precisely equal to that between the fixed mirror and the beamsplitter throughout the entire scan.

    In the instruments manufactured by Bomem, the fixed mirror is tilted to compensate for any variation in the tilt of the moving mirror during a scan. With this design, it is not even necessary to use air bearings in the drive; a mechanical carriage is used for the moving mirror in the Bomem spectrometer. Mirror alignment is optimized at power on and retained at all mirror positions during each scan. In fact, the angular deviation from optimum alignment is less that 10^{-6} rad in normal laboratory environments and about 10^{-5} rad under conditions of quite severe vibration [17,18]. It has been installed in aircraft and balloons, used in arctic conditions, and even been cooled to liquid ntirogen temperature.

It will be remembered from the earlier discussion in this chapter that the interferometers made by Beckman incorporate a Zeeman-split He–Ne laser with a beat frequency of about 250 kHz. The beam from this laser is expanded to a diameter of 2.5 mm and enters the interferometer coaxially with the collimated infrared beam. After modulation by the interferometer, the laser beam is passed to a detector array. The difference between the modulation frequency of the interferogram and the beat frequency of the laser is determined from the average of the signals from each detector. This average difference signal is then compared to the signal from each separate detector in a series of phase comparators. Error signals proportional to the phase difference between each sector of the detector array and the average signal are obtained, amplified, and sent to each of three piezoelectric transducers on the rear of the fixed mirror so that the alignment is maintained continuously. Although the Beckman interferometers are not designed for high-resolution operation, the dynamic alignment technique has the secondary advantage that the need for an air bearing is reduced and a less expensive glass bearing may be used.

(ii)   Cube-Corner and Cats-Eye Retroreflectors

If the moving mirror of a Michelson interferometer is tilted, the amplitude of the interferogram is reduced (see Chapter 1, Section IX). When a cube-corner retroreflector is used as the moving element, the effect of tilt is essentially eliminated. The principle behind the operation of a cube corner is shown in Fig. 4.12. All the FT–IR spectrometers made by Mattson Instruments, the Sirius 100, Cygnus 25 and the Alpha Centauri, incorporate cube corners. The Sirius 100 spectrometer permits spectra to be measured at remarkably high resolution (0.12 cm^{-1}) for an instrument that cost less than $50,000 at the start of 1984.

Even though cube corners eliminate the effects of mirror tilt, lateral displacements produce a shear that has a similar effect on the interferogram. In practice, it is easier to meet the tolerance on lateral displacements than on tilts, so that retroreflectors experimentally have a definite advantage over plane mirrors. Steel [19] has suggested a combination of a movable cube-corner and stationary plane mirrors that introduces neither shear nor tilts, shown in Fig. 4.13. Any interferometer using this arrangement would be fully compensated. This design is used by Analect in their "Multisept" interferometer, see Fig. 4.14.

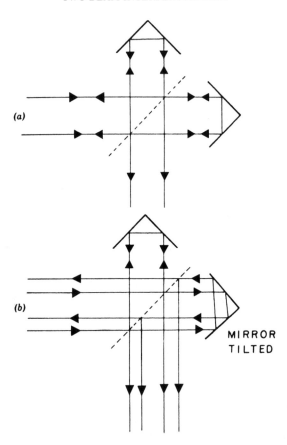

Fig. 4.12. Principle of the corner retroreflector: (*a*) Both retroreflectors are in good alignment so that there is no essential difference from the case of a plane mirror Michelson interferometer. (*b*) One of the retroreflectors is tilted, but the output beams from each retroreflector are still parallel, unlike the case for a conventional Michelson interferometer. The concept of the "roof" retroreflector shown in this figure can be extended into three dimensions through the use of cube corners.

The use of cube corners as retroreflectors has the disadvantage of introducing some polarization effects and requires quite delicate initial alignment. To get around these problems, *cats eyes* have been used as retroreflectors. These devices consist of a concave (usually paraboloidal) mirror with a convex sphere used to image the beamsplitter back on itself, usually at zero retardation. The principle of the interferometer is shown in Fig. 4.15.

Most of the highest-resolution Fourier spectrometers incorporate cats-eye retroreflectors. The first were built in France by the group led by Pierre Connes [20–22], and several descendents of these instruments have been developed in France [22,23], the United States [23,24], and elsewhere [25–27]. The most detailed description of a cats-eye interferometer has been given by Davis et al. [26], and interested readers are strongly recommended to read this paper for details of the construction of this highly sophisticated instrument. It is interesting to note that this instrument measures less than 1 m in any dimension, yet it generates 0.01 cm^{-1} resolution spectra. Since it is used predominantly for astronomy, the mirror drive has to be exceptionally stable, and a technique analogous, but not identical, to the dynamic alignment technique described above is incorporated.

Several of the cats-eye interferometers developed in France operate on the step-and-integrate principle. In some of them, the secondary mirror of the cats eye can be modulated so that PM interferograms are generated as described in Chapter 1, Section XI. A typical SNR for spectra measured at very high resolution is 10^3 (signal-to-rms noise), attesting to the high performance of these instruments. By using a rapid-scanning cats-eye interferometer with many levels of gain switching throughout the interferogram, Brault [27] has improved the performance of this type of in-

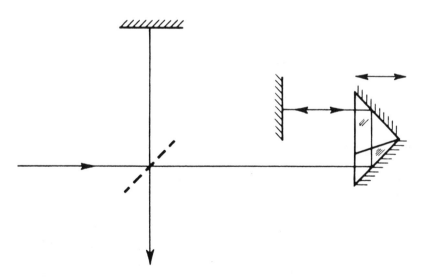

Fig. 4.13.   The mirror combination suggested by Steel [19] to compensate for both tilt and shear. (Reproduced from [19], by permission of the author.)

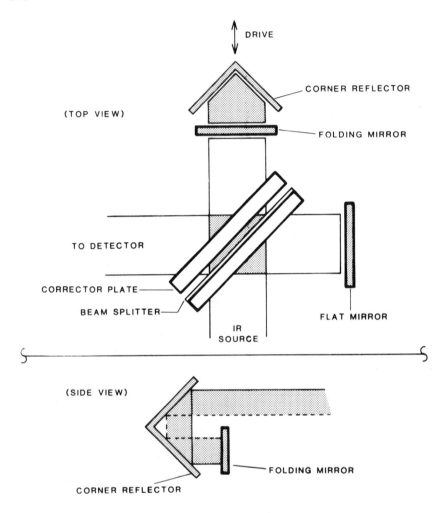

Fig. 4.14. Schematic representation shown both as a top view and a side view, of the Multisept® interferometer built by Analect Instruments, which is based on the original design of Steel. (Reproduced by permission of the Analect Instruments Division of Laser Precision Corporation.)

strument by a further order of magnitude. Few details of this instrument are available in easily accessible papers, however.

Most of the highest-quality FT–IR spectra measured at or near the Doppler limit have been acquired using cats-eye interferometers, but all those instruments have been constructed at the laboratory at which the

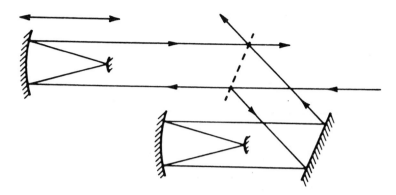

Fig. 4.15. Schematic representation of a "cats-eye" interferometer employed in many of the highest-resolution Fourier transform spectrometers. (Reproduced from [19], by permission of the author.)

measurements are to be made. Chemical spectroscopists who are unwilling to invest the large amount of time required to build one of these instruments are therefore advised to use interferometers that incorporate some type of dynamic alignment, and these are commercially available.

II. BEAMSPLITTERS FOR MICHELSON INTERFEROMETERS

One of the most important components governing the performance of a Michelson interferometer for infrared spectrometry is the beamsplitter. In this section, the theoretical factors governing the efficiency of beamsplitters are discussed, and the materials that are used in practice are evaluated.

Let us consider a beam of monochromatic radiation of intensity I entering a Michelson interferometer for which the beamsplitter has a reflectance R and a transmittance T. If there is no absorption of radiation,

$$R + T = 1 \tag{4.1}$$

Figure 4.16 shows the intensity of the various beams in the interferometer if no interference effects occurred. The intensity of the beam transmitted to the detector (the dc component of the interferogram) is $(2RT)I$, while that of the beam returning to the source is $(R^2 + T^2)I$. In order that there is no energy loss occurring at the beamsplitter, the sum of the intensities

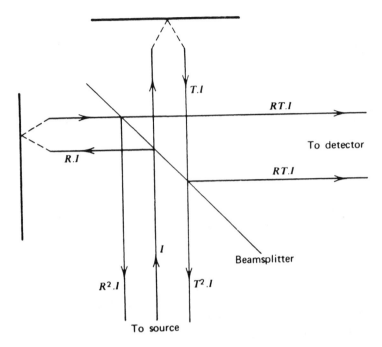

Fig. 4.16. Modified ray diagram of a Michelson interferometer in which beams traveling in opposite directions have been separated for clarity. If radiation of intensity I enters the interferometer, the average intensity of the output beam passing to the detector has an intensity $2RTI$, and the average intensity of the beam returning to the source is $(R^2 + T^2)I$.

of these two beams must be equal to I. This is easily verified:

$$(2RT)I + (R^2 + T^2)I = (R^2 + 2RT + T^2)I$$

$$= (R + T)^2 I$$

$$= I$$

The interferograms of the beams passing to the detector and returning to the source are 180° out of phase, and their amplitudes about the dc level are equal, but of opposite sign, at all moments during the scan. The amplitude of the ac portion of the interferogram is determined by the output beam with the lower intensity, which is always the beam transmitted to the detector. Figure 4.17 shows how $4RT$ varies with the reflectance of the beamsplitter, and it is seen that $2RT \le 0.5$ for all values of R, so that $2RT \le (R^2 + T^2)$.

Thus, for a nonideal beamsplitter, the ac component of the interferogram for monochromatic radiation is given by

$$I(\delta) = 2RTI(\bar{\nu}) \cos 2\pi\bar{\nu}\delta \qquad (4.2)$$

For an ideal beamsplitter, the interferogram is given by Eq. 1.4:

$$I(\delta) = 0.5I(\bar{\nu}) \cos 2\pi\bar{\nu}\delta \qquad (1.4)$$

The effect of nonideality of the beamsplitter material may be allowed for by multiplying the amplitude of the ac interferogram by a factor $\eta(\bar{\nu})$, which is less than 1, so that

$$I(\delta) = 0.5\eta(\bar{\nu})I(\bar{\nu}) \cos 2\pi\bar{\nu}\delta \qquad (4.3)$$

where $\eta(\bar{\nu})$ is equal to $4RT$ and is known as the *relative beamsplitter efficiency*. For infrared spectrometry, $\eta(\bar{\nu})$ should be as close to unity as possible over as wide a spectral range as possible.

Beamsplitters are generally thin films whose reflectance is determined by the refractive index of the material, n, the thickness of the film, d (in centimeters), the angle of incidence of the beam, θ, and the wavenumber of the radiation, $\bar{\nu}$ (reciprocal centimeters). Several factors have to be taken into account when calculating the reflectance of the film. In addition

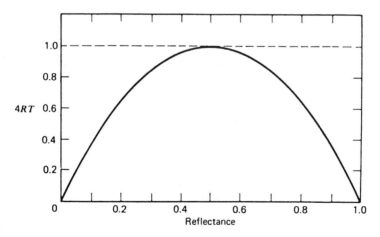

Fig. 4.17. Variation of the relative beamsplitter efficiency $4RT$ with reflectance R assuming that the beamsplitter has an absorbance of zero. Provided that $0.28 \le R \le 0.72$, the beamsplitter efficiency exceeds 80%.

to the single-surface reflectance, the effect of internal reflections must be considered, together with the fact that phase shifts of 180° occur at external reflection surfaces whereas no phase shifts occur for internal reflection. The ray that is reflected from the front surface of the beamsplitter can interfere with the parallel ray, which has undergone internal reflection.

To calculate the relative beamsplitter efficiency of any film, it is necessary to consider separately the reflectance for radiation whose electric field is polarized parallel to the plane of incidence (p-polarization) and perpendicular to the plane of incidence (s-polarization). The Fresnel equations [28] give the single-surface reflectivities for these two polarizations as

$$R_p = \frac{\sin^2(\theta - \theta')}{\sin^2(\theta + \theta')} \qquad (4.4)$$

$$R_s = \frac{\tan^2(\theta - \theta')}{\tan^2(\theta + \theta')} \qquad (4.5)$$

where θ' is the angle of refraction inside the film.

Chamberlain et al. [29] have shown that the relative beamsplitter efficiencies for parallel and perpendicular polarized light are given by the equations

$$\eta_p = \frac{4R_p T_p^2 E}{(T_p^2 + R_p E)^2} \qquad (4.6)$$

$$\eta_s = \frac{4R_s T_s^2 E}{(T_s^2 + R_s E)^2} \qquad (4.7)$$

where

$$T_p = 1 - R_p \qquad (4.8)$$

$$T_s = 1 - R_s \qquad (4.9)$$

and

$$E = 4 \sin^2(2\pi n \, d\bar{\nu} \cos \theta') \qquad (4.10)$$

$$\equiv 4 \sin^2 \epsilon$$

where   (4.11)

$$\epsilon = 2\pi n \, d\bar{\nu} \cos \theta'$$

The relative beamsplitter efficiency of a film for unpolarized incident radiation is simply given by the average of the values for parallel and perpendicular polarized radiation:

$$\eta = \tfrac{1}{2}(\eta_p + \eta_s) \tag{4.12}$$

From Eqs. 4.6 and 4.7, it is seen that η will take a zero value when E is zero, that is, when $\epsilon = m\pi$, where $m = 0, 1, 2, \ldots$. Similarly, η will exhibit a maximum when $m = \tfrac{1}{2}, \tfrac{3}{2}, \tfrac{5}{2}$, and so forth. The efficiency of any beamsplitter is zero at 0 cm^{-1} and returns to zero for the first time when $\epsilon = \pi$. Thus, the frequency at which this minimum value of $\eta(\bar{\nu})$ is seen, $2\bar{\nu}_0$ reciprocal centimeters, is given by

$$2\bar{\nu}_0 = (2n\, d \cos \theta')^{-1} \tag{4.13}$$

Between 0 and $2\bar{\nu}_0$, the variation of $\eta(\bar{\nu})$ is symmetrical about $\bar{\nu}_0$, although the actual shape of the curve is dependent on the values of n and θ, since it is these two parameters that determine the amount of internal reflection in the film.

For materials of low refractive index with a low angle of incidence, the proportion of the incident radiation that is internally reflected is small. The most commonly used beamsplitter material for far-infrared Fourier transform spectrometry is polyethylene terephthalate (Mylar in the United States, Melinex in Europe) for which the refractive index is 1.69 in the far infrared [29]. Figure 4.18 shows the efficiency of a Mylar film as a function of angle of incidence. It can be seen from the figure that for an angle of incidence of 45°, $\eta(\bar{\nu})$ never exceeds 0.7 for Mylar due primarily to the low single-surface reflectances at 45° incidence for parallel ($R_p = 0.02$) and perpendicular ($R_s = 0.14$) polarized radiation.

For a film to be useful as a beamsplitter for far-infrared spectrometry, $\eta(\bar{\nu})$ should be at least 0.3 and preferably greater than 0.5. If the former criterion is applied to Mylar, the spectral range for a film at 45° incidence is between $0.35\bar{\nu}_0$ and $1.65\bar{\nu}_0$. In practice, it is desirable to attain a wider spectral range than this if possible.

The effective spectral range may be increased either by increasing n or θ. If the angle of incidence of a beam of radiation on a Mylar film is increased to 70°, the shape of the efficiency curve changes such that although the maximum efficiency is slightly lower than for 45° incidence, the efficiency remains greater than 0.5 over a much wider frequency range. The value of $\eta(\bar{\nu})$ is greater than 0.5 for $0.25\bar{\nu}_0 < \bar{\nu} < 1.75\bar{\nu}_0$. In spite of the reduced maximum efficiency, Mylar is obviously preferable for spectrometry over wide spectral ranges when it is used with high angles

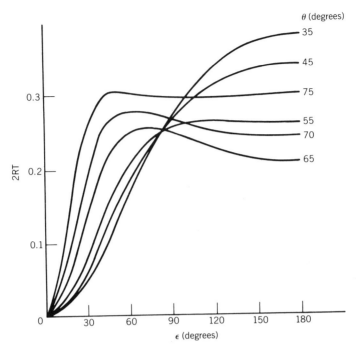

Fig. 4.18. Variation of $2RT$ with ϵ for different angles of incidence for a Mylar film. (Reproduced from [29], by permission of Pergamon Press; copyright © 1966.)

of incidence. On the other hand, if a high relative efficiency is desired over a small spectral range, it becomes preferable to reduce the angle of incidence below 45°. The minimum value of θ is determined by the Brewster angle, below which there would be total internal reflection.

The results for Mylar represent the typical case for a low-refractive-index material. If the angle of incidence is held constant at 45° and the refractive index is increased, the shape of the beamsplitter efficiency curve becomes similar to that of Mylar at 75° incidence, but the maximum value of $\eta(\bar{\nu})$ becomes closer to unity. The film therefore becomes better suited for mid-infrared spectrometry, where a high efficiency over a wide spectral range is required.

Very thin films are required for mid-infrared spectrometry, and thin films of high-index materials tend to be very brittle. They are therefore supported on a flat transmitting substrate. These substrates should have several important properties:

1. They should have a low refractive index to prevent a large reflection loss from the front surface of the plate.

2. They should have a high transmittance over the entire spectral range of interest and show no strong absorption bands.

3. They must be able to maintain a flatness equal to a quarter of the shortest wavelength being measured.

4. They should not be easily scratched or be susceptible to attack by any component of the atmosphere (especially water vapor).

To prevent severe chirping because of dispersion by the substrate plate (see Chapter 1, Section VI), a compensator plate of exactly the same thickness as the substrate must be placed on the other side of the film.

Germanium (n = 4.0) or silicon (n = 3.6) films are generally used for mid-infrared spectrometry. The plate on which the film is deposited depends on the spectral range to be studied. CsI and CsBr will allow the range to be extended into the far infrared, but both materials are soft and do not maintain their flatness well for measurements to high wavenumber. NaCl is a good material as far as hardness is concerned but does not transmit appreciably below 650 cm^{-1}. The best compromise between these two extremes for conventional mid-infrared spectrometry is KBr. For near-infrared spectrometry, a silicon or ferric oxide film (n = 3.0) film is usually deposited on a calcium fluoride or quartz flat.

Calculations of the beamsplitter efficiency of deposited films are somewhat more complex than for unsupported films since the refractive index of the substrate must be taken into account. Sakai [30] has calculated the efficiency of a typical film with the angle of incidence equal to 45°, the refractive index of the film equal to 3.6 and that of the substrate equal to 1.4 (Fig. 4.19). It can be seen that the relative efficiency is greater than 0.8 for wavenumbers between $0.3\bar{\nu}_0$ and $1.7\bar{\nu}_0$, but outside this range $\eta(\bar{\nu})$ drops off rapidly. For mid-infrared spectrometry ($\bar{\nu}_0$ = 2000 cm^{-1}) using this beamsplitter, good performance may be expected between 600 and 3400 cm^{-1}, but the performance will decrease at wavenumbers above 3400 cm^{-1} and below 600 cm^{-1}.

The Fresnel equations show that the relative beamsplitter efficiency of a given film is different for parallel and perpendicular polarized radiation. The reflectance for p-polarized light is always less than that for s-polarization, but this fact does not imply that $\eta(\bar{\nu})$ is always lower for parallel polarized radiation if the beamsplitter material has a high refractive index. For instance, Bell [31] has given the overall reflectance of germanium at $\bar{\nu}_0$ as 0.36 for p-polarized radiation and 0.62 for s-polarization. The beamsplitter efficiencies for these two polarizations are therefore given by $\eta_p(\bar{\nu}_0)$ = 0.92 and $\eta_s(\bar{\nu}_0)$ = 0.94. Thus, even though the reflectance values for the two polarizations are quite different, the fact that they differ from the ideal value of 0.5 by approximately the same

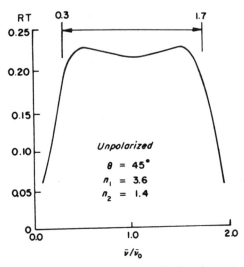

Fig. 4.19. Variation of RT with wavenumber, normalized to the center wavenumber for a silicon film ($n_1 = 3.6$) on a sodium chloride substrate ($n_2 = 1.4$) with a NaCl compensator plate of the same thickness, with $\theta = 45°$, calculated for the first "hoop" of the beamsplitter efficiency curve. (Reproduced from [30], by permission of the author.)

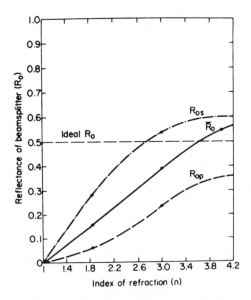

Fig. 4.20. Variation of the maximum reflectance R_0 of a beamsplitter film with refractive index n. Since the optimum value of $\eta(\bar{v})$ is found when $R = 0.5$, there is no "ideal" beamsplitter for p-polarized radiation, but the *average* reflectance R_0 of both Si and Ge approaches the optimal value. (Reproduced from [31], by permission of Academic Press; copyright © 1972.)

154

extent causes the relative efficiencies for each polarization to be approximately equal.

Figure 4.20 shows the variation of the overall maximum reflectance for materials of different refractive index. Since an ideal beamsplitter has a reflectance of 0.5, it is apparent from this diagram that the optimum

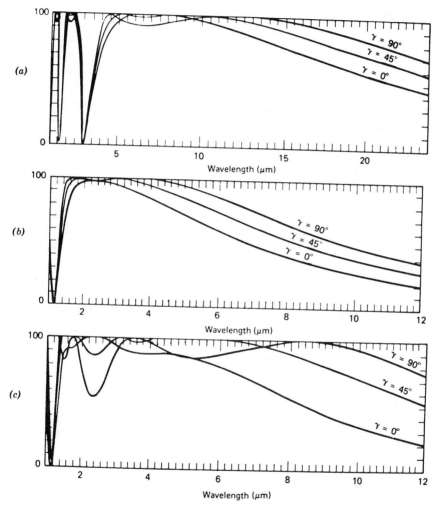

Fig. 4.21. Theoretically calculated beamsplitter efficiency curves for various combinations of materials at different incident angles: (*a*) a 0.4-μm film of Ge between NaCl plates; (*b*) a 0.17-μm film of Si between CaF$_2$ plates; (*c*) a 0.17-μm film of Si between CaF$_2$ plates with a 2.5-μm air gap between the film and the compensator plate to give improved long-wavelength performance. (Reproduced from [32], by permission of the author.)

beamsplitter material for an unpolarized beam should have a refractive index of about 3.6, whereas if only p-polarized radiation is being studied, the ideal refractive index is about 2.8.

Tescher [32] has described the results of a theoretical investigation into the effect of polarization on $\eta(\bar{\nu})$ for dielectric films on transmitting substrates, with compensator plates of the same material. His results are shown for a 0.4-μm film of germanium between NaCl plates in Fig. 4.21a and for a 0.17-μm film of silicon between CaF$_2$ plates in Fig. 4.21b. These systems show good performance over fairly limited spectral ranges, but the performance falls off at long wavelengths.

A substantial improvement in the long-wavelength performance can be achieved by adding an additional thin-film layer. If a CaF$_2$ compensator plate is separated from a silicon film by a 2.5-μm air gap, it is seen that $\eta(\bar{\nu})$ is increased at long wavelengths (Fig. 4.21c). Tescher also demonstrated that the width of the air gap can be varied to optimize the performance in different spectral regions.

In practice, it is difficult to control the thickness of an air gap, and Tescher has also described a system where 13 layers of different materials (Ge, TlBr, KRS-5) can be deposited to produce a beamsplitter with $\eta(\bar{\nu})$ > 0.5 from 7000 to 300 cm^{-1}. Although these beamsplitters would obviously be difficult to manufacture since the thickness of each layer must be carefully controlled, for any application where a large bandwidth is more important than good response over a small frequency range, systems such as this seem to hold the highest potential. Several manufacturers now use multilayer coatings for some of their beamsplitters to give improved performance at high wavenumbers. They are naturally unwilling to divulge the actual composition of the layers, but the success of this design is demonstrated by the comparison of single-beam spectra measured using the same source and detector with a conventional Ge–KBr beamsplitter and a multilayer beamsplitter (Fig. 4.22).

One type of material other than a Mylar film may become useful as a beamsplitter for far-infrared spectroscopy. Wire grids have been shown [33,34] to possess the properties of reflectance and transmittance for far-infrared radiation that give quite high relative beamsplitter efficiencies. The efficiency of a wire grid at any wavelength λ depends on the grating constant d of the mesh but is not strongly dependent on the width of the

$\longrightarrow$

Fig. 4.22. Ratio of single-beam spectra measured on a Nicolet 60-SX spectrometer equipped with a multilayer and a single-layer beamsplitter, with all other components unchanged. This curve demonstrates the increased efficiency of the multilayer beamsplitter both at higher and lower wavenumbers. (Reproduced by permission of the Nicolet Instrument Corporation.)

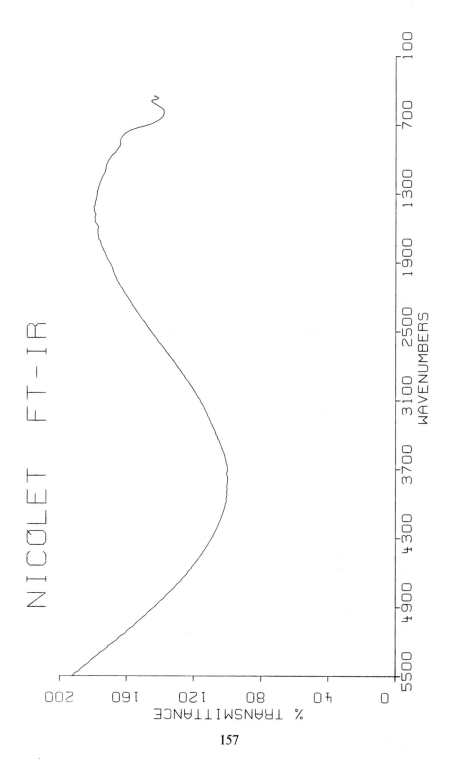

NICOLET FT-IR

157

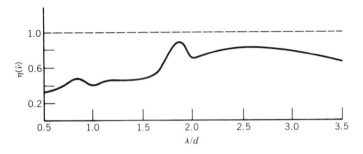

Fig. 4.23. Variation of the beamsplitter efficiency with λ/d for a metal mesh, where λ is the wavelength and d is the grating constant for the mesh.

wires. A plot of $\eta(\bar{\nu})$ against λ/d is shown in Fig. 4.23 that demonstrates that high efficiencies can be attained over reasonably long spectral ranges. Although this type of beamsplitter shows no advantages over high-refractive-index films of the type used in mid-infrared spectrometry (quite the reverse, in fact), wire grid beamsplitters appear to be superior to Mylar for far-infrared spectrometry over wide spectral ranges. It is difficult to maintain the flatness of wire grids over the entire 5-cm aperture of most conventional Michelson interferometers. However, the very small beamsplitters needed for Genzel interferometers permit wire grid beamsplitters to be used at fairly high efficiency.

III. LESS COMMON TYPES OF INTERFEROMETERS FOR FOURIER SPECTROMETRY

a. Lamellar Grating Interferometers

In spite of their obvious advantages over grating monochromators for far-infrared spectrometry, Michelson interferometers still exhibit certain disadvantages for very far infrared spectrometry. The main disadvantage of the Michelson interferometer as it is conventionally used for far-infrared spectrometry is the low efficiency and limited spectral range given by the Mylar beamsplitter. On the other hand, the type of interferometer known as a lamellar grating interferometer can have an efficiency close to 100% and can operate over a wide spectral range for far-infrared spectrometry [35].

Unlike the Michelson interferometer, in which the amplitude of the radiation is divided at the beamsplitter, the lamellar grating interferometer uses the principle of wavefront division. It consists of two sets of parallel interleaved mirrors, one set that can move in a direction perpendicular to the plane of the front facets, whereas the other set is fixed. Thus, when a beam of collimated radiation is incident onto this grating (Fig. 4.24), half the beam will be reflected from the front facets whereas the other half will be reflected from the back facets. By moving one set of mirrors, the path difference between the two beams can be varied so that an interferogram is generated. It can be seen from Fig. 4.24 that essentially all the radiation from the source reaches the detector, so that the efficiency of the lamellar grating interferometer can approach 100%.

In practice, the efficiency can only approach 100% through a limited wavenumber range, and it falls off at both high and low frequencies [35]. Figure 4.25 shows the theoretical efficiency of a lamellar grating interferometer compared with that of a Michelson interferometer using a Mylar beamsplitter. The cutoff at low wavenumber is caused by the cavity effect, by which the modulation of waves for which the electric vector is parallel to the sides of the cavity starts to decrease; $\bar{\nu}_L$ is equal to $(0.3a)^{-1}$ where

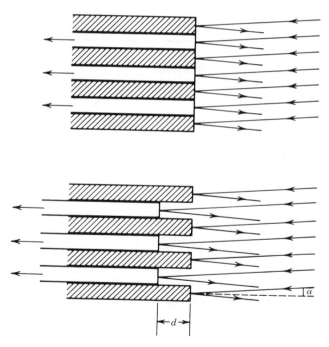

Fig. 4.24. Principle of the lamellar grating interferometer. Radiation reflected off the movable facets travels a distance of $2d/\cos \alpha$ further than the radiation reflected off the stationary (shaded) facets.

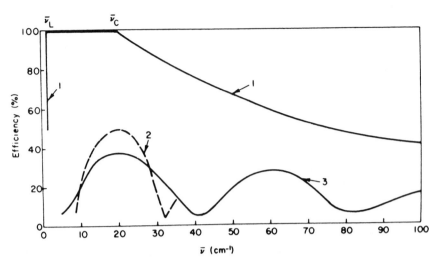

Fig. 4.25. Comparison of the theoretical efficiency of a lamellar grating interferometer (1) and a Michelson interferometer with (2) a metal mesh beamsplitter and (3) a Mylar beamsplitter. (Reproduced from [35], by permission of the Optical Society of America; copyright © 1964.)

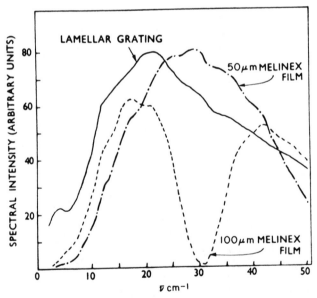

Fig. 4.26. Spectra measured with the same source and detector using a lamellar grating interferometer and a Michelson interferometer with 100- and 50-μm-thick beamsplitters. Note the better performance of the lamellar grating interferometer near the millimeter wavelength region, below 20 cm^{-1}.

160

a is the grating constant. The reduction in efficiency at high wavenumber is caused by waves diffracted at the grating (which can be considered as a series of long rectangular slits) being canceled at the exit aperture; $\bar{v}_c$ is equal to F/aS, where F is the focal length of the collimator and S is the diameter of the exit aperture. Although the optical and mechanical properties of lamellar grating interferometers preclude their use at wavenumbers much above 150 cm^{-1}, they have been successfully used from 150 cm^{-1} to as low as 1.5 cm^{-1}. For all measurements below 10 cm^{-1}, the lamellar grating interferometer is almost certainly the instrument of choice.

Figure 4.26 shows how the actual energy measured with a lamellar grating interferometer compares with the energy of the same source measured with a Michelson interferometer equipped with 100- and 50-μm-thick Mylar beamsplitters and using the same detector. Below 10 cm^{-1}, the energy in the single-beam spectrum measured with the lamellar grating interferometer is decidedly superior to that measured even with the thickest Mylar film. However, for use above 100 cm^{-1}, the lamellar grating interferometer requires an exceptionally high quality drive, and in practice it becomes preferable to use a Michelson interferometer. Although one lamellar grating interferometer, the Beckman–RIIC LR-100 was once available commercially, none is currently sold.

b.   Polarizing Interferometers for Far-Infrared Spectrometry

Another rather unusual far-infrared spectrometer, which is sold by Specac in England, is known as the polarizing interferometer. It is based on a concept originally proposed by Martin and Puplett [36] that can be most readily understood from the configuration shown in Fig. 4.27.

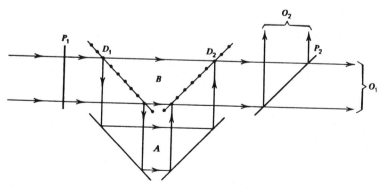

Fig. 4.27.   Diagram of the Martin–Puplett polarizing interferometer; see text for details.

A collimated beam is plane polarized at P_1 in the plane at 45° to the normal of this page. It is then divided by a flat wire grid polarizer, D_1, into beam A, polarized with its electric vector normal to the paper and beam B, polarized at 90° to A. Then A and B are recombined at the wire grid D_2, and the combined beam finally passes through polarizer P_2, the axis of which may be either parallel to that of P_1 or perpendicular to that direction.

For a monochromatic source, the beam is elliptically polarized after recombination at D_2 with an ellipticity varying periodically with increasing path difference between A and B. After P_2, the beam is plane polarized with an amplitude that varies periodically with path difference. If P_1 and P_2 are oriented in parallel and if the source is unpolarized, the intensity at the detector is given by

$$I_p(\delta) = 0.5I(\bar{\nu})[1 + \cos 2\pi\bar{\nu}\delta] \tag{4.14}$$

which is, of course, exactly the same expression as that found for a stan-

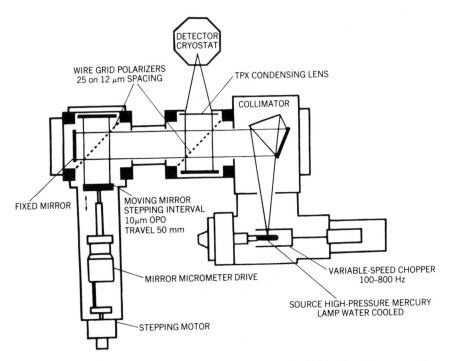

Fig. 4.28. Optical schematic of a commercial far-infrared polarizing interferometer. (Reproduced by permission of Specac Ltd., Orpington, Kent, U.K.)

dard Michelson interferometer. If P_1 and P_2 are crossed, the sign in Eq. 4.14 is changed, that is, the behavior is the same as for the beam that returns to the source for a Michelson interferometer (see Chapter 8).

Wire grids can have reflection and transmission coefficients approaching 100% for the appropriate planes of polarization, so that it is possible to combine P_1 and D_1 and P_2 and D_2. This is the approach taken in the Specac far-infrared polarizing interferometer shown in Fig. 4.28. The sample would usually be held in a module between the interferometer and the TPX (poly-4-methylpentene-1) lens. These instruments can incorporate a step-and-integrate drive (as shown in Fig. 4.28), with a piezoelectric modulator to give phase modulation. Alternatively, by increasing the scan speed and adding a laser reference, it is possible to convert it to a rapid-scanning interferometer. A theoretical analysis by Lambert and Richards [38] attests to the power of these instruments. Interferometers using grids of 0.0125 mm spacing and 0.005-mm-diameter wire have been used successfully in broadband spectrometry from 2 to 700 cm^{-1} [39].

c.   Special-Purpose Interferometers

A wide variety of special-purpose interferometers have been designed for an enormous number of applications. Interferometers have been designed to operate at liquid nitrogen or even liquid helium temperatures. They have been flown on spacecraft, satellites, rockets, and aircraft, to say nothing of the many interferometers that have been mounted in vans and trucks or on the end of telescopes in an observatory. A detailed discussion of all these instruments is beyond the scope of this book, but each is based on one of the types of interferometers described in this chapter.

REFERENCES

1. L. Mertz, *Astron. J.*, **70**, 548 (1965).
2. P. R. Griffiths, R. Curbelo, C. T. Foskett, and S. T. Dunn, *Analytical Instrumentation* (Inst. Soc. Am.), **8**, II-4 (1970).
3. A. S. Zachor, *Appl. Opt.*, **16**, 1412 (1977).
4. "The Midac Interferometer Spectrometer System," Midac Corp., Costa Mesa, CA (1982).
5. R. P. Walker and J. D. Rex, *Proc. Soc. Photo-Opt. Instrum. Eng.*, **191**, 88 (1979).
6. H. R. Chandrasekhar, L. Genzel, and J. Kuhl, *Opt. Commun.*, **17**, 106 (1976).
7. L. Genzel and J. Kuhl, *Infrared Phys.*, **18**, 113 (1978).
8. L. Genzel and J. Kuhl, *Appl. Opt.*, **17**, 3304 (1978).

9.   P. Bouchareine and P. Connes, *J. Phys. Radium.*, **24**, 134 (1963).

10.   A. M. Despain, F. R. Brown, A. J. Steed, and D. J. Baker, Aspen Int. Conf. on Fourier Spectrosc., 1970 (G. A. Vanasse, A. T. Stair, and D. J. Baker, eds.), AFCRL-71-0019, p. 293 (1971).

11.   A. M. Despain, D. J. Baker, A. J. Steed, and T. Tohmatsu, *Appl. Opt.*, **10**, 1870 (1971).

12.   A. J. Steed, *Proc. Soc. Photo-Opt. Instrum. Eng.*, **191**, 2 (1979).

13.   W. M. Doyle, B. C. McIntosh, and W. L. Clarke, *Appl. Spectrosc.*, **34**, 599 (1980).

14.   W. M. Doyle and B. C. McIntosh, *Proc. Soc. Photo-Opt. Instrum. Eng.*, **289**, 322 (1981).

15.   W. L. Truett, J. P. Dybwad, G. D. Propster, C. D. Prozzo, and J. Szappanos, *Amer. Lab.*, **16**(3), 115, (1984).

16.   R. S. Sternberg and J. F. James, *J. Sci. Instrum.*, **41**, 225 (1964).

17.   H. Buijs, *Proc. Soc. Photo-Opt. Instrum. Eng.*, **191**, 116 (1979).

18.   H. Buijs, D. J. W. Kendall, G. Vail, and J.-N. Bérubé, *Proc. Soc. Photo-Opt. Instrum. Eng.*, **289**, 322 (1981).

19.   W. H. Steel, Aspen Int. Conf. on Fourier Spectrosc., 1970 (G. A. Vanasse, A. T. Stair, and D. J. Baker, eds.), AFCRL-71-0019, p. 43 (1971).

20.   J. Connes and P. Connes, *J. Opt. Soc. Am.*, **56**, 896 (1966).

21.   J. Connes, H. Delouis, P. Connes, G. Guelachvili, J.-P. Maillard, and G. Michel, *Nouv. Rev. Opt.*, **1**, 3 (1970).

22.   P. Connes and G. Michel, *Appl. Opt.*, **14**, 2067 (1975).

23.   M. Cuisinier and J. Pinard, *J. Phys.*, **28**, C2:97 (1967).

24.   G. Guelachvili and J.-P. Maillard, Aspen Int. Conf. on Fourier Spectrosc., 1970 (G. A. Vanasse, A. T. Stair, and D. J. Baker, eds.), AFCRL-71-0019, p. 151 (1971).

25.   R. B. Sanderson and H. E. Scott, Aspen Int. Conf. on Fourier Spectrosc., 1970 (G. A. Vanasse, A. T. Stair, and D. J. Baker, eds.), AFCRL-71-0019, p. 167 (1971).

26.   D. S. Davis, H. P. Larson, M. Williams, G. Michel, and P. Connes, *Appl. Opt.*, **19**, 4138 (1980).

27.   J. W. Brault, *J. Opt. Soc. Am.*, **66**, 1081 (1976).

28.   M. Born and E. Wolf, *Principles of Optics*, 2nd ed., Pergamon Press, Oxford (1964).

29.   J. E. Chamberlain, G. W. Chantry, F. D. Findlay, H. A. Gebbie, J. E. Gibbs, N. W. B. Stone, and A. J. Wright, *Infrared Phys.*, **6**, 195 (1966).

30.   H. Sakai, Aspen Int. Conf. on Fourier Spectrosc., 1970 (G. A. Vanasse, A. T. Stair, and D. J. Baker, eds.), AFCRL-0019-71, p. 19 (1971).

31.   R. J. Bell, *Introductory Fourier Transform Spectroscopy*, Academic Press, New York (1972).

32.   A. G. Tescher, Aspen Int. Conf. on Fourier Spectrosc., 1970, (G. A. Vanesse, A. T. Stair, and D. J. Baker, eds.), AFCRL-0019-71, p. 225 (1971).

33.   P. Vogel and L. Genzel, *Infrared Phys.*, **4**, 257 (1964).

34. W. G. Chambers, C. L. Mok, and T. J. Parker, *J. Phys. D. Appl. Phys.,* **13,** 515 (1980).

35. P. L. Richards, *J. Opt. Soc. Am.,* **54,** 1474 (1964).

36. D. H. Martin and E. Puplett, *Infrared Phys.,* **10,** 105 (1969).

37. D. E. Martin, in *Infrared Detection Techniques for Space Research,* (V. Manno and J. Ring, eds.), D. Reidel Publishing, Dordrecht, Holland (1972).

38. D. K. Lambert and P. L. Richards, *Appl. Opt.,* **7,** 1595 (1978).

39. P. A. R. Ade, A. E. Costley, C. T. Cunningham, C. L. Mok, G. F. Neill, and T. J. Parker, *Infrared Phys.,* **19,** 599 (1979).

CHAPTER

5

AUXILIARY OPTICS
FOR FT–IR SPECTROMETRY

I. SPECTROMETER DESIGN

a. Introduction

Fourier transform–infrared spectrometers measure single-beam spectra directly, but the most common applications of infrared spectrometry require either the transmittance or absorbance spectrum of the sample to be measured. The transmittance $\tau(\bar{\nu})$ is calculated by taking the ratio of single-beam spectra of a broadband source measured with and without the sample present. The absorbance spectrum $A(\bar{\nu})$ is subsequently calculated as $-\log_{10}\tau(\bar{\nu})$.

The method by which transmittance spectra are obtained on a double-beam grating spectrophotometer is quite different from the way they are measured on an FT–IR spectrometer. These authors choose to use the name *spectrometer* for the FT–IR instrument, although it is technically a spectrophotometer. This distinction is drawn to indicate that FT–IR instruments always produce single-beam spectra. On modern double-beam grating spectrophotometers, $\tau(\bar{\nu})$ is obtained in real time. In Fourier spectrometry, this type of output cannot be obtained since an interferogram contains multiplexed information from all spectral elements. A typical procedure for measuring $\tau(\bar{\nu})$ on a Fourier transform spectrometer is to measure the reference interferogram first and to store it in the memory of the computer (usually, but not necessarily, after transforming it to the spectrum). The sample is then inserted and the sample interferogram is measured. The ratio of the sample and reference spectra is calculated to yield $\tau(\bar{\nu})$, which is then usually converted to $A(\bar{\nu})$. Finally, the spectrum is displayed on the CRT and/or plotted as hard copy.

All of the early far-infrared FT–IR spectrometers operated in this single-beam mode. Many of the less expensive contemporary mid-infrared FT–IR spectrometers are also single-beam instruments. Because of the simplicity of the optical systems required for single-beam measurements, these instruments are often quite compact, and several bench-top FT–IR spectrometers are now available.

166

b. Slow-Scanning Far-Infrared Fourier Spectrometers

In the late 1960s and early 1970s far-infrared FT–IR spectrometers were available from several companies, including Research and Industrial Instruments Company (subsequently Beckman–RIIC) and Grubb-Parsons in the United Kingdom, Coderg in France, and Polytec in Germany. Although none of these organizations still markets far-infrared spectrometers, it is instructive to study the design of one of these instruments, the Grubb-Parsons Mark II, chosen because it was directly based on the original National Physical Laboratory "cube" interferometer, which is still being used for high-sensitivity far-infrared measurements in many laboratories. A similar instrument is marketed today by Imperial College Instruments in England.

Radiation from a mercury lamp is modulated by a rotating chopper, collimated by an off-axis paraboloid, and passed through the slow-scan

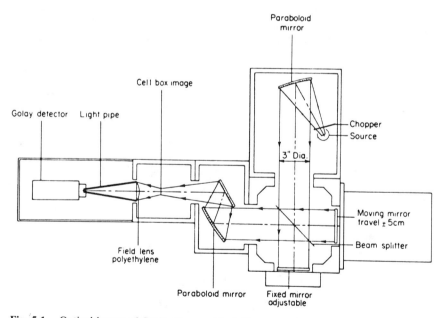

Fig. 5.1. Optical layout of Grubb-Parsons Mark II spectrometer. Like many of the spectrometers discussed in this section, this instrument incorporates off-axis paraboloidal mirrors to collect and collimate the source radiation and to focus the beam in the sample compartment. Unlike the other instruments, which are designed for both far- and mid-infrared spectrometry, the beam emerging from the sample compartment in this spectrometer is focused onto the detector by a field lens and light cone. (Reproduced by permission of Grubb-Parsons, Ltd.)

interferometer. The beam is then focused onto the sample by another paraboloid (of longer focal length than the one used to collimate the source radiation) and refocused on a Golay detector by a polyethylene field lens and conical light pipe, see Fig. 5.1. The modular nature of this instrument can readily be seen. The sample compartment shown in the figure was designed for the measurement of transmittance spectra of large condensed-phase samples or gases held in standard (10-cm-pathlength) cells. This module could be unbolted from the spectrometer and replaced by microsampling optics, a multiple-pass gas cell, a cryostat, or a specular reflectance attachment, with provision for a polarizer if necessary.

c. Single-Beam Rapid-Scanning FT–IR Spectrometers

Many of the less expensive contemporary mid-infrared FT–IR spectrometers are also single-beam instruments. These include the Beckman 1100 and 2100, Digilab FTS-40, FTS-50, FTS-60 and Qualimatic (see Fig. 5.2), the IBM IR/85 and IR/32 and the Nicolet 5-MX, 5-DX, 10-MX (see Fig. 5.3), 10-DX, 20-MX, 20-DX (see Fig. 5.4), and 60-SX. The evolution of

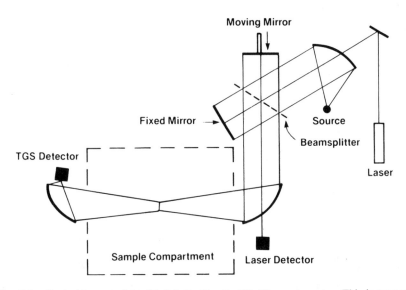

Fig. 5.2. Optical layout of the Digilab Qualimatic FT–IR spectrometer. This instrument contains the minimum number of mirrors required for a Fourier transform spectrometer. The beam from the sample compartment is focused onto the detector by an aspherical mirror with a 6:1 ratio of the effective focal lengths so that the diameter of the image in the sample compartment (12 mm diameter) is reduced to 2 mm at the detector. (Reproduced by permission of the Digilab Division of Bio-Rad Laboratories, Inc.)

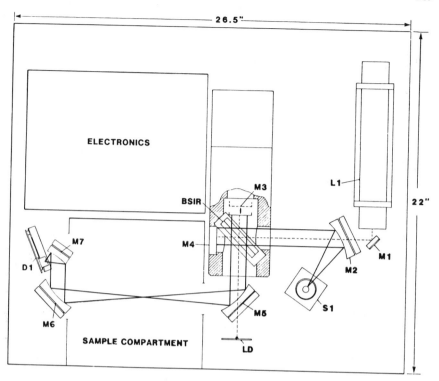

Fig. 5.3. Optical layout of the Nicolet 10-MX spectrometer. The only substantive difference between the optics of this spectrometer and the Qualimatic shown in Fig. 5.2 is the use of two paraboloidal mirrors, one (M6) of long effective focal length (e.f.l.) and the other one (M7) of short e.f.l., to collect and focus the beam from the sample compartment onto the detector (D1). S1, source; L1, He–Ne laser; M1, plane mirror; M3, M4, interferometer mirrors; M2, M5, paraboloidal mirrors; BSIR, infrared beamsplitter; LD, laser detector. (Reproduced by permission of the Nicolet Instrument Corporation.)

their optics from the early far-infrared instruments may be seen by comparing their optics to the Grubb-Parsons Mark II shown in Fig. 5.1. There is, of course, no need for a separate chopper in the optics of mid-infrared Fourier spectrometers because the rapid-scanning interferometer itself modulates the source radiation at audio frequencies. Since the maximum optical throughput of mid-infrared Fourier spectrometers is about one-tenth the throughput allowed for far-infrared measurements at the same resolution, the lens–light-cone combination used as the detector for optics of the Grubb-Parsons Mark II is replaced by mirrors in all mid-infrared instruments. Either a single ellipsoid or a combination of a long and short-focal length paraboloidal mirror is used to focus the beam on the detector in most mid-infrared spectrometers.

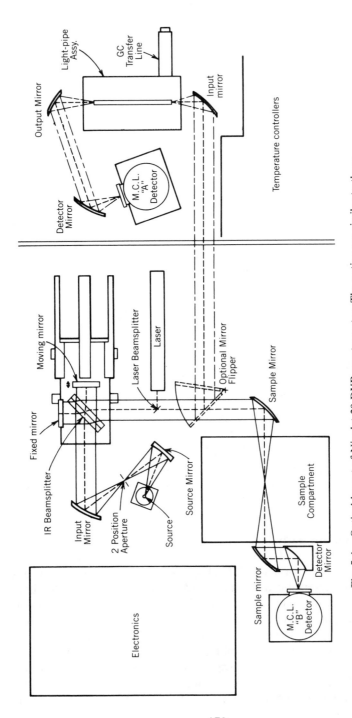

Fig. 5.4. Optical layout of Nicolet 20-DXB spectrometer. The optics are similar to those of the 10-MX shown in Fig. 5.3, with two exceptions. The source optics incorporate a two-position aperture for resolution control (see Section I.e, this chapter), and a flip mirror has been installed to permit the collimated beam to be passed to an accessory located outside the main optical bench. In this case, the accessory is a GC–IR interface (see Chapter 18). (Reproduced by permission of the Nicolet Instrument Corporation.)

170

A few of the more expensive mid-infrared Fourier spectrometers also incorporate single-beam optics but are more flexible than the small benchtop instruments. For example, in the evacuable Nicolet 200SX, the operator is allowed to choose between two sources (S1 and S2) or two detectors (D1 and D2) and switch between them without breaking the vacuum, see Fig. 5.5. It is very convenient to be able to switch between a room temperature pyroelectric detector for high-energy measurements and a liquid-nitrogen-cooled photodetector for samples that transmit less energy. Nevertheless, it should be recognized that in many cases if different sources are to be used to investigate different spectral regions the

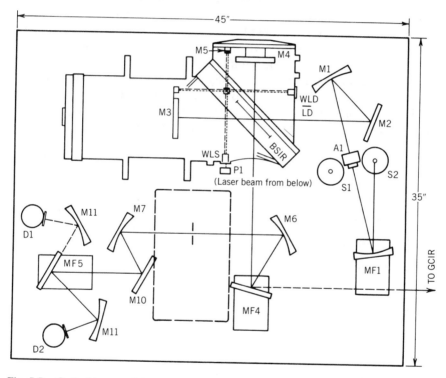

Fig. 5.5. Optical layout of the Nicolet 200-SXV spectrometer. Radiation from one of two sources (S1 and S2) may be selected by the flip mirror, MF1, passed through an aperture for resolution control, and collimated by the paraboloid, M1. The collimated beam emerging from the interferometer may be either passed into the sample compartment or outside the main bench by the flip mirror, MF4. The paraboloids, M6 and M7, focus the beam at the sample and recollimate the beam, respectively. Either of two detectors, D1 or D2, can be selected by the flip mirror, MF5. M2 and M10 are plane mirrors, and the M11 mirrors are both short e.f.l. paraboloids. All symbols for the interferometer are explained in the caption for Fig. 5.3. (Reproduced by permission of the Nicolet Instrument Corporation.)

beamsplitter must also be changed. On most contemporary FT–IR spectrometers, this cannot be achieved without breaking the vacuum or purge. Another feature of the 200SX that is not to be seen in any other Nicolet FT–IR spectrometer is the externally mounted He–Ne reference laser. By holding the laser at an unevacuated location, the possibility of the laser arcing in a high vacuum is eliminated. Unlike several of the other evacuable instruments, the 200SX incorporates sealed air bearings, thereby permitting a high vacuum to be attained.

Most mid-infrared spectrometers are purged with dry air to avoid atmospheric interferences rather than being evacuated. Certainly, purging with dry air is a much more convenient means of operation, provided that water vapor can be rigorously excluded from the path of the infrared beam. Since new samples must be changed many times each day, laboratory (wet) air is introduced very easily into the optical path unless an "air lock" is provided at the sample compartment. After introducing a new sample, the operator must usually wait for at least a minute for the partial pressure of water vapor to return to an acceptable level before initiating the measurement. For many measurements, the presence of interfering lines absorbing a few tenths of a percent due to the rotation–

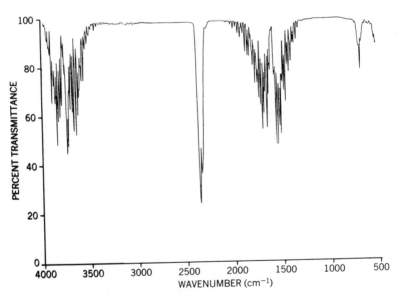

Fig. 5.6. Transmittance spectrum of water and carbon dioxide at levels typically observed in poorly purged or unpurged FT–IR spectrometers. The rotation–vibration bands of water are centered near 3750 and 1650 cm^{-1}, and the bands of CO_2 are located near 2350 and 668 cm^{-1}. This spectrum was measured at 4 cm^{-1} resolution with triangular apodization.

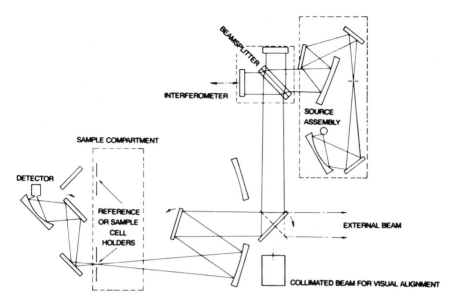

Fig. 5.7. Optical layout of the Digilab FTS-90 spectrometer. This instrument contains three flip mirrors. The first serves the dual purpose of directing the beam either to the sample compartment or to an externally mounted accessory; when this mirror is in the latter position, the beam from a source of collimated visible light is directed through the sample compartment to aid in the alignment of sampling accessories. The second and third flip mirrors work in tandem to direct the infrared beam to either the front or rear channels in the sample compartment. (Reproduced by permission of the Digilab Division of Bio-Rad Laboratories, Inc.)

vibration spectrum of atmospheric water may not be important. If, however, large ordinate expansions or spectral subtraction experiments are to be performed, the presence of this much interference in the spectrum may represent the difference between success and failure for the measurement. The spectrum of atmospheric H_2O and CO_2 are shown for reference in Fig. 5.6.

d. Double- or Multiple-Beam FT–IR Spectrometers

To overcome the problem of too much atmospheric interference, several manufacturers have designed double-beam optical systems, which have two possible beam paths. The sample is held in one path and the reference (if any) in the other; see, for example, Figs. 5.7 and 5.8. With double-beam instruments, it is possible to minimize the effect of water vapor by signal averaging a few (say 20) scans in one beam, storing the interferogram, switching to the other beam, and again signal averaging 20 scans.

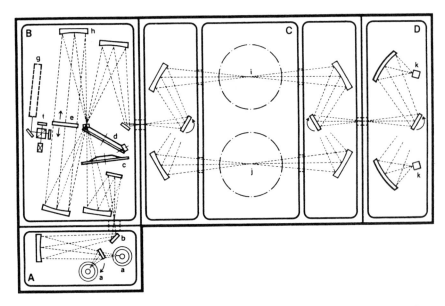

Fig. 5.8. Optical layout of the IBM Instruments IR/90 series and Bruker IFS 113 and 114 spectrometers. Radiation from either of two sources (a) can be selected and passed through a variable aperture (b) to the interferometer (see Fig. 4.5). Two flip mirrors direct the beam to the front (i) or rear (j) channel of the sample compartment, and another flip mirror directs the beam to one of two detectors (k). The evacuable versions of these instruments are modular in a similar manner to the far-infrared spectrometer shown in Fig. 5.1. (Reproduced by permission of IBM Instruments, Inc.)

The process is continued until the desired SNR has been attained. Then the sample and reference interferograms are transformed and the single-beam spectra ratioed to obtain the transmittance spectrum.

Although the effect of residual water vapor is reduced by taking measurements in this fashion, the operator is still advised to wait a short time after introducing the sample before initiating the measurement, especially if the flowrate of the purge gas is low or on hot, humid days. Another point should also be mentioned. Many rapid-scan single-beam or double-beam FT–IR spectrometers are equipped with a deuterated triglycine sulfate (DTGS) detector. When the beam of a double-beam spectrometer is switched from one channel to the other, radiation from the source is prevented from hitting the detector for a few tenths of a second. The temperature of the detector element therefore falls temporarily, and the response of the detector drops dramatically. Usually it only takes a couple of seconds for it to recover, but data should not be acquired during this recovery period. Therefore, the computer of the spectrometer's data system should be instructed to create a delay of about 4 sec after the mirrors to the sample optics have been switched.

One final practical point concerning double-beam measurements may also be noted here. Although the beam paths for the front and rear channels are supposedly identical, in practice this is rarely found to be the case. Thus, if no sample is held in either beam, it is often found that the 100% line may neither be flat nor exactly at 100% transmittance. This behavior is caused by inconsistent alignment of the optics in the front and back beams. In some of the more recent instruments, such as the Perkin–Elmer Model 1800, careful manufacturing practices have reduced this optical mismatch. In many instances, a much better 100% line is found in "single-beam" measurements, where only one channel is used for both sample and reference spectra.

One practical advantage of a double-beam FT–IR spectrometer is that it may be used as two single-beam instruments. For example, if many measurements are performed each week in a certain laboratory using a beam condenser, a reflectance accessory, or a long-path gas cell, the accessory may be permanently mounted in the instrument (provided that it does not obscure the other beam). Spectra of solutions, KBr disks, mulls, 10-cm gas cells, and so on, may then be measured in the other beam, and the time-consuming process of checking the alignment of the accessory each time it is replaced in the sample compartment is eliminated.

Sometimes it is convenient to mount accessories outside of the spectrometer optics. This is particularly important for bulky accessories such as the interface to a gas chromatograph (GC/FT–IR). Many instruments incorporate a flipping mirror to deflect the collimated beam from the interferometer outside the spectrometer. Top-of-the-line FT–IR spectrometers made by several companies usually incorporate such a feature, as may be seen in Figs. 5.4, 5.5, 5.7, and 5.9.

Every commercial FT–IR spectrometer has the sample compartment located between the interferometer and the detector. This is particularly important for the measurement of hot samples, since radiation emitted from these samples is unmodulated and hence not observed in the ac interferogram passed by the high-pass electrical filter. In this way an important potential source of stray radiation is eliminated. An additional benefit is that the heating effect of near-infrared radiation emitted from the source is reduced, since germanium beamsplitters reflect most of the near-infrared and visible radiation back to the source so that it does not reach the sample.

One of the disadvantages of locating the sample compartment between the source and the detector is that it becomes impossible to align sampling accessories visually using the light emitted from the source (as is normally done on grating spectrometers). Several instrument manufacturers have incorporated an auxiliary visible-light source that can replace the infrared beam by switching one mirror in the system. Several Digilab FT–IR spectrometers use a source of "collimated" white light, whose divergence angle exactly reproduces the divergence of the infrared beam through the interferometer at its maximum allowed throughput. In this way the amount of light loss taking place in any accessory can be estimated directly. In spectrometers made by other instrument companies, a He–Ne laser is commonly used for accessory alignment. Due to multiple reflections of the laser beam at the beamsplitter, it is often difficult to follow the central ray when aligning an accessory. The use of a white-light alignment source is therefore preferred over the use of a laser.

After the mirrors in the accessory have been aligned roughly using the auxiliary visible-light source, the accessory should be aligned accurately using the infrared beam. The interferometer is set scanning repetitively across the centerburst, and the signal is observed on an oscilloscope. The mirrors in the accessory are then fine tuned to maximize the infrared interferogram at the centerburst.

Since 1982, several companies have introduced instruments with compartments in which special sampling accessories can be mounted. With the Nicolet 60-SX, for example, the operator can divert the beam away from the conventional sample chamber either to optics mounted externally (such as a GC/FT–IR light pipe) or into an auxiliary chamber in which any one of several accessories can be mounted. These include a beam condenser, micro-ATR, or specular reflectance accessory. The mirrors in these optics are held on microcomputer-controlled translation stages. Energy at the detector is maximized automatically by feedback to piezoelectric transducers controlling the position of each mirror. The optical design of this instrument is shown in Fig. 5.9.

With the design of the 60-SX, three large accessories can be mounted with some degree of permanency, one at the "GC/FT–IR position," one in the auxiliary chamber, and one in the sample compartment. It is very convenient that an expensive research grade spectrometer does not have to be tied up with a single experiment but can be used for several purposes simultaneously.

An alternative but related approach has been taken by Mattson Instruments for their Sirius 100 spectrometer. A "universal sampling optics stage" can be kinematically mounted in the sample compartment of this spectrometer. The optics are mounted on precision orthogonal scales, and the positions of the various mirrors can be adjusted by micrometers to form a transmittance beam condenser or a specular, diffuse, or atten-

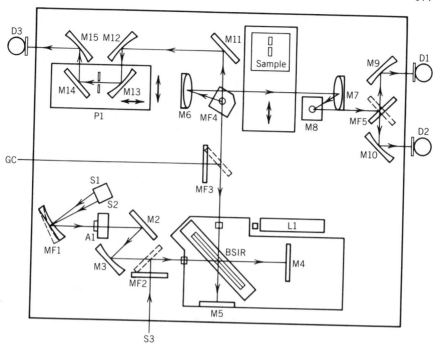

Fig. 5.9. Optical plan of the Nicolet 60-SX spectrometer. This instrument offers a selection of sources (S1 and S2 mounted in the same housing, or a remote source, S3) and detectors (D1, D2, or D3, together with the detector on the GC–IR accessory), which are selectable by flip mirrors MF1, MF2, MF3, MF4, and MF5. The sample and reference cells may be held on a computer-controlled translation stage. An accessory, such as the beam condenser shown, may be mounted permanently in the chamber shown in the upper right-hand corner; the position of several of the mirrors in this accessory are under computer control to permit automated alignment. (Reproduced by permission of the Nicolet Instrument Corporation.)

uated total reflectance accessory [1]. Once the micrometer scale readings have been determined for a particular experimental configuration, the settings can be reused each time that particular experiment is performed.

e. Optical Throughput

It is interesting and instructive to study the optics of mid-infrared Fourier spectrometers in some detail, since different manufacturers have differing design philosophies for their instruments. From Eq. 1.41, we see that the maximum allowed solid angle of the beam passing through the interferometer depends on the ratio of the resolution, $\Delta \bar{\nu}$, to the maximum wavenumber in the spectral range, $\bar{\nu}_{max}$. Optical throughput is defined as the product of the area of a beam at a focus and its solid angle. The effect of throughput on signal-to-noise ratio is discussed in some detail in Chapter 7. At low resolution (i.e. large $\Delta \bar{\nu}$), the allowed solid angle, and hence throughput, may become so large that it becomes impossible to focus all the light permitted to pass through the interferometer onto the rather small (1- or 2-mm-square) detectors used in these spectrometers. Thus, a decision has to be made on the practical upper limit for optical throughput. An example will be used to illustrate this point.

Extreme values for the maximum throughput of commercial FT–IR spectrometers may be found for the Digilab FTS-20 and the Nicolet NIC-7199B, the top-of-the-line instruments sold by these two companies in 1977. These instruments are approximately equivalent to the present FTS-90/20 and 170-SX, respectively. The largest allowed throughput of the FTS-20 was about eight times larger than that of the NIC-7199B. The standard detector on both instruments was DTGS, with the Digilab instrument having a 2-mm detector and the Nicolet spectrometer a 1-mm detector. All else being equal, this would result in an increased SNR of the FTS-20 of a factor of 4 over the NIC-7199B (since the noise equivalent power of a detector depends on the square root of its area, see Chapter 7). Part of this advantage is offset by the fact that beam intensity is not perfectly uniform across the image at a focus, especially for large images found in cases of high optical throughput. Thus, the practical gain in SNR for the FTS-20 over the NIC-7199B when both instruments were equipped with a DTGS detector was typically about a factor of 3. This gain could always be realized when an appreciable portion of the incident radiation was lost by scattering or absorption.

The smaller throughput of the NIC-7199B led to two benefits that offset this sensitivity disadvantage in certain circumstances. First, the fact that the diameter of the beam at its sample focus was four times smaller than that of the FTS-20 offsets the advantage of the FTS-20 for small samples. Second, the DTGS detector could be replaced by a wide-range mercury cadmium telluride (MCT) detector, which is about four times more sensitive than DTGS, without the interferogram exceeding the dynamic range of the ADC, even for samples of high average transmittance.

Users of FT–IR spectrometers therefore have a choice. Spectroscopists who tend to deal with samples of small diameter are advised to use a spectrometer with low throughput optics, or at least a small detector, whereas spectroscopists who deal more commonly with samples that scatter or absorb much of the incident energy are better advised to use a spectrometer with higher throughput. The throughput of most commercial FT–IR spectrometers falls between the two extremes discussed above.

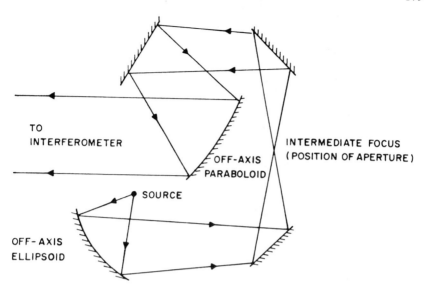

TO
INTERFEROMETER

INTERMEDIATE FOCUS
(POSITION OF APERTURE)

OFF-AXIS
PARABOLOID

SOURCE

OFF-AXIS
ELLIPSOID

Fig. 5.10. Typical configuration of the source optics of an FT–IR spectrometer. The size of the aperture, which is held at the intermediate focus (sometimes known as the Jacquinot stop), determines the resolution at all wavenumbers below a certain value.

The maximum allowed solid angle for a given spectral range and resolution is given by Eq. 1.41. In medium and high-resolution instruments, an aperture stop (sometimes known as the Jacquinot stop) is installed in the source optics to control the solid angle. This aperture has the same function as the entrance slit of a monochromator. A typical optical configuration to produce a beam with the correct solid angle is shown in Fig. 5.10. For most spectrometers, the source is imaged at the Jacquinot stop, at the focus in the sample compartment, and at the detector. Examples of optics in this category are seen in Figs. 5.5 and 5.7–5.9. Although a design of this type is highly efficient, it does have one major disadvantage. In Chapter 1, Section VII, it was shown that if the optical throughput is changed or if the location of a sample that is smaller than the focus is altered, a shift in the wavenumber scale results.

The optical design of the instruments made by Perkin-Elmer are designed to minimize spectral shifts while retaining a high optical throughput. Two series of conjugate images have been incorporated in the optics of these instruments. The first is a series of Jacquinot stops, which are located at the usual places in the source optics, in the sample compartment, and at the detector. The second is a series of images of the beam-

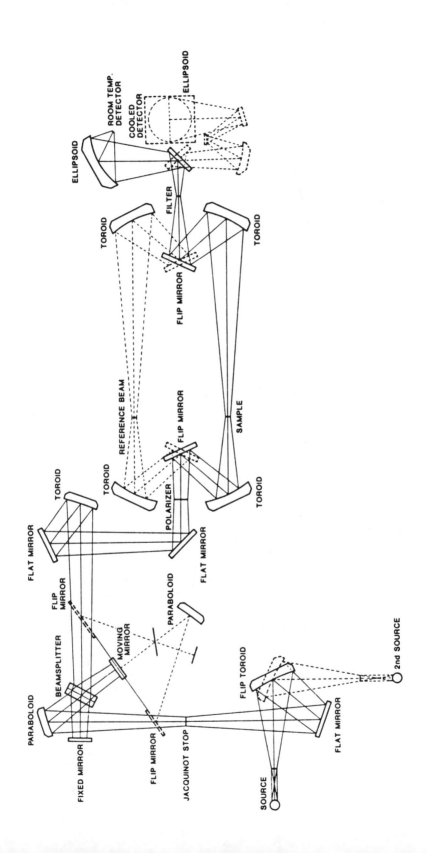

splitter, which are located at the source and at a slightly different location in the sample compartment than the Jacquinot stop. The optics of the Model 1800 are shown in Fig. 5.11. This design not only minimizes shifts in the wavenumber scale when the sample or sampling accessory vignettes the beam but also leads to improved baseline flatness in double-beam operation.

II. CELLS AND ACCESSORIES FOR ABSORPTION SPECTROMETRY

a. Cells

Many sample cells designed for infrared grating spectrophotometers can by used at high efficiency on FT–IR spectrometers. The main practical difference between cells optimized for use on each type of spectrometer is that cells designed for grating spectrometers usually have rectangular apertures (reflecting the shape of the slits of the monochromator) whereas cells for FT–IR spectrometers usually have round apertures (which is the shape of the Jacquinot stop). Nevertheless, most holders for mulls, solutions, 13-mm KBr discs, and 10-cm-pathlength gas cells have open apertures of at least 10 mm in the shortest dimension. These cells can therefore all be installed directly in FT–IR spectrometers since it is rare that a significant portion of the beam is lost by vignetting.

One important point should be noted. If a sample with a high front surface reflectance is mounted so that its plane is exactly perpendicular to the direction of the beam, the reflected beam passes back into the interferometer. This component of the beam will be remodulated by the interferometer, and part of it will be transmitted by the sample and measured at the detector. In this case, spurious weak spectral features will be observed in the measured spectrum at twice the wavenumber of strong features in the true spectrum. This effect is particularly important when narrow-bandpass filters are to be characterized by FT–IR spectrometry. A sample where this effect is appreciable should always be mounted

Fig. 5.11.   Optical layout of the Perkin-Elmer Model 1800 spectrometer. This instrument has two unusual features. First, the sample is located at a position intermediate between images of the Jacquinot stop and the beamsplitter, thereby minimizing wavenumber shifts caused when the sample or an accessory is misaligned or causes vignetting. The other unique feature is the capability of switching the beam to a second interferometer, the moving mirror of which is the back side of the moving mirror of the first interferometer, thus allowing rapid interchange between two spectral regions. (Reproduced by permission of the Perkin-Elmer Corporation.)

slightly tilted so that the beam reflected from its front surface does not ultimately reach the detector.

Many accessories designed for use on grating spectrometers can also be used at fairly high efficiencies on FT–IR spectrometers, especially after relatively minor adjustment of the positions of one or two mirrors. It should be noted, however, that the optimal performance will always be found using accessories designed for use on a particular instrument. This is particularly true for the more complex accessories such as 6× beam condensers or folded-path gas cells.

b. Beam Condensers and Microscopes

The circular shape of the beam at the sample focus often has distinct benefits over a rectangular image for the measurement of the spectra of small samples. For example, when 0.5-mm micropellets or high-pressure diamond anvil cells are being used, the aperture is approximately circular and therefore matches the image of the FT–IR spectrometer. Beam condensers must be configured to match the spectrometer optics. Like the grating spectrometers available in the late 1960s, the sample focus in many of the earlier FT–IR spectrometers was located at the side of the sample compartment. A center focus is somewhat more versatile than a side focus if several different accessories are to be employed, and most modern

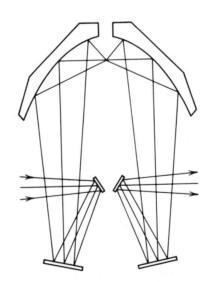

Fig. 5.12. Optical configuration of a 6× beam condenser for center-focus spectrometers. (Reproduced by permission of Harrick Scientific Corporation.)

infrared spectrometers (both FT–IR and grating) are designed with center-focusing optics. Typical optics for a beam condenser for a center focus are shown in Fig. 5.12.

Beam condensers are, of course, used for the study of samples that are constrained in a small aperture. When very small regions of a larger sample, such as a spot on a polymer film, are to be investigated nondestructively, the sample must be mounted behind a pinhole drilled through a thin metal sheet. The area of the pinhole should be approximately the same as the area of the spot. Mounting the film so that the spot of interest is located directly behind the pinhole can be quite difficult.

Recently, several microscope accessories have been designed for FT–IR spectrometers to circumvent this problem. In these accessories, the size of the image at the sample is controlled by an aperture located at a focus in the microscope where the solid angle of the beam is small rather than by an aperture at the sample focus where the solid angle is large. Microscope accessories are usually equipped with an MCT detector with a small element (typically 0.1 × 0.1 or 0.2 × 0.2 mm) to minimize the noise equivalent power. The smallest sample that can be examined on infrared microscopes is about 8–20 μm diameter and is controlled by diffraction effects rather than by the signal-to-noise ratio.

The first microscope accessory for an FT–IR spectrometer was introduced by Digilab. The optical configuration for this device is shown in Fig. 5.13. Samples can be mounted on a computer-controlled translation stage, and spectral profiling can be achieved with a resolution of about 20 μm.

c. Long-Path Gas Cells

20- and 40-m folded path gas cells based on the three mirror optical arrangement originally described by White [2] have been used on grating spectrometers for many years. The body of these cells can be used for the measurement of spectra on an FT–IR spectrometer, but only with transfer optics specifically designed for the particular spectrometer being used. Although these cells enable the stronger bands of most species present at parts per million (ppm) levels to be obtained with some ease, the observation of absorption bands from species present at the low parts per billion (ppb) level is much more difficult and requires a considerably longer pathlength cell.

The design of eight-mirror multiple-pass cells for very long pathlength measurements has been described by Hanst [3], and their application for trace atmospheric analysis using an FT–IR spectrometer has been described [4–9]. A cell based on Hanst's eight-mirror design capable of

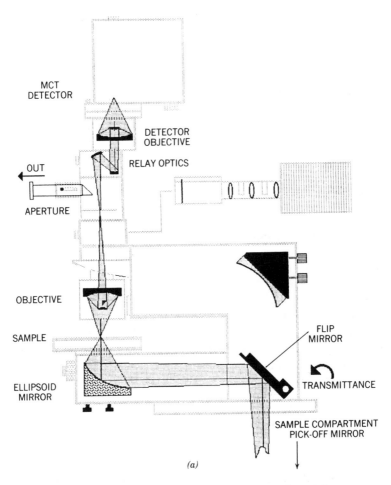

MCT
DETECTOR

DETECTOR
OBJECTIVE

OUT

RELAY OPTICS

APERTURE

OBJECTIVE

SAMPLE

FLIP
MIRROR

ELLIPSOID
MIRROR

TRANSMITTANCE

SAMPLE COMPARTMENT
PICK-OFF MIRROR

(a)

Fig. 5.13. (*a*) Optical diagram of the Digilab microscope configured for measurement of transmission spectra. Radiation is passed to the microscope by a pick-off mirror located in the sample compartment. The size of the image at the sample is determined by the size of the aperture. The sample is first aligned visually using a low-power refracting objective, which is then removed for the IR measurement. (*b*) Configuration of the Digilab microscope required for microreflectance measurements. The position of a flip mirror determines whether reflection or transmission spectra will be measured. (Reproduced by permission of the Digilab Division of Bio-Rad Laboratories, Inc.)

operating with a total pathlength of more than 1 km is shown in Fig. 5.14. Four rectangular mirrors cut from a single-plane Pyrex blank comprise the in-focus end of the cell. Three of these mirrors are 4 cm wide and 32 cm long, whereas the fourth is slightly shorter (28 cm) to provide space for the entrance and exit apertures. The assembly of collimating mirrors

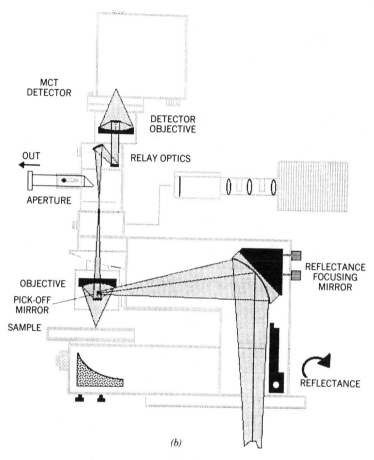

MCT
DETECTOR

DETECTOR
OBJECTIVE

OUT

RELAY OPTICS

APERTURE

OBJECTIVE

PICK-OFF
MIRROR

SAMPLE

REFLECTANCE
FOCUSING
MIRROR

REFLECTANCE

(b)

Fig. 5.13. (continued)

consists of four 30-cm-diameter spherical mirrors 22.5 m from the rectangular plane mirrors. As many as 48 passes of the beam through the cell are possible, giving a total pathlength for the cell shown in Fig. 5.14 of 1080 m.

Hanst [9] has published a detailed description of various configurations for long-path measurements including a discussion of how multiple-pass cells are interfaced to Michelson interferometers. Readers interested in this area are strongly advised to read this article. Not mentioned in this article, however, is the design of Horn and Pimentel [10], which has also been used for several long-path measurements using FT–IR spectrometers [11,12].

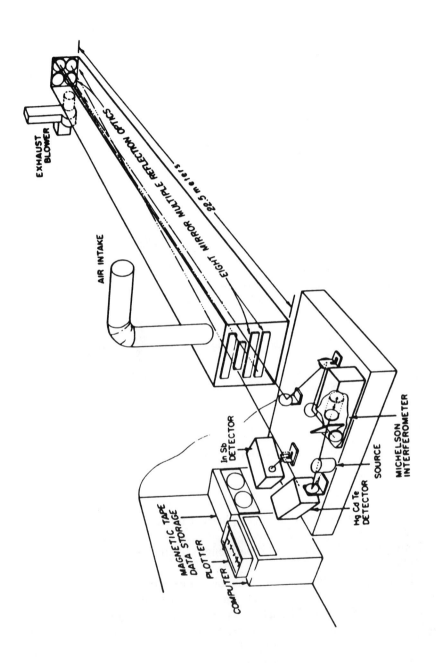

EXHAUST BLOWER

AIR INTAKE

EIGHT MIRROR MULTIPLE REFLECTION OPTICS

22.5 meters

In Sb DETECTOR

Hg Cd Te DETECTOR

SOURCE

MICHELSON INTERFEROMETER

MAGNETIC TAPE DATA STORAGE

PLOTTER

COMPUTER

III. ACCESSORIES FOR REFLECTION MEASUREMENTS

a. Specular Reflection Spectrometry

Commercially available accessories for FT–IR measurements of specular reflectance and attenuated total reflectance (ATR) spectra fall into the same general category as beam condensers for absorption spectrometry in that many accessories designed for grating spectrometers may be used on FT–IR spectrometers. Nevertheless, it should still be recognized that optimal data are obtained using accessories specifically designed for a particular instrument.

In the past, the term *specular reflectance* has been applied to two types of measurements. One applies to the measurement of the reflectance spectrum of a flat, clean surface of an absorbing material (either inorganic or organic). The other involves a double pass through a thin surface film on a reflective (typically metallic) surface. It is now becoming common practice to call this type of measurement reflection–absorption (R–A) spectrometry, whereas the former measurement is still known as specular reflectance spectrometry.

Specular reflectance spectra are generally rather easily measured for samples as small as 1 mm in diameter with incidence angles from about 15° to 75°. For small samples, it is often convenient to stop the beam down by placing a suitable aperture at the intermediate focus of the source optics, see Figs. 5.5, 5.7, and 5.9. Incidence angles much less than 15° are difficult to attain with conventional optics since it becomes difficult to separate the incident and reflected rays. Specular reflectance spectra at normal incidence can be measured by incorporating a beamsplitter in the accessory. The design shown in Fig. 5.15 can be effectively used for measurements at normal incidence. Although this optical arrangement only results in a maximum of 25% of the energy that leaves the interferometer and reaches the detector, even for reflection from a specular mirror, this signal is still entirely adequate for an FT–IR spectrometer. The problem with measuring specular reflectance spectra at high incidence (*i.e.*, near grazing) angles is somewhat different and will be described in the following discussion of R–A spectrometry.

Fig. 5.14. The eight-mirror multiple reflection cell developed by Hanst for use with a Fourier transform spectrometer and either an InSb or MCT detector. The base path of this cell was 22.5 m and the maximum pathlength attainable was over 1 km. (Reproduced from [8], by permission of the Optical Society of America; © 1978.)

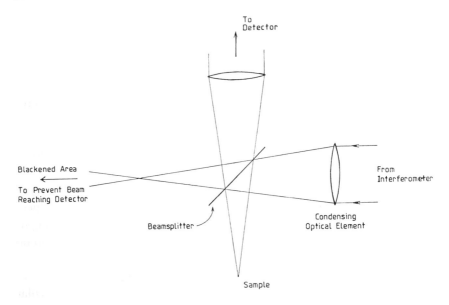

Fig. 5.15. A potentially useful optical configuration for the measurement of reflection spectra at normal incidence. It may be noted that radiation reflected from the sample and then reflected by the beamsplitter can be remodulated by the interferometer and reflected again by the sample. If this beam is measured at the detector along with the singly reflected beam, artifacts at double the wavenumber of any real spectral feature can be observed in the spectrum. This situation may be eliminated by masking off one-half of the input beam using a "half-shade."

b.   Reflection–Absorption Spectrometry

Reflection–absorption measurements of fairly thick surface films (*i.e.*, thicknesses between 0.2 and 20 μm) are easily achieved using standard accessories for specular reflectance spectrometry. Typical of the type of samples that may be measured in this way are coatings on beverage containers and epitaxial layers on silicon wafers of the kind used in the manufacture of semiconductor devices. (It might be mentioned here that a special-purpose instrument based on a Michelson interferometer has been developed for the characterization of epitaxial layers.) The measurement of the R–A spectra of thin layers, especially monomolecular layers on metallic surfaces, presents a more difficult problem, as can be seen by the following example.

Most organic samples measured in transmittance require a thickness of about 10 μm to yield a transmittance spectrum requiring little or no ordinate expansion, that is, where the absorbance of the stronger bands is between 0.1 and 1 a.u. (The thickness of the polystyrene film commonly used to calibrate infrared spectrophotometers is about 50 μm.) The thickness d of a film on a metal substrate required to yield an R–A spectrum of the same intensity as the transmittance spectrum of a 10-μm film would therefore be given by

$$2d = 10 \cos \alpha \qquad \mu m \qquad (5.1)$$

where α is the angle of incidence. For a sample measured with $\alpha = 45°$, the calculated pathlength d is equal to 3.5 μm. As α is increased by approaching closer to grazing incidence, the effective pathlength, $2d/\cos \alpha$, also increases, so that the film thickness required to produce an intense spectrum is decreased. However, even for $\alpha = 85°$, a thickness of 0.43 μm is still required to produce band absorbances equal to those produced by a 3.5-μm sample using 45° incidence. It is apparent that for films much thinner than 1 μm, R–A measurements should always be performed using high incidence angles. An accessory to achieve this goal, made in Japan by JASCO, has been described by Ishitani et al. [13] and is shown in Fig. 5.16.

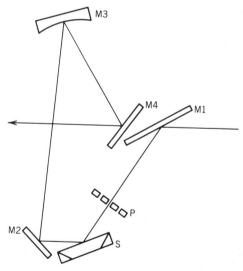

Fig. 5.16. A simple optical configuration for measuring polarized reflection–absorption spectra of species on the surface of a sample S at near grazing incidence. The polarizer P is located as close as possible to the sample to avoid the polarization changes that could occur at mirrors. This device was designed to be mounted in the sample compartment of a commercial FT–IR spectrometer. (Reproduced from [13], by permission of the American Chemical Society; © 1982.)

The thickness of a monomolecular layer can be assumed to be approximately 1 nm. Therefore, applying Eq. 5.1 and the Beer–Lambert law to the material used in the example above, we might expect band absorbances measured with 85° incidence to be between 2×10^{-3} and 2×10^{-4} a.u. Reflection–absorption spectra with peak absorbances of this order of magnitude have been measured on commercial FT–IR spectrometers (although more than 1000 scans must usually be signal averaged) [14,15]. Nevertheless, it should be recognized that the experimental realization of a beam at an incidence angle of 85° on a small sample is not a trivial matter, as can be seen from the following discussion.

Consider first the sample size required if a perfectly collimated beam of 5 cm diameter emerging from the inteferometer is to be reflected from a sample at an incidence angle of exactly 85°: the sample should be 57 × 5 cm to avoid any losses. Such large samples are rarely encountered in practice, of course, so that either a smaller sample must be used—in which case some fraction of the beam does not hit the sample—or the criterion of using a collimated beam to define the angle of incidence precisely must be relaxed. If the sample is held at the approximate focus of a beam emanating from a point source converging with a half-angle of only 6° and the incidence angle of the central ray is 85°, the incidence angle of one extreme ray would be 79° whereas that of the other would be greater than 90° (*i.e.*, it would miss the sample completely). The problem is compounded when it is realized that there is no such thing as a point source (since by definition it would emit no radiation); thus, the range of incidence angles would be further increased.

The use of a multiple-reflection cell does not in theory lead to large advantages even though such cells have been used in several earlier experiments for R–A spectrometry. For example the sample size required for 10 reflections at 45° incidence is about the same as that required for 1 reflection at a mean angle of 85°. Since, cos 45° is about 10 times larger that cos 85°, the measured absorbance (in the absence of polarization effects, to be discussed in the next paragraph) would be about the same in each case.

One additional and very important advantage may be cited for the application of large incidence angles. Greenler [16,17] has shown that if a beam polarized parallel to the surface is reflected at close to grazing incidence, interference between the incident and reflected rays can set up an intense standing field at the surface. This gives an intensity enhancement for absorbing species on the surface by a factor of up to about 25 relative to the absorption of the perpendicular polarized ray. Surface chemists have made use of this property and employed parallel polarized radiation at high incidence angles (but rarely as high as the value of 87°

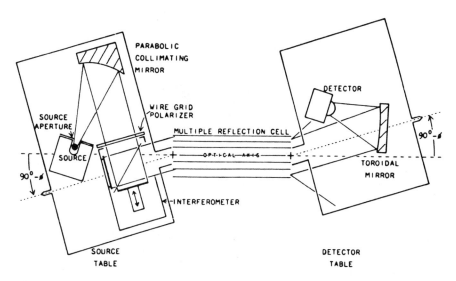

Fig. 5.17. Optical arrangement designed by Harkness [18] for IRRAS measurements of adsorbates on thin metal foils. Multiple reflections are achieved even with a high incidence angle, θ, by separating several parallel foils by a very small distance as shown schematically above. This system is suitable for studying the adsorption of gases onto metal surfaces but is not useful for studying monolayers on large substrates.

calculated by Greenler to give the largest enhancement) for R–A measurements of monomolecular layers and even partial monolayers. The difference in absorption by surface films of parallel and perpendicular polarized radiation is the basis of the important technique of polarization modulation for surface characterization, which will be described in Chapter 8.

One optical configuration combines the advantages of using a high incidence angle and multiple reflection simultaneously. Harkness developed a multiple-pass cell consisting of many parallel layers of palladium foil in a configuration such that several reflections occur between each facet, see Fig. 5.17 [18]. Surprisingly, even though this system was described over a decade ago, no similar experiments have since been performed using an analogous system.

c. Attenuated Total Reflectance Spectrometry

ATR spectrometry has been one of the more popular sampling techniques employed by FT–IR spectroscopists. Many other texts [*e.g.*, 19-21] describe ATR in some detail, and ATR accessories for grating and FT–IR

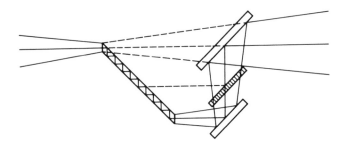

Fig. 5.18. Simple accessory for ATR spectrometry comprising two plane mirrors and the IRE. The exit plane of the IRE is located at the normal sample focus in the spectrometer. A short or long IRE can be accommodated by adjusting the position of the second mirror. (Reproduced by permission of Harrick Scientific Corporation.)

spectrometers are available from almost all manufacturers of accessories for infrared spectrometry. A schematic diagram of a particularly simple ATR accessory that can be used on several FT–IR spectrometers is shown in Fig. 5.18.

The internal reflection elements (IREs) of most ATR accessories are rectangular in cross section, matching the shape of the slits of the grating spectrometer for which they were first developed. However, special-purpose IREs with square and circular cross sections have been designed for FT–IR spectrometers and are available commercially from Spectra-Tech and Harrick Scientific Corp. ATR spectrometry may be used for

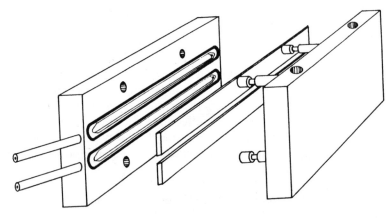

Fig. 5.19. Holder for two parallel thin rectangular IREs in contact with a flowing liquid stream. Using this arrangement, two experiments can be completed before the cell is dismantled. This unit is used for the protein deposition experiments described in Chapter 14. (Reproduced by permission of R. J. Jakobsen, Battelle Columbus Laboratories.)

sampling flexible flat or woven sheets and for fibers by wrapping them around the IRE and compressing them to ensure good surface contact. It is also possible to deposit materials onto the surface of the IRE by applying drops and then evaporating the solvent or by means of suitable plumbing to study liquids directly. For example, Gendreau *et al.* [22] have described how an ATR cell can be used for dynamic studies of protein deposition occurring during blood clotting. In this work, a germanium IRE 50 × 5 × 2 mm was mounted as shown in Fig. 5.19. The results of this study are described in some detail in Chapter 14.

Possibly the best shape for an IRE for FT–IR spectrometry is a cylinder with conical ends, in view of the excellent matching of its cross section with the round beam [23]. A particularly useful ATR accessory for FT–IR spectrometry is the Spectra-Tech Circle cell shown in Fig. 5.20. The energy transmitted through this accessory is high, greater than 50% when a ZnSe element is used. A cylindrical IRE is also very suitable for the study of flowing solutions since its smooth surface is more easily sealed with O-rings than is the surface of a rectangular IRE. These cells have already been used for many different applications including on-line and off-line process monitoring, characterization of peaks eluting from a high-performance liquid chromatograph, and the study of homogeneous catalysis.

In an internal reflectance element, a standing wave is established near the surface of the crystal. The amplitude of the electric field decreases exponentially with distance from the surface of the IRE. The distance at which the field reaches 1/*e* of its initial magnitude is defined as the pen-

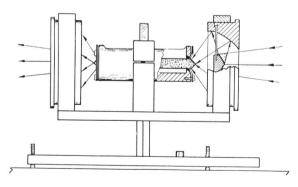

Fig. 5.20. Schematic diagram of a cylindrical internal reflection accessory. The round cross section of the IRE has the dual advantages of matching the round beam of an FT–IR spectrometer and being easier to seal than a rectangular element for measurement of liquid samples. (Reproduced by permission of Spectra-Tech Corporation.)

etration depth d_p. The magnitude of d_p depends on the wavelength of the radiation, λ, the refractive index of the prism, n_p, the refractive index of the sample, n_s, and the angle of incidence of the beam at the surface of the IRE, θ, according to the equation

$$d_p = \frac{\lambda}{2\pi n_p(\sin^2 \theta - n_{sp}^2)^{1/2}} \tag{5.2}$$

where $n_{sp} = n_s/n_p$ ($n_s < n_p$). For most materials, $d_p \sim 0.1\lambda$ and is smallest for materials of high refractive index and for high incidence angles.

Because both θ and n_p can be readily changed, it is possible to vary d_p from about 0.12λ (for a KRS-5 IRE, with $\theta = 45°$) to 0.05λ (for a germanium IRE, with $\theta = 60°$), assuming $n_s = 1.5$. Thus, several workers have claimed that under favorable circumstances it is possible to obtain a depth profile of a surface [24,25].

It is also possible to lay a very thin layer of a substrate on the surface of the IRE. If the thickness of this layer is much less than d_p, the beam can penetrate through the entire layer and the ATR spectrum of species on the outer surface of the layer can be measured. This surface layer may be a polymer, in which case its thickness may be as great as 100 nm, or a metal. Because of the optical properties of metals, however, the thickness of a metallic layer should only be about 1–3 nm.

d. Diffuse Reflectance Spectrometry

A few years ago, diffuse reflectance infrared spectra of powdered samples were thought to be extremely difficult to measure with an adequate SNR even though a variety of optical designs had been tested [26–29]. It was not until the mid-1970s, when a special-purpose diffuse reflectance Fourier transform spectrometer was introduced, that the measurement of these spectra became feasible on a routine basis [30]. This instrument was the Willey Model 318 Total Reflectance Infrared Spectrophotometer, and its optical layout is shown in Fig. 5.21. The beam from an infrared source is passed first through a slow-scanning Michelson interferometer and then

$\longrightarrow$

Fig. 5.21. Optical diagram of the Willey Model 318 Total Reflectance Infrared Spectrophotometer (*a*) showing the path of the He–Ne laser beam; (*b*) showing the path of the IR beam passing through the open quadrant of the rotating chopper mirror and hence to the sample, which is mounted at a port on the surface of the integrating sphere; and (*c*) showing the IR beam reflected from a reflective quadrant of the chopper to the diffuse reflectance reference. Specularly reflected radiation may be eliminated by removing a port on the surface of the integrating sphere. The two quadrants between the open and reflective positions are blackened to give a zero energy signal at the infrared detector.

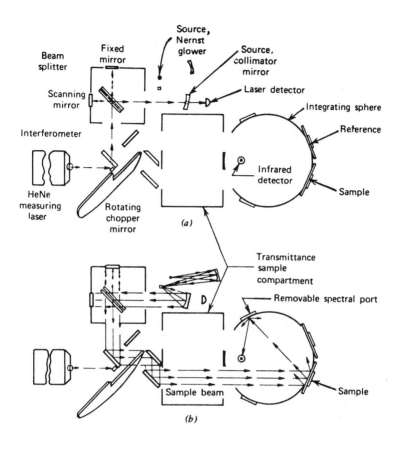

Beam splitter

Fixed mirror

Source, Nernst glower

Source, collimator mirror

Scanning mirror

Laser detector

Integrating sphere

Interferometer

Reference

Infrared detector

Sample

HeNe measuring laser

Rotating chopper mirror

(a)

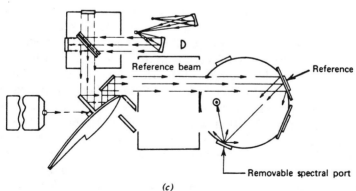

Transmittance sample compartment

Removable spectral port

D

Sample beam

Sample

(b)

D

Reference beam

Reference

Removable spectral port

(c)

195

to a rotating chopper that directs the beam to one of two paths. The chopper itself is divided into four equal sections. One quadrant is a specular mirror, and the opposite quadrant is removed so that the beam is transmitted. The other two quadrants are made of a nonreflecting material that is known to closely approximate a blackbody in the infrared. When the beam passes through the open portion of the chopper, it is then directed to the position where the sample under study is held. When the chopper rotates a full 180° from this position, the beam is reflected from the mirror quadrant of the chopper on to the sample held at the reference position. When the chopper rotates 90° from either the sample or reference position, the beam impinges on the blackbody and is totally absorbed. The signal measured when the chopper is at this position gives the "zero" energy, which is subtracted from the sample and reference signals so that real-time double-beam spectra may be obtained. The interferometer used for these measurements has to be of the slow-scanning type so that the sample, reference, and zero energies are all measured at effectively the same retardation.

Both the sample and reference are held at the surface of an integrating sphere constructed with a diffusely reflecting gold surface. Using a flux-averaging device such as the integrating sphere, the collecting optics are equally efficient for measuring the reflection spectrum of a flat specular mirror or a highly scattering diffuse reflector. However, specularly reflected radiation from either the sample or the reference can be prevented from reaching the detector by removing ports on the wall where specularly reflected radiation from a flat sample would first hit the surface of the sphere. Thus, both the total and the diffuse reflectance of any sample can be measured using this instrument. The capability of accurately measuring the total reflectance, even of rough samples, enables the optical properties of all types of materials to be readily calculated.

The integrating sphere used in this spectrometer has the advantage of collecting rays reflected at any angle, so that equal proportions of the rays reflected from the sample in all directions reach the detector. Unfortunately, the efficiency of the integrating sphere is, to a first approximation, proportional to the ratio of the area of the detector to the surface area of the sphere, so that for large integrating spheres with small detectors the device becomes rather inefficient. It was for this reason that until the development of FT-IR spectrometers all measurements taken using an integrating sphere had to be of rather low resolution to allow a sufficient signal-to-noise ratio to be attained. Although the Willey 318 could be used to measure the spectra of a variety of samples [30] and had excellent photometric accuracy [31], its sensitivity was rather low (even at 8 cm^{-1} resolution) and it did not meet with wide acceptance in the chemical

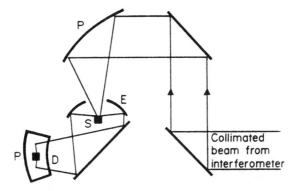

Fig. 5.22. Optics for diffuse reflectance FT–IR spectrometry originally described by Fuller and Griffiths. Collimated radiation from the interferometer is focused onto the sample (S) by a paraboloidal mirror. Diffusely reflected radiation is collected by an ellipsoidal mirror (E) and focused on the detector (D) by the paraboloid (P). Subsequently, the ellipsoid was replaced by a 12.5-mm-focal-length on-axis paraboloidal mirror. (Reproduced from [32], by permission of the American Chemical Society; © 1978.)

community. However, this instrument did give an indication of the potential of diffuse reflectance techniques for obtaining the infrared spectrum of powdered samples.

In 1978, Fuller and Griffiths [32] described an optical arrangement that could be used in conjunction with commercial rapid-scanning interferometers for the measurement of diffuse reflectance infrared spectra at high SNR. Their design, which is shown in Fig. 5.22, led to the rapid acceptance of diffuse reflectance as an important sampling technique by FT–IR spectroscopists. The optical configuration is rather simple. The sample is held at the common focus of a fast off-axis paraboloid (the *focusing* mirror, used to focus the collimated beam from the interferometer onto the sample) and a short focal length aspherical mirror used on-axis (the *collecting* mirror, which collects radiation diffusely reflected from the sample). A hole drilled through the axis of the collecting mirror allows the input beam from the focusing mirror to pass to the sample; it also prevents specularly reflected radiation from reaching the detector. Initially, an ellipsoidal mirror with a large ellipticity was used as the collecting mirror. This mirror was replaced subsequently by a 12-mm focal length paraboloid to produce an approximately collimated beam of 5 cm diameter [33]. This beam is then focused onto an MCT detector by an off-axis paraboloid identical to the one used to focus the beam onto the sample. These optics permitted about 15% of the radiation that was passed through the interferometer to be measured by the detector provided a nonabsorbing powder (such as KCl) is held in the sample cup. Although

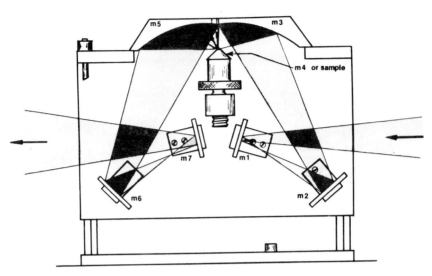

Fig. 5.23. The Collector for DR spectrometry designed by Spectra-Tech. Unlike the configuration shown in Fig. 5.22, this accessory fits in the sample compartment of commercial FT–IR spectrometers. Mirrors m1, m2, m6, and m7 are plane, and the focusing mirror, m3, and the collecting mirror, m5, are ellipsoids. The sample and reference are installed by pulling m3 and m5 apart. (Reproduced by permission of Spectra-Tech Corporation.)

many diffuse reflectance measurements are performed using an MCT detector, the high collection efficiency of this optical configuration, and that of others to be described in the rest of this section, allows good spectra to be measured even when a DTGS detector is employed.

After some of the early applications using the Fuller–Griffiths design had been described, several companies (including Spectra-Tech, Harrick Scientific, and JASCO) introduced accessories for diffuse reflectance spectrometry that could fit into the sample compartment of standard FT–IR spectrometers. The optics of the Spectra-Tech Collector and the Harrick Praying Mantis accessories are shown in Figs. 5.23 and 5.24, respectively. Analect Instruments have introduced a reflectance accessory for their fX-6200 spectrometer that requires replacing the paraboloidal mirrors used to focus and collect the beam at the sample compartment by plane mirrors, see Fig. 5.25. In their simplest configuration, each of these accessories measures the "total reflectance" of the sample (actually all of the specularly reflected component and part of the diffusely reflected component). In the Harrick and Analect designs, the specular component can be eliminated by rotating either the plane of the sample cup (in the Harrick accessory) or the plane of the mirrors (in the Analect design). In the Spectra-Tech design, a post can be introduced that prevents the specularly reflected component from reaching the detector.

One of the more important potential applications of diffuse reflectance spectrometry is the study of molecules adsorbed on powdered adsorbents. For these measurements, it is necessary not only to have efficient optics (for maximum SNR) but also to be able to enclose the sample in a chamber so that it may be heated, evacuated, or subjected to a controlled atmosphere.

The first commercially available system for studying gas–solid reactions was designed by Harrick Scientific for installation in the Praying Mantis reflectance accessory shown in Fig. 5.24. This unit consisted of a small cell, with a single port for gas inlet or evacuation; the sample may be heated from below by a small cartridge. The input and exit beams are not in the same plane, so that the specularly reflected component is dis-

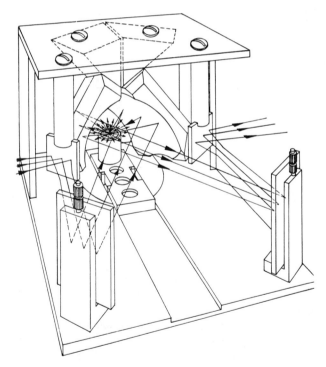

Fig. 5.24. Praying Mantis accessory for diffuse reflectance spectrometry. Like the accessory shown in Fig. 5.23, this design incorporates ellipsoidal focusing and collection mirrors and fits in the sample compartment of commercial FT–IR spectrometers. (Reproduced by permission of Harrick Scientific Corporation.)

(A)

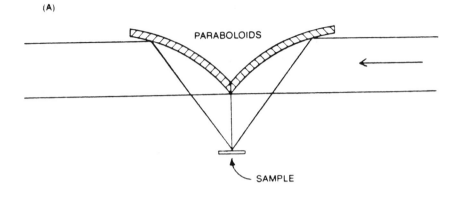

(B)

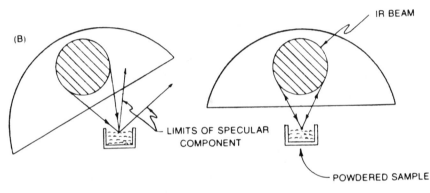

Fig. 5.25. (*a*) Simple paraboloidal mirror reflectometer designed to focus the collimated beam from an interferometer onto the sample and to collect and recollimate both diffusely and specularly reflected radiation. (*b*) The specular component can be eliminated by tilting the two paraboloids. (Reproduced by permission of Analect Instruments Division of Laser Precision Corporation.)

criminated against. A later design had an inlet and an outlet port and allowed gas to be drawn through the sample to avoid condensation of the products on the windows of the cell and to permit rapid equilibration, see Fig. 5.26. Subsequently, Spectra-Tech introduced a similar cell.

Recently, a special-purpose system for measuring the diffuse reflectance spectra of powders in a controlled atmosphere was described by Hamadeh et al. [34]. The optics and cell are illustrated in Fig. 5.27. Although this device is much larger than the Harrick and Spectra-Tech design, it has the advantage of remarkably high sensitivity, especially for the study of species adsorbed on the surface of adsorbents with a high

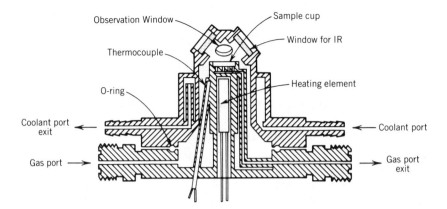

Fig. 5.26. Heatable evacuable cell designed to fit at the sample position of a Praying Mantis DR accessory, allowing the study of gas–solid reactions at elevated temperature. (Reproduced by permission of Harrick Scientific Corporation.)

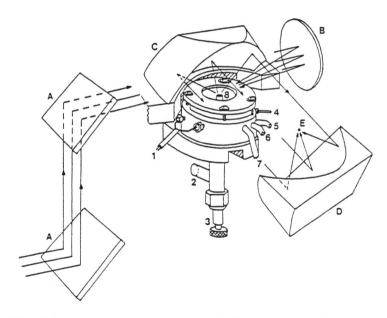

Fig. 5.27. Cell and optical configuration to permit DR measurements of adsorbed species at very high SNR . Optics: A, beam-steering plane mirrors; B, paraboloidal mirror to focus the collimated beam onto the sample; C, paraboloidal collection mirror; D, paraboloidal mirror to focus the collimated beam from C onto the downward-looking MCT detector, E. Cell: 1, heater leads; 2, vacuum line; 3, knurled screw to permit rapid sample removal and installation, 4, thermocouple leads; 5,6, cooling water lines; 7, line to manometer; 8, sample. (Reproduced from [34], by permission of Academic Press; © 1984.)

201

surface area. It has been claimed that a surface coverage of as little of 10^{-4} monolayer of CO adsorbed on certain supported metal catalysts can be observed using this device in a measurement time of less than 1 min.

IV.   INFRARED EMISSION SPECTROMETRY

a.   Laboratory Measurements

For practical purposes, infrared emission spectrometry can be divided into two categories: measurements where the source is located in the laboratory, or at least close to the spectrometer, and those where the source is remote. For laboratory studies, most workers need to measure the emittance (or emissivity) spectrum $\epsilon(\bar{\nu})$, which is the ratio of the energy emitted by the sample at a given temperature to the energy emitted by a blackbody of the same geometry at the same temperature. For gaseous samples or condensed phase samples on a metallic support or in a non-absorbing matrix [where $\epsilon(\bar{\nu}) \simeq 0$ for both cases], the emittance spectrum is given to a good approximation by the relation

$$\epsilon(\bar{\nu}) = 1 - \tau(\bar{\nu}) \tag{5.3}$$

where $\tau(\bar{\nu})$ is the transmittance spectrum of the sample.

For condensed phase samples, the direct measurement of $\epsilon(\bar{\nu})$ is difficult since the single-beam spectrum $S(\bar{\nu}, T)$ of a sample at a temperature T has contributions from several sources and is given by [35]

$$S(\bar{\nu}, T) = R(\bar{\nu})[\epsilon(\bar{\nu})H(\bar{\nu}, T) + B(\bar{\nu}) + I(\bar{\nu})\rho(\bar{\nu})] \tag{5.4}$$

where
$R(\bar{\nu})$ = instrument response function
$\epsilon(\bar{\nu})$ = emittance of sample
$H(\bar{\nu}, T)$ = Planck function
$B(\bar{\nu})$ = background radiation reaching the detector directly
$I(\bar{\nu})\rho(\bar{\nu})$ = radiation reflected from sample
$\rho(\bar{\nu})$ = reflectance of sample

To eliminate the effects of $R(\bar{\nu})$, $H(\bar{\nu}, T)$, $B(\bar{\nu})$, and $I(\bar{\nu})\rho(\bar{\nu})$, Kember et al. [36] and Chase [35] have suggested that measurements of the sample and a blackbody should each be made at two temperatures T_1 and T_2. For the blackbody reference, $\epsilon(\bar{\nu}) = 1$ and $\rho(\bar{\nu}) = 0$, so that

$$S_1(\bar{\nu}, T_1) = R(\bar{\nu})[H(\bar{\nu}, T_1) + B(\bar{\nu})] \tag{5.5}$$

$$S_3(\bar{\nu}, T_2) = R(\bar{\nu})[H(\bar{\nu}, T_2) + B(\bar{\nu})] \tag{5.6}$$

For the spectra of the sample measured at T_1 and T_2,

$$S_2(\bar{\nu}, T_1) = R(\bar{\nu})[\epsilon(\bar{\nu})H(\bar{\nu}, T_1) + B(\bar{\nu}) + I(\bar{\nu})\rho(\bar{\nu})] \qquad (5.7)$$

$$S_4(\bar{\nu}, T_2) = R(\bar{\nu})[\epsilon(\bar{\nu})H(\bar{\nu}, T_2) + B(\bar{\nu}) + I(\bar{\nu})\rho(\bar{\nu})] \qquad (5.8)$$

Assuming $\epsilon(\bar{\nu})$ and $\rho(\bar{\nu})$ are independent of temperature, we have

$$\epsilon(\bar{\nu}) = \frac{S_4 - S_2}{S_3 - S_1} \qquad (5.9)$$

In practice, when cooled detectors and/or interferometers thermostated above ambient temperature are used for these measurements, the interferograms contain a contribution due to emission from the interferometer (primarily from the beamsplitter), which is out of phase with the emission from the sample or blackbody reference. The standard Mertz-type phase correction routines do not treat these contributions independently so that spectra are improperly phase corrected. However, since the instrumental emission is identical for all four measurements, the problem is eliminated by subtracting interferogram $I_2(\delta, T_1)$ from $I_4(\delta, T_2)$ and $I_1(\delta, T_1)$ from $I_3(\delta, T_3)$ before computing the spectra $[S_4 - S_2]$ and $[S_3 - S_1]$ from the difference interferogram using standard software. At least one manufacturer (Nicolet Instrument Corp.) sells an accessory for measuring emittance spectra in this fashion. Chase [35] has noted that if both the interferometer and detector are at ambient temperature, the measurements at T_2 are unnecessary, and $\epsilon(\bar{\nu})$ is given simply by the ratio of S_2 and S_1.

Discrete contributions to $S(\bar{\nu}, T)$ due to emission from the beamsplitter thermostated at 38°C have been observed directly by Kember et al. [36], see Fig. 5.28. The main feature is a broad band at 850 cm^{-1} observed previously in dual-beam FT–IR measurements [37] (Chapter 8) that is due to GeO on the outer surface of the germanium film forming the beamsplitter.

For the actual measurements, radiation emitted by the sample may be collimated using specially designed optics, passed through the interferometer, and focused directly onto the detector. Alternatively, the emitted radiation may be passed into the source optics of a standard FT–IR spectrometer directly (i.e., without precollimation), either letting the detector optics limit the optical throughput (for low resolution measurements) or by restricting the optical throughput needed to achieve a given resolution by placing a suitable aperture at the focus in the sample compartment.

An alternative method of measuring the emittance spectrum of a sample held at a single temperature T_1 is to measure interferograms from the sample, $I_1(\delta, T_1)$ and blackbody reference, $I_2(\delta, T_1)$, which are each held

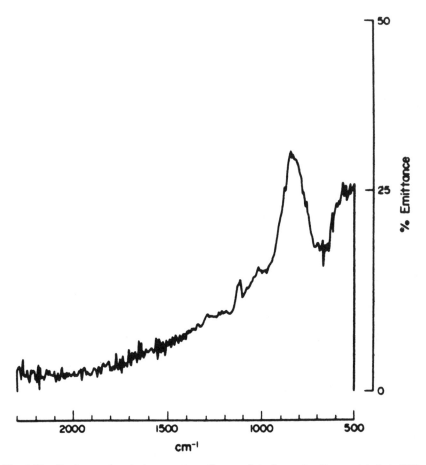

Fig. 5.28. Background emission spectrum from an interferometer thermostated at 40°C. The band centered near 850 cm^{-1} is probably caused by germanium oxides on the surface of the beamsplitter. (Reproduced from [36], by permission of Pergamon Press; © 1979.)

in a cooled box with blackened walls. The interferogram from the walls, $I_3(\delta)$, is then measured and subtracted from both $I_1(\delta, T_1)$ and $I_2(\delta, T_1)$. The ratio of the spectra computed from $[I_1(\delta, T_1) - I_3(\delta)]$ and $[I_2(\delta, T_1) - I_3(\delta)]$ is $\epsilon(\bar{\nu})$. The fact that samples are held in a box with cooled walls may give rise to temperature gradients in the sample and reference, so that this method is less preferable than the method involving measurements at two temperatures.

One problem with the measurement of emission spectra of samples with discrete bands (especially gaseous samples) is that the Mertz method of phase correction requires that the spectrum has energy at all wavelengths in the spectrum of interest. If this is not the case, the phase spectrum computed from the interferogram of a blackbody source (after subtraction of background contributions) must be stored and used in subsequent contributions.

One type of sample for which laboratory measurement of the emission spectrum may become quite important is a metal with a thin film on its surface. It is interesting to note that Greenler [38] has shown theoretically that the intensity of emission bands from surface species will increase as the angle of the normal to the sample is increased with respect to the optical axis of the spectrometer, to a limit of about 80°. This has been verified experimentally [36], but no accessory for this purpose is available commercially.

For the measurement of extremely weak sources, it may become necessary to use a liquid-helium-cooled germanium bolometer (Section VI.c of this chapter) to achieve an adequate SNR. In this case, the photon flux on the detector from an interferometer and the surroundings at ambient temperature will usually be much greater than the photon flux from the sample. It then becomes necessary to cool the interferometer and surroundings to liquid nitrogen temperature. Although this is by no means easily achieved, several workers have managed to modify commercially available interferometers to liquid nitrogen temperature [39–41]. Although it is unlikely ever to be necessary for chemical FT–IR spectrometry, it is still interesting to note that for astronomical measurements, liquid-helium-cooled interferometers have also been designed and built [42].

b. Measurements of Remote Sources

The size of remote sources can vary substantially. The image can be extensive (as in the case of airglow studies) or very small (e.g., when the spectrum of a remote star is measured). An intermediate case is found when a terrestrial source of 1 or 2 m diameter is to be monitored from a distance of about 1 km; this would be the case if the emission spectrum of the hot gaseous effluent from a smokestack is to be measured. Each type of measurement will usually require a telescope to collect the radiation and optics, which may or may not be an integral part of the telescope, to collimate the radiation before it is passed into the interferometer. In the case of extended sources, the solid angle of the beam passing through the interferometer may be greater than the maximum allowed angle for a given resolution on a standard Michelson interferometer (see Eq. 1.41). In this case, an aperture stop must be placed at a focus somewhere between the telescope and the detector. One simple design for this purpose based on a modified Cassegrain telescope is shown in Fig. 5.29.

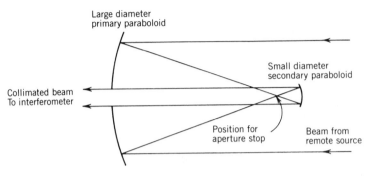

Fig. 5.29. Optics of a simple telescope suitable for remote emission measurements by FT–IR spectrometry. The field of view and the spectral resolution can both be controlled by the size of the aperture place between the primary and secondary mirrors.

A superior design based on a Dall-Kirkham telescope forms the basis of the U.S. Environmental Protection Agency's ROSE (Remote Optical Sensing of Emissions) system [43] depicted in Fig. 5.30. The ROSE system is actually separated into two physically distinct parts: (a) the remote light source and source telescope and (b) the receiver telescope, interferometer, detector, and calibration unit, which are mounted in a large van. Radiation from the emission source (the plume in Fig. 5.30) is reflected into the receiver telescope by an elliptical tracking mirror, mounted outside the van, and then focused at an aperture whose diameter can be varied to allow the field of view to be controlled. From this aperture, the radiation is collimated and passed through the interferometer to the detector.

To calibrate the intensity of measured emission spectra from remote sources, radiation from a blackbody source can be passed into the interferometer (through the use of a two-position flip mirror) so that it fills the field of view of the detector to the same extent as the remote source. For this measurement, the calibration cell shown in Fig. 5.30 is removed.

The same system may also be used for absorption measurements over long paths by mounting the remote light source (a 1000-W quartz-iodine

————————————————————————————→

Fig. 5.30. Schematic representation of the ROSE system for remote absorption and emission FT–IR spectrometry developed by the U.S. Environmental Protection Agency. These optics can be installed in a van for mobility. Absorption spectra are measured by locating a light source and collimating telescope on the side of the area being monitored, which is remote from the van, and collecting this radiation by the receiver telescope. The receiver telescope can also collect radiation emitted from a hot plume and pass it to the interferometer. A flip mirror can be used to direct the radiation from a light source and cell mounted next to the interferometer in the van to permit rapid calibration measurements. (Reproduced from [43], by permission of Academic Press; © 1979.)

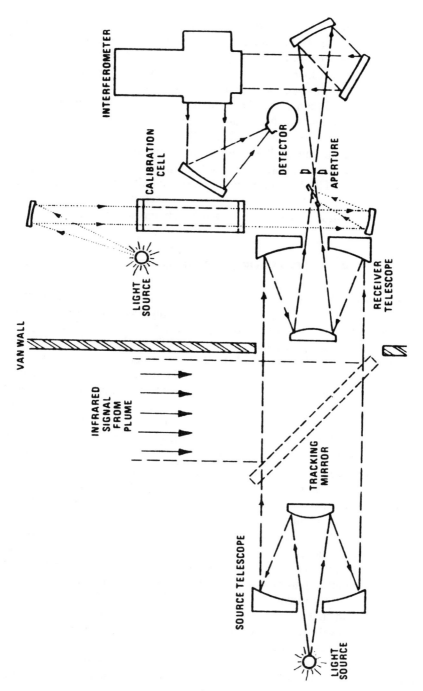

lamp) and source telescope on a pick-up truck at a location situated at one side of the area to be sampled. The van with the receiver optics and FT–IR spectrometer is located on the other side of this area. The calibration cell shown in Fig. 5.30 may be used to obtain quantitative data from the absorption spectra measured in this way.

At this point, it can certainly be stated that emission spectrometry is the least popular of the three major branches of infrared spectrometry being studied today by chemists. Nevertheless, we should recognize that it was the measurement of the weak emission spectra of very remote objects by astronomers that provided the impetus for many of the original developments in FT–IR spectrometry. Undoubtedly, more measurements in such fields as surface analysis, infrared chemiluminescence, and atmospheric monitoring will be reported by chemists in the next decade.

V. SOURCES FOR FOURIER SPECTROMETRY

There is virtually no difference between the types of sources that are used for infrared spectrometry performed using a monochromator and a Fourier spectrometer. For mid-infrared spectrometry, Globars, Nernst glowers, and nichrome coils have all been used. The Nernst glower has the highest operating temperature of these three detectors, but its poor emissivity at high wavenumber offsets many of the advantages that might be thought to accrue because of the high blackbody emission of a source operating at about 1500°C. Many of the less expensive instruments incorporate nichrome coils operating at between 1000 and 1100°C. These sources have fairly good emissivity across the mid-infrared spectrum and, like the Nernst glower, the advantage of being air cooled. However, their low operating temperature and poor thermal stability have precluded their use in most top-of-the-line instruments. Perkin-Elmer claim to have improved the stability of the wire source and incorporate it on their Model 1800 FT–IR spectrometer in preference to a Globar.

Most research-grade instruments incorporate the Globar (silicon carbide) source, which is usually, but not always, water cooled. The operating temperature and sensitivity of the Globar are slightly higher than the corresponding parameters for a nichrome coil. The Globar also has one other very significant advantage over the Nernst glower and the nichrome coil in that it has a fairly high emissivity down to 80 cm^{-1}. It is therefore possible to use the same source for many far-infrared measurements of organic and inorganic compounds.

For measurements below 100 cm^{-1}, the use of a high-pressure mercury lamp is to be preferred over a Globar, and below about 75 cm^{-1}, the mercury lamp is mandatory. Emission of continuous radiation from the plasma exceeds the spectral energy density calculated from the Planck equation for a source operating at the temperature of this source. Mercury lamps have been used for far-infrared spectrometry with slow-, stepped- and rapid-scanning interferometers. Although an ac power supply is generally used for the slow- and stepped-scan instruments, a dc power supply must be incorporated for rapid-scanning interferometers, or else 60- and 120-Hz glitches (see Chapter 7, Section III.c) may fall right in the middle of the spectral range.

For near-infrared and visible Fourier transform spectrometry, the same sources used in grating spectrometers are again used. For the wavelength region between 320 and 2500 nm, the source is almost invariably a tungsten filament lamp. Little absorption spectrometry in the ultraviolet region has been reported in which a Fourier spectrometer has been used in conjunction with a source of continuous radiation, such as hydrogen or deuterium lamps, largely because of the effects of shot noise (see Chapter 7, Section III.f). Atomic emission measurements using flame and inductively coupled plasma sources have been reported in the near-infrared, visible, and near-ultraviolet regions. This work is discussed in Chapter 16.

VI. INFRARED DETECTORS

a. Thermal Detectors

A detailed discussion of infrared detectors is beyond the scope of this monograph. Nevertheless, spectroscopists interested in getting the most out of their instruments should understand the principles behind the more commonly encountered detectors used in FT–IR spectrometers, and a brief summary encompassing many of the more important types of detectors used in Fourier spectrometry will be given in this section.

Infrared detectors can be divided into two types: thermal detectors and quantum detectors. Thermal detectors operate by sensing the change of temperature of an absorbing material. The output obtained from them may be in the form of a thermal electromotive force (e.g., thermocouples), a change in resistance of a conductor (bolometer) or semiconductor (thermistor bolometer), or the movement of a diaphragm caused by the expansion of a gas (pneumatic detector), which may lead to the change in illumination of a subsidiary photocell (Golay detector). Thermal detectors often respond to radiation over a wide range of wavelengths. However, because the temperature of the element must change, they are usually slow, typically taking from 0.01 to 0.1 sec to respond.

Because of their slow response, thermocouples and bolometers have rarely been used for mid-infrared Fourier spectrometry, although several of the early low-resolution interferometer spectrometers made by Block Engineering incorporated a thermistor bolometer detector. Two types of thermal detectors have been used extensively, however: the Golay detector and the pyroelectric bolometer.

Golay detectors have been used for far-infrared Fourier spectrometry with slow-scanning and stepped-scan interferometers. A schematic of a Golay cell is shown in Figs. 5.31 and 5.32. A chamber containing a gas of low thermal conductivity is sealed at one end with an infrared transparent window (A) through which radiation reaches a thin absorbing film (B). This film has a low thermal capacity, so that its temperature rises

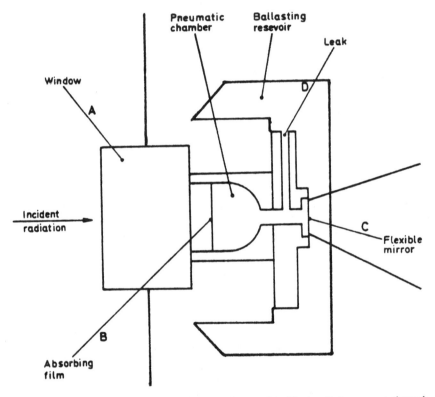

Fig. 5.31.   Front section of a Golay detector. Chopped incident radiation passes through the window onto a blackened film, causing the pressure of the gas in the pneumatic chamber to fluctuate and a flexible mirror to deform with the same modulation frequency as the chopper. A pinhole leak minimizes the effect of long-term temperature variations. (Reproduced by permission of Cathodeon, Ltd.)

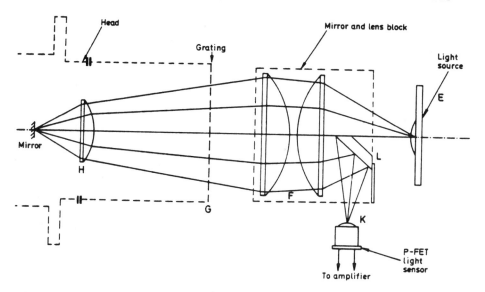

Fig. 5.32. Rear section of a Golay detector. Visible radiation is passed through a grating and focused onto the flexible mirror shown in Fig. 5.31. The reflected radiation passes back through the grating and is measured by a P-FET light sensor. When the flexible mirror is undistorted, the grating prevents any light from reaching the P-FET detector. The greater the distortion of the flexible mirror, the greater the measured signal. (Reproduced by permission of Cathodeon, Ltd.)

rapidly, in turn warming the gas with which it is in contact. A rise in temperature of the gas in the chamber produces a corresponding rise in pressure and therefore a distortion of the mirror membrane (C) with which the other end of the gas chamber is sealed. A fine leak (D) connects the detection chamber with a ballasting reservoir of gas on the other side of the mirror membrane to prevent slow changes in ambient temperature from affecting the detector.

The temperature of the absorbing film follows the modulation frequency. Light from a source (E) passes through a condensing lens (F) to a line grid (G) and is focused on the flexible mirror. A meniscus lens (H) between the line grid and mirror focuses the beam so that in the absence of any deformation an image of one part of the line grid is superimposed on another part of the same grid. If the image of a gap between lines coincides with a gap in the grid, light will be transmitted and hence detected by a sensor (K). Deformation of the mirror membrane produces a corresponding change in the relative positions of the image and grid and hence a change in the intensity of light reaching the sensor.

Golay detectors are considerably more sensitive than thermocouples and thermistor bolometers. At one stage, they were extremely fragile, but the application of solid-state technology has greatly improved their reliability. These detectors are slow, however, and may only be used with slow-scanning or stepped-scan interferometers, where the infrared radiation is modulated by an external chopper. Radiation modulated in the audio-frequency regime must be detected by pyroelectric or quantum detectors.

Pyroelectric bolometers incorporate as their heat-sensing element ferroelectric materials, which exhibit a large spontaneous electrical polarization at temperatures below their Curie point. If the temperature of these materials is changed, the degree of polarization is changed. The change in polarization may be observed as an electrical signal if electrodes are placed on opposite faces of a thin slice of material to form a capacitor. When the polarization changes, the charges induced on the electrodes can either flow as a current through a relatively low external impedance or produce a voltage across the slice if the external impedance is comparatively high. The detector will only give a signal when the temperature of the element changes.

The most commonly used material for pyroelectric detectors is deuterated triglycine sulfate. The DTGS element is usually mounted so that the thermal resistance between the element and its surroundings is large and the thermal time constant is consequently long. The thermal circuit can be represented by an electrical analog comprising a capacitance shunted by a large resistance. The voltage responsivity of a pyroelectric detector at frequency f is given by

$$R_v = \frac{p(T)}{\rho C_p d} \frac{R_E}{[1 + (2\pi f R_E C_E)^2]^{1/2}} \tag{5.10}$$

where ρ is the density of the crystal, C_p is the specific heat, d is the spacing between the electrode surfaces, $p(T)$ is the pyroelectric coefficient at temperature T, R_E is the feedback or load resistance, and C_E is the effective capacitance.

To a good approximation, R_v is proportional to $1/f$. For a rapid-scanning interferometer, therefore, the scan speed v should be as low as possible. The capacitive element is essentially noiseless, and the long thermal time constant leads to low thermal noise. In fact, the noise performance attainable with DTGS pyroelectric bolometers is dominated by the noise performance of its associated amplifiers. For wide-frequency bandwidths such as those encountered in rapid-scanning FT–IR spectrometers, Johnson noise in the load resistor dominates.

b. Quantum Detectors

The alternative method of detecting infrared radiation depends on the interaction of radiation with the electrons in a solid, causing the electrons to be excited to a higher energy state. These effects depend on the quantum nature of radiation, and detectors using them are called quantum detectors. The energy of each photon is directly proportional to its wavenumber. A transition of electrons from one state to another will only occur if the wavelength is less than some critical value.

The idealized response of a quantum detector is shown in Fig. 5.33. The rising form of the curve is due to the fact that as the wavelength increases, the number of quanta for a given amount of energy increases. Since the output is controlled by the number of electrons excited, it also rises. For thermal detectors, on the other hand, the response per watt is constant.

To use this quantum effect, it is necessary to be able to excite electrons from one state to another in which their electrical properties are different. One way is by photoemission, where electrons are given enough energy to escape from the surface and flow through a vacuum to give a current. Phototubes and photomultipliers fall into this category. Since surfaces require high energies for the release of electrons, this effect can only be used in the ultraviolet, visible, and near-infrared (to about 1 μm) regions.

Inside a semiconductor, however, the properties of an electron when it is in the valence band are very different from when it is in the conduction band. In an n-type semiconductor, the electrons in the valence band are unable to contribute to the conduction. If excited to the conduction band, they can act as current carriers, as can the hole left behind in the valence band. To detect mid-infrared wavelengths, a very small amount of excitation energy is needed. Some detectors make use of the release of an electron from an added impurity atom, which needs less energy than for an excitation across a band gap; PbS and PbSe fall into this category. Others, such as mercury cadmium telluride (MCT), make use of the properties of a mixture of two semiconductors.

The very low energy that must be sensed by mid-infrared photodetectors has the important consequence that electrons can be excited by thermal agitation of the solid. This is a random effect giving rise to electrical noise in the output. For good sensitivity, therefore, it is necessary to cool the element. For certain detectors, such as PbS and PbSe, it may be sufficient to use a thermoelectric cooler to maintain their temperature just below ambient. Mercury cadmium telluride detectors must be maintained at liquid nitrogen temperatures (77 K), whereas others may require cooling to the temperature of liquid helium (4.2 K).

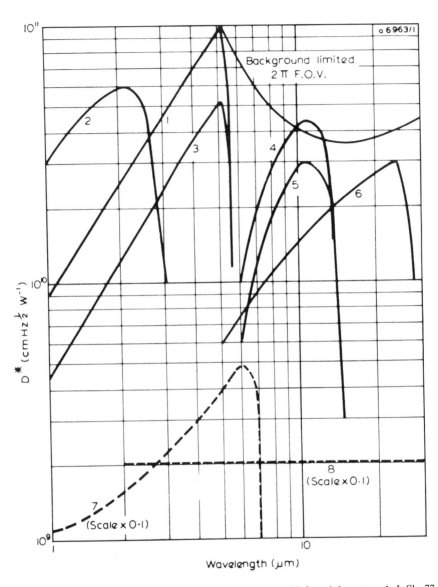

Fig. 5.33. The variation of D^* with wavelength for several infrared detectors. 1, InSb, 77 K, 60° field of view (FOV); 2, PbS, 300 K, 2π FOV; 3, InSb, 77 K, 2π FOV; 4, CdHgTe, 77 K, 2π FOV; 5, Ge–Hg, 35 K, 60° FOV; 6, Ge–Cu, 4.2 K, 60° FOV; 7, (scale $\times 0.1$) InSb, 300 K, 2π FOV; 8, (scale $\times 0.1$) TGS pyroelectric bolometer. Note that D^* increases with the FOV. For any wavelength and FOV, there is a theoretical limit to the D^* of any detector, known as the background limit for infrared photons (BLIP). The BLIP (idealized response) for a 2π FOV is indicated on this curve. (Reproduced by permission of Mullard, Ltd.)

By far the most commonly used photodetector for mid-infrared Fourier spectrometry is the MCT detector. The specific detectivity D^* of these detectors (see Chapter 7, Section I) is strongly dependent on their composition. MCT detectors with a certain composition have a very high D^*, and their peak response is at about 800 cm^{-1}. By 750 cm^{-1}, however, the response of these detectors is approximately zero. If response is required at longer wavelengths, some sacrifice must be made in the maximum value of D^*. Mercury cadmium telluride detectors are often classified as high-sensitivity (or narrow-band) or low-sensitivity (broadband, or wide-range) detectors. However, it should be realized that the composition of the HgTe–CdTe mixture can be varied continuously to produce detectors with the desired response characteristics, see Fig. 5.34.

Unlike the case for pyroelectric bolometers, the response of most quantum detectors *increases* with modulation frequency. For MCT detectors, D^* usually increases with f, to a limit of about 1 kHz. Then D^* remains approximately invariant with modulation frequency to about 1 MHz, see Fig. 5.35. At a still higher frequency, D^* starts to decrease again. When an MCT detector is used in conjunction with a rapid-scanning interferometer, the mirror velocity v should therefore be set as high as possible.

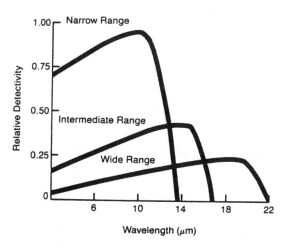

Fig. 5.34. The relative D^* values of a narrow-range, intermediate-range, and wide-range MCT detector. The narrow-range detector is the most sensitive, but only has good response to about 750 cm^{-1} and generally has poorer linearity at high signal levels than the wider-range detectors. (Reproduced from [44], by permission of the American Chemical Society; © 1983.)

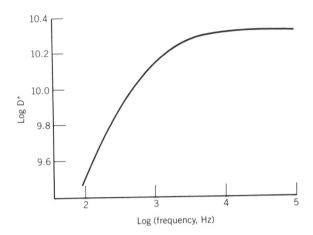

Fig. 5.35. Typical variation of the D^* of MCT detectors with modulation frequency. The reduction in D^* at modulation frequencies below about 1 kHz indicates that for most FT–IR measurements, the moving mirror of the interferometer should be translated at a fairly high speed so that radiation above 800 cm^{-1} is modulated at a frequency greater than 1 kHz. (Reproduced from [44], by permission of the American Chemical Society; © 1983.)

c. Liquid-Helium-Cooled Detectors

For operation in the far infrared, where photon energies are very low, quantum detectors cannot be used at all. The sensitivity of thermal detectors is often not sufficient to permit spectra to be measured at sufficiently high signal-to-noise ratios in view of the very low spectral energy densities of sources in this spectral region. For improved sensitivity, two types of detectors, each operating at liquid helium temperatures, have been used.

The first is the germanium bolometer. These detectors are often doped with a low level of copper, gallium, or antimony. They operate close to the Johnson noise limit with fast (0.8-msec) thermal time constants. For increased responsivity, these detectors may be operated at 1.5 K by pumping the detector cryostat to 3.6 torr, that is, the detector is operated below the lambda point of liquid helium. The typical wavenumber range of these detectors is usually quoted as 5–400 cm^{-1}, but many may be operated through the entire mid infrared. In this case, however, the photon flux from surroundings at ambient temperature is so great that is is necessary to cool all surfaces from which thermal emission can take place. Filters cooled by the liquid helium cryogen are placed directly in front of the detector. For optimal operation in the mid infrared, these detectors are best used when the interferometer and all its surroundings are cooled to 77 K.

The second type of far-infrared detector is the InSb hot-electron detector. Absorption of radiation by free carrier electrons causes their mean temperature T_e to rise above that of the host lattice. Electron mobility is proportional to $T_e^{3/2}$. This temperature rise can therefore be sensed as a change in conductivity. The very low thermal mass of the free electrons, with short energy relaxation times of about 10^{-7} sec, leads to very fast, sensitive submillimeter and millimeter wave detectors. Detectors of this type have been satisfactorily matched to the preamplifier either by using a transformer (Rollin type), which restricts the bandwidth and introduces noise, or by using a magnetic field to increase impedance (Putley type), which restricts the bandwidth. A new InSb detector has been introduced by QMC Instruments in England, which uses a specially shaped high-purity n-type InSb crystal. This detector can be coupled directly to a low-noise preamplifier without sacrificing bandwidth or introducing noise. The wavenumber response of these detectors is from 2 to 50 cm^{-1}.

Many other infrared detectors have been described. It is often possible to find a detector with the desired characteristics for a particular measurement. A summary of the D^* values for several common (and some less frequently encountered) detectors for Fourier transform spectrometry is shown in Fig. 5.33.

REFERENCES

1. R. L. White, P. J. Coffey, J. P. Covey, and D. R. Mattson, *Amer. Lab.*, **15**(11), 90 (1983).

2. J. U. White, *J. Opt. Soc. Am.*, **32**, 285 (1942).

3. P. L. Hanst, *Adv. Environ. Sci. Technol.*, **2**, 91 (1971).

4. P. L. Hanst, A. S. Lefohn, and B. W. Gay, *Appl. Spectrosc.*, **27**, 188 (1973).

5. P. L. Hanst, W. E. Wilson, R. K. Patterson, B. W. Gay, L. W. Chaney, and C. S. Burton, Report No. EPA-650/4-75-006, U.S. Environ. Prot. Agency, Research Triangle Park, NC (1975).

6. J. N. Pitts, J. M. McAfee, W. D. Long, and A. M. Winer, *Environ. Sci. Technol.*, **10**, 787 (1976).

7. E. C. Tuazon, R. A. Graham, A. M. Winer, R. R. Easton, J. N. Pitts, and P. L. Hanst, *Atmos. Environ.*, **12**, 865 (1978).

8. P. L. Hanst, *Appl. Opt.*, **17**, 1360 (1978).

9. P. L. Hanst, in *Fourier Transform Infrared Spectroscopy: Applications to Chemical Systems*, Vol 2 (J. R. Ferraro and L. J. Basile, eds), Academic Press, New York, pp. 79–110 (1979).

10. D. Horn and G. C. Pimentel, *Appl. Opt.*, **10**, 1892 (1971).

11. E. C. Tuazon, W. P. L. Carter, A. N. Winer, and J. N. Pitts, *Env. Sci. Technol.*, **15**, 823 (1981).

12. K. C. Herr, L. J. Ortega, R. G. Robins, S. S. Durso, R. M. Young, D. S. Urevig, and C. J. Rice, Report DS-TR-82-60, Aerospace Corp. El Segundo, CA (1982).

13. A. Ishitani, H. Ishida, F. Soeda, and Y. Nagasawa, *Anal. Chem.*, **54**, 682 (1982).

14. D. L. Allara, A. Baca, and C. A. Pryde, *Macromolecules*, **11**, 1215 (1978).

15. J. F. Rabolt, F. C. Burns, N. E. Schlotter, and J. D. Swalen, *J. Chem. Phys.*, **78**, 946 (1983).

16. R. G. Greenler, *J. Chem. Phys.*, **44**, 10 (1966).

17. R. G. Greenler, *J. Chem. Phys.*, **50**, 1963 (1969).

18. J. B. L. Harkness, Ph.D. dissertation, Massachusetts Institute of Technology, Cambridge, MA (1970).

19. N. J. Harrick, *Internal Reflection Spectroscopy*, Wiley-Interscience, New York (1967), now available from Harrick Scientific, Ossining, NY.

20. American Society for Testing and Materials, 1977 Annual Book of ASTM Standards, Part 42, p. 435 *ff*, ASTM, Philadelphia (1977); an updated version of this document was being prepared in 1985.

21. A. L. Smith, *Applied Infrared Spectroscopy*, Wiley-Interscience, New York, pp. 84–95 (1979).

22. R. M. Gendreau, S. Winters, R. I. Leininger, D. Fink, C. R. Hassler, and R. J. Jakobsen, *Appl. Spectrosc.*, **35**, 353 (1981).

23. A. Rein and P. Wilks, *Amer. Lab.*, **14**(10), 152 (1982).

24. H. G. Tompkins, *Appl. Spectrosc.*, **28**, 335 (1974).

25. T. Hirschfeld, *Appl. Spectrosc.*, **31**, 289 (1977).

26. J. T. Gier, R. V. Dunkle, and J. T. Bevans, *J. Opt. Soc. Am.*, **44**, 558 (1954).

27. J. U. White, *J. Opt. Soc. Am.*, **54**, 1332 (1964).

28. G. Kortüm and H. Delfs, *Spectrochim. Acta*, **20**, 405 (1964).

29. B. E. Wood, J. G. Pipes, A. M. Smith, and J. A. Roux, *Appl. Opt.*, **15**, 940 (1976).

30. R. R. Willey, *Appl. Spectrosc.*, **30**, 593 (1976).

31. R. R. Willey, *Appl. Opt.*, **15**, 1124 (1976).

32. M. P. Fuller and P. R. Griffiths, *Anal. Chem.*, **50**, 1906 (1978).

33. M. P. Fuller and P. R. Griffiths, *Appl. Spectrosc.*, **34**, 533 (1980).

34. I. M. Hamadeh, D. King, and P. R. Griffiths, J. Catal., **88**, 264 (1984).

35. D. B. Chase, *Appl. Spectrosc.*, **35**, 77 (1981).

36. D. Kember, D. H. Chenery, N. Sheppard, and J. Fell, *Spectrochim. Acta*, **35A**, 455 (1979).

37. D. Kuehl and P. R. Griffiths, *Anal. Chem.*, **50**, 418 (1978).

38. R. G. Greenler, *Surf. Sci.*, **69**, 647 (1977).

39. J. Engel, G. Wijntjes, and A. Potter, Aspen Int. Conf. on Fourier Spectrosc., 1970 (G. A. Vanasse, A. T. Stair, and D. J. Baker, eds.), AFCRL-71-0019, p. 289 (1971).

40. J. G. Moehlmann, J. F. Gleaves, J. W. Hudgens, and J. D. McDonald, *J. Chem. Phys.*, **60,** 4790 (1974).

41. D. L. Allara, D. Teicher, and J. F. Durana, *Chem. Phys. Lett.*, **84,** 20 (1981).

42. G. C. Augason and N. Young, Aspen Int. Conf. on Fourier Spectrosc., 1970 (G. A. Vanasse, A. T. Stair, and D. J. Baker, eds.), AFCRL-71-0019, p. 281 (1971).

43. W. F. Herget, in *Fourier Transform Infrared Spectroscopy: Applications to Chemical Systems*, Vol. 2 (J. R. Ferraro and L. J. Basile, eds.), pp. 79–110 (1979).

44. P. R. Griffiths, J. A. de Haseth, and L. V. Azarraga, *Anal. Chem.*, **55,** 1361A (1983).

CHAPTER

6

DATA SYSTEMS

I. HISTORICAL PERSPECTIVES

Until the early 1970s, the tremendous time saving gained through using an interferometer because of Fellgett's advantage had not been realized to its fullest extent because of the time delays involved in the computation of the spectrum from the interferogram. In the early days of interferometry, all interferograms had to be recorded, generally using a paper tape punch or magnetic tape recorder, carried to the computing center, read into the computer, and then transformed into spectra. This method not only had the disadvantage of requiring several visits to the computer center every week but also frequently necessitated long waiting periods while other computer users' jobs finished. Periods of as long as two or three weeks before the computed data were returned to the spectroscopist were not uncommon.

Most of the early interferometers used by chemists were the slow-scanning type used for far-infrared spectrometry. For these instruments, the data rate was sufficiently slow that a paper tape punch could be used to record the digitized interferogram. Since the data rate is slow, time sharing on a large computer was successfully used with these interferometers, although it should be noted that this method did require installation of a terminal close to the interferometer.

When rapid-scanning interferometers are used for Fourier transform spectrometry, it is usually less economical to perform the Fourier transform on each interferogram individually than to signal-average interferograms and transform the result, since the only difference between successive interferograms should be noise. There is, of course, no difference between a spectrum produced from the transform of a signal-averaged interferogram and the resultant spectrum produced after individual interferograms are transformed and their spectra averaged, provided that nothing changes instrumentally between successive scans. The most important factor affecting the validity of signal averaging is the necessity for coherent addition, that is, sampling each interferogram at precisely

220

the same retardation values, and in particular the need to start data collection at the same point on the scan each time.

For interferometers with a white-light reference and a laser reference interferometer or with fringe counting, coherent signal averaging should take place automatically, but many early interferometers did not possess such fringe-referencing features. Many of these early rapid-scanning interferometers used hard-wired signal averagers for data collection. For example, all the early Block Engineering [1] interferometers were equipped with either a specially designed Block Co-Adder or a Fabri-Tek Inc. [2] Model 1062 BE signal processor into which the signal was fed. The Fabri-Tek processor could average up to 20,000 interferogram points per second if necessary. It contained up to 4096 18-bit memory units, and its 10-bit analog-to-digital converter (ADC) was sufficient for most applications for which the interferometer was used at that time.

At the end of the signal-averaging process, the interferogram could be punched out on a paper tape for processing on a large computer. If spectral data were required immediately after data collection, at somewhat reduced accuracy than if a digital Fourier transform were performed, the interferogram could also be analyzed with an audio-frequency wave analyzer. Several low-energy measurements were performed using these interferometers in a field environment, where the interferogram and clock signals were recorded on two-track magnetic tape. After the measurement was completed, the tape recordings were played back to the signal averager in the laboratory, so the amount of equipment required for field measurements was quite small.

Audio-frequency wave analyzers can be used to determine the amplitude of each frequency component in the interferogram. The analog interferogram is rapidly and repeatedly presented to the wave analyzer at such a rate that the component frequencies in the interferogram fall in the audio-frequency spectrum, and a signal is produced that can be recorded conveniently on an XY plotter. A plot of intensity versus audio-frequency results, and since there is a linear relationship between the infrared and audio frequencies, the output of the wave analyzer is a single-beam spectrum, see Fig. 6.1.

This technique represents the cheapest method of transforming the interferogram to a spectrum, but at the cost of a certain degree of accuracy, sensitivity, and flexibility.

A commercially available unit for slow-scanning interferometers that used an audio-frequency wave analyzer was the FTC-100 Fourier transform computer manufactured by Beckman–RIIC [3,4]. Analog information from the amplifier was digitized and stored serially in a ferrite core matrix memory as binary words of 12 bits. Two sections of 1024-word

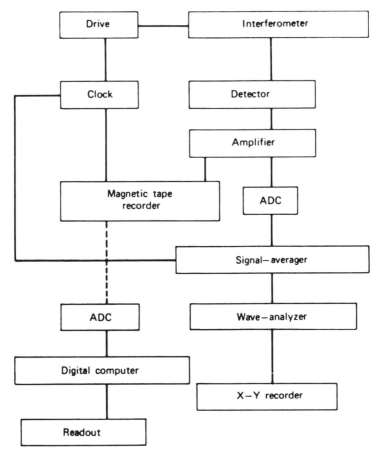

Fig. 6.1. Block diagram of an early type of Fourier transform spectrometer in which the interferogram and clock signals could be recorded on two-track tape for later computation or signal averaged and transformed directly using an audio-frequency wave analyzer.

capacity were available to enable a ratio of two spectra to be obtained as the two interferograms were simultaneously analyzed. Computation consisted of cycling the store, converting back to an analog waveform, and passing this output waveform through the wave analyzer. The program provided for either section of the store to be transformed and displayed separately and the ratio of the two to be displayed to eliminate background effects. The pen carriage was coupled directly to the timing system so that the frequency accuracy is independent of the operator. A diagram of the FTC-100 computer in combination with the FS-620 interferometer is shown in Fig. 6.2.

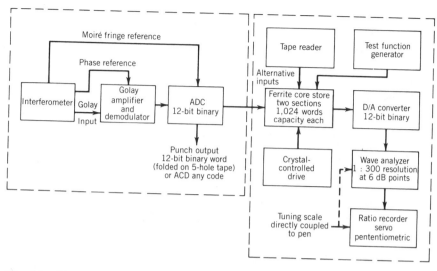

Fig. 6.2. Block diagram of the combination of the Beckman-RIIC FS-620 interferometer and the FTC-100 data system. (Courtesy of Beckman Instruments.)

The same manufacturer went one stage further with their FTC-300 system. Here the interferogram was stored in a 20-k-word memory, so that ratio-recorded spectra were able to be computed at higher resolution or over a greater frequency range than was possible using the FTC-100 computer.

II. MINICOMPUTER DATA SYSTEMS

a. Scope and Capabilities

It is no surprise that stand-alone FT–IR spectrometers became commercially available shortly after moderately priced minicomputers were marketed. By present-day standards, these early minicomputers were rather unsophisticated, which is a function of the hardware technology and the software available. The earliest stand-alone systems were supplied by Digilab in 1969 [5]. These early systems consisted of a Data General Nova minicomputer with 4-k words (16 bits per word) of core memory (~2 μsec cycle time), an ASR-33 teletypewriter, a digital plotter, and an Alphadata 128-k-word fixed-head disk. The spectrometer was not controlled by the data system, as is common today, but the spectrometer optical bench was a self-contained unit that caused the moving mirror to scan, monitored scan velocity, initiated data collection, sampled the signal, and performed the analog-to-digital conversions. Functions such as signal gain and maximum mirror retardation were set manually by the operator with switches on the optical head controller. As a result, the optical head constantly supplied data in terms of scan initiation, detector signal, retrace, and so on, to the computer. The functions the computer controlled were minimal, such as retracing the moving mirror after the desired number of data points had been collected, controlling the aperture in the source optics, and flipping two mirrors on the optical bench so that sample or reference data were collected.

The computer software consisted of a program that allowed the operator to select the number of sample and reference scans to be collected and to initiate data collection. The program also monitored various functions of the system, such as the air pressure in the interferometer drive mechanism and the intensity of the laser reference signal, and checked to see that the infrared beam was passing through the correct channel (that is, front or back sample position) to the detector. Interferometric data from the infrared detector were passed directly to computer memory, from which the data were transferred to disk for storage. Coaddition of the data was accomplished in real time, and the data were constantly monitored for word overflow. This original system was limited to a 16-bit data word and a 14-bit ADC, so that the size of the data word could be easily exceeded at the centerburst upon the coaddition of a few scans. This obstacle was overcome by eliminating the least significant bit when overflow was imminent by dividing the signal-averaged data by 2 (simply shifting the data one bit to the right). As more data were collected, they were added to the accumulated data. The hypothesis for this method of signal averaging was that only the least significant bit was lost, which should have been attributed to noise. Nevertheless, when the energy passing through the sample or reference cell was high, the SNR could become limited by digitization noise after fewer than 50 scans.

Before the scan was initiated, the appropriate information was typed in at the data console by the operator, setting the apodization, phase correction and plotting parameters. After data acquisition was complete, the sample (and reference) interferograms were transformed and the spectrum was plotted according to the parameters supplied by the operator prior to data acquisition. Because the early spectrometers did not have display oscilloscopes, it was necessary to obtain hard copy plots to assess the quality of the spectra. Although from an operating standpoint this system was awkward and primitive by today's standards, the spectrometer system was an extremely complex apparatus that was successful in chemical applications. It can be considered the forerunner of the present-day instrument. Modern instruments have capitalized on the early design ideas of this first widely accepted commercial spectrometer.

The evolution of the Fourier transform infrared spectrometer has been linked closely to the development of advanced electronics and data-handling hardware as well as the evolution of software to a sophisticated level. Unlike the earliest commercial FT–IR systems, most present-day spectrometers operate under a file management system. A file management system is one in which all data and programs reside in a mass storage unit, usually a magnetic disk. In such a system, the computer functions under an operating system that performs most operations, for example, the transfer of data between devices and peripherals, and is used to invoke all programs. An operating system allows a user to have many different programs and data files on mass storage peripherals, which can be called and put into execution at the direction of the user. Hence, the FT–IR tasks are often incorporated into a single program that is invoked by the operator. This program (or in at least one case, a series of programs) allows the spectroscopist to communicate with the computer system through a keyboard console or display 'scope and light pen. The spectroscopist is able to control virtually all aspects of data collection, processing, and presentation through the keyboard.

The primary advantage of a sophisticated file management operating system is that it is more flexible than a stand-alone system. This flexibility leads to features such as *multitasking* and *spooling*. In systems that permit multitasking, several jobs or tasks can be completed seemingly simultaneously. For example, it is possible with some systems to collect data from the spectrometer and perform the Fourier transform while simultaneously viewing and plotting spectra from previous data collections. Spooling is a method by which time-consuming tasks such as printing and plotting that require the use of relatively slow peripherals are removed from direct monitoring by the computer. If a large plot is to be produced, the data can be formatted appropriately and stored on a disk file. The computer then reads a portion of the disk file and sends it to the plotter where it is stored in buffer memory. The plotter is started by the computer whereupon it reads from the buffer memory and begins to produce the plot. The computer detaches itself from the plotter and is free to assume other tasks. When the buffer memory for the plotter is emptied, an interrupt is sent to the computer that causes the buffer memory to be filled again. These interrupts and data transfers are extremely brief, and their occurrence is essentially transparent to the user.

Instrumental control varies among today's FT-IR spectrometers. There are two basic philosophies: either the computer controls all spectrometer functions directly or the optical head has its own computer that controls the parameters associated with driving and monitoring the interferometer as well as sampling the data to be supplied to the main computer. The second of the two design philosophies has become feasible with the availability of low-cost microcomputers that can be dedicated to the limited task of running the spectrometer. In either case, the computational overhead required of the main computer is not great and the efficiency of operation is not largely affected. Of course, it is advantageous to keep computer overhead to a minimum so that each separate task can be done as efficiently as possible without interference from other tasks. Regardless, during data collection, the interferometer must be controlled to produce scans of the correct retardation. Therefore, the laser or sample clock signal must be monitored to ensure correct data collection. For spectrometers that have these components, the bearing air pressure, interferometer temperature, and purge rate may also be monitored. If the computer controls the spectrometer directly, its full resources cannot be devoted to alternate tasks since the main computer must swap between monitoring the operation of the spectrometer and the other tasks, that is, multitasking. When a separate processor controls the spectrometer, this processor can be detached from the main unit, and the majority of the resources of the main unit can be devoted to alternate tasks.

Regardless of the philosophy of system design, both approaches provide the user with an extremely flexible instrument. Many of the major vendors that employ minicomputers in their instruments permit the user to perform several tasks simultaneously. As stated above, the user may wish to collect data with the spectrometer and perform a Fourier transform of the interferogram at the end of data collection. This can be a time-consuming procedure and may be considered wasted time by the spectroscopist. With a system that supports multitasking, the spectroscopist may initiate data collection and then proceed to plot previously collected data or display it and perform various mathematical functions on the data in order to extract information. Under such conditions, available time may be more efficiently used and resources more effectively exploited.

In practice, most minicomputers are not fully utilized even during multitasking operations. As a result, it is possible to interface several spectrometers to the single computer. In this way, not only may data from several samples be collected simultaneously, but additional data processing may also be effected at the same time. Hence, several spectroscopists may use a single instrument cluster without seriously impeding the work of others.

b. Hardware

The computer hardware available to FT–IR manufacturers, and hence spectroscopists, has changed dramatically since the early commercial stand-alone instruments. The capabilities of the minicomputer itself have increased significantly. The speed with which computations are completed and the speed at which memory is accessed have improved substantially. An internal "clock" or crystal oscillator determines the cycle time of the processor. Simple operations are completed in one clock cycle. For example, memory access time for the early Digilab spectrometers, which used a Data General Nova 1200 minicomputer (which superceded the original Nova minicomputers), exceeded 1200 nsec. Several of the more modern Digilab instruments incorporate a Data General Nova 4/X minicomputer, which has a memory access cycle time of only 400 nsec. Other innovations have been made such as *instruction prefetch*, which means that the central processing unit (CPU) looks ahead to store future instructions in a special buffer while simultaneously executing an instruction. This eliminates a "fetch" cycle. In other words, the CPU does not access an instruction, execute it, access another, and so on; rather it executes instructions virtually sequentially because accessing (or fetch) is done simultaneously to execution. To draw another example from the Nova series of computers, the Nova 1200 (early 1970s) completed a simple *add* instruction in 1.35 μsec, but at least an additional 0.6 μsec must be added to account for the fetch cycle. The Nova 4 series (late 1970s) can complete an add instruction in about 400 nsec, and virtually no time is required for the fetch cycle. Consequently, computers are now capable of completing more tasks per unit time, which provides a faster response for the spectroscopist.

Besides the change in speed of minicomputers, the computer memory that is avilable has increased. The size of computer memories was at one time restricted by two parameters, the cost of the memory and the ability of computers to address individual data words in the memory. The cost problem has been solved by going from ferrite core memories to metal oxide semiconductor (MOS) high-speed memories. Core memories are still available for many computers but are generally an order of magnitude more expensive than semiconductor memory. The addressing problem is related to the computer word size. In a 16-bit-word computer, the maximum number of addressable units is 2^{15}, or 32,768. Thus, many 16-bit-word computers were restricted to 32,768 words of computer memory. This has been solved by *memory mapping*, which increases the bit length of the address register so that memories of up to 128-k words are commonly available. For computers that have a longer word, such as the 20-bit Nicolet 1280, memories of up to 512-k words (actually 524,228 words) are available [6]. The potential memory size of the Aspect 24-bit computers in the IBM/Bruker instruments is potentially even larger (up to 8.2 × 10^6 words), but memories of up to 80-k words are supplied to the present time [7].

A significant function of any data system is the input–output (I/O) operations required for communicating with the spectroscopist. On most minicomputer-based FT–IR spectrometer systems, spectroscopist input is via a keyboard. Keyboard systems sometimes use instructions that are mnemonics restricted to just a few characters. It is possible to program a series of input instructions into a single command or, in some cases, to a separate function key on the console front panel to reduce time in certain operations. All manufacturers supply an oscilloscope or video display as well as a hard copy plotter for output. These display systems incorporate graphics for the presentation of spectra. Storage oscilloscopes are not used since this necessitates the erasure of the entire screen if only a small portion is to be removed; hence all displays use refresh technology. These displays are either monochrome or color depending on the vendor. For example, the Nicolet systems have an eight-color oscilloscope that has 560 × 380 pixels (discrete picture elements). Once data have been collected and processed, spectra (or interferograms) can be transmitted to a hard copy plotter. The spectra can be plotted and labeled in the format desired by the operator. In addition to these output devices, hard copy printers are available for the printing of alphanumeric data.

Other I/O devices are primarily magnetic storage devices. Most of the commercial FT–IR minicomputer systems are disk based. That is, the main peripheral is a hard disk that stores the operating system and other programs as well as all the collected data. Some of these hard disks employ moving-head technology and range in size from 10 to 300 Mbytes (1 byte = 8 bits). As explained in the preceding section, data are usually collected from the spectrometer and stored directly on disk. If the interferogram is sufficiently small and memory size permits it, coadding may be done in the memory; otherwise data are written onto the disk and coadding of the interferogram is done in sections.

Moving-head disks are cheaper to produce than older fixed-head technology. A disk is somewhat akin to a phonograph record except that it has a series of magnetically coded concentric tracks rather than a single spiral groove. A moving-head disk has one set of magnetic heads that can be positioned over any track to read or write the data. A fixed-head disk requires a head for each track and is hence more expensive as the heads tend to be very costly. Moving-head disks also often have removable cartridges, so various operating configurations or different sets of data can be stored on separate disk cartridges.

Several manufacturers are beginning to supply Winchester disks with their systems. These disks employ moving-head technology with very high encoding densities on the magnetic medium. This leads to faster, higher-capacity disk drives that require less space and power to operate. Some

Winchester disks come with removable cartridges, and storage capacities of 50–300 Mbytes are not uncommon. The Winchester drives are less expensive, per byte, than the older hard disks. It is expected that all vendors will shift to this or similar technology for the main storage device.

It was stated above that modern minicomputers are faster than the earlier models, but it is often found that the newer spectrometers are not remarkably faster in computation than the earlier systems. This is because the data systems are generally rate limited by the hard or Winchester disk. Data access times on these disks are on the order of tens of milliseconds, which is far slower than the computational speed of the computer. (However, Winchester disks transfer data 8–10 times faster than the hard disks.) In fact, the earlier fixed head disks had faster access times because the heads did not have to jump tracks. Present-day data systems are faster than their predecessors due to CPU and software enhancements; however, this speed advantage rarely exceeds a factor of 4.

Other magnetic storage media are available, such as floppy (or flexible) diskettes and magnetic tape. Flexible disks usually have rather small capacities, up to about 512 kbytes or 1Mbyte, and are rather slow, about three times slower than the hard disks. These devices are useful for the archival storage of small data sets. Nine-track magnetic tape is also an excellent archival storage medium, since here the capacity can be rather large (on the order of 20 Mbytes per 2400-ft reel). The major drawback to magnetic tape is that the data must be accessed sequentially, and it may take minutes to access data written far along the tape. Magnetic tape is sometimes the storage medium of choice for GC/FT–IR single-scan interferograms due to its large storage capacity and relatively low cost (see Chapter 18). Magnetic tape also has certain advantages in multiuser environments, where each user may have his or her own tape(s).

It is a common misconception that computers are inherently capable of many arithmetic operations. In practice, computers are generally capable of only a few select operations, and arithmetically these involve addition and bit shifting. All other arithmetic operations are comprised of these basic functions. Consequently, operations such as multiplication and division can be speeded up by employing electronic hardware that is constructed specifically for these tasks. Some FT–IR instrument manufacturers have included hardware multiply/divide processors for the minicomputers in their instruments. These devices reduce computation times by factors of 10–20 over the standard software-controlled multiply/divide functions in the computer. A hardware multiply/divide option should not be confused with an array (or vector) processor, which is discussed in the last section of this chapter.

c. Software

On most Fourier transform spectrometers designed for chemical spectro-metry, interferograms may be measured either from a single channel or from two separate beams sequentially, in which case either the front or the back beam may be specified as the sample beam. Before data collection is initiated, the resolution and the number of scans in the sample and reference beams must be specified, and apodization and phase correction parameters are entered before the computation of the spectrum from the interferogram. In most Fourier transform spectrometers, all the parameters for data collection, computation, and plotting are entered before the measurement, and the complete operation of measuring the spectrum is initiated with one command.

Once interferograms have been collected, signal averaged, and stored, the next step is usually the transformation of the data to a spectrum. The algorithm has been explained in detail in Chapter 3, so the procedure will not be reiterated here. Because all FT–IR spectrometers are in effect single-beam instruments, reference and sample spectra are produced and stored independently. Once the data have been collected and transformed, the remainder of the software is designed to present the data and to extract information by various manipulations.

A transmittance spectrum is produced by ratioing a sample to a reference spectrum and is known as a ratio-recorded transmittance spectrum. A characteristic of this type of spectrum is that in wavenumber regions where the energy in the single-beam background spectrum is low, either due to bandpass or solvent effects, the spectrum may become very noisy. When the relative spectral energy $B(\bar{\nu})$ is very low in both the reference and sample spectra, the relative error upon ratioing can be quite large. It is possible to monitor the absolute intensity of the reference or background spectrum, and if it falls below a certain percentage of the maximum energy, no ratioed data are produced. This produces gaps, or dead regions, in the spectrum, but those regions where the noise may be excessive can be eliminated. It should be noted that a similar effect is produced in optical-null dispersive spectrometers, but it is a mechanical limitation of the system known as a *dead pen.*

Ratio-recorded spectra can be presented in several ways, linear transmittance, linear absorbance, or linear log absorbance. Conversion from one form of output to another is rather straightforward. As the output format is changed, the ordinate scale of the display is modified to match the data. Plotting routines have become quite sophisticated as far as data presentation is concerned. The routines can mark abscissa and ordinate scales at operator-selected intervals and scale the data in either axis. The plots can be labeled with the file information and the spectrometer parameter settings at the time of data collection. Plotted data can be previewed on a display scope prior to hard copy plotting to ensure that the data are in the desired format. This is often preferable because it generally takes considerably less time to display data than to produce a paper copy. If necessary, plot files can be stored on disk or magnetic tape to eliminate the need to reproduce all the appropriate plot parameters. Thus digital copies of important data can be stored and reproduced as required.

Other features associated with the presentation of data include *baseline correction, plot smoothing*, and *interpolation*. Occasionally, spectra may be found to have baselines with a continuous drift or slope. Most FT–IR instrument manufacturers supply software whereby the user can correct these conditions. One method by which baseline correction may be accomplished has been employed by Digilab [5]. Two parameters are defined, the tilt (T) and the curvature (C). The curvature is the slope of the baseline, and the tilt is a parameter to account for any constant offset in the baseline. The baseline is corrected at each wavenumber in the spectrum by the equation

$$Y'(\bar{v}) = Y(\bar{v}) - (T + C \cdot \bar{v}) \qquad (6.1)$$

where $Y(\bar{v})$ is the original ordinate value at wavenumber $\bar{v}$, T is the tilt, C is the curvature, and $Y'(\bar{v})$ is the corrected ordinate value. The curvature is empirically determined by the spectroscopist until all baseline slope is removed, but from Eq. 6.1 it can be seen that the curvature offsets the baseline at all wavenumbers except zero. The tilt parameter value is then determined to correct for the offset. The determination of these parameters can be completed rapidly by plotting the spectrum on the display scope and examining the spectrum after each modification in the parameter values is made.

Smoothing the spectrum reduces the effects of noise but degrades the spectral resolution. Many smoothing algorithms have been reported in the literature, but all have the same basic premise. Several data surrounding a single datum are used to calculate a new value for that datum. One of the simplest smoothing functions is boxcar integration. Boxcar integration is accomplished by taking each ordinate value in the spectrum and calculating a new value by computing the mean of the value plus the ordinate values from a small number (usually ≤4) of preceding and succeeding data. Other algorithms perform similar operations, but the surrounding data are weighted by various functions such as Lorentzian, Gaussian, Savitsky–Golay, or triangular functions. When these other functions are employed, data displaced from the datum to be corrected

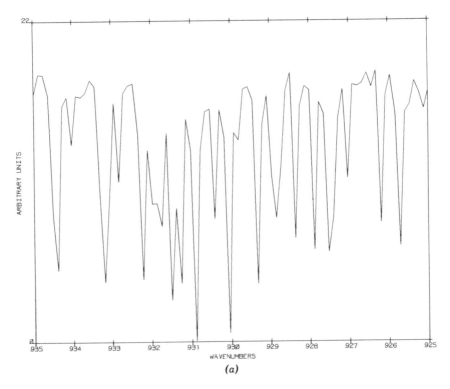

Fig. 6.3. (a) Spectrum of ammonia gas at a resolution of 0.125 cm^{-1} and no zero-filling. (b) The same spectrum using a quadratic interpolation function with nine points interpolated between each original datum.

are given less weight. Smoothing may also be effected in the Fourier domain by computing the Fourier transform of the spectrum, multiplying by a suitable apodization function, and computing the inverse transform. Although smoothing a spectrum may make the data more aesthetically pleasing, resolution can be reduced by a factor of 2 or more.

It may be advantageous to smooth data selectively within specific wavenumber regions, such as near the bandwidth limits of the spectrometer where the energy throughput is low and the SNR decreases. The effect of the smoothing function can be gradually increased from zero to a maximum value in regions where the SNR begins to decrease. When this is done, there is no abrupt change in the resolution or the effective SNR. This type of smoothing may be compared with the application of a slit program on a grating spectrophotometer.

The effect of interpolation of data may at first seem quite similar to

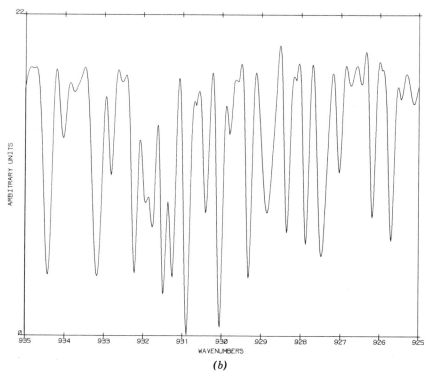

ARBITRARY UNITS

WAVENUMBERS

(b)

Fig. 6.3. (continued)

smoothing. Interpolation leaves the original values unchanged but adds data between the original discrete data points. It was shown in Chapter 3 that zero filling an interferogram interpolated the spectral data, but this is an expensive operation computationally. To interpolate one datum between two successive original data by zero filling requires that the interferogram be doubled in length, which results in a Fourier transform that theoretically requires four times as much time. If a limited region is of interest, it is often beneficial to interpolate the spectral data by applying a polynomial fit to the spectrum to calculate interpolated points. Different numbers of points may be interpolated between the original data, with the number of points ranging from 1 to 25. Interpolation does not reduce resolution but can change the contour of a spectrum so that features are more aesthetically pleasing. Figure 6.3 illustrates this point by showing spectral data before and after interpolation.

Additional special functions may be incorporated into FT–IR operating systems. One that has had wide acceptance and utility is spectral subtraction. This is a technique where two absorbance spectra are subtracted in order to ascertain specific information about one of the spectra. Spectral subtraction may be used qualitatively to determine if an unknown spectrum and a known spectrum are identical. Although this may initially appear to be a trivial example, absorbance spectra can be different in appearance when the difference in concentration is large. Low SNRs may also add uncertainty. Spectral subtraction routines permit data to be scaled to account for concentration differences. The values of the scaling factors are usually selected by monitoring deviations from the baseline. If two spectra are of the same compound the spectral subtraction will produce a flat baseline (within the limits of the noise). If the unknown is suspected of being a mixture, spectral subtraction can be used to isolate components. In quantitative analysis, this technique can be used to determine the concentration of each component in a mixture by examination of the scaling factors necessary for appropriate subtraction. (Spectral subtraction is covered in more detail in Chapter 10.)

Wavenumber assignment of spectral features is often of importance in spectrometry. If a stable sample is left in a Fourier spectrometer and several spectra are measured consecutively (without varying any parameters), the spectra will usually be reproducible within 0.01 cm^{-1}. Residual features in their difference spectra will be very small. This repeatability may not always be observed, however, when the cell is removed and reinserted between spectra. Several reasons may be advanced to explain this observation. First, the temperature of the sample may change, leading to small shifts of spectral features. Second, it was noted in Chapters 1 and 5 that changes in the effective solid angle of the beam can lead to small spectral shifts. Therefore, unless the cell is replaced in exactly the same position for each measurement, the great wavenumber precision of the interferometer can be lost. A mount in which the cell fits loosely obviously should be avoided for any measurement involving spectral subtraction or estimation of band centers to high precision.

The method by which the wavenumber of a band or its width is determined is also important. In several early reports, the wavenumber of the actual data point with the highest ordinate was listed, often to several decimal places, as the band center. Since spectra of samples in solution are often measured at 4 cm^{-1} resolution, data points are typically computed at 2-cm^{-1} intervals so that shifts of, say, 0.5 cm^{-1} would not be readily detected.

Cameron et al. [8] have illustrated the pitfalls of this approach by considering the effect of shifting a Lorentzian band with a FWHH of 10 by 0.5 cm^{-1} when its data points are separated by 4 cm^{-1}, see Fig. 6.4a. The datum giving the band maximum in this example is the same for both

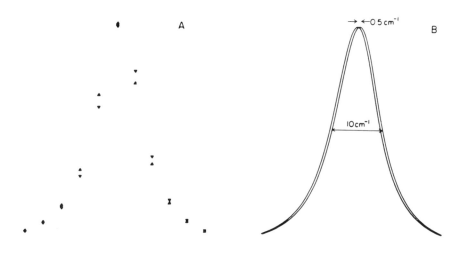

Fig. 6.4. Effect of shifting a Lorentzian band with a FWHH of 10 cm^{-1} by 0.5 cm^{-1} when the band is digitized every 4 cm^{-1}. (A) The two bands have the same maximum and only a small ordinate difference is seen. (B) The errors in the band maxima become apparent only after extensive interpolation. (Reproduced from [8], by permission of the Society for Applied Spectroscopy; copyright © 1982.)

spectra, and even the ordinate values of the points are approximately the same. Extensive interpolation would indicate that a shift has occurred, as seen in Fig. 6.4*b*. Nevertheless, there is still an inherent error in this method, and to obtain a peak wavenumber in error by only 0.01 cm^{-1}, 200 interpolated points must be calculated between each linearly independent datum.

Two more efficient ways of calculating the peak wavenumber are the center of gravity and least-squares methods [8]. In the center-of-gravity method, the spectrum in the region where the absorbance $E(\bar{v})$ of the peak exceeds a certain fraction f of the maximum peak height is used, see Fig. 6.5. The center of gravity in this region, $\bar{v}_{CG}$, is given for a continuous spectrum by

$$\bar{v}_{CG} = \frac{\int_{\bar{v}_j}^{\bar{v}_k} \bar{v}[E(\bar{v}) - E_f]\, d\bar{v}}{\int_{\bar{v}_j}^{\bar{v}_k} [E(\bar{v}) - E_f]\, d\bar{v}} \tag{6.2}$$

where $\bar{v}_j$ and $\bar{v}_k$ are the wavenumbers at which $E(\bar{v}) = E_f$.

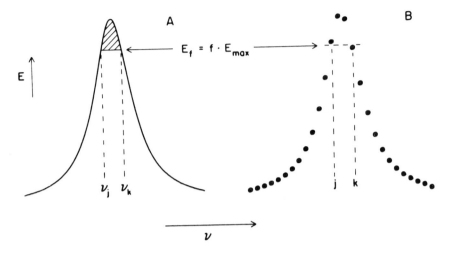

Fig. 6.5. Center-of-gravity computation data. Fraction of the peak height, f, for (A) a continuous function and (B) the same function as in (a) digitized at constant-frequency intervals. (Reproduced from [8], by permission of the Society for Applied Spectroscopy; copyright © 1982.)

For a spectrum that is digitized at constant intervals, $\bar{\nu}_{CG}$ can be computed by the following summation:

$$\bar{\nu}_{CG} = 0.5 \left(\frac{\sum\limits_{i=j}^{k} \bar{\nu}_i(E_i - E_k)}{\sum\limits_{i=j}^{k} (E_i - E_k)} + \frac{\sum\limits_{i=j}^{k-1} \bar{\nu}_i(E_i - E_j)}{\sum\limits_{i=j}^{k-1} (E_i - E_j)} \right) \qquad (6.3)$$

where $E_j \geq E_k$, and E_k is the value of $E(\bar{\nu})$ closest to E_f. If the peak is symmetric, $\bar{\nu}_{CG}$ will be the wavenumber of the peak maximum. For asymmetric peaks, $\bar{\nu}_{CG}$ will differ from the true peak maximum by an amount dependent on f and the degree of asymmetry. The choice of f is not very critical if the peak is fairly symmetrical. For a symmetrical peak with a FWHH of 10 cm^{-1} measured with a SNR of 1000, the maximum uncertainty is less than 0.01 cm^{-1} for $0.2 < f < 0.9$, as shown in Fig. 6.6.

An alternative approach is to fit a polynomial, $y = f(x)$ to the top of the peak and then to solve for $dy/d\bar{\nu} = 0$. Generally, quadratic equations are used; higher-order polynomials do not appear to lead to increased precision. It can be shown that these two approaches are quite similar, with the principal difference being that in the center-of-gravity approach all points are equally weighted whereas in the least-squares approach the weighting of a given point depends on its position relative to the band center. In view of the computational simplicity of the center-of-gravity method, it will probably become the method of choice for FT–IR measurements.

The accurate determination of bandwidths is also very important, for example, in biochemical FT–IR spectrometry. Although bandwidths are often quoted at half-height, it may often be more informative to make the determination at some other fraction f of the peak height. The peak height may be defined as $E_{max} - E_{ref}$, where E_{ref} is an arbitrarily selected zero point and $\bar{v}_{max} = \bar{v}_{ref}$. The fractional width $\Delta \bar{v}_f$ may be calculated as

$$\Delta \bar{v}_f = \left| \bar{v}_{k-1} + \frac{\Delta \bar{v}[f(E_{max} - E_{ref}) - E_{k-1}]}{E_k - E_{k-1}} \right.$$
$$\left. - \bar{v}_j + \frac{\Delta \bar{v}[f(E_{max} - E_{ref}) - E_j]}{E_{j+1} - E_j} \right| \quad (6.4)$$

where $E_j < f(E_{max} - E_{ref}) < E_{j+1}$, and $E_k < f(E_{max} - E_{ref}) < E_{k-1}$.

The judicious application of Eqs. 6.3 and 6.4 can allow an enormous increase in the amount of information that can be obtained from spectra.

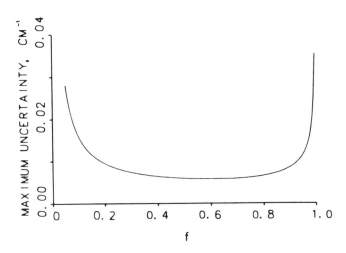

Fig. 6.6. The maximum uncertainty in the band maximum using the center-of-gravity method versus f when the band has a FWHH of 10 cm^{-1} and the SNR is 1000. (Reproduced from [8], by permission of the Society for Applied Spectroscopy; copyright © 1982.)

The use of sophisticated computers that support high-level languages has led to the development of applications packages. Occasionally, users find the software supplied with their spectrometer inadequate for their needs, so they write programs to solve specific problems. Instrument manufacturers have exploited this resource of new software by collecting, documenting, debugging, and distributing the programs to other users. The contents of these applications packages include plotting routines, deconvolution and bandfit software, and gas chromatography/Fourier transform infrared software.

III. MICROCOMPUTERS

At one time, the discrepancy between microcomputers and minicomputers in terms of performance was rather large, but the gap is narrowing rapidly. Technically, a microcomputer is a device that has the CPU isolated to a single integrated circuit. A minicomputer has the CPU constructed from many discrete elements (including integrated circuits) and is generally distributed on one printed circuit board. However, the distinctions between mini- and microcomputers based on these criteria often do not hold. The actual differences have become somewhat blurred. When microcomputers first began to appear, they were restricted by available semiconductor technology and were rather slow compared to minicomputers. At the present time, microcomputers are available that have word sizes equal to those of most minicomputers (up to 32 bits/word) and are able to address large memories directly (up to 16 Mbytes.)

The operating cycle time of microcomputers may be very fast; for example, the Motorola 68000 microcomputer operates at 8 MHz. This corresponds to a cycle time of 125 nsec, which is faster than a Data General Nova 4/X minicomputer. For the most part, microcomputers remain less expensive than minicomputers.

Clearly, the shift to microcomputers is appealing when one examines their performance with respect to their cost. Several manufacturers have begun to incorporate microcomputers in FT–IR systems, initially in the less expensive models, but now in some top-of-the-line systems. The first manufacturer to incorporate a microcomputer in spectrometers to the knowledge of these authors was Bomem [9]. In fact, two microcomputers were employed, one (an Intel 8085) to control the spectrometer, the second (an Intel 8086) to process data through the array (or vector) processor. (Array processors are discussed in Section IV.) This system is not low cost by any means but represents a true multiprocessor spectrometer. The host computer is used to retrieve and display data, but the optical head is a stand-alone system. Commands are passed from the host computer to the spectrometer control microprocessor, which then detaches itself and collects the required data. The interferograms are passed to the vector processor for phase correction and coaddition and finally transformation to a spectrum. At this time, a link to the host computer is reestablished and the data are passed to the host. The host computer can be any mini- or microcomputer that can support an IEEE-488 general-purpose interface bus (GPIB) and that can perform the necessary data storage and presentation operations. Bomem supplies Digital Equipment Corporation (DEC) mini- and microcomputers with their systems; however, these interferometers have also been interfaced to such diverse systems as an LSI-11/03 microcomputer and a VAX 730 mainframe computer [10]. The operating system used is written in FORTRAN and can therefore be installed on different DEC systems with little difficulty. A schematic of the Bomem computer system is shown in Fig. 6.7.

An example of a stand-alone microcomputer system is the Analect fX-6200 FT–IR spectrometer [11]. This system actually consists of three microcomputers, two in the CPU and one for the display unit. The CPU acts as an executive and delegates tasks to different sections of the spectrometer data system. The CPU directs the acquisition of data and sends the raw interferogram to an ADC, where it is digitized and then transformed. Fourier transformation is performed with a hard-wired TTL (transistor-transistor logic) device rather than by conventional software. When the data have been transformed and signal averaged, the spectra are stored or displayed. If display is desired, the data are transferred to the display processor, which contains a separate microcomputer. A block diagram for the fX-6200 data system is given in Fig. 6.8. It is possible to display spectra in real time with this spectrometer, that is, a spectrum can be displayed after each single scan.

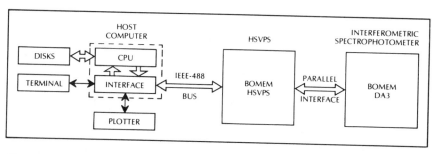

Fig. 6.7. Schematic of the Bomem computer system. HSVPS is the High Speed Vector Processor System, *vide infra*. (Courtesy of Bomem, Inc.)

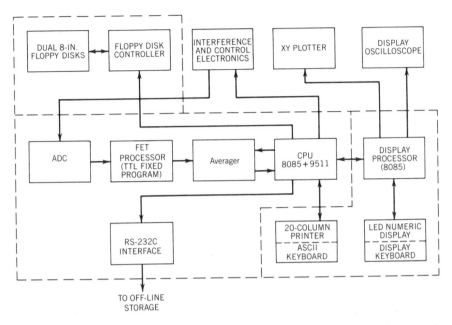

Fig. 6.8. Block diagram of the Analect fX-6200 series spectrometer system. (Courtesy of Analect Instruments.)

Until 1984, the Perkin-Elmer Corporation marketed the Model 1500 FT–IR spectrometer [12]. This instrument is essentially an Analect fX-6200 spectrometer interfaced to a Perkin-Elmer Model 3600 or a Model 7500 data station. The interferometer has been modified to meet Perkin-Elmer's specifications, but the spectrometer control microcomputer and hard-wired FFT device remain the same. The Perkin-Elmer Model 3600 incorporates a Motorola 6800 8-bit microprocessor that can support up to 64 kbytes of memory. The data station supports two 160-kbyte micro-floppy disks that serve as storage for data and application program storage. The operating system resides in 12 kbytes of read-only memory (ROM). The data station controls all plotting and data presentation as well as performs searches and characterizations of unknown organic compound spectra.

Other manufacturers have begun to market systems that incorporate microcomputers. Nicolet uses a Z80 microprocessor in their 60SX system to help control the spectrometer, but the majority of the processing is done by the host computer, a Nicolet 1280 minicomputer [6]. Digilab employs three microprocessors in the Qualimatic FT–IR spectrometer; a Motorola 68000 is used as the primary processor with the other two being 8-bit microprocessing devices [5]. More recently, Digilab has begun marketing a research-grade FT–IR spectrometer controlled by a Motorola-68000-based processor. IBM Instruments uses a Motorola 68000 microcomputer to control the IR/32 spectrometer [7]. Beckman Instruments has also introduced a microcomputer-controlled FT–IR spectrometer series, the FT-1000 and FT-2000, which also incorporates a Motorola 68000 processor [13].

In a very real sense, Motorola-68000-based microcomputers are beginning to dominate the FT–IR field. As noted above, several manufacturers have introduced this processor to their product line. In addition to IBM and Digilab, Perkin-Elmer uses the Motorola 68000 microprocessor in its Model 7300 and 7500 data station. The 7500 data station is the standard computer for the Model 1800 FT–IR spectrometer. This processor can be interfaced to the lower-cost Models 1500 series and 1700 FT–IR spectrometers. Mattson Instruments uses the Motorola 68000 in their Starlab computer system [14]. This system can be interfaced to the Mattson Sirius, Cygnus and Alpha Centauri FT–IR spectrometers as well as many other competitors' systems. The Mattson computer not only controls the Mattson line of instruments but can also be used to communicate between different vendors' systems and process the data from a competitor's spectrometer.

All Motorola-68000-based microprocessor systems have multiple coprocessors. Generally, a second processor is used to coadd the data and to compute the fast Fourier transform. (Sometimes another Motorola 68000 is used.) A third processor is used to handle data communications, and additional processors may be used to transfer data to and from storage devices and to operate the interferometer. Despite the large number of processors, these systems tend to be more efficient and less expensive than their predecessors, the minicomputers.

Three of the FT–IR spectrometer manufacturers that use the Motorola 68000 microprocessor have adopted the UNIX or a UNIX-like real-time operating system. UNIX is a general-purpose, highly flexible standard operating system developed by Bell Laboratories. Mattson Instruments uses UNIX, and Digilab and Perkin-Elmer use the UNIX-like operating system IDRIS.

Coincident with the large-scale introduction of microcomputers to the FT–IR spectrometer marketplace is a shift in the method of user interaction. For the most part, earlier systems have used mnemonic-driven systems to control the spectrometers. That is, the spectroscopist was required to recall abbreviated commands and parameters to execute tasks. Systems are becoming menu-driven so that the user is always presented with a logical sequence of commands or tasks. A menu-driven system makes it easier to recall the appropriate commands and parameters, and often less time is spent setting up an operation. The user interaction is reduced further by the fact that many of the commands or parameter value changes can be invoked by pressing a single key or touching the menu annotation on the display screen with a light pen. The introduction of microprocessors, including those in the display terminal and keyboard, have led to programmable keys, or "soft" keys. Depending on the menu or task, these programmable keys can have assigned functions and can be changed at any time. As the menus change for different operations, the purpose of each of the programmable keys may change. Menu-driven systems clearly require less training on the part of the operator to master the system, they are user friendly, and they are as flexible as the older mnemonic systems. It is expected that menu-driven operating systems will dominate the field in the future.

Clearly, the differences between mini- and microcomputers are disappearing quickly. It is undoubtedly true that as minicomputers and microcomputers incorporate more of the same technology, the differences will disappear altogether. The low cost and ease of operation of spectrometers with on-line computers make the dedicated computer data system a highly desirable component of a Fourier transform spectrometer. The time savings in data processing and the capability of the operator to interact with the data system certainly account in no small measure for the growing popularity of infrared Fourier transform spectrometry. In fact, this capability has undoubtedly spurred the development of computer-assisted dispersive infrared spectrometers.

IV. REAL-TIME DATA SYSTEMS AND ARRAY PROCESSORS

Most of the data systems of the type described above have been designed to collect and store the digitized interferogram first and to compute the Fourier transform using the FFT algorithm at the end of the data collection step. It was noted in Chapter 3, Section IV, that the technique of real-time computation can be used to give a readout of the spectrum after each interferogram point has been collected. This technique has been used with several slow-scanning interferometers. The first real-time data systems to be described in the literature did not use general-purpose computers. The first such system was described by Yoshinaga et al. [15] in 1966 and used a special-purpose digital computer. In 1969, Hoffman [16] described an analog data system for this purpose. However, the real-time computing technique did not appear to be of great importance until the first real-time data system using a general-purpose minicomputer became commercially available.

The system was described by Levy et al. [17] and formed the basis of the Coderg [18] Fourierspec 2000 Fourier transform spectrometer. The computer used by Levy et al. was a Varian 620i minicomputer with 4-k words of core. With this memory size, up to 1500 frequency points were able to be computed for a single-beam spectrum, or 750 frequency points if the ratio between two spectra was being found. The remaining locations were used for programs and instructions (1572 words) and a table of cosine values (1024 words). The time of computation of this system was found to be almost directly proportional to the number of frequency points calculated.

The time required to calculate 1500 frequency points was a little under 2 sec. Thus, it is apparent that the scan speed of the interferometer had to be quite slow if the full 1500 frequency points were to be calculated. If only a limited frequency range was of interest, the speed of this data system placed few restrictions on the operating speed of a typical slow-scanning far-infrared interferometer.

The designer of the Coderg data system considered it unlikely that phase correction procedures (which must necessarily take place after several points after the retardation point) could be carried out without endangering the speed of the real-time method of computation or substantially reducing the number of frequency points that can be calculated. Thus, it was difficult to measure photometrically accurate spectra using this instrument. However, the designers of a second commercially available far-infrared Fourier transform spectrometer that also used a real-time computing technique, the Polytec FIR-30 [19], claimed to have overcome this problem. In this system, the position of the fixed mirror could be changed to coincide with the zero retardation point by means of a very fine translational adjustment. Any residual phase errors were then corrected by fitting a parabola to the largest three sampled points around zero retardation. The interferogram retardation values were then adjusted accordingly.

The two real-time computing systems just described were designed around slow-scanning, far-infrared interferometers with a slow data rate and relatively short frequency range. It is obvious that the real-time computing method will show less compatibility with rapid-scanning interferometers where signal averaging of interferograms is used to increase the final SNR. Nevertheless, several very high resolution interferometers have been described for mid-infrared spectrometry where signal-averaging techniques are not applied. For these interferometers, up to 10^6 data points may be collected at a relatively high data rate (>10 points/sec). In view of their wide frequency range, high resolution, and high data rate, it would seem on first inspection that the real-time computing technique would have very little application for this type of measurement. It is just in this field, however, that some of the most significant advances in real-time computing techniques have been made. For high-resolution measurements where one scan can take several hours to complete, it is particularly important to check whether the spectral bands of interest to the operator are in fact being measured during that measurement. If they are not, the run can be aborted before several further hours of measurement are wasted. By only looking at a very small fraction of the total frequency range, but in a region in which important spectral bands are known to absorb, Connes and Michel [20] developed a method of computing 1000 output points at the rate of 50 input points/sec. This represented a gain in speed of almost a factor of 20 over the Polytec data system described previously.

The data system that was used to achieve this increase involved no radically new principle. It was simply a digital computer with a hard-wired classical Fourier transform algorithm. Analog devices were used only for apodization and at the output for interpolation between spectral points to present a continuous curve on the oscilloscope. The two features of this data system that differ from general-purpose computers were the elimination of all programming by the use of hard wiring and storing the spectrum in a sequential-access circulating memory. The real-time processor devised by Connes and Michel was a design predecessor of the hard-wired FFT processor produced by Analect [11]. The Analect system departs from the earlier system in that it uses random-access memory (RAM) rather than a magnetostrictive drum. The Analect processor is able to transform a 3.6-cm^{-1} resolution double-sided interferogram in 200 msec.

Another type of processor, other than the hard-wired FFT processors, has come into prominence in recent years. This is the array processor, or, as it is sometimes called, a vector processor. An array processor is a programmable computer (or device) that has been constructed to perform high-speed data computations. Technically, most array processors are high-speed multipliers and adders. These operations are achieved by constructing hard-wired logic that breaks down the multiplication or addition process into a few discrete steps. An array processor can complete a single multiplication appreciably faster than a standard software multiplication and gains further speed by doing operations simultaneously. The data to be processed are loaded into a dedicated RAM in the array processor and fed into the computation devices. For example, it may be desired to take the dot product of two arrays. The first two multiplicands are fed into the first step of the multiplier, and after this step is complete, the data are passed to the second step. Simultaneously, the second data values from the two arrays are passed into the first step of the multiplier.

As the operations are completed, these two sets of data are passed one step further down the line and new data are fed into the system. As products emerge from the processor, they are added as the final step of the dot product. If the multiplication process has three steps, three multiplications are being run simultaneously. Such an operation is sometimes referred to as "pipelining" the data.

The first array processor developed for FT–IR spectrometers was manufactured by Bomem [9,21]. Bomem's High Speed Vector Processor System (HSVPS) performs high-speed calculations on floating point data. Sixteen-bit floating point arrays are input to the HSVPS, where they are converted to fixed point. The multiplier can complete a high-speed precision multiplication in 0.5 μsec for a 40-bit product or 0.8 μsec for a 64-bit product. The data are converted to floating point, upon output. The array processor has 4k × 10-bit on-board first-in-first-out (FIFO) memory, another 4k × 10-bit memory for the storage of numerical filter coefficients, and a sine–cosine generator for up to 2×10^6 different values. Associated with the HSVPS is random-access memory for coadding and storage that can vary in size from 16k × 32 bits to 240k × 32 bits. This system is a general-purpose programmable array processor, and software is currently available to perform phase corrections by the Forman (convolution) method [22], numerical filtering, fast Fourier transforms, and discrete Fourier transforms. The Bomem HSVPS is designed to phase-correct and numerically filter incoming raw interferograms and then coadd processed interferograms as the spectrometer is collecting data. The spectrometer can collect data at a mirror velocity of up to 4 cm/sec (optical) without overrunning the data processing speed of the HSVPS. The HSVPS can also interpolate spectral points when an FFT is performed. A 4k × 32-bit interferogram can be transformed, without interpolation, in 10.3 sec, and a 240k × 32-bit interferogram with 25-point interpolation, which yields a 6M × 32 bit spectrum, can be transformed in 214 min.

In a very real sense, the Bomem spectrometer is built around the array processor, but this yields very impressive specifications. Other FT–IR manufacturers have begun to offer array processors as optional equipment. Digilab markets the HI-COMP-32® microprogrammable processor for use with its Data General computers. This device is a 32-bit fixed point processor and has been programmed to perform FFTs and dot products for spectral search systems. The HI-COMP-32 is a direct memory access (DMA) device to the host computer main memory. The array processor includes 4k × 32-bit 300-nsec cycle time MOS memory. This is a single computer board device and fits directly in a Nova 4/X chassis. (It may be attached to other Nova computers via an expansion chassis.) A 2k × 32-bit fixed point interferogram can be transformed from a disk file in 420 msec, a 256k × 32-bit interferogram can be transformed in $8\frac{1}{2}$ min. The array processor provides a gain in computation time of a factor of 10–20.

Nicolet has a 24-bit precision array processor. This is also programmable and has 128k × 24-bit on-board memory. This on-board memory is directly addressable by the Nicolet 1280 host minicomputer and is thus in effect DMA memory. A 64k × 20-bit interferogram can be transformed in about 2 sec. Beckman instruments has also announced a 24-bit array processor, but no details are available at this writing.

Other high-speed devices are being constructed for FT–IR spectrometers. For example, Digilab has included a Special Function Multiplier (SFM) in its Qualimatic spectrometer data system. The SFM is not an array processor, but it does provide near array processor speeds. Apparently, this is a hard-wired device to calculate the products in the fast Fourier transform. Other manufacturers, such as Perkin-Elmer, offer similar devices because of the excellent performance and low cost.

Real-time Fourier transforms for rapid-scanning interferometers are not yet a reality; however, array processors are drastically reducing the computation times in Fourier spectrometry. These devices represent time savings by a factor equal to that realized by the fast Fourier transform algorithm. As technological advances in computer and semiconductor technology push the prices of components down, spectroscopists will see the development and availability of extremely powerful computation devices. The Digilab and Nicolet array processors currently sell for approximately $10,000 each. It can be expected that in the future devices with similar computational specifications will cost less than this amount, and the cost of even more powerful devices will not greatly exceed this figure.

REFERENCES

1. Block Engineering, A Division of Bio-Rad, Inc., 19 Blackstone Street, Cambridge, MA 02139.
2. Now Nicolet Analytical Instruments, 5225-1 Verona Road, P.O. Box 4508, Madison, WI 53711-0508.
3. Beckman-RIIC Ltd., Eastfield Industrial Estate, Glenrothes, Fife, KY7 4NG, Scotland.
4. J. N. A. Ridyard, *J. Phys.*, **28**, C2:62 (1967).
5. Digilab, A Division of Bio-Rad Inc., 237 Putnam Avenue, Cambridge, MA 02139.
6. Nicolet Analytical Instruments, 5225-1 Verona Road, P.O. Box 4508, Madison, WI 53711-0508.

7. IBM Instruments Inc., Orchard Park, P.O. Box 332, Danbury, CT 06810.
8. D. G. Cameron, J. K. Kauppinen, D. J. Moffatt, and H. H. Mantsch, *Appl. Spectrosc.*, **36**, 245 (1982).
9. Bomem Inc., 625 Marais Street, Vanier, Quebec, Canada G1M 2Y2.
10. H. L. Buijs, Bomem Inc., 625 Marais Street, Vanier, Quebec G1M 2Y2, Canada, personal communication.
11. Analect Instruments, A Division of Laser Precision Corporation, 1731 Reynolds Avenue, Irvine, CA 92714.
12. Perkin-Elmer Corporation, Main Avenue (MS-12), Norwalk, CT 06856.
13. Beckman Instrument, Inc., Scientific Instruments Division, Campus Drive at Jamboree Boulevard, P.O. Box C-19600, Irvine, CA 92713.
14. Mattson Instruments, Inc., 6333 Odana Road, Madison, WI 53719.
15. H. Yoshinaga, S. Fujita, S. Minami, Y. Suemoto, M. Inoue, K. Chiba, K. Nakano, S. Yoshida, and H. Sugimari, *Appl. Optics,* **5,** 1159 (1966).
16. J. E. Hoffman, *Appl. Optics,* **8,** 323 (1969).
17. F. Levy, R. C. Milward, S. Bras, and R. Letoullec, Aspen Int. Conf. on Fourier Spectrosc., 1970 (G. A. Vanasse, A. T. Stair, and D. J. Baker, eds.), AFCRL-71-0019, p. 331 (1971).
18. Societe Coderg, 15 Impasse Barbier, 92 Clichy, France.
19. Polytec Gmbh., 7501 Wettersback-Karlsruhe, W. Germany.
20. P. Connes and G. Michel, Aspen Int. Conf. on Fourier Spectrosc., 1970 (G. A. Vanasse, A. T. Stair, and D. J. Baker, eds.), AFCRL-71-0091, p. 313 (1971).
21. J. N. Berbe and H. L. Buijs, Description of a High Speed Vector Processor, in *Minicomputers and Large Scale Computations*, (Peter Lykos, ed.), ACS Symposium Series, **57**, 106 (1977).
22. M. L. Forman, W. H. Steele, and G. A. Vanasse, *J. Opt. Soc. Am.,* **56,** 59 (1966).

SIGNAL-TO-NOISE RATIO

I. DETECTOR NOISE

The most basic and unavoidable of all types of noise in a spectrum measured using a Fourier transform spectrometer is detector noise. Every instrument should be designed so that detector noise is greater than noise from all other sources combined.

The sensitivity of infrared detectors is commonly expressed in terms of the *noise equivalent power* (NEP) of the detector, which is the ratio of the root mean square (rms) detector noise voltage, V_n, in volts $Hz^{-1/2}$, to the voltage responsivity, R_v, in volts per watt, that is,

$$\text{NEP} = \frac{V_n}{R_v} \quad \text{W Hz}^{-1/2} \tag{7.1}$$

The NEP is dependent on the area of the detector, A_D, and its *specific detectivity* D^*, where D^* is a measure of the sensitivity of a given detector and is, to a first approximation, independent of the area of the element. Generally,

$$D^* = \frac{(A_D)^{1/2}}{\text{NEP}} \quad \text{cm Hz}^{1/2} \text{ W}^{-1} \tag{7.2}$$

The *noise power* observed in a measurement time t (in seconds) is given by

$$N' = \frac{\text{NEP}}{t^{1/2}} \quad \text{W} \tag{7.3}$$

To determine the SNR obtainable in any measurement, we must know not only the noise power but also the power of the signal. The spectral energy density $U_{\bar{v}}(T)$ at wavenumber $\bar{v}$ from a blackbody source at a temperature T is given by the Planck equation

$$U_{\bar{v}}(T) = \frac{C_1 \bar{v}^3}{\exp(C_2 \bar{v}/T) - 1} \quad \text{W/sr cm}^2 \text{ cm}^{-1} \tag{7.4}$$

where C_1 and C_2 are the first and second radiation constants, having the values

$$C_1 = 2hc^2 = 1.191 \times 10^{-12} \text{ W/cm}^2 \text{ sr } (\text{cm}^{-1})^4 \qquad (7.5)$$

and

$$C_2 = \frac{hc}{k} = 1.439 \text{ K cm} \qquad (7.6)$$

The power received at a detector through any optical system is determined by the *throughput* of that system, that is, the product of the area of the beam and its solid angle at any focus. For an optimally designed FT–IR spectrometer, the throughput is determined by the area of the mirrors of the interferometer and the maximum allowed solid angle, which may be calculated using Eq. 1.41. The power received at a detector through an interferometer having a throughput Θ, a resolution $\Delta\bar{\nu}$, and an efficiency ξ, in unit wavenumber interval, is given by

$$S' = U_{\bar{\nu}}(T) \cdot \Theta \cdot \xi \cdot \Delta\bar{\nu} \qquad \text{W} \qquad (7.7)$$

Thus, the SNR of a spectrum measured using a Michelson interferometer is given by

$$\text{SNR} = \frac{S'}{N'} = \frac{U_{\bar{\nu}}(T) \cdot \Theta \cdot \Delta\bar{\nu} \cdot t^{1/2} \cdot \xi}{\text{NEP}} \qquad (7.8)$$

$$= \frac{U_{\bar{\nu}}(T) \cdot \Theta \cdot \Delta\bar{\nu} \cdot t^{1/2} \cdot \xi \cdot D^*}{A_D^{1/2}} \qquad (7.9)$$

It has been noted in Chapters 1 and 5 that the throughput can, in practice, be limited either by the maximum allowed solid angle of the beam or by the physical constraints of the optics, especially the detector size or the f-number of the detector foreoptics. The latter criterion is often encountered for low-resolution measurements, where a large throughput is allowed for the interferometer. In this case, the throughput is limited by the area of the detector and is given the symbol Θ_D, where Θ_D is equal to the product of the solid angle of the beam being focused on the detector, Ω_D (in steradians), and the detector area A_D; that is,

$$\Theta_D = A_D \Omega_D \qquad \text{cm}^2 \text{ sr} \qquad (7.10)$$

where Ω_D is given to a good approximation by [1]

$$\Omega_D = 2\pi\{1 - \cos \alpha_M\} \qquad (7.11)$$

where α_M is the maximum half-angle of convergence achievable from the detector foreoptics and rarely exceeds 45°, so that Ω is usually less than 2 sr.

When the solid angle of the beam through the interferometer is determined by the maximum wavenumber in the spectrum, $\bar{\nu}_{max}$, and the desired resolution $\Delta\bar{\nu}$, the solid angle of the beam through the interferometer, Ω_I, is given by

$$\Omega_I = \frac{2\pi \, \Delta\bar{\nu}}{\bar{\nu}_{max}} \qquad \text{sr} \qquad (7.12)$$

so that

$$\Theta_I = \frac{2\pi A_M \, \Delta\bar{\nu}}{\bar{\nu}_{max}} \qquad \text{cm}^2 \text{ sr} \qquad (7.13)$$

where A_M is the area of the mirrors in the interferometer being illuminated.

To determine whether Θ_I or Θ_D should be used in Eq. 7.7, both parameters should be calculated and the *smaller* one should be used. This is fairly easily done, and the procedure will be illustrated by using the parameters found on many Digilab FT–IR spectrometers:

$$A_D = 0.04 \text{ cm}^2 \text{ (2-mm-square detector)}$$

$$A_M = 20.3 \text{ cm}^2 \text{ (2-in.-diameter mirrors)}$$

$$\Delta\bar{\nu} = 1 \text{ cm}^{-1} \text{ (arbitrarily)}$$

$$\bar{\nu}_{max} = 4000 \text{ cm}^{-1} \text{ (for mid-infrared spectroscopy)}$$

$$\Omega_D = 1.5 \text{ sr } (f/1 \text{ condensing optics})$$

Thus,

$$\Theta_D = 0.04 \times 1.5 = 0.06 \text{ cm}^2 \text{ sr}$$

and

$$\Theta_I = \frac{2\pi \times 20.3 \times 1}{4000} = 0.03 \text{ cm}^2 \text{ sr}$$

Since Θ_I is less than Θ_D, the throughput for 1-cm^{-1}-resolution measurements is limited by the maximum allowed divergence of the beam through the interferometer rather than by the detector foreoptics. For measurements at 2 cm^{-1} resolution, Θ_D and Θ_I are both approximately equal to 0.06 cm^2 sr; this value is used in the calculations performed in the following paragraph. For lower-resolution spectrometry, Θ_D should be used in Eq. 7.7.

It is instructive to compare a reported SNR value measured using a commercial FT–IR spectrometer with the SNR calculated by inserting appropriate values for each parameter in Eq. 7.7. Foskett and Hirschfeld [2] reported that the SNR of the transmittance spectrum of a bandpass filter centered at 1690 cm^{-1} measured at 2 cm^{-1} resolution by averaging 1500 scans (0.79 sec/scan) on an early Digilab FTS-14 spectrometer was 61,000, see Fig. 7.1. This instrument incorporated a 2-mm-square TGS detector ($D^* \sim 1 \times 10^8$ cm Hz$^{1/2}$ W^{-1}) and a nichrome wire source operating at approximately 1000°C. From Eq. 7.4, $U_{1690}(1273) = 1.0 \times 10^{-3}$ W/sr cm^2 cm^{-1}. Since the reported spectrum was actually the ratio of a sample and reference spectrum measured using equal numbers of scans, the SNR of the single-beam spectrum, or of a transmittance spectrum measured using a completely noise free reference, would be $\sqrt{2}$ times the reported value, or 8.6×10^4.

The SNR calculated from Eq. 7.9, assuming an efficiency ξ of 0.1 (*vide infra* Section IV.a.(v) of this chapter) is

$$\text{SNR}_{\text{calc}} = \frac{1.0 \times 10^{-3} \times 0.06 \times 2 \times (1500 \times 0.79)^{1/2} \times 0.1 \times 1 \times 10^8}{(0.04)^{1/2}}$$

$$= 2.0 \times 10^5$$

The measured SNR was therefore 43% of the calculated SNR, which suggests that this early instrument was operating quite close to its theoretical performance. Mattson [3] later measured the SNR of the spectrum of another bandpass filter using a Nicolet 7199 spectrometer equipped with a Globar source operating at 1250°C and a mercury cadmium telluride detector. This instrument was operating under very low throughput conditions ($\Theta = 0.0024$ cm^2 sr) to avoid exceeding the dynamic range of the ADC, and Mattson found that the experimentally measured and calculated values of SNR were within about 25%.

Thus, Eq. 7.9 represents a fairly good way of rapidly estimating the SNR of any measurement. It should be noted that the reported noise is the rms value, which is about five times smaller than the peak-to-peak noise level.

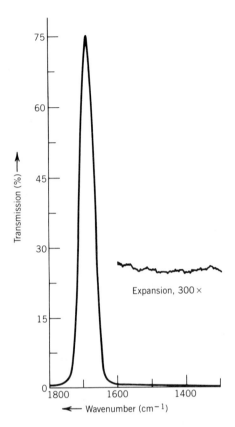

Fig. 7.1. Transmittance spectrum of a narrow-bandpass filter measured at 2 cm^{-1} resolution using a TGS detector and nichrome wire source obtained by averaging 1500 scans, indicating a SNR of 61,000 (rms). This result indicates a SNR (peak-to-peak) of about 400 for a single-scan spectrum in this region. It may be noted that this performance is quite difficult to match, even on contemporary instruments, unless an energy-limiting device, such as this filter, is placed in the beam. (Reproduced from [2], by permission of the Society for Applied Spectroscopy; copyright © 1977.)

For an instrument operating with just one detector of a given area (which is usually the case), the optimum performance is found when Θ_D = Θ_I, that is, when the detector is exactly filled by the image of the source while the beam passes through the interferometer with the maximum allowed solid angle. At this point, the optics are said to be *throughput matched*. Any further increase in throughput would result in the area of the image of the source at the detector exceeding A_D, so that the effective throughput would remain constant at Θ_D. The throughput of most commercial low-resolution FT–IR spectrometers, such as the Analect fX-6200, the Digilab FTS-50, and the Nicolet 5-DXB, is limited by the detector foreoptics. These instruments always operate with a *constant throughput* Θ_D, with $\Theta_D \leq \Theta_I$ for all resolution settings. For higher-resolution measurements, on the other hand, instruments are provided with interchangeable aperture stops (usually located at an intermediate focus in the source optics, see Figs. 5.7 and 5.8) to reduce the solid angle of the beam passing through the interferometer to the value given by Eq. 7.11. Measurements taken at high resolution are therefore measured with a *variable throughput* Θ_I, which must be decreased as the spectral resolution increases (or $\Delta\bar{\nu}$ decreases).

Hirschfeld has discussed the variation of SNR with throughput when the condition of retaining a detector of constant size is relaxed [4]. His conclusions are summarized in Fig. 7.2. For a detector of a certain size, the SNR will increase linearly with Θ until the diameter of the focused beam, d, equals the area of the smallest detector available. At this point, Eq. 7.11 gives that

$$d = \frac{1}{\pi} \left\{ \frac{2\theta}{1 - \cos \alpha_M} \right\}^{1/2} \tag{7.14}$$

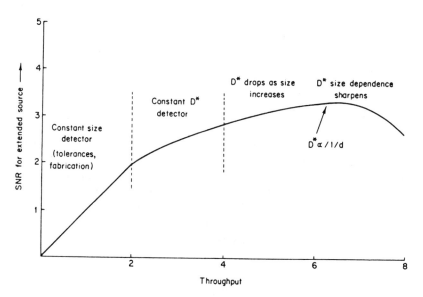

Fig. 7.2. Simulated variation of SNR with optical throughput for interferograms measured using an extended source; see text for details. (Reproduced from [4], by permission of Academic Press; copyright © 1979.)

Beyond this point, the SNR can be improved by using a detector of larger area. Since the NEP increases with $(A_D)^{1/2}$ (see Eq. 7.2), the SNR will increase with $\Theta^{1/2}$ if the diameter of the focused beam equals the diameter of the detector. The size of a detector cannot be increased ad infinitum, however, while maintaining a constant D^*. Above a limiting value of A_D, D^* starts to decrease, and for *very* large detectors,

$$D^* \propto \frac{1}{A_D^{1/2}} \tag{7.15}$$

At this point, there is actually a degradation in SNR on increasing the detector size to match the optical throughput.

In practice, pyroelectric detectors can be made with a constant D^* up to sizes in excess of 2 mm. This was not the case with MCT detectors before 1980; the D^* of most 2-mm MCT detectors was less than that of a 1-mm detector with the same λ_{max}. Recently, at least one detector manufacturer (Infrared Associates) has fabricated 2-mm MCT detectors with D^*'s equal to the D^*'s of 1-mm detectors with the same λ_{max}.

It should be pointed out that Hirschfeld's calculations are based on the assumption that the spectral energy density of a beam is uniform across a focal plane. This is probably not true in practice, since the energy density will be greater at the center of the focus than near an outer edge (cf. Chapter 5, Section I.e). In any event, the qualitative behavior illustrated in Fig. 7.2 is still useful in predicting the correct detector size to select for a given measurement.

II. "TRADING RULES" IN FT–IR SPECTROMETRY

The quantitative relationships between SNR, resolution, and measurement time are commonly referred to as "trading rules." The trading rules for FT–IR spectrometry are fairly simple for measurements made on a rapid-scanning interferometer operating with a constant mirror velocity. The effect of changing the mirror velocity and detector size may be somewhat more difficult to calculate. In the next sections, we will attempt to show how SNR may vary with parameters such as measurement time, resolution, throughput, and scan speed.

a. Effect of Measurement Time

For any spectrometer, conventional or interferometric, the SNR of a spectrum measured at a given resolution is proportional to the square root of the measurement time. The dependence of SNR on $t^{1/2}$ shown in Eq. 7.9 demonstrates that this relationship holds in FT–IR spectrometry. For measurements made with a rapid-scanning interferometer operating with a certain mirror velocity at a given resolution, SNR therefore increases with the *square root* of the number of scans being signal averaged. Note that the SNR of a spectrum measured using a single 1-sec scan may be increased by an order of magnitude by averaging 100 scans in a time of less than 2 min. To increase the SNR by another order of magnitude necessitates averaging 10,000 scans, which takes about 3 hr. A further 10-fold increase in SNR would require averaging one million scans and would take 11.5 days. (By this time several other factors would almost certainly limit the SNR.) If a measurement necessitates averaging more than 20,000 scans, there is probably a more efficient way for it to be performed, for example, by substituting a detector of higher D^* or possibly by reconfiguring the sample by multiple passing to yield a greater absorbance.

For measurements made on a slow-scanning interferometer, the measurement time may be increased either by signal averaging (usually spectra, rather than interferograms) or by decreasing the scan speed. When the scan speed is decreased, the time constant of the lock-in amplifier should, of course, be increased proportionately (see Section I.d of this chapter).

b. Effect of Resolution and Throughput

Consider now the effect on SNR of doubling the retardation for measurements made using a rapid-scanning interferometer. Let us assume that the measurement time t, the optical throughput Θ, the velocity of the moving mirror, v, and the duty cycle efficiency (see Chapter 4, Section I.c) all remain constant for the measurements at higher and lower resolution. These conditions imply that the lower-resolution measurement requires twice the number of scans needed for the higher-resolution scan. For this discussion, we define SNR as $100/N$, where N is the rms noise level on the 100% line produced by ratioing two single-beam spectra measured in the same times under identical conditions. From Eq. 7.9, we see that the SNR is proportional to $\Delta\bar{\nu}$; thus, the SNR is halved on doubling Δ_{max}. Since SNR $\propto t^{1/2}$, the measurement time required to achieve a certain baseline noise level must therefore be quadrupled each time Δ_{max} is doubled for measurements made at a constant optical throughput. The experimental verification of this conclusion is illustrated in Fig. 7.3 [5].

If measurements are made under the *variable-throughput* criterion, the throughput Θ_I is also halved when Δ_{max} is doubled (or $\Delta\bar{\nu}$ is halved), see

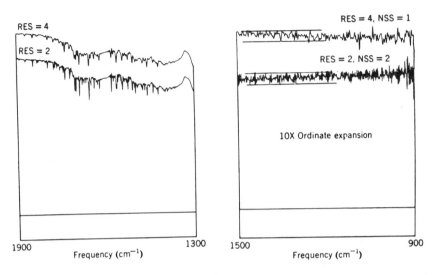

Fig. 7.3. (*Left*) Single-beam spectra measured at 2 and 4 cm^{-1} resolution (RES) and plotted in the region between 1900 and 1300 cm^{-1} to show the effect of resolution on the vibration–rotation spectrum of water vapor. Interferograms from which these spectra were calculated were measured with the same optical throughout, which for these measurements was limited by the area of the detector. (*Right*) Ordinate-expanded "100% lines" measured at 2 and 4 cm^{-1} resolution with the number of scans (NSS) indicated; the same number of scans was used for the sample and reference spectrum in each case. These data indicate an identical peak-to-peak noise level, as forecast by Eq. 7.9. (Reproduced from [5], by permission of the American Chemical Society; copyright © 1972.)

Eq. 7.13. This results in a further halving of the SNR, necessitating an *additional* fourfold increase in the number of scans to recover the original SNR. Thus, for FT–IR spectrometers operating under the variable-throughput criterion, the overall measurement time is increased by a factor of 16 when the retardation is doubled [5]. This conclusion has also been verified experimentally, see Fig. 7.4.

It is interesting to note that this is the same trading rule that applies for a grating spectrometer. In this case, when the spectral slit width is halved, the widths of both the entrance and exit slits of the monochromator must be halved, resulting in a fourfold reduction in energy at the detector. Thus, an increase in measurement time by a factor of 16 is again required to recover the SNR of the original measurement.

So far we have only considered the effect on the noise level of the baseline (or 100% line). The signal being measured is the peak *absorptance*, i.e. the percentage of the incident radiation being absorbed at the band center. The resolution may also affect the absorptance of the bands being measured. For example, the peak absorptance of a weak spectral feature whose FWHH is much less than the instrumental resolution will approximately double on doubling the retardation. If this feature were measured under the constant-throughput criterion, therefore, its effective SNR would be the same for measurements taken at the higher and lower resolution in equal measurement times. However, the degree of overlap by nearby spectral features will be reduced when the measurement is taken at a higher resolution, so the higher-resolution measurement is pref-

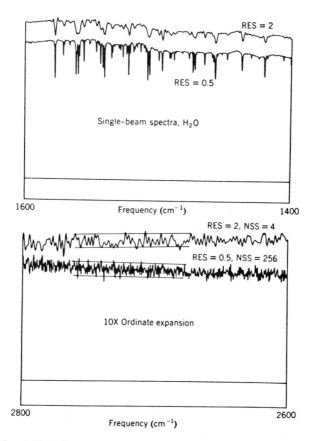

Fig. 7.4. (*Above*) Single-beam spectra measured at 0.5 and 2 cm^{-1} resolution and plotted between 1900 and 1300 cm^{-1}; the optical throughput was changed by a factor of 4 between these measurements. (*Below*) Ordinate-expanded "100% lines" measured at 2 cm^{-1} resolution with four scans and at 0.5 cm^{-1} resolution with 256 scans, indicating identical peak-to-peak noise levels as forecast by Eq. 7.9. (Reproduced from [5], by permission of the American Chemical Society; copyright © 1972.)

erable in this case. For weak *broad* spectral features whose peak absorptance does not change as the resolution is altered, measurement at the lower resolution is obviously preferable when the optical throughput is not varied. For strongly absorbing bands, whether they are broad or narrow, it is necessary to refer to Fig. 1.11 or 1.12 to determine how their apparent absorbance will be affected as the instrumental resolution is varied.

It is rather easy to show, using arguments of the type applied above, that the optimum SNR for *isolated* lines measured using the variable-throughput criterion is achieved in the measurement made at the lowest resolution. However, the usual reason for measuring a spectrum at high resolution (where the variable-resolution criterion almost invariably applies) is to effect the separation of the spectral lines of interest from nearby overlapping lines. Under these circumstances, the maximum analytical sensitivity is found for measurements made at the lowest resolution that adequately resolves the lines of interest from the interfering lines. Increasing the resolution *beyond* this point degrades the SNR.

One final point may be raised concerning the identification of weak bands in a spectrum. When the FWHH of a band is less than or equal to $\Delta \bar{\nu}$, the band will appear in the measured spectrum with a width approximately equal to the width of the ILS function. The period of the noise on the baseline cannot be any higher than the resolution (unlike the case for measurements made on a grating spectrometer) so that it becomes quite difficult to distinguish arbitrarily a weak real feature from a noise spike. Digitally smoothing the spectrum is of no help in this case, and the only way to ensure that the feature is real is to remeasure the spectrum using at least four times the number of scans, thereby reducing the baseline noise by a factor of 2 while leaving the signal unchanged.

c.   Effect of Apodization Function

For all the results derived above, it has been assumed that the apodization function has not been changed. However, it should be recognized that the process of apodization affects both the resolution and the noise of any spectrum. Perhaps the simplest example of how apodization affects the SNR of a spectrum is to compare the effect of changing from boxcar truncation to triangular apodization. The information contained in the interferogram around the centerburst determines the profile of the single-beam spectrum, and this low-resolution spectrum is changed little when the interferogram is multiplied by any apodization function. Thus, the signal and low-frequency noise are essentially the same whether the interferogram is apodized or not. On the average, the rms noise on this spectrum is decreased by a factor of $\sqrt{3}$ for triangular apodization [4].

The apparent absorbance of narrow bands will also be affected by the choice of apodization function, as shown on Figs. 1.11 and 1.12, with the bands in a spectrum computed with boxcar truncation always being more intense than bands in the spectrum of the same sample computed from the same interferogram after applying a triangular apodization function. The greatest difference for weak bands measured with and without apodization is found when the true FWHH of the bands is narrow. When the FWHH of the bands is much greater than the FWHH of the ILS function, the measured absorbance is about the same in apodized and unapodized spectra. In general, to obtain the optimum SNR for spectra of small molecules with resolvable fine structure, boxcar truncation is preferable if side lobes from neighboring intense lines do not present an interference; if they do, apodization definitely becomes necessary.

There is also no general rule to govern which particular apodization function is preferable. The effect of any apodization function is to decrease the noise on the baseline, with the extent of the noise suppression being determined by the shape of the apodization function. If the three preferred functions of Norton and Beer given in Eqs. 1.22 and Table 1.1 are considered, the greatest noise suppression will be obtained with the strong function, but the resolution (and hence the band intensities) will be greatest for the weak function.

d. Effect of Changing Mirror Velocity

The effect of changing the scan speed of a rapid-scanning interferometer may be inferred from Eq. 7.9. Let us first assume that the D^* of the detector does not change with modulation frequency (a situation that in practice rarely occurs). In this case, the measurement time *per scan* will be inversely proportional to the scan speed for a given retardation. Thus, if the scan speed is doubled, the SNR will decrease by a factor of $2^{1/2}$, and the SNR of the measurement made at the faster scan speed may be recovered by signal averaging two scans. The actual time required for the measurement will more than double because the *duty cycle efficiency* of the interferometer, that is, the ratio of the time between the start and end of data acquisition for one scan to the time between the start of successive scans, will decrease as the scan speed increases. This effect must be included in the calculation of the efficiency ξ.

At first glance, therefore, little might appear to be gained by increasing the mirror velocity for any measurement. In practice, however, the assumption that D^* does not vary with modulation frequency is not true, and for most photodetectors, D^* increases with modulation frequency. Thus, measurements made with an MCT or InSb detector are better carried out with a higher mirror velocity (to raise the modulation frequencies, $f_{\tilde{v}} = 2v\bar{v}$, in the interferogram). The variation of D^* with modulation frequency for a mercury cadmium telluride detector is shown in Fig. 5.35. This behavior may be contrasted with the behavior of most thermal detectors, including the DTGS pyroelectric bolometer, for which D^* *decreases* as the modulation frequency increases, see Eq. 5.10. Measurements made with photodetectors are therefore optimally performed with a high mirror velocity, giving Fourier frequencies well above 1 kHz. The limit on the scan speed is usually determined by the maximum data acquisition rate of the ADC or data system but can also be set by the reduced duty cycle efficiency obtainable at high scan speeds.

Using the converse of these arguments, it can be seen that measurements made with pyroelectric detectors should be made with as low scan speeds as possible, with the lower limit being determined by

1. the SNR of the interferogram building up to the point that it exceeds the dynamic range of the ADC,
2. the low-frequency ($1/f$) noise from the amplifier becoming dominant, and
3. other low-frequency interferences (e.g., building vibrations or vibrations of the mirrors in the optical train) being found in the same frequency range as the Fourier frequencies generated by the interferometer.

Because of 1, many mid-infrared absorption spectra measured with an intense Globar source, high throughput optics, and a DTGS detector cannot be made with a mirror velocity of less than a few millimeters per second. A commonly used mirror velocity for standard Michelson interferometers is 1.58 mm sec^{-1}, which leads to a data acquisition rate of 5 kHz if each wave of a 632.8-nm He–Ne laser is used to trigger data acquisition. For *far*-infrared spectrometry performed using a DTGS detector, on the other hand, the energy emitted by a mercury lamp source is low enough that one is rarely limited by the dynamic range of the ADC, and substantial gains have been reported when the mirror velocity is reduced well below 1.58 mm/sec.

III. OTHER SOURCES OF NOISE

a. Digitization Noise

As a general rule, the dynamic range of the A/D converter should always exceed the SNR of the interferogram in the region of the centerburst. This is particularly important for rapid-scanning FT–IR spectrometers, otherwise efficient signal averaging cannot be performed (see Chapter 2). Since many measurements on modern mid-infrared Fourier spectrometers are made with the peak-to-peak noise level changing only the least significant bit, it is often beneficial to increase the gain of the amplifier by at least a factor of 2 when the interferogram of a sample that does not have a high overall transmittance (either because of absorption, scattering, or vignetting) is being measured.

The use of gain-ranging amplifiers is prevalent in modern FT–IR spectrometers. In these devices, the gain may be switched in the time interval between two sample points by a factor g (which is usually 2^N, where N is an integer), a user-specified number of data points on either side of the centerburst. At the end of the run, that portion of the interferogram measured with the increased gain is divided by g. Since the amplitude of the interferogram of broadband sources dies down rapidly on either side of the centerburst, most of the interferogram is measured with the increased gain, only the few points around the centerburst being recorded with less than the optimum gain. Hirschfeld [4] has suggested that the best way of using gain ranging is to use a series of steps approximating the envelope of the interferogram. For the spectrum of a broadband source, gain ranging has been stated [4] to give an improvement in SNR over the measurement where no gain ranging was applied equal to

$$\frac{\text{SNR}_{\text{GR}}}{\text{SNR}} = 2^{1/2}[\text{int}(\log_2 g) + 1] \tag{7.16}$$

This principle has been applied in the Perkin-Elmer Model 1800 FT–IR spectrometer. When this instrument is equipped with a narrow-range mercury cadmium telluride detector, the SNR of the interferogram certainly exceeds the dynamic range of the ADC. However, when the amplifier is equipped with seven stages of gain ranging, a remarkably high performance is achieved. When the Jacquinot stop (see Chapter 5) is set to its minimum value and the mirror velocity is set to 0.75 cm (mechanical) $\sec^{-1}$, the rms noise level between 1100 and 1200 cm^{-1} on a 2-cm^{-1}-resolution spectrum calculated by ratioing two single-scan single-beam spectra was measured at 0.012%. The peak-to-peak noise level on this spectrum was 0.059%. The corresponding values measured for 100 scan spectra were 0.0011% (rms) and 0.0063% (peak-to-peak). The active data acquisition time was 0.35 sec/scan, so that the total acquisition time for 100 scans was less than 1 min. Assuming a band must have an absorbance of three times the peak-to-peak noise level, these data imply that a band of intensity 8×10^{-5} absorbance units is detectable in a measurement time of 1 min at this resolution. With a stored reference spectrum measured with 400 scans and a resolution of 4 cm^{-1}, the weakest band measurable in 1 min should have an absorbance of less than 3×10^{-5}. At the time of this writing, this is the highest performance reported for any commercial FT–IR spectrometer.

b.   Sampling Noise

If the interferogram is not sampled at precisely equal intervals of retardation, the signal that is measured is different from the signal that should have been sampled according to information theory, thereby increasing the noise level of the calculated spectrum. This source of noise is illustrated in Fig. 7.5 for a sinusoidal wave; it can be seen that errors in

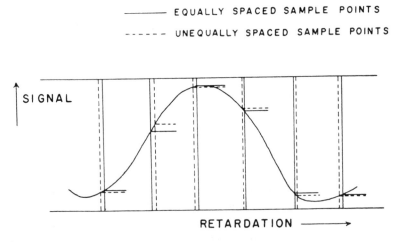

Fig. 7.5.   Simulated interferograms sampled at equally spaced (solid vertical lines) and unequally spaced sampling intervals (broken lines). The difference between the ordinate values of the correctly and incorrectly sampled interferograms is equivalent to noise superimposed on the true interferogram. The error is dependent both on the difference between the correct and the actual sampling positions and on the gradient of the interferogram at each point.

sampling of only a few percent can lead to substantial errors in the measured intensity of the wave.

The maximum SNR allowed for by a positional error Δl is given by [4]

$$SNR_{max} = \frac{4}{\Delta l \, \bar{\nu}_{max}} \qquad (7.17)$$

For $SNR_{max} = 10^3$ and $\bar{\nu}_{max} = 4000$ cm^{-1}, $\Delta l = 10$ nm. The need for the He–Ne reference laser found in most modern FT–IR spectrometers becomes obvious from this calculation. It may also be noted here that small temperature variations in the interferometer itself may lead to changes in the refractive index of the air and hence to small effective positional errors. Hirschfeld [4] has suggested that this is a good reason why interferometers should be thermostated. This source of noise is probably worse for slow-scanning interferometers than for rapid-scanning interferometers since signal averaging will compensate for errors of this type.

A second-order phenomenon may become important if the velocity of the mirror is not precisely constant even though sampling occurs at exactly equal intervals of retardation. In most interferometers, a servoloop senses variations in the frequency of the reference sinusoidal interferogram and adjusts the voltage to the drive transducer to restore the frequency to its set value. The infrared detector and preamplifier have a bandwidth just high enough to pass the infrared interferogram and prevent the high-frequency noise from being folded into the spectrum (*vide infra*). The response time of the infrared signal is therefore usually slower than that for the reference laser photodiode. If the mirror velocity is precisely constant, the time delay between the laser photodiode and the infrared detector does not introduce any sampling errors; however, if the velocity varies even by as little as 2%, a significant sampling error can be introduced that can degrade the SNR of the spectrum.

A technique suggested by Logan [6] for reducing the effect of this time delay involves matching the delays and offsets of filters for the laser reference and infrared signal channels. The filters must, of course, provide the optimum bandwidth for noise removal and at the same time have a uniform time delay as a function of signal frequency. A good way of testing system stability is to subtract consecutively measured interferograms [7]. The result of subtracting consecutive interferograms measured by Logan on an interferometer equipped with the standard electronics is shown in Figs. 7.6a, b. The residual difference interferogram has about 3% of the amplitude of the measured interferogram around the centerburst and is

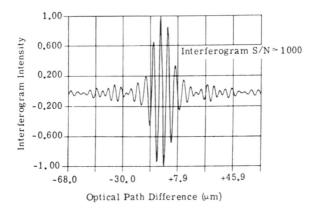

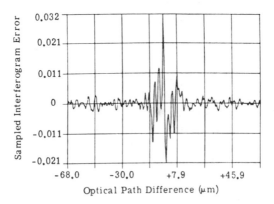

Fig. 7.6. (*Above*) Interferogram of a broadband source with a 1400–2500-cm^{-1} bandpass filter measured using conventional filtering, that is, a Butterworth four-pole electrical filter with break frequency of 800 Hz, a sampling rate of 1600 Hz, and a sampling interval of 0.3164 μm. The analog-peak-signal-to-peak-noise is approximately 1000. (*Below*) Difference between consecutive single-scan interferograms. This error is about 15 times greater than the intrinsic detector noise. (Reproduced from [6], by permission of the author.)

at least 15 times greater than the detector noise level. When the same system (source, interferometer, detector, and optical filter) was equipped with linear phase electronics and delays matched for dynamic compensation of velocity errors, the peak-to-peak amplitude of the difference interferogram was about four times smaller than the case when "standard" electronics were used, see Fig. 7.7.

When spectra were computed from interferograms measured with and without electrical filter compensation, and then the *spectra* computed from successive interferograms were subtracted, the difference spectra computed from interferograms measured with dynamic filter compensation were much flatter than difference spectra measured using standard electronics, see Figs. 7.8*a*, *b*. The differential-shaped features due to subtracting slightly frequency shifted water vapor lines seen in Fig. 7.8*a* are not observed in Fig. 7.8*b*, indicating that phase correction is superior in the latter case.

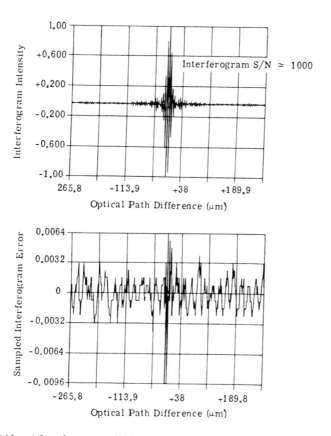

Fig. 7.7. (*Above*) Interferogram of the same source, filter, and detector used for Fig. 7.6 but different electronics; Bessel four-pole filter with break frequency of 800 Hz, delay line on laser detector sampling reference matched to Bessel filter delay, a sampling rate of 1600 Hz, and a sampling interval of 1.2656 μm. (*Below*) Difference between consecutive single-scan interferograms. Compared to Fig. 7.6, the error is substantially reduced by filter compensation and is comparable to the digitization noise. A low-frequency component with higher amplitude than the noise at these frequencies is also evident. (Reproduced from [6], by permission of the author.)

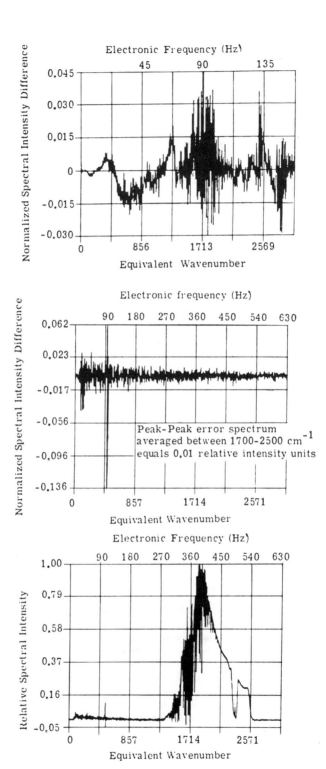

One final potential source of sampling noise can arise from the fact that most He–Ne lasers used to generate the reference interferogram do not emit a single mode, so that a very small sampling error may result. Multimode lasers will cause the wavenumber precision of most FT–IR spectrometers to be limited to about ± 0.002 cm^{-1}. This is quite adequate for many chemical FT–IR measurements but not quite enough for ultra-high-resolution spectrometry and possibly some very difficult measurements involving spectral subtraction of intense sharp features. In these cases, a single-mode laser should be used.

c.   Vibrations and Microphonics

Several detectors used in FT–IR spectrometry are microphonic and can pick up vibrations rather easily. If a single event occurs during a scan, a sinusoidal interference will be seen across the spectrum (see Chapter 2). On the other hand, if a sinusoidal vibration of frequency f is picked up throughout a scan (e.g., by a poorly damped mirror vibrating at its resonant frequency), then a spike or "glitch" will be seen at a wavenumber $\bar{v}$ equal to $f/2v$ (reciprocal centimeters) for a standard Michelson interferometer and $f/4v$ for a Genzel interferometer, see Fig. 7.9. If the phase of this vibration is not coherent with the phase of the interferogram, which is usually the case, its amplitude will decrease in the same way as the noise level, that is, with the square root of the number of scans being averaged. If the vibration is not purely sinusoidal but has some square-wave characteristics, glitches may also be observed at harmonics of $\bar{v}$, with the odd harmonics being more intense than the even harmonics.

The electrical line frequency (60 Hz in the United States) often falls in the range of modulated spectral frequencies, especially when far-infrared spectra are being measured. Thus, if cables are not well shielded or the instrument is not well grounded, a glitch may often be seen in the spectrum, not only at the fundamental frequency but also at its harmonics. A particularly severe example is observed if an ac power source is used.

Fig. 7.8. (a) Difference between two single-scan spectra computed from consecutive interferograms measured using the electronics described in the caption to Fig. 7.6. (b) Difference between two single-scan spectra computed from consecutive interferograms measured using the electronics described in the caption to Fig. 7.7. (c) Transform of the interferogram shown in Fig. 7.7. (Reproduced from [6], by permission of the author.)

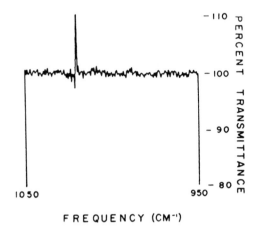

FREQUENCY (CM⁻¹)

Fig. 7.9. Typical appearance of a "glitch" in the spectrum. When the scan speed remains uniform and the oscillation being picked up has a constant frequency, the glitch can be very sharp (as in this case). It will often have a derivative shape, that is, upward going on one side and downward going on the other, since the phase of the imposed oscillation and that of the interferogram are usually different.

d. Folding

Excess noise in the high-frequency region of the spectrum can arise because of folding. It was stressed in Chapter 2 that for a sampling frequency of $2F$ hertz, not only should all *optical* modulations above the Nyquist frequency F be filtered out, either electronically or optically, but the *noise* frequencies above F should also be filtered out electronically. If electrical filtering is not performed efficiently, noise will be folded back into the spectrum below F.

In early rapid-scanning FT–IR spectrometers whose data storage capacity was usually limited, the Nyquist frequency for mid-infrared measurements was selected to correspond with one-quarter of the wavenumber of the He–Ne laser (3951 cm^{-1}). A sharp low-pass filter transmitting greater than 75% at 3800 cm^{-1} and less than 0.01% at 4000 cm^{-1} was used to prevent folding of the infrared signal. However, an electrical filter this sharp cannot be used without introducing a very rapid change in the phase spectrum around 3800 cm^{-1}. Therefore, in practice, a slower electrical filter had to be used, so that some high-frequency noise was folded back into the spectrum.

Modern FT–IR spectrometers have much more memory, so that the sampling frequency is usually chosen to be twice as great as it was a decade ago. In this case, twice as many interferogram points must be acquired for a spectrum of a given resolution. Since the Nyquist wavenumber is now 7902 cm^{-1}, both optical modulations and electrical noise are easily filtered out electronically and there is no need for sharp optical filters to be installed.

e. Fluctuation Noise

If the intensity of a signal varies, either due to fluctuations of the source output or detector response or to scintillation in the atmosphere between the source and detector, noise will be seen in the spectrum over the frequency range in which the variations occur. The effect is particularly important in remote-sensing studies and for measurements taken with slow-scanning interferometers using an external chopper when the source intensity varies slowly during the measurement. Contrary to the types of noise we have considered previously, which are additive, fluctuation noise is multiplicative and much more serious for Fourier spectrometry.

When fluctuation noise is much greater than detector noise, the changes caused by the fluctuation may negate the multiplex gain [8]. The magnitude of the effect of fluctuation noise has been given by Hirschfeld [4] as

$$\text{SNR} = \frac{f(\bar{v})\, \Delta\bar{v}\, \xi t^{1/2}}{[a^2 + b^2(\bar{v}_{\max} - \bar{v}_{\min})\overline{f^2}]} \tag{7.18}$$

where a is the detector noise, b is the relative fluctuation amplitude, $f(\bar{v})$ is the spectral energy density function, $\Delta\bar{v}$ is the resolution, $\bar{f}$ is the average value of $f(\bar{v})$, ξ is the spectrometer efficiency, and t is the measurement time (cf. the discussion at the start of this chapter).

For a boxcar spectrum, where $\bar{f} = f(\bar{v})$, the Fellgett advantage in the presence of fluctuation noise is given by

$$F_g = \left[\frac{1 + g^2}{M^{-1} + Mg^2}\right]^{1/2} \tag{7.19}$$

where g is the ratio of fluctuation to detector noise in each resolution element; that is,

$$g = \frac{b\bar{f}\, \Delta\bar{v}}{a} \tag{7.20}$$

and M is the number of resolution elements. The multiplex advantage is completely canceled when $F_g = 1$, which occurs when $g = M^{-1/2}$.

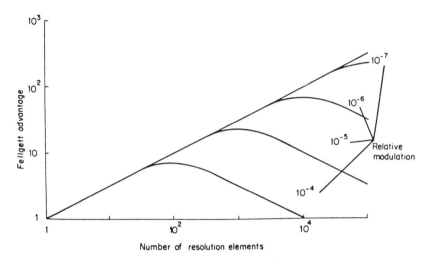

Number of resolution elements

Fig. 7.10. Effect of multiplicative noise on the Fellgett advantage as a function of the number of resolution elements for a spectral SNR of 100 and several different values of the relative fluctuation amplitude. (Reproduced from [4], by permission of Academic Press; copyright © 1979.)

Hirschfeld has shown the effects of various values of b on spectra measured with an SNR of 100 as M varies; his results are summarized in Fig. 7.10. It can be seen that even very small fluctuations (so small that they cannot be conveniently measured electronically) cause degradation in the SNR. Hirschfeld has suggested that the reduction in fluctuation noise is a major cause of the "5 o'clock effect," that is, the performance of some FT–IR spectrometers improving in the evening when many external perturbations are shut down.

No single technique will completely compensate the effects of fluctuation noise. The best way of minimizing this effect is to use a scan speed that causes the infrared wavelengths to be modulated at frequencies outside the frequency range of the perturbations. Another method, which is much less frequently used, is for both output beams from the interferometer, which are 180° out of phase, to be measured using two detectors [9]. The use of an analogous technique for reducing the effect of digitization noise will be discussed in the next chapter, but here we will discuss the application of dual-beam FT–IR spectrometry for reducing the effect of fluctuation noise.

We must first note that fluctuation noise affects both the ac and dc components of the interferogram, and its effect may be represented by multiplying both the ac and dc components of $I'(\delta)$ by a randomly varying function, $E(\delta)$. The general form of an interferogram exhibiting both fluctuation noise and detector noise [represented by $D(\delta)$] is

$$J(\delta) = I'(\delta)E(\delta) + D(\delta) \tag{7.21}$$

For a single-input beam, let the portion that is transmitted through the interferometer be referred to with the subscript t, whereas that reflected back to the source will be referred to with the subscript r. Thus, when no noise is present,

$$
\begin{aligned}
I'_t(\delta) &= \int_{-\infty}^{+\infty} 2R_{\bar{\nu}}T_{\bar{\nu}}I(\bar{\nu}) \, d_{\bar{\nu}} + \int_{-\infty}^{+\infty} 2R_{\bar{\nu}}T_{\bar{\nu}}I(\bar{\nu}) \exp(2\pi i \bar{\nu}\delta) \, d\bar{\nu} \\
&\equiv I_t + F(\delta)
\end{aligned}
\tag{7.22}
$$

where I_t is the constant component of the signal transmitted through the interferometer and $F(\delta)$ is the ac component. Similarly, for the reflected beam,

$$I'_r(\delta) = I_r - F(\delta) \tag{7.23}$$

where

$$I_r = \int_{-\infty}^{+\infty} (R_{\bar{\nu}}^2 + T_{\bar{\nu}}^2)I(\bar{\nu}) \, d\bar{\nu} \tag{7.24}$$

When additive noise and multiplicative noise are both present, the resulting interferograms are given by

$$J_t(\delta) = I_t E(\delta) + F(\delta)E(\delta) + D_t(\delta) \tag{7.25}$$

$$J_r(\delta) = I_r E(\delta) - F(\delta)E(\delta) + D_r(\delta) \tag{7.26}$$

The difference between $J_t(\delta)$ and $J_r(\delta)$ augments the ac interferogram but reduces the value of the constant terms; that is,

$$J_d(\delta) = J_t(\delta) - J_r(\delta) \tag{7.27}$$

$$= (I_t - I_r)E(\delta) + 2F(\delta)E(\delta) + D_t(\delta) - D_r(\delta) \tag{7.28}$$

For an ideal beamsplitter, $I_t = I_r$, so that the first term of Eq. 7.28 disappears, giving a substantial increase in the SNR. Even when a beamsplitter with quite low efficiency is used, this background term becomes much smaller. For an ideal beamsplitter,

$$J_d(\delta) = 2F(\delta)E(\delta) + D_t(\delta) - D_r(\delta) \tag{7.29}$$

If the NEPs of both detectors are equal, the detector noise level $[D_t(\delta)$ $- D_r(\delta)]$ will be greater than $D_t(\delta)$ or $D_r(\delta)$ by a factor of $\sqrt{2}$, while the amplitude of the ac component of the interferogram has doubled. Thus, even if fluctuation noise is small with respect to detector noise, the use of dual-beam techniques with two detectors will increase the SNR of the spectrum by a factor of $\sqrt{2}$.

f. Shot Noise

The primary benefit of Fourier spectrometry over the use of a scanning monochromator, the multiplex advantage, is only fully valid when the noise level is independent of the signal level. When the detector is photon shot noise limited (as it generally is for a photomultiplier), the noise level will be proportional to the square root of the total signal. For a boxcar spectrum, this means that shot noise is proportional to the square root of the number of resolution elements. This disadvantage therefore precisely offsets Fellgett's advantage when continuous broadband sources are employed. It may also be noted that the Jacquinot advantage is likewise reduced to the square root of the ratio of the throughputs of the interferometer and monochromator. These considerations help to explain why Fourier spectrometry has not become popular for the measurement of molecular spectra in the ultraviolet–visible region.

Hirschfeld [10] has proposed that for measurements of atomic emission spectra, where discrete lines are observed, shot noise will be distributed across the entire spectrum. Thus, for emission spectrometry of weak emission lines in the presence of several stronger lines, the SNR will almost certainly be degraded when compared to the same measurement made in the absence of the strong lines. This is also very important for Raman spectrometry, where shot noise caused by the very intense Rayleigh line would almost certainly mask all but the strongest Raman bands.

For the measurement of an emission spectrum consisting of few weak lines and no very intense features, the distribution of shot noise would appear to be an advantage. However, in one of the few experiments in which this hypothesis might be checked Horlick, Hall, and Yuen [11] found that the noise level in the region of an isolated emission line (as estimated by the standard deviation of several measurements of the peak intensity of the line) was substantially greater than the baseline noise. The source of this noise is probably related to source instability. This effect is also discussed in Chapter 16.

g. Sampling Errors

The rapid development of interferometers and data systems over the past decade has meant that sampling errors due to missed data points are now

rarely encountered during data acquisition. A missed point in the interferogram of a monochromatic source results in a very rapid change of
phase angle for the remaining points of the interferogram, see Fig. 7.11.
Roland [12] has computed how the theoretical ILS function will change
when an abrupt phase change occurs. For discrete emission spectra, it is
possible that two peaks could be observed instead of one, with one of the
peaks being negative under certain circumstances. The effect is greatest
if the missed point is close to the centerburst.

Horlick and Malmstadt [13] demonstrated the effect of missed data
points experimentally using an interferometer operating in the visible region of the spectrum. Their spectra were rather noisy, making it difficult
to distinguish between noise generated because of sampling errors and
detector noise. Thus, a detailed comparison of their results with Roland's
predictions was not possible, but it is certainly true that the deterioration
of measured spectra was considerably worse if sampling errors were made
close to the centerburst than if the points were missed close to the end
of the scan.

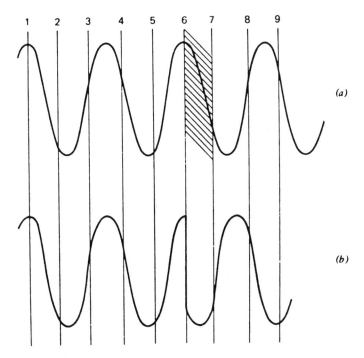

Fig. 7.11. The effect of missing one sample point. A sharp discontinuity is imposed on the
interferogram, which leads to both a phase error and an increased noise level. In general,
the closer the error occurs to the centerburst, the greater is the effect on the spectrum.

IV.   INTERFEROMETERS VERSUS GRATING SPECTROMETERS

a.   Theoretical Calculations

As mentioned earlier in this chapter, several studies have demonstrated that commercial FT–IR spectrometers are operating close to their theoretically calculated performance. Perhaps more important to the spectroscopist who has to decide whether to purchase an FT–IR or a grating spectrometer is the *relative* performance of these instruments. In 1977–1978, several comparisons of the theoretical and practical performance of commercial grating and FT–IR spectrometers were published [3,14–16], and the conclusions will be summarized below both from a theoretical and a practical viewpoint.

(i)   Jacquinot's Advantage

The maximum allowed throughput of a Fourier spectrometer is

$$\Theta_I = 2\pi A^1 \frac{\Delta \bar{\nu}}{\bar{\nu}_{max}} \qquad cm^2 \ sr \qquad (7.30)$$

where A^1 is the area of the interferometer mirrors being illuminated. The throughput of a grating spectrometer is

$$\Theta_G = \frac{hA^G \Delta \bar{\nu}}{f a \bar{\nu}^2} \qquad cm^2 \ sr \qquad (7.31)$$

where A^G is the area of the grating being illuminated, f is the focal length of the collimating mirror, h is the slit height, and a is the grating constant. Jacquinot's advantage is therefore given by

$$\frac{\Theta_I}{\Theta_G} = \frac{2\pi A^1 f a \bar{\nu}^2}{A^G h \bar{\nu}_{max}} \qquad (7.32)$$

This equation shows that there is a $\bar{\nu}^2$ dependence for Jacquinot's advantage if neither the beamsplitter nor the grating is changed during the measurement. Since most mid-infrared grating spectrometers have automatic grating interchanges (typically at 2000 and 667 cm^{-1}), the grating constant a may change as $\bar{\nu}$ is decreased, so that a plot of Θ_G versus $\bar{\nu}$ will show several discrete changes. In Fig. 7.12 the ratio of the calculated throughput of the Digilab FTS-14 and Beckman 4240 spectrometers operating at 2 cm^{-1} resolution is plotted as a function of wavenumber. It can be seen that over much of the spectrum, Θ_I is substantially superior

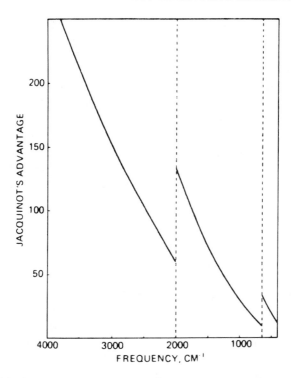

Fig. 7.12. Calculated variation of Jacquinot's advantage between a Digilab FTS-14 Fourier transform spectrometer and a Beckman Instruments Model 4240 grating spectrometer, both operating at 2 cm^{-1} resolution. The broken lines indicate grating changes for the monochromator. (Reproduced from [15], by permission of the Society for Applied Spectroscopy; copyright © 1977.)

to Θ_G, but at low wavenumbers, the use of an optimized grating makes Jacquinot's advantage quite small [15].

For FT–IR spectrometers where the throughput is limited by the detector foreoptics (the constant-throughput case discussed in Section II.b of this chapter), Jacquinot's advantage may be considerably less than the value shown in Fig. 7.12. For example, if the resolution is changed from 2 to 8 cm^{-1}, the throughput of the grating spectrometer can increase by a factor of 16, while Θ_I remains unchanged.

(ii) Fellgett's Advantage

The multiplex advantage of Fourier spectrometers can be expressed in the following way: The SNR of spectra measured on a Fourier spectrometer will be greater than the SNR of the spectrum measured in the same time and at the same resolution on a grating spectrometer with the same source, detector, optical throughput, optical efficiency, and modulation frequency by a factor equal to $\sqrt{M}$, where M is the number of resolution elements. For mid-infrared spectra, with $\bar{\nu}_{max} = 4000$ cm^{-1} and $\bar{\nu}_{min} = 400$ cm^{-1}, measured in equal times at a resolution of 2 cm^{-1}, Fellgett's advantage will be $\sqrt{1800}$, or 42.

The time advantage of a Fourier spectrometer should be even larger than the sensitivity advantage, being *directly* proportional to M. Thus, all other parameters being equal, a 2-cm^{-1}-resolution spectrum taking 30 min to measure on a grating spectrometer should be measured at equal sensitivity in 1 sec on a Fourier spectrometer.

(iii)   Detector Performance

Combining the values of Jacquinot's advantage and Fellgett's advantage given above, we should find commercial FT–IR spectrometers to be about 2000 times more sensitive than grating spectrometers, but smaller advantages are usually found in practice. To understand why, we must consider other components of the spectrometers. The most important among these is the detector.

The NEP of thermocouples presently used in most grating spectrometers is typically 1×10^{-10} W Hz$^{-1/2}$ at the modulation frequency of 15 Hz imposed by the chopper in these instruments [15]. This number increases rapidly as the modulation frequency is increased, so higher chopping rates are rarely used. For this reason, thermocouples are never used in modern rapid-scanning FT–IR spectrometers, where modulation frequencies can exceed 1 kHz.

The only ambient temperature detectors covering the entire mid-infrared spectrum with a rapid enough response time to be used with modern FT–IR spectrometers are the pyroelectric bolometers. The most sensitive pyroelectric detector is the DTGS, but even this detector is not as sensitive as a thermocouple. The typical NEP of a DTGS detector operating at 15 Hz is about 2×10^{-9} W Hz$^{-1/2}$, that is, about 20 times greater than the thermocouple. The NEP of the DTGS detector increases with modulation frequency, but not as rapidly as the NEP of a thermocouple. Assuming a mirror velocity of 0.158 cm sec^{-1}, the variation in the NEP of a DTGS detector with wavenumber is shown in Fig. 7.13. Comparing Figs. 7.12 and 7.13, it can be seen that the value of Jacquinot's advantage is almost exactly compensated by the detector *disadvantage* for a standard commercial Fourier spectrometer.

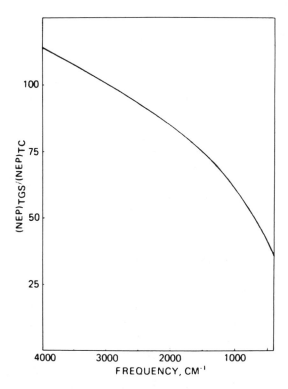

Fig. 7.13. Ratio of the noise equivalent powers of a TGS pyroelectric bolometer of the ′ type used on FT–IR spectrometers and a thermocouple of the type used on grating spectrometers. The mirror velocity for the interferometer was chosen as 1.58 mm sec^{-1} (5 kHz He–Ne laser frequency), which is a commonly used scan speed for interferometers equipped with TGS detectors. Note that this disadvantage of FT–IR spectrometers equipped with TGS detectors largely offsets the magnitude of Jacquinot's advantage shown in Fig. 7.12. (Reproduced from [15], by permission of the Society for Applied Spectroscopy; copyright © 1977.)

(iv)   Sources

The most commonly used source in both grating and FT–IR spectrometers is the Globar, so that this component has little effect on the relative performance of the two instruments. It may be noted, however, that before about 1975 many commercial FT–IR spectrometers were supplied with nichrome wire sources, which operate at lower temperatures than the Globar and also have lower emissivity. The performance of instruments equipped with nichrome wire sources was considerably poorer than that of modern FT–IR spectrometers equipped with a Globar source.

(v) Efficiency

One very important factor contributing to the relative efficiencies of FT–IR and grating spectrometers is sometimes included in the Fellgett advantages as $\sqrt{M/x}$, where x has been reported to be anywhere from 8 [17] to 1 [3]. The origins and magnitude of x is still a subject of considerable controversy [4]. The situation is made even more complicated because some workers include a contribution from the beamsplitter efficiency in the value of x by defining beamsplitter efficiency as $2RT$ rather than $4RT$ (see Chapter 4, Section II).

A good description of all the factors contributing to the efficiency ξ of an FT–IR spectrometer has been given by Mattson [3]. He measured the effect of many different parameters separately and found the overall value of ξ to be 0.096 for a Nicolet 7199. Factors included in Mattson's study were beamsplitter efficiency, Fresnel losses at the substrate and compensator, reflection losses at the mirrors, radiation obscured by the mounting hardware for the reference laser, the emissivity of the Globar source, and losses due to imperfect optical alignment.

The calculated efficiency of a perfectly flat Ge–KBr beamsplitter is superior to the efficiency of gratings used in commercial grating spectrometer above 800 cm^{-1} by anywhere from a few percent (at the maximum of the grating efficiency curves) to a factor of 4 (at the grating interchange at 2000 cm^{-1}). Below 600 cm^{-1}, the efficiency of a Ge–KBr beamsplitter designed to cover the entire mid-infrared spectrum falls below the efficiency of the grating being used in this region. Below 400 cm^{-1}, KBr can no longer be used as the substrate and CsBr or CsI must be used instead. These materials are often softer than KBr and can deform so that their efficiency may fall off at high wavenumbers.

(vi) Other Factors

A few other relatively minor parameters also affect the relative performance. For example, optical null grating spectrometers measure transmittance spectra in real time by alternating the beam between sample and reference channels while the grating is being slowly scanned. In FT–IR spectrometers, single-beam sample and reference spectra must be measured separately and ratioed subsequently. Duty cycle efficiencies of rapid-scanning FT–IR spectrometers can exceed 80% for measurements made with low scan speeds and at high resolutions. However, the duty cycle efficiency of several early low-resolution FT–IR spectrometers operating at high scan speeds was reported to be worse than 20%. By acquiring data during both directions of mirror travel (as described in Chapter 4, Section I.c), the duty cycle efficiency of modern instruments operating at very high speeds has been improved better than 80%. The duty cycle efficiency of most grating spectrometers, even when they are used in a signal-averaging mode, is generally better than 90% because of this low scan speed.

b. Practical Comparison

In 1977, the SNR of spectra measured at 2 cm^{-1} resolution on a Digilab FTS-14 was compared with that of a Beckman 4240 grating spectrophotometer [15]. The results of this study are summarized in Fig. 7.14. The theoretical advantage shown in Fig. 7.14a was calculated taking into account Fellgett's and Jacquinot's advantages, the relative performance of the detectors, the calculated beamsplitter and grating efficiencies, duty cycle efficiency of the FTS-14, and optical null advantage of the 4240. It did not take into account practical factors such as actual beamsplitter efficiency, optical alignment errors, and so on, and whether the factor x discussed above should be included with $\sqrt{M}$, the value assumed for Fellgett's advantage. The advantage curve determined experimentally, shown in Fig. 7.14b, is always about a factor of 4 below the theoretical curve.

If a value of $x = 8$ is assumed, as suggested by Tai and Harwit [17], the practical advantage then comes close to matching the theoretical advantage, as demonstrated by Fig. 7.14c. It may be noted, however, that Mattson [3] found that many practical factors contribute to the magnitude of ξ which, when combined, yield a value of less than 0.25. Therefore, we are still not convinced that a value of $x = 8$ should be assigned to the Fellgett advantage. A more probable value for x is 4, reflecting the fact that a chopper generates a square-wave modulation and an interferometer generates a sinusoidal modulation [18].

The practical performance of commercial mid-infrared Fourier spectrometers increased tremendously between 1969 (when the FTS-14 was introduced) and about 1977. This was in part because of the use of improved components, such as better sources and higher dynamic range ADCs, but also because the experience gained by the design engineers enabled several improvements in electronic circuitry to be effected that are not obvious to the chemical spectroscopist and rarely discussed in manufacturers' literature. Subsequent improvements have been relatively minor, indicating that we have reached the plateau of the learning curve. We can therefore predict that future improvements will be derived from advances in the design of specific components such as sources and detectors. In the next two or three years no major developments appear likely.

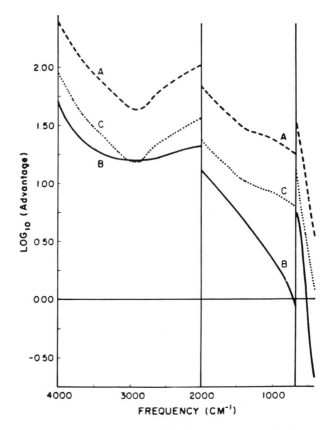

Fig. 7.14. (a) Theoretical advantage of an FT–IR spectrometer with the same optical parameters as a Digilab FTS-14 over an optical null grating spectrometer with the same optical parameters as a Beckman 4240, with both instruments operating at 2 cm^{-1} resolution. (b) Measured advantage of a Digilab FTS-14 over a Beckman 4240, with both instruments operating at 2 cm^{-1} resolution. (c) Theoretical advantage of an FT–IR spectrometer over a grating spectrometer, with the same instrumental parameters as curve (a), assuming that Fellgett's advantage is $(M/8)^{0.5}$ rather than simply $M^{0.5}$. (Reproduced from [15], by permission of the Society for Applied Spectroscopy; copyright © 1977.)

Finally, it should be noted that the comparison described above was made on the basis of equal data acquisition times, that is, equal times to complete the measurement of *spectra* on a grating spectrometer and *interferograms* on a Fourier spectrometer. The time required to compute a spectrum from an interferogram is no longer a significant fraction of the data acquisition time except in experiments where time resolution is needed and many interferograms must be measured sequentially (e.g., in kinetics and GC/FT–IR). In this case, the interferograms must be stored in some way and transformed at the end of the experiment unless the spectrometer is equipped with an array processor so that the transform times are very rapid.

In many laboratories, plotting the spectrum on a digital plotter *does* represent a time-consuming operation, especially if labeled axes and other parameters are to be plotted with the spectrum. On many instruments, computed spectra may be spooled and plotted when the data system is not tied up with high-priority operations such as data acquisition or computation, making the overall time required from insertion of the sample into the spectrometer to completion of a hard copy plot as long as it takes to measure a grating spectrum on preprinted paper. For laboratories involved in many rapid routine measurements, the use of preprinted chart paper on their FT–IR spectrometer is a very cost effective measure.

c. Far-Infrared Spectrometry

Since the frequency range covered by a single beamsplitter for far-infrared spectrometry is smaller than the range covered in mid-infrared measurements, Fellgett's advantage for a Michelson interferometer will be smaller than the factor of about 40 that was derived in the previous section. For far-infrared spectrometry at 2 cm^{-1} resolution between 450 and 80 cm^{-1}, Fellgett's advantage for FTS is approximately 14. If the interferometer is used in conjunction with a mechanical chopper (which is the case with slow-scanning far-infrared interferometers), this numerical factor would be reduced to about 10, since the measurement time is effectively halved.

Another factor working against the Michelson interferometer for far-infrared spectrometry is the poor efficiency of Mylar beamsplitters. It was seen in Chapter 4 that the relative beamsplitter efficiency of Mylar is greater than 50% only over a very small portion of the full frequency range usually covered in practice with a single beamsplitter. For a 6.25-μm-thick Mylar beamsplitter, whose peak efficiency is at 250 cm^{-1}, the efficiency is greater than 20% only between 110 and 390 cm^{-1} (for 45° incidence). This problem is offset in practice by the fact that grating efficiency curves in the far infrared are not much better than beamsplitter efficiency curves. The recent development of interferometers with an automatic beamsplitter interchange has been of greater benefit for far-infrared measurements than for any spectral region. With these instruments, the poor efficiency of Mylar beamsplitters no longer degrades the performance of far-infrared Fourier spectrometers significantly relative to that of grating monochromators operating over the same range.

Since it is difficult to focus the full throughput allowed for far-infrared spectrometry onto a small detector without being limited by the detector foreoptics, Jacquinot's advantage is also much smaller in this spectral region. It is for this reason that light cones are often used in conjunction with far-infrared FT–IR spectrometers, since they generally permit greater solid angles to be collected than do aspherical mirrors such as fast ellipsoids or paraboloids.

In spite of the fact that most of the individual advantages of interferometers for far-infrared spectrometry are significantly smaller than the values for mid-infrared spectrometry, the combination of Fellgett's and Jacquinot's advantages still allows an order of magnitude greater sensitivity than the corresponding grating monochromator. In such an energy-starved spectral region, this is still a very significant factor. In fact, when liquid-helium-cooled germanium bolometers are used, the SNR of far-infrared interferograms can approach the dynamic range of a 15-bit ADC.

One final advantage of interferometry over grating spectrometry that is particularly applicable to far-infrared spectrometry is the ease with which stray radiation can be filtered out. This factor is particularly important for very far infrared measurements for which second-order radiation from a grating may be only a few wavenumbers higher than radiation diffracted in the first order, so that second- and higher-order radiation is extremely difficult to filter out without severely attenuating the first-order radiation. With an interferometer, however, high-frequency radiation need not be filtered out provided the sampling interval is sufficiently short that high-frequency light is not folded back into the low-frequency spectrum; thus, all that is needed is electronic filtering to attenuate the high audio frequencies in the interferogram. It should be noted that the high-frequency information outside the frequency range of interest should not be so intense that the low-frequency components of the interferogram are not adequately digitized or that shot noise is generated. For this reason, black polyethylene filters are usually placed before the detector of far-infrared interferometers to ensure that most of the intense mid-infrared radiation emitted from the source is removed, and further attenuation is achieved through the use of electronic filtering for both rapid- and slow-scanning interferometers.

Black polyethylene has a fairly high transmittance in the far infrared, whereas the combined transmittance of the several filters required in a grating spectrometer is generally much lower. Among the filters that are used with a grating monochromator, the most common are transmission filters, such as the Yoshinaga [19] type or alkali halide plates for very far infrared spectroscopy, and reflection filters, such as scatter plates and *reststrahlen* plates; when used in combination, these filters rarely have a transmittance above 40%. The stray light in most far-infrared grating spectrometers is still usually rather high (especially when very long wavelengths are being measured), whereas the effect of stray light in far-infrared FT–IR spectrometry is generally negligible.

In summary, the combination of all the advantages of interferometry for far-infrared spectrometry are so great that very few dispersive spectrometers are used today for spectral measurements below 200 cm^{-1}.

REFERENCES

1. T. Hirschfeld, *Appl. Spectrosc.*, **31**, 471 (1977).
2. C. T. Foskett and T. Hirschfeld, *Appl. Spectrosc.*, **31**, 239 (1977).
3. D. R. Mattson, *Appl. Spectrosc.*, **32**, 335 (1978).
4. T. Hirschfeld, Chapter 6 in *Fourier Transform Infrared Spectroscopy: Applications to Chemical Systems*, Vol. 2 (J. R. Ferraro and L. J. Basile, eds.), Academic Press, New York (1979).
5. P. R. Griffiths, *Anal. Chem.*, **44**, 1909 (1972).
6. L. Logan, in *Multiplex and/or High Throughput Spectroscopy*, (G. A. Vanasse, ed.), *Proc. Soc. Photo-Opt. Instrum. Eng.*, **191**, 110 (1979).
7. J. A. de Haseth, *Appl. Spectrosc.*, **36**, 544 (1982).
8. T. Hirschfeld, *Appl. Spectrosc.*, **30**, 234 (1976).
9. W. J. Burroughs and J. Chamberlain, *Infrared Phys.*, **11**, 1 (1971).
10. T. Hirschfeld, *Appl. Spectrosc.*, **30**, 68 (1976).
11. G. Horlick, R. H. Hall, and W. K. Yuen, Chapter 2 in *Fourier Transform Infrared Spectroscopy: Applications to Chemical Systems*, Vol. 2 (J. R. Ferraro and L. J. Basile, eds.), Academic Press, New York (1979).
12. G. Roland, Ph.D. thesis, University of Liege, Liege, Belgium (1965).
13. G. Horlick and H. V. Malmstadt, *Anal. Chem.*, **42**, 1361 (1970).
14. P. R. Griffiths, R. G. Greenler, and N. Sheppard, *Appl. Spectrosc.*, **31**, 448 (1977).
15. P. R. Griffiths, H. J. Sloane, and R. W. Hannah, *Appl. Spectrosc.*, **31**, 485 (1977).
16. D. H. Chenery and N. Sheppard, *Appl. Spectrosc.*, **32**, 79 (1978).
17. M. H. Tai and M. Harwit, *Appl. Opt.*, **15**, 2664 (1976).
18. T. Hirschfeld, personal communication (1981).
19. Y. Yamada, A. Mitsuishi, and H. Yoshinaga, *J. Opt. Soc. Am.*, **52**, 17 (1962).

CHAPTER

8

REDUCTION OF DYNAMIC RANGE
IN FT–IR SPECTROMETRY

I. THE DYNAMIC RANGE PROBLEM

One of the more difficult measurements to make with an FT–IR spectrometer is the rapid detection of very weak bands, that is, those with an absorptance $[1 - \tau(\bar{\nu})]$ of 0.01% or less. If we approximate the single-beam spectrum of a broadband source as a boxcar (see Fig. 1.26) and limit the noise to this bandwidth, the signal-to-noise ratio of the spectrum, SNR_S, is given by

$$SNR_S = \frac{SNR_I}{\sqrt{M}} \tag{8.1}$$

where SNR_I is the SNR of the interferogram at the centerburst and M is the number of resolution elements. The maximum value of SNR_I for a single scan is determined by the dynamic range of the ADC. For a 16-bit ADC (15 bits + sign), SNR_I should not exceed about 2^{14}, or 16,384 (peak-to-peak rather than rms noise). For a spectrum measured at 4 cm^{-1} resolution between 4000 and 400 cm^{-1}, $\sqrt{M} = 30$, so that SNR_S is approximately 500.

Since the signal, that is, the band to be observed, must be about three times the peak-to-peak noise level, the minimum absorptance observable in a single scan is about 0.6%, which corresponds to an absorbance of 2.6×10^{-3}. The use of a gain-ranging amplifier (see Chapter 7) might reduce this value somewhat, but not by much more than a factor of 2. To reduce the minimum detectable absorbance to 1×10^{-4} a.u., it would be necessary, therefore, to average $(2.6 \times 10^{-3}/1.0 \times 10^{-4})^2$, or about 650, scans.

It is important to recognize that once the maximum value of SNR_I allowed by the ADC has been attained, no further improvement can be gained by increasing the intensity of the source or the detectivity of the detector. In fact, quite the reverse effect is observed, since if SNR_I exceeds the dynamic range of the ADC, signal averaging no longer causes the SNR to increase with the square root of the number of scans. Many experiments for which, at first glance at least, FT–IR spectrometry would appear to be well suited are in practice limited by the dynamic range of the ADC. Among these are the interface between a gas chromatograph and a Fourier spectrometer (GC/FT–IR), measurement of the spectra of monomolecular layers and partial monolayers of adsorbed molecules on flat metal surfaces, and the measurement of vibrational circular dichroism (VCD) spectra.

One method of increasing SNR_S is to reduce the spectral range by inserting an appropriate optical filter into the beam. Of course, the use of spectral filtering reduces Fellgett's advantage, but in cases where only a limited spectral range is of interest, filtering is a useful method of increasing the sensitivity of FT–IR measurements.

An alternative approach is to reconfigure the experiment in order to reduce or eliminate this dynamic range problem. As an example, let us consider the infrared characterization of trace quantities of a material prepared as a KBr micropellet. If a 1.5-mm-diameter pellet is prepared and mounted in a $4\times$ beam condenser, the SNR of an interferogram measured using a narrow-range MCT detector will usually exceed the dynamic range of a 16-bit ADC. If the same quantity of sample is pressed into a 0.5-mm-diameter pellet, the area of the sample (and hence the energy flux at the detector) is reduced by a factor of 9. In this case, the narrow-range MCT detector could be used without SNR_I exceeding the dynamic range of the ADC and signal averaging can be used to increase the SNR still further. In addition, the absorbance of the bands due to the analyte is increased almost 10-fold so that less ordinate expansion is needed and atmospheric interferences are reduced. For this experiment, it is apparent that the spectroscopist gains in just about every respect by reconfiguring the experiment.

For some measurements, however, it is not possible to overcome the problem of limited dynamic range so easily. In these cases it is necessary to resort to unusual optical techniques to suppress the centerburst of the interferogram. Several techniques allowing an interferogram to be measured that is due only to the small amount of radiation *absorbed* by the sample, rather than by the large amount of radiation *emitted* by the source, have been developed. Two techniques designed toward this end have received the most attention, although neither can yet be called popular. One is *polarization modulation* and the other is *optical subtraction* (or dual-beam) FT–IR spectrometry. Since in the opinion of the authors both these techniques have the potential to become much more widely used in the next decade, they will be described in some detail in the following sections.

II. POLARIZATION MODULATION

a. Vibrational Circular Dichroism

The measurement of difference spectra is often needed to study phenomena that cause only very minor perturbations to the spectra. As an example, the measurement of vibrational circular dichroism spectra may be cited. The VCD spectrum, ΔA_{CD}, is the difference in the absorbance of an optically active molecule to right and left circularly polarized radiation:

$$\Delta A_{CD} = A_R(\bar{\nu}) - A_L(\bar{\nu}) \qquad (8.2)$$

For most molecules exhibiting the VCD effect, ΔA_{CD} is typically four or five orders of magnitude less than either A_L or A_R. Thus, if ΔA_{CD} was to be measured by subtracting absorbance spectra measured using first right and then left circularly polarized radiation with a mid-infrared Fourier spectrometer operating at close to the digitization noise limit, at least 100,000 scans would have to be measured for all of the single-beam spectra needed for $A_R(\bar{\nu})$ and $A_L(\bar{\nu})$. This would take most of the day if the SNR of the VCD spectrum was determined by detector noise. Unfortunately, other problems can also manifest themselves that make the measurement of ΔA_{CD} impossible no matter how many scans are averaged. For example, if any component (such as the sample or polarizer) changes in position only marginally between the measurement of $A_R(\bar{\nu})$ and $A_L(\bar{\nu})$, a wavenumber shift can be caused (see Eq. 1.47). This shift would impose derivative-shaped features on the difference spectrum whose magnitude would in all likelihood exceed the VCD spectrum. Also, the use of He–Ne reference lasers, which are not mode locked in most commercial Fourier spectrometers, can lead to small shifts in the wavenumber scale (by as much as 0.005 cm^{-1}). Even such a small shift as this can lead to derivative-shaped features in the difference spectrum that are as large or larger than ΔA_{CD}. Thus, the measurement of VCD spectra using an FT–IR spectrometer in its conventional single-beam mode of operation is not feasible.

A way to avoid such problems is through the use of *differential spectrometry*, where the beam is rapidly modulated between the parent states. For VCD modulation between right and left circular polarization would be imposed, so that the drawback of measuring $A_R(\bar{\nu})$ and $A_L(\bar{\nu})$ at different times is eliminated. Such measurements make use of the well-known *ac advantage*, that is, that a small difference signal can be measured more accurately as the amplitude of a periodically varying ac signal than as the difference between two time-independent dc signals. The measurement of such ac signals is best achieved with a lock-in amplifier (LIA) tuned to the frequency of the modulation so that all contributions (random or coherent) at other frequencies are not detected.

Modulation spectrometry was first applied successfully for the measurement of VCD spectra on a grating spectrometer in 1974 [1]. In this and subsequent [2] experiments, radiation emerging from the exit slit of a monochromator was first linearly polarized and then passed into a *photoelastic modulator* (PEM) to create circularly polarized radiation in the manner originally proposed by Grosjean and Legrand [3,4]. This device consists of an isotropic optical element (ZnSe is commonly used in the infrared) that is periodically compressed and expanded by one or more piezoelectric transducers (PZT). The linear polarizer is set at 45° with respect to the orthogonal stress axes of the PEM so that equal intensities of polarized light lie along these axes. When the ZnSe optical element is stretched and compressed by the PZT, the indices of refraction of the ZnSe element along the stress axes vary sinusoidally. This causes the components of polarized light along the axes to be phase advanced or retarded with respect to one another. When the phase retardation limits are $\pm 90°$, modulation between left and right circular polarization is achieved. The beam emerging from the PEM is passed through the sample and is then focused on a photodetector. The VCD spectrum is measured as the monochromator is scanned.

The same type of experiment can be performed on an FT–IR spectrometer; in this case, Fellgett's and Jacquinot's advantages should improve the SNR above the best obtainable with a monochromator. Radiation from the source is modulated by a Michelson interferometer at frequencies between $2v\bar{v}_{max}$ and $2v\bar{v}_{min}$ (see Eq. 1.10); these frequencies are called the *Fourier frequencies* (see Chapter 1). As the radiation is modulated twice, once by the interferometer and again by the PEM, this is sometimes referred to as *double-modulation* spectrometry. The frequency of the PEM should be *at least* 10 times higher than the highest Fourier frequency, $2v\bar{v}_{max}$ [5], and it is preferable to separate them by an even greater amount. In a paper in which some excellent VCD spectra were shown, Nafie et al. [6] indicated that mirror velocities of 0.37 and 0.185 cm sec^{-1} were used, so that the Fourier frequencies for $\bar{v}_{max} = 4000$ cm^{-1} were 3.0 and 1.5 kHz, respectively. The modulation frequency f_m of the PEM used for these measurements was 70 kHz. It can also be seen from these numbers that the time constant τ of the LIA must be very short—preferably less than 1 msec—in order that the Nyquist sampling criterion is obeyed.

The low-frequency interferogram that would be measured in the conventional FT–IR experiment, called $V_{dc}(\delta)$ by Nafie [6,7], is given by (cf. Eq. 1.32)

$$V_{dc}(\delta) = \int_0^\infty B(\bar{v}) \frac{[10^{-A_R(\bar{v})} + 10^{-A_L(\bar{v})}]}{2} \exp(-2\pi i\bar{v}\delta) \, d\bar{v} \quad (8.3)$$

The PEM induces a wavenumber and time-dependent phase retardation angle, $\alpha(\bar{v}, t)$, on the beam. The magnitude of the intensity variation depends on the sine of $\alpha(\bar{v},t)$, given by [5]

$$\sin[\alpha(\bar{v},t)] = \sin[\alpha_0(\bar{v}) \sin(2\pi f_m t)] \quad (8.4)$$

where $\alpha_0(\bar{v})$ is the polarization phase shift at wavenumber $\bar{v}$. This equation can be expanded in a series of odd-order spherical Bessel functions:

$$\sin[\alpha(\bar{v},t)] = 2 \sum_{n,\text{odd}} J_n[\alpha_0(\bar{v})] \sin(2\pi n f_m t)] \quad (8.5)$$

The first term in the summation, $2J_1[\alpha_0(\bar{v})]$, corresponds to the fundamental frequency of the modulator, so that the ac interferogram measured as the output of the LIA is given by

$$V_{ac}(f) = \int_0^\infty \frac{B(\bar{v})}{2} [10^{-A_R(\bar{v})} - 10^{-A_L(\bar{v})}]2J_1[\alpha_0(\bar{v})]$$

$$\times \exp(-2v\bar{v}\tau) \exp(-2\pi i\bar{v}\delta) \, d\bar{v} \quad (8.6)$$

The function $\exp(-2v\bar{v}\tau)$ represents the decrease in signal imposed by the time constant of the LIA, τ.

Once the interferograms, $V_{ac}(\delta)$ and $V_{dc}(\delta)$, have been properly transformed to yield the spectra, $B_{ac}(\bar{v})$ and $B_{dc}(\bar{v})$, these spectra are ratioed to give

$$\frac{B_{ac}(\bar{v})}{B_{dc}(\bar{v})} = 2J_1[\alpha_0(\bar{v})] \exp(-2v\bar{v}\tau) \tanh\{\tfrac{1}{2} \ln 10[A_R(\bar{v}) - A_L(\bar{v})]\} \quad (8.7)$$

If ΔA_{CD} is small, which it always is for VCD spectra, and τ is sufficiently small that $\exp(-2v\bar{v}\tau)$ is approximately unity, we have

$$\frac{B_{ac}(\bar{v})}{B_{dc}(\bar{v})} = 2.3J_1[\alpha_0(\bar{v})] \Delta A_{CD} \quad (8.8)$$

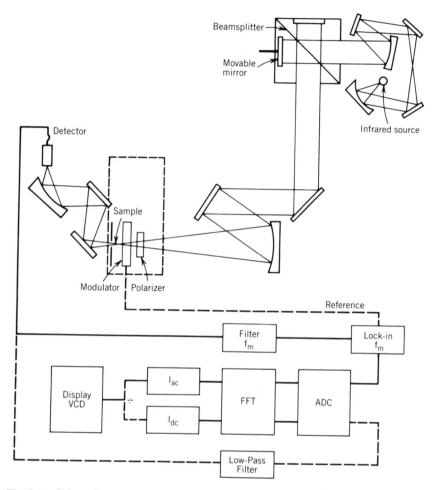

Fig. 8.1. Schematic representation of the vibrational circular dichroism FT–IR spectrometer proposed by Nafie and Diem. (Reproduced from [5], by permission of the Society for Applied Spectroscopy; copyright © 1979.)

The term $2J_1[\alpha_0(\bar{\nu})]$ accounts for the effectiveness of the PEM in producing circularly polarized radiation as a function of wavenumber. Its value can be determined experimentally by a calibration procedure based on the combined use of a birefringent plate followed by a polarizer at the sample position [7]. The phase lag imposed by the lock-in amplifier can also be determined using the same equipment. A block diagram of a Fourier transform VCD spectrometer is shown in Fig. 8.1.

Since the ac interferogram is composed of both positive and negative

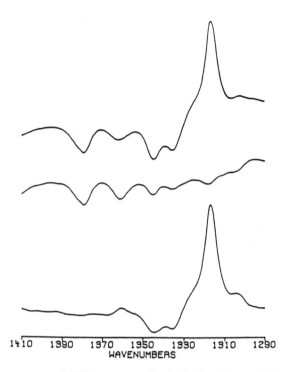

Fig. 8.2. (*Lower spectrum*) VCD spectrum of a 0.22 *M* solution of (+)-3-methylcyclo-hexanone in CCl₄ measured by subtracting the raw spectrum of the racemic mixture (*middle spectrum*) from that of the enantiomer (*upper spectrum*). (Reproduced from [8], by permission of the Society for Applied Spectroscopy; copyright © 1984.)

spectral intensities, the zero retardation point need not correspond to an intensity maximum in the interferogram. Under this circumstance, the standard Mertz and Forman methods of phase correction are usually believed to be inapplicable. One way of phase correction that was applied initially involved a separate measurement of the interferogram of a statically stressed ZnSe plate followed by a linear polarizer. The phase spectrum calculated from this interferogram was stored and used for the phase correction of VCD interferograms measured subsequently. Surprisingly, perhaps, VCD spectra computed in this way exhibited large derivative-shaped artifacts indicative of poor phase correction [7]. Even more surprisingly, use of the standard Mertz method for phase correction yields spectra in which the artifacts were much smaller [8].

The success of this approach appears to be due to the presence of a distinct centerburst in the ac interferogram. Lipp and Nafie [8] have attributed this feature to a polarization sensitivity of the optical system, which apparently displays a preference for one state of circular polarization over the other. The resultant spectra therefore exhibit a baseline offset, which must be corrected by subtracting the measured VCD spectrum of a racemic mixture of the same compound.

One of the most troublesome problems in VCD spectrometry is the appearance of absorption artifacts, which appear at the wavenumbers of all bands in the spectrum. These artifacts are observed in VCD spectra measured both using dispersive and FT–IR spectra. Like the baseline offset, they are eliminated by subtracting the spectrum of a racemic mixture from the spectrum of the enantiomer of interest, with both samples present at the same concentration [6,7], see Fig. 8.2. Many approaches have been investigated for the removal of these artifacts [8,9]. The most successful has involved the subtraction of a measured VCD spectrum of an enantiomer at concentration M from the spectrum of the same enantiomer at concentration $2M$. A comparison of a VCD spectrum measured using the two-concentration method with one measured using the racemic mixture is shown in Fig. 8.3.

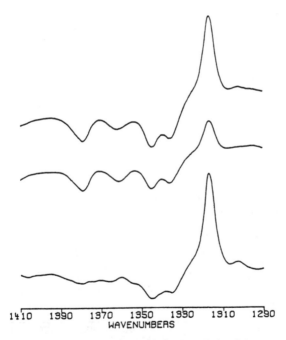

Fig. 8.3. (*Lower spectrum*) VCD spectrum of (+)-3-methylcyclohexanone measured by subtracting the raw spectrum of a 0.11 M solution (*middle spectrum*) from that of a 0.22 M solution (*upper spectrum*). (Reproduced from [8], by permission of the Society for Applied Spectroscopy; copyright © 1984.)

Spectra down to 800 cm^{-1} at resolutions up to 2 cm^{-1} have been measured by Lipp and Nafie using a BaF$_2$ polarizer [8]. Typical data are shown in Fig. 8.4. Polavarapu [10] has reported VCD spectra down to 600 cm^{-1} using a KRS-5 polarizer, see Fig. 8.5. In this work, the low-wavenumber limit was determined by the transmission of the ZnSe modulator.

Thus, it can be seen that the state of the art for VCD spectrometry has advanced to the point that spectra of single enantiomers can be measured

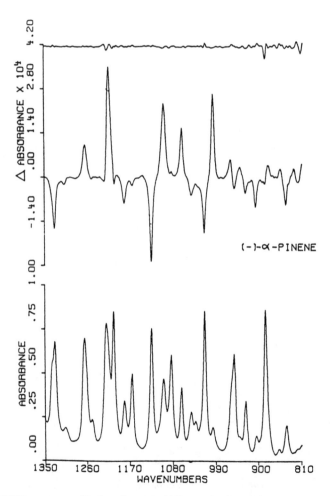

Fig. 8.4.   VCD spectrum of (−)-α-pinene (*middle trace*), shown along with the absorbance spectrum of this compound (*lower trace*) and an estimate of the noise (*upper trace*). (Reproduced from [8], by permission of the Society for Applied Spectroscopy; copyright © 1984.)

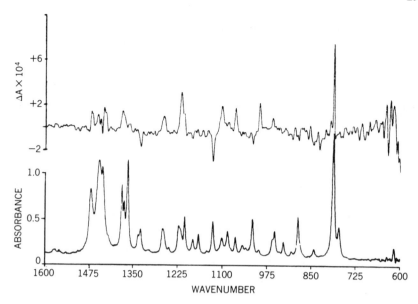

Fig. 8.5. (*Upper trace*) VCD spectrum of ($-$)-α-pinene measured to 600 cm^{-1} by Pola-varapu et al. (*Lower trace*) Absorbance spectrum of same compound. (Reproduced from [9], by permission of the Society for Applied Spectroscopy; copyright © 1984.)

with little ambiguity. Much of the early VCD data were on small molecules at high concentration, but now larger biochemically significant molecules are starting to be investigated. Back and Polavarapu [11] have reported the first spectra–structure correlations for sugars prepared as 1 M solutions in DMSO-d_6. It is surely only a matter of time before VCD spectra of biologically important molecules in H$_2$O will be reported.

b.   Linear Dichroism Spectrometry

The PEM used in the measurement of VCD spectra can also be used in the measurement of *linear dichroism* spectra. The magnitude of the intensity variation in the linear dichroism experiment depends on $\cos[\alpha(\bar{v}, t)]$ [5,7], where

$$\cos[\alpha(\bar{v},t)] = \cos[\alpha_0(\bar{v}) \sin(2\pi f_m t)] \tag{8.9}$$

On expansion, we have

$$\cos[\alpha_0(\bar{v}) \sin(2\pi f_m t)]$$

$$= J_0[\alpha_0(\bar{v})] + \sum_{n,\text{even}} 2J_n[\alpha_0(\bar{v})] \cos(2\pi n f_m t) \tag{8.10}$$

The term in Eq. 8.10 corresponding to $n = 0$ does not vary in time. The first term in the expansion, corresponding to $n = 2$, provides modulation of the difference spectrum at the frequency $2f_m$. The terms corresponding to higher values of n are rejected by the LIA. Thus, for measurement of linear dichroism spectra, the modulator must be set at a higher retardation level than for VCD measurements, corresponding to a greater value of $\alpha_0(\bar{\nu})$, and the LIA must be tuned to $2f_m$.

Although linear dichroism spectra of DNA measured using a PEM in conjunction with a monochromator have been reported [12], we believe that by far the most important application of linear dichroism spectrometry will be the measurement of the spectra of species adsorbed onto flat metallic surfaces. As discussed in Chapter 5, for radiation incident at a very high angle, the component of the electric field vector polarized parallel to the plane of incidence is absorbed to a much greater extent than the component polarized perpendicular to the plane of incidence. If a PEM is used to generate alternately parallel and perpendicular polarized radiation, the ac signal detected at $2f_m$ will be due to the difference in the absorption of radiation of each polarization.

The combined use of a PEM and an FT–IR spectrometer for surface analysis was first shown by Dowrey and Marcott [13] to be feasible. The optical arrangement used by these workers was relatively simple, consisting of a slightly modified reflectance accessory held in the sample compartment of an FTS-14 spectrometer, see Fig. 8.6. Spectra of a 1-nm-thick layer of cellulose acetate on a copper slide measured using a conventional single-beam technique and using a PEM are shown in Fig. 8.7. In each case, the angle of incidence was 75°. The SNR of the spectrum measured using a PEM was only slightly superior to the conventional R–A spectrum in this early work, with much of the advantage arising from the discrimination against atmospheric water vapor interference, but later reports are starting to indicate that greater improvements are possible.

For example, Golden et al. [14,15] have shown that the spectrum of monolayer coverages of cadmium arachidate on silver can be measured using a ZnSe PEM. For this work, a switching circuit was designed so that both the difference interferogram ($I_p - I_s$) and the sum interferogram ($I_p + I_s$) could be measured, where I_p and I_s are the interferograms measured with parallel and perpendicular polarized radiation, respectively, and $I_p - I_s$ is the interferogram due to the difference in absorption of surface species to parallel and perpendicular polarized radiation. The normalized spectrum computed by dividing the FT of $I_p - I_s$ by the FT of $I_p + I_s$ is proportional to the absorptance of the surface species.

The early data reported by Golden et al. were measured on an unpurged spectrometer and were consequently a little less than optimal. Neverthe-

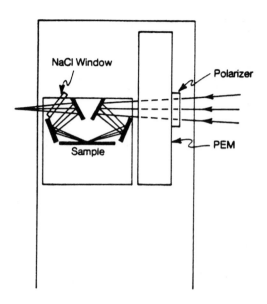

Fig. 8.6. Optical schematic diagram of the polarization modulation R-A experiment. As in the VCD experiment, the polarizer, PEM, and cell (in this case a small reflection accessory) all fit in the sample compartment of the spectrometer. The NaCl window is used to null the offset signal at the 74 kHz modulation frequency. (Reproduced from [13], by permission of the Society for Applied Spectroscopy; copyright © 1982.)

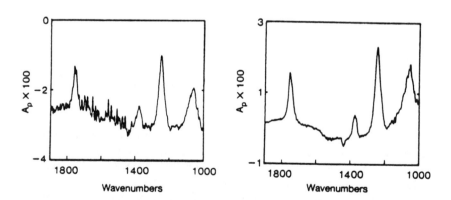

Fig. 8.7. (*Left*) Normal R-A spectrum of a 10-Å film of cellulose acetate on a polished copper slide versus the clean slide. (*Right*) Polarization modulation spectrum of the same sample. (Reproduced from [13], by permission of the Society for Applied Spectroscopy; copyright © 1982.)

295

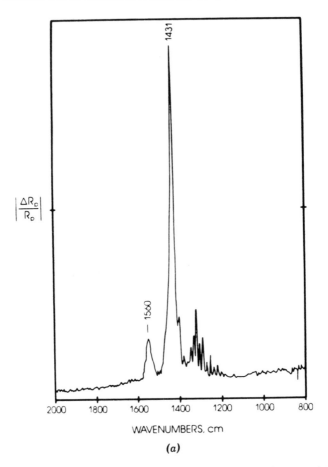

$\left|\dfrac{\Delta R_{\mathrm{c}}}{R_{\mathrm{D}}}\right|$

WAVENUMBERS, cm

(a)

Fig. 8.8. Polarization modulation R-A spectra of (*left*) six monolayers and (*right*) one monolayer of cadmium arachidate on silver measured at 4 cm^{-1} resolution. (Reproduced from [15], by permission of the American Chemical Society; copyright © 1984.)

less, they indicate the feasibility of characterizing monolayers of adsorbed species on flat substrates. The spectra of six layers and one monolayer of cadmium arachidate on silver are shown in Figs. 8.8a and *b*, respectively. In light of the remarkable improvement in VCD data shown by Nafie and Polavarapu over a period of only one or two years, we forecast that the spectrometry of surface species measured by polarization modulation techniques will show a corresponding improvement over the next two or three years.

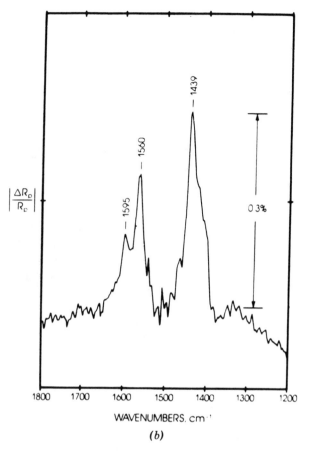

Fig. 8.8. (continued)

c. Sample Modulation Spectrometry

It should be noted here that the idea for superimposing a modulation of high frequency on a beam where the Fourier frequencies are low is not new. We have seen in Chapter 4 that many of the first FT–IR spectrometers used exactly this principle. For example, mechanical choppers were generally used to modulate radiation passing through slow-scanning far-infrared interferometers. Phase modulation was achieved by modulating the fixed mirror of the interferometer over a small amplitude with a PZT. Several other potentially interesting means of modulating the spectrum besides changing the polarization have been forecast [7]. In theory, "sample modulation'' could be achieved either by switching the beam between different samples, by changing the absorption of the sample by a perturbation such as an electric field, or by periodically initiating photochemical or electrochemical events in a cycle that involves relaxation back to equilibrium. Doubtless, many applications of this type will be reported in the next few years.

III. OPTICAL SUBTRACTION

Although polarization modulation represents a powerful way of overcoming the dynamic range problem of single-beam FT–IR spectrometers, it can only be used when some means of modulating the signal to discriminate between the two states of interest is available. If, for example, one needs to measure the spectrum of a molecule present in the vapor or liquid phase at low concentration, polarization modulation cannot be used. In this case, an alternative form of differential FT–IR spectrometry designed to suppress the centerburst of interferograms due to continuous sources may become applicable. This technique goes under a variety of names, such as *optical subtraction*, *dual-beam*, and *double-output* FT–IR spectrometry.

In its usual mode of operation, the conventional Michelson interferometer accepts one input beam of radiation, and two modulated output beams 180° out of phase emerge. One of the output beams (the transmitted beam) is passed to the detector along an axis perpendicular to the input beam; the other (the reflected beam) generally returns to the source along the same axis as the input beam. This latter beam is, of course, rarely measured because of the experimental difficulties in separating it from the input beam. Optical subtraction measurements may be performed in one of two ways. In one, the beams from two sources are passed into each arm of the interferometer, and one or both output beams are measured. In the other, both of the output beams from a single input beam are measured, usually at the same detector. A brief summary of the theory of optical subtraction FT–IR spectrometry is presented below. A more rigorous treatment of double-input and/or double-output FT–IR spectrometry has been given by Genzel et al. [16].

Let us consider only the case of a single input beam and two output beams. Using the nomenclature introduced in Chapters 1 and 7, we have

$$I'_t(\delta) = 2 \int_0^\infty R_{\bar{\nu}} T_{\bar{\nu}} I(\bar{\nu}) \, d\bar{\nu} + 2 \int_0^\infty R_{\bar{\nu}} T_{\bar{\nu}} I(\bar{\nu}) \, \exp(2\pi i \bar{\nu} \delta) \, d\bar{\nu} \quad (8.11)$$

and

$$I'_r(\delta) = \int_0^\infty (R_{\bar{\nu}}^2 + T_{\bar{\nu}}^2)I(\bar{\nu})\ d\bar{\nu} - 2 \int_0^\infty R_{\bar{\nu}}T_{\bar{\nu}}I(\bar{\nu})\ \exp(2\pi i\bar{\nu}\delta)\ d\bar{\nu} \qquad (8.12)$$

If both beams are measured at the same detector, we have

$$I'_t(\delta) + I'_r(\delta) = \int_0^\infty (R_{\bar{\nu}}^2 + T_{\bar{\nu}}^2 + 2R_{\bar{\nu}}T_{\bar{\nu}})I(\bar{\nu})\ d\bar{\nu}$$

$$= \int_0^\infty (R_{\bar{\nu}} + T_{\bar{\nu}})^2 I(\bar{\nu})\ d\bar{\nu} \qquad (8.13)$$

If there is no absorption by the beamsplitter, so that $R_{\bar{\nu}} + T_{\bar{\nu}} = 1$, we have the logical result

$$I'_t(\delta) + I'_r(\delta) = \int_0^\infty I(\bar{\nu})\ d\bar{\nu} \qquad (8.14)$$

that is, the total energy output is equal to the total energy input. Note that the amplitude of the modulated component of the interferogram is reduced to zero. In rapid-scanning FT–IR spectrometers, only the modulated component of $I'(\delta)$ is measured, so that in this case the centerburst would not be observed.

If a sample having a transmittance $\tau(\bar{\nu})$ is held in the reflected output beam and no sample is held in the transmitted beam, the amplitude of the ac component of the combined interferogram measured at the detector is

$$I_t(\delta) + I_r(\delta) = 2 \int_0^\infty R_{\bar{\nu}}T_{\bar{\nu}}[1 - \tau(\bar{\nu})]I(\bar{\nu})\ \exp(2\pi i\bar{\nu}\delta)\ d\bar{\nu} \qquad (8.15)$$

Therefore, the signal measured at the detector is

$$V_t(\delta) + V_r(\delta) = \int_0^\infty B(\nu)\ [1 - \tau(\bar{\nu})]\ \exp(2\pi i\bar{\nu}\delta)\ d\bar{\nu} \qquad (8.16)$$

The spectrum computed from this interferogram is $B(\bar{\nu})\ [1 - \tau(\bar{\nu})]$, where $B(\bar{\nu})$ is the single-beam spectrum computed from $V_t(\delta)$. The ratio of the spectra computed from the dual-beam interferogram and the single-beam interferogram is therefore the absorptance spectrum of the sample $[1 - \tau(\bar{\nu})]$ (cf. Eq. 8.8 for polarization modulation).

The weaker is the absorptance of the sample, the smaller is the dynamic range of the ADC required to digitize it, which is, of course, the reverse of the case encountered for single-beam FT–IR spectrometry. From Eqs. 8.15 and 8.16, it might be thought that the single instrumental criterion for optical subtraction spectrometry is that the transmitted and reflected beams must be identical in all respects except the presence of the sample. However, we will see later that a few other problems, such as detector response going nonlinear at the high radiation flux encountered in optical subtraction measurements, rarely allow the full advantages of the technique to be realized in practice.

Many different optical configurations have been suggested for optical subtraction FT–IR spectrometry. One of the first was described by Bar-Lev [17] who reported a system using a single source and two detectors measuring $V_t(\delta)$ and $V_r(\delta)$ individually, with the two signals being passed into a summing amplifier. Low [18] described an experimental arrangement designed for GC/FT–IR with two input beams and two measurable output beams, only one of which was actually measured. This configuration is reproduced in Fig. 8.9. Several problems with this design are evident. For example, the mirrors used to collect the output beams block a substantial fraction of each input beam. The beams passing into the interferometer converge at a fairly high angle, so that this arrangement could only be used for very low resolution measurements. Finally, the gas cells (light pipes) used for GC/FT–IR measurements are hot; they should therefore be located between the interferometer and the detector so that radiation emitted from the cells is not modulated by the interferometer and hence not detected by the audio-frequency amplifier of the detector.

In view of the importance of GC/FT–IR measurements and the limitations imposed on them by the dynamic range of the ADC [19], several other designs for optical subtraction GC/FT–IR measurements have been reported. In 1974, a short-lived company called Spectrotherm introduced a special-purpose instrument [20] shown schematically in Fig. 8.10. Unfortunately, this company folded before the development of this instrument was completed, but several interesting features of this optical design could still be incorporated in a low-cost special-purpose GC/FT–IR instrument. It is also noteworthy that the drive of the moving mirror of the interferometer involved a "porch-swing" mechanism—an early indication of the potential of this type of drive for low-cost FT–IR spectrometers (see Chapter 4).

The first configuration for optical subtraction FT–IR spectrometry based on a medium-resolution commercial Michelson interferometer was described by Kuehl and Griffiths [21]; this design is shown in Fig. 8.11. These optics were used in a slightly modified configuration for GC/FT–IR by Gomez-Taylor and Griffiths [22] by locating two parallel light pipes

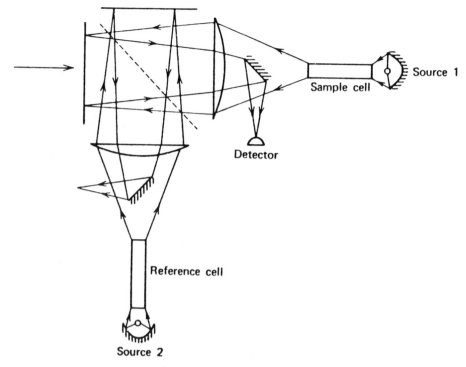

Fig. 8.9. Optical layout of the dual-beam system designed by Low for measuring the spectra of GC peaks. (Reproduced from [18], by permission of Marcel Dekker, Inc.; copyright © 1968.)

at the sample and reference positions. This design allowed many of the problems of the early optics for dual-beam FT–IR spectrometry to be overcome. For example, no pick-off mirrors blocked the beam and a high optical throughput (0.06 cm² sr) was attained. However, two new problems appeared that were caused by the effects of instabilities in the interferometer drive and detector saturation. The first problem was subsequently eliminated when the manufacturer redesigned the drive mechanism to permit a smoother and more rapid turnaround at the end of the retrace cycle [23], but the second problem appears to be more fundamental.

In optical subtraction measurements, the radiation flux at the detector is very high—much higher than can usually be tolerated in single-beam measurements using sensitive photodetectors. The linear range of detectors and their amplifiers used in single-beam measurements has to be

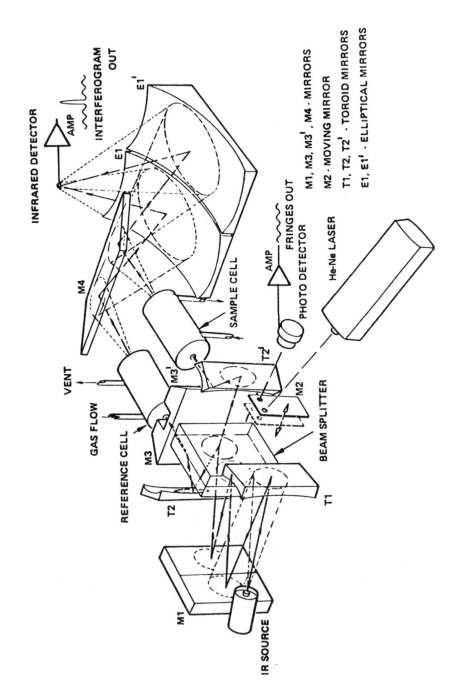

INFRARED DETECTOR

AMP

INTERFEROGRAM OUT

E1'

E1

M4

SAMPLE CELL

VENT

M3'

GAS FLOW

REFERENCE CELL

M3

T2

M1

IR SOURCE

T1

BEAM SPLITTER

M2

T2'

PHOTO DETECTOR

AMP

FRINGES OUT

He-Ne LASER

M1, M3, M3', M4 - MIRRORS

M2 - MOVING MIRROR

T1, T2, T2' - TOROID MIRRORS

E1, E1' - ELLIPTICAL MIRRORS

somewhat greater than 10^4, and few detectors have a smaller linear range than this. However, it should be noted that only about 15% of the full allowed throughput (Eq. 1.41) from an intense Globar source is needed before the SNR_I of interferograms measured using a narrow-range MCT detector becomes equal to the maximum dynamic range of a 16-bit ADC. If the *full* allowed throughput were allowed to fall on the detector, the response of the detector and amplifier could become nonlinear.

The linear ranges of a TGS and a narrow-range MCT detector for optical subtraction FT–IR measurements made using a Nernst glower source are illustrated in Fig. 8.12. The scale is compressed by a factor of 10 for the MCT curve relative to that for the TGS detector. In this experiment, therefore, the sensitivity of the measurement made with the MCT detector is about 10 times better than the measurement made with a TGS detector. At high radiation flux, the relative sensitivity of the MCT detector was found to be only 10-fold better than that of the TGS detector even though NEP measured at very low photon flux was about 50 times better. It should be noted that the limited linear range of the MCT detector rarely causes spectral artifacts on the spectra of weakly absorbing samples measured using the optical subtraction technique. This is because over the small modulation amplitude of the interferogram caused by the absorbing sample, the response curve is approximately linear. However, because the *slope* of the response curve is reduced by a factor of 5 at high radiation flux, the gain in sensitivity of the optical subtraction experiment over the corresponding single-beam measurement is reduced.

It is probable that an analogous result should probably be found in polarization modulation experiments for which the radiation flux at the detector is also high and the modulation amplitude is low. For both optical subtraction and polarization modulation measurements, the flux could be reduced through a judicious use of optical filters so that the sensitivity in limited spectral regions might be increased at the expense of spectral range.

Although the problems of dynamic range encountered in the far infrared are much less severe than those of mid-infrared Fourier spectrometry, several designs for double input and/or double output for far-infrared spectrometry have been described. Some have been designed to reduce the effects of fluctuation noise as discussed in Chapter 7. For example, Burroughs and Chamberlain [24] reported a design for a far-infrared interferometer based on the use of corner retroreflectors that allow the re-

←——————————————————————————

Fig. 8.10. Three-dimensional optical diagram of the dual-beam FT–IR spectrometer designed by Spectrotherm Corp. for the on-line measurement of the spectra of GC peaks. (Note that Spectrotherm Corp. no longer manufactures FT–IR spectrometers.)

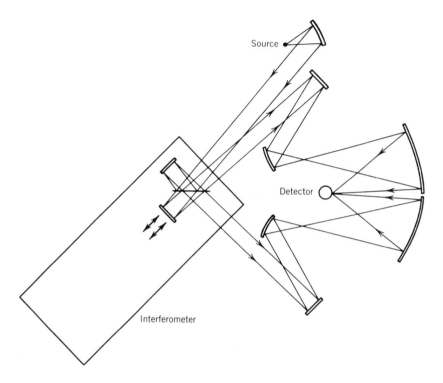

Fig. 8.11.   Optical diagram of the dual-beam FT–IR spectrometer designed by Kuehl and Griffiths. (Reproduced from [21], by permission of the American Chemical Society; copyright © 1978.)

flected and transmitted beams to be separated easily, see Fig. 8.13. Shepherd et al. [25] and Zehnpfennig et al. [26] have described a very similar design for near-infrared work using a PbS detector. The only disadvantage to designs involving the use of retroreflectors is that either two beamsplitters or one very large one is required. Overall, this design is very efficient, and its advantages appear to outweigh its disadvantages.

A far-infrared Fourier spectrometer based on a similar principle to the Spectrotherm instrument shown in Fig. 8.10 has been reported by Chandrasekhar et al. [27]. A schematic of this instrument is shown in Fig. 8.14. It can be seen that it is very similar to the design of the Genzel interferometer shown in Chapter 4. The main difference is that the two spherical collimating mirrors, S_2 and S_3, are set at an angle that displaces the input and output beams enough to enable them to be differentiated. Because of the small area of the focused beam at the beamsplitter, the image separation is accomplished by only a small tilt of S_2 and S_3, resulting in only minor distortion due to aberrations. It was claimed that this design could be modified readily for mid-infrared operation by removing the chopper, C, and speeding up the moving mirror. No mid-infrared data have been reported at the time of this writing.

An instrument designed for far-infrared spectrometry that also shows promise for mid-infrared measurements has been reported by Genzel and Kuhl [28]. This instrument is shown schematically in Fig. 8.15. The interferometer is based on a crossed version of a Czerny–Turner system to minimize distortions caused by the spherical mirrors. Two small beam-splitters are used, and the authors remark that these could be made out of different materials if desired. For far-infrared measurements, the two input beams are usually chopped out of phase. In the design shown in Fig. 8.15, the two output beams are focused onto a sample (S) and a

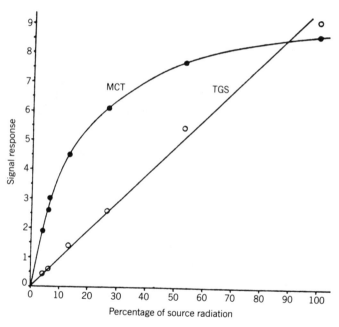

Fig. 8.12. Indication of the nonlinearity of an MCT detector and its amplifier when used at high signal levels in dual-beam FT–IR spectrometry. The maximum intensity of the strong 2875-cm^{-1} band of polyethylene is plotted against the percent transmittance of a neutral density filter placed just after the collimating mirror in the source optics. The corresponding data for a TGS detector are also shown, expanded by a factor of 10 for clarity, to show the linearity of the response of this detector. (Reproduced from [21], by permission of the American Chemical Society; copyright © 1978.)

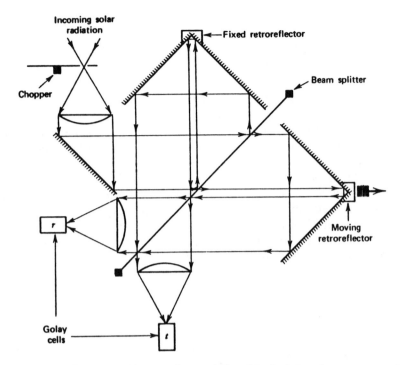

Fig. 8.13. Interferometer with retroreflectors designed for far-infrared measurements. This design enables the reflected output beam to be separated easily from the input beam. (Reproduced from [24], by permission of Pergamon Press; copyright © 1971.)

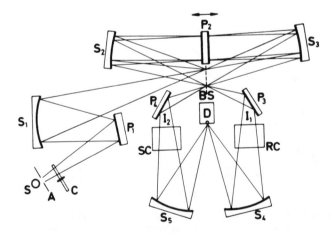

Fig. 8.14. Optical layout of a dual-beam FT–IR spectrometer based on a Genzel-type interferometer. S_1–S_5, spherical mirrors; P_1–P_4, plane mirrors; BS, beamsplitter; S, source; A, aperture; C, chopper; D, detector; RC and SC, reference and sample chambers, respectively. (Reproduced from [27], by permission of Elsevier Science Publishers, North Holland Publishing Division; copyright © 1976.)

306

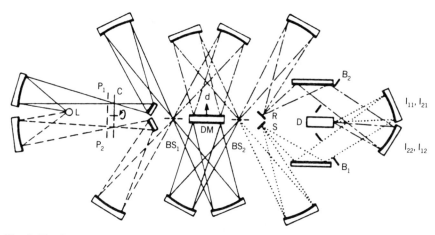

Fig. 8.15. Optical layout of far-infrared Fourier spectrometer with double-input, double-output interferometer and input chopping. L, source; p_1 and p_2, input power attenuators; C, chopper; BS_1 and BS_2, metal-mesh beamsplitters; DM, movable double mirror; S, sample position and, R, reference position for reflection spectrometry; B_1 and B_2, iris diaphragms; D, detector. (Reproduced from [17], by permission of Pergamon Press; copyright © 1978.)

reference (R) for reflectance measurements, but it is easy to visualize how this design could be modified to permit measurement of differential absorbance. A related design was recently reported by Krenn et al. [29].

The initial optical alignment of all dual-output interferometers is much more difficult than that of single-beam Fourier spectrometers. One way of testing how well the optics are aligned is to look for a complete disappearance of the modulated component of the interferogram, and this technique works well in the far and near infrared [25–28]. For mid-infrared measurements made using Ge–KBr beamsplitters, there is usually a thin oxide layer on the surface of the germanium dielectric next to the compensator. In this case, $R_{\bar{\nu}} + T_{\bar{\nu}} \neq 1$, and a discrete absorption band at 780 cm^{-1} is seen in the transform of $[V_t(\delta) + V_r(\delta)]$, see Fig. 8.16. It therefore becomes impossible to nullify the modulated component of the interferogram. The *nulling ratio* may be defined as the peak-to-peak amplitude of the single-beam interferogram $[V_t(\delta)]$ divided by the peak-to-peak amplitude of the dual-beam interferogram $[V_t(\delta) + V_r(\delta)]$ in the region of the centerburst. On well-aligned mid-infrared systems, it is usually possible to achieve nulling ratios between 20 and 100. Even for the lower value, SNR$_I$ is reduced well below the dynamic range of the ADC.

The phase of the spectral features in the background interferogram, i.e., the dual-beam interferogram measured with no sample present, is

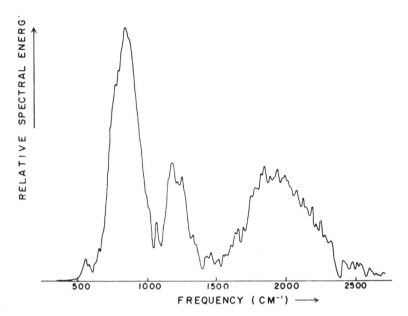

Fig. 8.16. Transform of an interferogram measured on a dual-beam optical subtraction FT–IR spectrometer. The two bands at about 850 and 1250 cm^{-1} appear to be caused by germanium oxides on the surface of the beamsplitter, and the energy at the higher wavenumber is probably a result of imperfect matching of the two beams.

unlikely to be the same as the phase of the weak features due to the absorbing sample. Therefore, it is necessary to take some care over the phase correction procedure used for optical subtraction Fourier spectrometry. Perhaps the best method of achieving accurate phase correction is first to subtract the background interferogram from the sample absorption bands. At this point, we have an interferogram that contains a limited number of discrete sinusoidal components, and commonly used phase correction algorithms, such as the Mertz method, are not applicable to interferograms of this type. To get around this problem, the phase spectrum of a sample that exhibits weak continuous "absorption" (such as a screen with 85–95% transmittance) should be measured prior to the experiment, stored, and used in all subsequent phase correction computations.

The typical procedure used in the calculation of a transmittance spectrum by optical subtraction FT–IR spectrometry [21] is illustrated in Fig. 8.17. It is noteworthy that for weakly absorbing samples, the absorptance spectrum [$1 - \tau(\bar{\nu})$] is equal to $2.303A(\bar{\nu})$. Thus, the absorptance spectrum can be calculated to a good approximation by dividing the optical subtraction spectrum $[1 - \tau(\bar{\nu})]B(\bar{\nu})$ by the single-beam spectrum $B(\bar{\nu})$ and dividing by 2.303. The success of this approximation can be seen by the fact that GC/FT–IR spectra calculated in this way correspond sufficiently closely to the absorbance spectra of reference standards so that good matches are found when spectral search programs are applied [30].

Optical subtraction techniques have been applied primarily to GC/FT–IR but have also been shown to be applicable to surface measurements

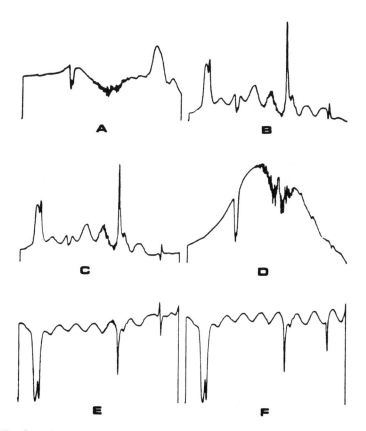

Fig. 8.17. Steps in the calculation of a transmittance spectrum by dual-beam FT–IR spectrometry using a polyethylene sheet as the sample. (a) Dual-beam background with no sample in either beam; (b) uncorrected dual-beam spectrum of polyethylene; (c) result of subtracting (a) from (b); (d) single-beam instrumental background spectrum; (e) result of dividing (c) by (d), subtracting the resulting spectrum from unity and multiplying by -1; (f) transmittance spectrum of polyethylene measured using a conventional single-beam FT–IR spectrometer. Only the sharp band at 720 cm^{-1} is different in the dual-beam spectrum, suggesting poor phase correction in this region. (Reproduced from [21], by permission of the American Chemical Society, copyright © 1978.)

[31, 32] and (with less success) to VCD [33]. Double-output Fourier spectrometry is not readily achievable on most commercial FT–IR spectrometers, but for certain applications this technique seems to hold considerable promise in getting around the problems of digitization noise imposed by the limited dynamic range of 15- or 16-bit ADCs.

REFERENCES

1. G. Holzwarth, E. C. Hsu, H. S. Mosher, T. R. Faulkner, and A. Moskowitz, *J. Am. Chem. Soc.*, **96**, 251 (1974).
2. L. A. Nafie, T. A. Keiderling, and P. J. Stephens, *J. Am. Chem. Soc.*, **98**, 2715 (1976).
3. M. Grosjean and M. Legrand, *C. R. Acad. Sci. (Paris)*, **251**, 2150 (1960).
4. L. Velluz, M. Grosjean, and M. Legrand, *Optical Circular Dichroism*, Academic Press, New York (1965).
5. L. A. Nafie and M. Diem, *Appl. Spectrosc.*, **33**, 130 (1979).
6. L. A. Nafie, E. D. Lipp, and C. G. Zimba, *Proc. Soc. Photo-Opt. Instrum. Eng.*, **289**, 457 (1981).
7. L. A. Nafie and D. W. Vidrine, in *Fourier Transform Spectroscopy: Applications to Chemical Systems*, Vol. 3 (J. R. Ferraro and L. J. Basile, eds.), Academic Press, New York (1982).
8. E. D. Lipp and L. A. Nafie, *Appl. Spectrosc.*, **38**, 20 (1984).
9. P. L. Polavarapu, D. F. Michalska, and D. M. Back, *Appl. Spectrosc.*, **38**, 438 (1984).
10. P. L. Polavarapu, *Appl. Spectrosc.*, **38**, 26 (1984).
11. D. M. Back and P. L. Polavarapu, *Carbohyd. Res.*, **133**, 163 (1984).
12. T. Kusan and G. Holzwarth, *Biochemistry*, **15**, 3352 (1976).
13. A. E. Dowrey and C. Marcott, *Appl. Spectrosc.*, **36**, 414 (1982).
14. W. G. Golden and D. D. Saperstein, *J. Electron. Spec.*, **30**, 43 (1983).
15. W. G. Golden, D. D. Saperstein, M. W. Severson, and J. Overend, *J. Phys. Chem.*, **88**, 572 (1984).
16. L. Genzel, H. R. Chandrasekhar, and J. Kuhl, *Opt. Commun.*, **18**, 381 (1976).
17. H. Bar-Lev, *Infrared Phys.*, **7**, 93 (1967).
18. M. J. D. Low, *Anal. Letters*, **1**, 819 (1968).
19. P. R. Griffiths, *Appl. Spectrosc.*, **31**, 284 (1977).
20. J. N. Willis, *Infrared Phys.*, **16**, 299 (1976).
21. D. Kuehl and P. R. Griffiths, *Anal. Chem.*, **50**, 418 (1978).
22. M. M. Gomez-Taylor and P. R. Griffiths, *Anal. Chem.*, **50**, 422 (1978).
23. G. J. Kemeny and P. R. Griffiths, *Appl. Spectrosc.*, **34**, 95 (1980).
24. W. J. Burroughs and J. Chamberlain, *Infrared Phys.*, **11**, 1 (1971).
25. O. Shepherd, W. Reidy, and G. A. Vanasse, *Proc. Soc. Photo-Opt. Instrum. Eng.*, **191**, 64 (1979).

26. T. Zehnpfennig, S. Rappaport, and G. A. Vanasse, *Proc. Soc. Photo-Opt. Instrum. Eng.*, **191**, 71 (1979).

27. H. R. Chandrasekhar, L. Genzel, and J. Kuhl, *Opt. Commun.*, **17**, 106 (1976).

28. L. Genzel and J. Kuhl, *Infrared Phys.*, **18**, 113 (1978).

29. H. Krenn, I. Roschger, and G. Bauer, *Appl. Opt.*, **23**, 3065 (1984).

30. W. J. Yang and P. R. Griffiths, unpublished data (1982).

31. M. J. D. Low and H. Mark, *J. Paint Technol.*, **43**(553), 31 (1971).

32. G. J. Kemeny and P. R. Griffiths, *Appl. Spectrosc.*, **35**, 128 (1981).

33. W. J. Yang, P. R. Griffiths, and G. J. Kemeny, *Appl. Spectrosc.*, **38**, 337 (1984).

CHAPTER
9

PHOTOTHERMAL FOURIER
SPECTROMETRY

I. PHOTOACOUSTIC SPECTROMETRY OF GASES

When radiation is absorbed by a gas, liquid, or solid, the energy absorbed is entirely or partially converted to kinetic energy. For gases held in an enclosed chamber, the increase in temperature gives rise to a corresponding increase in pressure. If the input radiation is modulated at a frequency in the audio range, the pressure fluctuations occurring at this modulation frequency can be detected with a microphone. The first experiments into the *photoacoustic*, or optoacoustic, effect were performed a century ago by Alexander Graham Bell and others [1–3]. However, little interest in this effect was shown by physicists and analytical chemists until almost a century after Bell's original experiments.

In 1971, Kreuzer and Patel [4] reported that by using a tunable infrared laser in conjunction with a sensitive microphone, the photoacoustic effect could be applied to the detection of very low concentrations of gases. Subsequently, photoacoustic spectrometry with infrared laser sources has been used to provide information about deexcitation processes and photochemical and kinetic effects in gases [5,6] and the determination of components in fluids at very low concentrations [7,8].

Until the late 1970s, most infrared photoacoustic experiments were performed using high-intensity tunable laser sources. Due to the monochromaticity of the source, this type of experiment is only useful for monitoring a single line in the spectrum of one molecule. For multicomponent analysis, however, a polychromatic source is required. If a broadband incandescent source is used in conjunction with a monochromator, the radiation flux at any wavenumber is very low, and photoacoustic spectra measured in this way have a rather poor SNR [9]. The multiplex and throughput advantages of Fourier spectrometry suggest that a Michelson interferometer could well be used for the measurement of infrared photoacoustic spectra, and recent work has indicated this to be true.

Recall that for a rapid-scanning Michelson interferometer with a mirror

Fig. 9.1. Representation of the system designed by Busse and Bullemer for the measurement of both photoacoustic spectra (spectrophone) and conventional absorption spectra of gases. (Reproduced from [10], by permission of Pergamon Press; copyright © 1978.)

velocity of v (cm sec^{-1}), radiation of wavenumber $\bar{v}$ (cm^{-1}) is modulated with a frequency of $2v\bar{v}$ (Hz). For mirror velocities on the order of 1–2 mm sec^{-1}, the wavelengths in the mid-infrared spectrum are modulated at audio frequencies, ideally suited for detection with a microphone. Since the photoacoustic interferogram is only due to radiation *absorbed* by the sample, this technique represents an alternative means of suppressing the centerburst of the interferogram caused by the intense emission of blackbody sources (see Chapter 8).

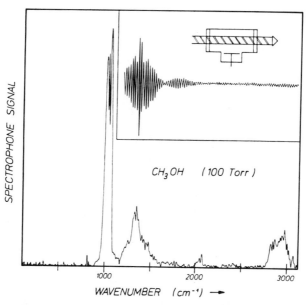

Fig. 9.2. Photoacoustic spectrum of 100 torr of methanol vapor measured using the spectrophone system shown schematically in Fig. 9.1. (*Inset*) The interferogram from which this spectrum was computed. (Reproduced from [10], by permission of Pergamon Press; copyright © 1978.)

The first experiment showing the feasibility of Fourier transform photoacoustic spectrometry was reported by Busse and Bullemer [10]. Their experimental layout is illustrated diagrammatically in Fig. 9.1. The modulated beam from a rapid-scanning interferometer is passed through a short gas cell in which a microphone (or spectrophone as it was called by these authors) is mounted. With this configuration, both the photoacoustic interferogram and the conventional interferogram can be measured simultaneously, see Figs. 9.2 and 9.3. The suppression of the single-beam background spectrum in the photoacoustic spectrum is readily observed.

These experiments suffered from a lack of sensitivity because the cell design and microphone had not been optimized. Subsequently, Krishnan [11] measured the photoacoustic spectrum of gaseous ethylene at moderate resolution and appreciably higher sensitivity, see Fig. 9.4. For this work, the sample was held in a very small cell designed for holding solids in contact with a nonabsorbing gas (*vide infra*, Section II of this chapter).

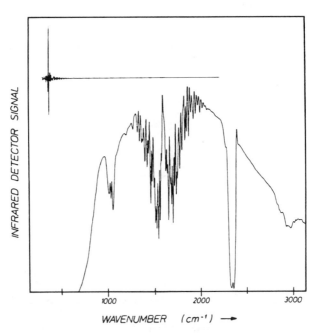

Fig. 9.3. Single-beam spectrum measured through the same cell used for Fig. 9.2, showing methanol, water, and CO_2 absorption bands. (*Inset*) The interferogram from which this spectrum was computed. (Reproduced from [10], by permission of Pergamon Press; copyright © 1978.)

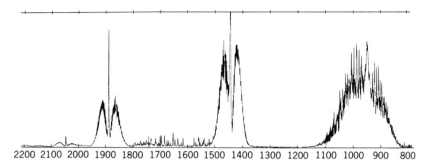

2200 2100 2000 1900 1800 1700 1600 1500 1400 1300 1200 1100 1000 900 800

Fig. 9.4.   Spectrum of gaseous ethylene contained in the Digilab photoacoustic cell designed for measuring the spectra of solids, which is shown schematically in Fig. 9.5, measured at moderate resolution. (Reproduced by permission of Digilab Division of Bio-Rad Laboratories.)

Although much more work has been published on the photoacoustic spectrometry of *solid* samples using Fourier spectrometers, the ethylene spectrum well illustrates the application of this technique to the monitoring of small volumes of *gases*.

II.   PHOTOTHERMAL INFRARED SPECTROMETRY OF SOLIDS

The most common method of measuring the photoacoustic (or, more generally, *photothermal*) spectrum of a solid sample is to hold the sample in contact with a nonabsorbing gas. The sample is illuminated with the modulated beam emerging from the monochromator or interferometer. At wavelengths where the sample absorbs some fraction of the incident radiation, a modulated temperature fluctuation will be generated at the same frequency (but not necessarily with the same phase) as that of the incident radiation. If the gas is held in an enclosed cell, the layer of gas next to the surface of the solid sample is heated so that an acoustic pressure wave is created. At the same time, the refractive index of the gas varies sinusoidally. Thus, photothermal spectra can be measured in two ways.

a.   Microphone Detection

By far the most common technique is to use a microphone to detect the acoustic signal. The first infrared spectra measured using a Fourier spectrometer in this fashion were reported by Rockley [12,13]. Depending on its geometry, the photoacoustic cell may be used in either a resonant or

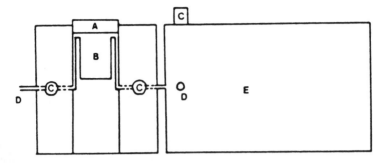

Fig. 9.5. Schematic representation of typical cell used for measuring the photoacoustic spectra of solid samples. A, KBr window; B, sample cup; C, inlet or outlet valves; D, inlet or outlet ports; E, microphone. (Reproduced from [21], by permission of the Society for Applied Spectroscopy; copyright © 1981.)

a nonresonant mode. A typical cell configuration is shown in Fig. 9.5. One key factor in the design of cells with microphonic detection is the degree of isolation from external sound and other vibrations, and good shock absorption is critical if adequate spectra are to be obtained with this type of cell [14].

The size of the cup can determine whether or not the cell exhibits resonance, as demonstrated in Fig. 9.6 for carbon black. Since carbon black is essentially totally absorbing, its photoacoustic spectrum is analogous to the single-beam reference spectrum measured in conventional absorption spectrometry. As a result, most early photoacoustic spectra were ratioed against a carbon black reference to remove the effects of the instrumental background. Certain carbon black samples have been observed to yield photoacoustic spectra with appreciable spectral structure. It is therefore becoming more common to use a conventional single-beam background spectrum measured with a DTGS detector as a reference for photoacoustic spectra [15]. Since the noise level of spectra measured using a DTGS detector is usually much lower than that of photoacoustic spectra of carbon black, the sensitivity of the measurement will also exhibit a small increase when a DTGS background spectrum is employed.

b.   Beam Deflection

An alternative method of measuring the temperature fluctuations introduced by the absorption of modulated radiation involves the refraction of a laser beam because of the "mirage" effect [16]. For *beam deflection* photothermal FT–IR spectrometry, a laser beam is passed over the surface of the sample at the same time as the modulated beam from the interferometer interacts with the sample. The thermal gradients and consequent refractive index gradients in the contact gas cause the laser beam to be deflected. Since the sample temperature is modulated at the frequency of the incident radiation, the deflection of the laser beam is also modulated. The photothermal beam deflection can be measured either by a single position-sensing detector or, better, by a dual-detector system to compensate the effects of fluctuations in the laser beam intensity and external vibrations. Several experimental designs for photothermal FT–IR spectrometry have been summarized by Low et al. [17–19] and are illustrated schematically in Fig. 9.7. The single-detector systems (I and III) are quite sensitive to the effects of acoustic noise and mechanical

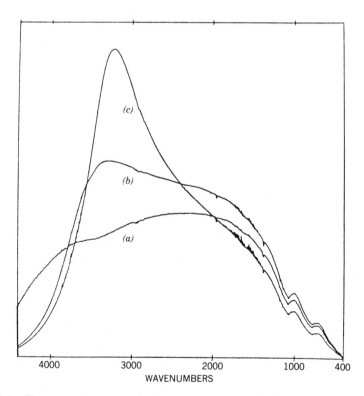

WAVENUMBERS

Fig. 9.6. Photoacoustic spectra of carbon black held in Digilab photoacoustic accessory using cups of three different depths: (a) 3 mm; (b) 6 mm; (c) 9 mm. The appearance of a resonance near 3000 cm^{-1} as the cup depth is increased is easily observed. (Reproduced from [21], by permission of the Society for Applied Spectroscopy; copyright © 1981.)

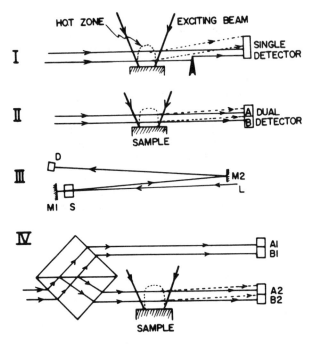

Fig. 9.7. Several different schemes for measuring beam deflection photothermal FT–IR spectra. (I) Simplest system in which half the undeflected beam is blocked before it reaches the detector; on photothermal deflection, the signal at the detector increases. (II) Dual-detector system in which the signal at the upper detector increases while that at the lower detector decreases on photothermal deflection. (III) Optical system for measuring beam deflection of transparent samples by double passing the laser beam through the sample. (IV) Device for compensating acoustic noise and mechanical vibrations, in which the laser beam is split into two essentially identical paths by a prism. Each of the beams hits a dual detector as in (II), or better yet a quadrant-following detector for compensating vibrations in two directions. External interferences and noise are compensated by subtracting the signal of A1 from that of A2, and B1 from B2. (Reproduced from [17], by permission of Marcel Dekker Publishing Company; copyright © 1982.)

vibrations. These effects can be minimized using a dual-detector system with "sample" and "reference" beams that are both affected to the same extent by mechanical or acoustic noise such as the one shown as IV in Fig. 9.7. The difference signal then compensates for beam deflection in both beams and gives the signal due to beam deflection because of infrared absorption. In all cases, the sample is placed on a leveling platform that is raised or lowered so that the probe laser beam grazes the surface of the sample.

It is perhaps too early to compare photoacoustic and beam deflection techniques for photothermal FT–IR spectrometry. The first photoacoustic spectra of solids [12–14] had a very poor SNR, but more recent data [20–23] are much better. It is difficult to assess at what point on the learning curve for each technique we are today. The first beam deflection photothermal FT–IR spectra were measured in late 1981 [17–19] and had substantially better SNR than the first photoacoustic FT–IR spectra. Assuming that the sensitivity of mirage detection improves at the same rate as that of photoacoustic FT–IR spectrometry over the initial years of its existence, this technique could evolve into a very useful technique for the analysis of solid samples.

It has been claimed that one of the advantages of photoacoustic FT–IR spectrometry is that it can be used for the measurement of condensed phase samples of a variety of types—solid chunks and powders, gels,

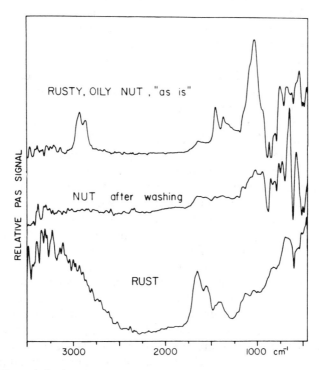

Fig. 9.8. Beam deflection photothermal spectra of a massive object, in fact, a rusty, oily nut. The upper curve shows the spectrum of the nut as received, and the middle trace shows the spectrum of the nut after washing to remove the oil. The lower trace shows the spectrum of rust. (Reproduced from [24], by permission of Marcel Dekker Publishing Company; copyright © 1982.)

foams, and liquids [20]—and undoubtedly this property is one of the main advantages of the technique. Even so, certain types of solid samples cannot be studied by photoacoustic FT–IR spectrometry. For example, large samples cannot be investigated, since the size of the sample must be small enough that it fits in the cell and cells must be of low volume to maximize sensitivity. One of the largest advantages of beam deflection techniques for photothermal FT–IR spectrometry is that massive objects can be measured [19,24] as illustrated in Fig. 9.8.

III. UNCONVENTIONAL PHOTOTHERMAL MEASUREMENT TECHNIQUES

It is interesting to speculate whether future instrumental developments will lead to improved sensitivity for the photothermal FT–IR spectrometry of solids. Some less conventional designs of interferometers and detection systems for photoacoustic Fourier transform *visible* spectrometry of solids have been described. For example, Farrow et al. [25] constructed a special-purpose interferometer whose moving mirror could be scanned slowly and modulated sinusoidally, giving rise to a phase-modulated photoacoustic interferogram. The SNR of spectra measured using this early instrument was not high. A subsequent publication from the same laboratory described a modification to this instrument in which the "fixed" mirror was the phase-modulating mirror and the moving mirror was translated at a uniform velocity [26]. Spectra were measured at a more acceptable sensitivity using this interferometer. It may be noted that photoacoustic spectrometry performed in this way may be considered to be another manifestation of double-modulation spectrometry (see Chapter 8). A third generation instrument from Eyring's group [27] involved the use of a phase-modulated step-and-integrate interferometer in a similar manner to their earlier slow-scan interferometers.

Débarre et al. [28] have also described an interferometer for PA measurements in the visible region; this instrument is shown schematically in Fig. 9.9. The movable mirror of this stepped-scan interferometer was driven using a flex pivot drive (see Chapter 4), and phase modulation techniques were again used on this instrument. Spectra measured by Débarre et al. using a xenon arc source compare very favorably with data obtained using either a monochromator with a xenon arc or a tunable dye laser source. The resolution obtained using the Fourier transform spectrometer was about an order of magnitude better than that obtained with the monochromator at comparable SNR and measurement time. The resolution of spectra obtained with the interferometer was comparable to

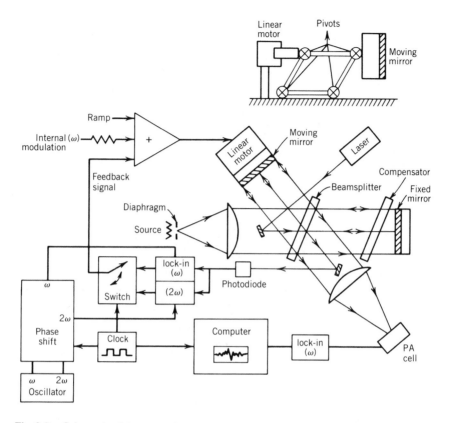

Fig. 9.9. Schematic of the stepped-scan flex-pivot drive interferometer and associated electronics designed by Débarre et al. for photoacoustic spectrometry in the visible region of the spectrum. (Reproduced from [28], by permission of the Optical Society of America; copyright © 1981.)

that obtained with the dye laser, and whereas the SNR was slightly better for spectra measured with the dye laser, the spectral range covered using the interferometer was about 10-fold greater than could be achieved with a dye laser.

Although microphone and mirage detection have been the two most commonly applied techniques for Fourier transform photothermal spectrometry, an alternative detection technique using a PZT has been proposed [29,30]. For visible spectrometry, the sample must be immersed in a transparent epoxy or cyanoacrylate matrix that attaches the sample to a front surface mirror. A $BaTiO_3$ or $PbTiO_3$ PZT is then glued to the back of the mirror and used to sense small motions induced in the sample.

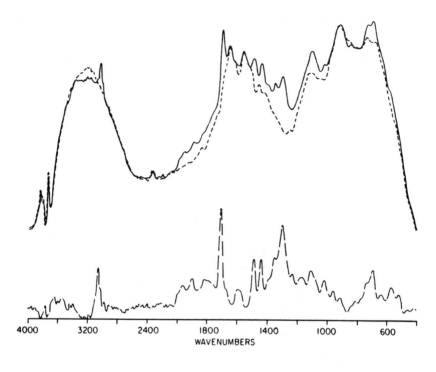

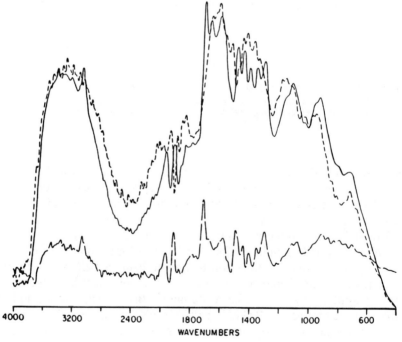

Since there is no such thing as an infrared transparent adhesive, it might be thought that this technique could not be applied to mid-infrared measurements, but this is not the case. For samples that are optically opaque, the effective penetration depth of the beam is usually less than 100 μm. For solid samples falling into this category, the PZT may be glued to the back of the sample since the modulated input radiation does not reach the adhesive. Lloyd et al. [30] have demonstrated that photothermal FT–IR spectra of samples such as thick compressed pellets of organics and thin-layer chromatographic plates can be measured by gluing a $BaTiO_3$ PZT to the back of the sample. Spectra of a ketone adsorbed on an aluminum-backed alumina TLC plate using PZT and microphonic detection are shown in Fig. 9.10. The SNR of the spectra measured with PZT detection appears to be comparable to that of spectra measured microphonically, but the relative intensities of certain bands due to the adsorbate are different. Since this paper [30] was published, few photoacoustic FT–IR spectra measured with PZT detection have been reported. Nevertheless, there seem to be certain definite advantages of this technique over microphonic detection, and we expect to read of further developments in this area in the future.

IV. QUANTITATIVE CONSIDERATIONS

For transmission spectrometry at a certain wavelength, Beer's law may be written as

$$I = I_0 \exp(-\beta x) \tag{9.1}$$

where I_0 is the incident energy on the sample, I is the transmitted energy, β is the absorption coefficient, and x is the sample thickness. When a source of radiation modulated sinusoidally at a frequency ω (rad sec^{-1}) is incident on a solid sample, the radiation penetrates to a depth that depends on β, with the intensity at a depth x being given by

$$I = 0.5I_0 \exp(-\beta x)(1 + \cos \omega t) \tag{9.2}$$

Fig. 9.10. Photoacoustic spectra of (solid line) 300 μg of tetraphenylcyclopentadienone on an aluminum-backed alumina TLC plate, (broken line) the same TLC plate with no sample, and (below, broken line) the difference spectrum. (*Above*) Spectra measured using microphonic detection. (*Below*) Spectra measured with PZT detection. (Reproduced from [30], by permission of the American Chemical Society; copyright © 1982.)

The power supplied per unit volume (known as the heat density H_d) is given by βI. This heat must then be dissipated to the gas that is in contact with the solid sample.

Rosencwaig and Gersho [31] have discussed the generation of photo-acoustic signals from isotropic solids (note, *not powders*) in terms of the thickness of the sample, l, the *optical absorption depth* μ_β (which is equal to β^{-1}), and the *thermal diffusion depth* μ_s, defined as

$$\mu_s = \left(\frac{2K_s}{\rho_s C_s \omega} \right)^{1/2} \tag{9.3}$$

where K_s is the thermal conductivity, ρ_s is the density, and C_s is the specific heat of the solid sample. The incremental change in pressure of the gas above the sample is given by

$$\delta P(t) = Q \exp[i(\omega t - \tfrac{1}{4}\pi)] \tag{9.4}$$

The explicit form of the coefficient Q depends on the nature of the sample. The intensity of the photoacoustic signal is proportional to Q and is shifted in phase relative to the phase of the modulated input signal.

Rosencwaig and Gersho have divided solids into six categories with different relative values of l, μ_β, and μ_s and have evaluated the magnitude of Q for each category. The categories are:

1a. Optically transparent and thermally thin ($\mu_s \gg l$, $\mu_s > \mu_\beta$).
1b. Optically transparent and thermally thin ($\mu_s > l$, $\mu_s < \mu_\beta$)

For both these cases, the photoacoustic signal is proportional to βl and has an ω^{-1} dependence.

1c. Optically transparent and thermally thick ($\mu_s < l$, $\mu_s \ll \mu_\beta$).

In this case, the signal is proportional to $\beta \mu_s$ and varies as $\omega^{-3/2}$ so that only the radiation adsorbed within the first μ_s contributes to the signal. Since μ_s has a $\omega^{-1/2}$ dependence, changing the modulation frequency of the incident radiation will change the thickness of the sample contributing to the photoacoustic signal. When rapid-scanning interferometers are used, the effective penetration depth μ_s is therefore inversely proportional to the velocity of the moving mirror and varies with the wavenumber across the spectrum. When phase modulation techniques are used to generate the interferogram, the effective penetration depth is inversely proportional to the square root of the modulation frequency of the dither applied to the interferometer mirror but is independent of the wavenumber. Either technique can therefore (in theory at least) be used for depth profiling studies.

2a.   Optically opaque and thermally thin ($\mu_s \gg l$, $\mu_s \gg \mu_\beta$).
2b.   Optically opaque and thermally thick ($\mu_s < l$, $\mu_s < \mu_\beta$).

In each case, the photoacoustic signal is independent of β and varies as ω^{-1}. Since most of the incident radiation is adsorbed within a length that is smaller than μ_s, *photoacoustic saturation* has occurred.

2c.   Optically opaque and thermally thick ($\mu_s \ll l$, $\mu_s < \mu_\beta$).

The signal is proportional to $\beta\mu_s$ and varies as $\omega^{-3/2}$, as in case 1c.
Carbon black falls in category 2a. The expression for Q in this case is given by Rosencwaig and Gersho as

$$Q = \frac{(1 - i)}{4\sqrt{2}} \frac{P_0 I_0 \gamma}{l_g T_0} \frac{\mu_s \mu_g}{K_s} \qquad (9.5)$$

where the subscripts s and g refer to the solid sample and the gas with which it is in contact, respectively; that is,

$$\mu_g = \left(\frac{2K_g}{\omega \rho_g C_g}\right)^{1/2} \qquad (9.6)$$

Here P_0 and T_0 are the equilibrium pressure and temperature of the gas, l_g is the thickness of the layer of gas above the sample, and γ is the ratio of the principal specific heats.

The frequency response of a Nicolet photoacoustic cell has been reported by Vidrine [20], who measured the signal at a particular wavenumber (780 cm^{-1}) for a carbon black sample studied with interferometer mirror velocities between about 0.05 and 2 cm sec^{-1}. His results are illustrated in Fig. 9.11. For frequencies between 60 and 400 Hz, these data show the predicted frequency response of exactly ω^{-1}. At a frequency of 1.25 kHz, a resonance is shown, so that the response on either side of this response is higher than would be predicted from a simple ω^{-1} frequency dependence. For isotropic solid samples, such as polymer films, l may be greater than μ_β in spectral regions where β is large (case 2) and smaller than μ_β in regions where β is small (case 1). In this case,

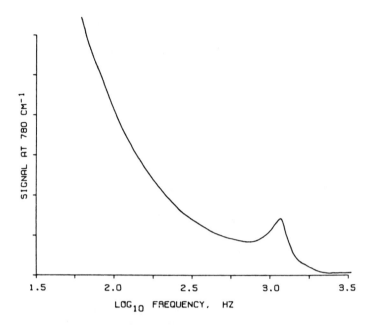

Fig. 9.11. Frequency response of Nicolet photoacoustic cell. Spectra of carbon black were measured at several different interferometer mirror velocities. In each case, the signal at 780 cm^{-1} was recorded and plotted against the modulation frequency of this wavenumber. The resonance just above 1 kHz is readily observed. (Reproduced from [20], by permission of the Society for Applied Spectroscopy; copyright © 1980.)

samples may be photoacoustically saturated in some regions and not in others. This behavior can make quantitative determinations extremely difficult. Krishnan [21] has shown examples where a transition between these cases can be observed. He prepared samples of Plexiglas deposited on salt plates at different thicknesses from 0.3 to 13.8 μm. The photoacoustic spectra of these samples measured using a rapid-scanning interferometer together with the absorbance spectrum of the thickest sample are shown in Fig. 9.12.

As the sample thickness l is increased from 0.3 to 3.7 μm, the relative intensity of all bands in the spectrum remains approximately constant and the intensity of each band increases linearly with l. These samples are therefore optically transparent and thermally thin. At greater sample thicknesses, saturation effects start to become apparent. For the strongest band in the spectrum (at 1730 cm^{-1}), $\mu_\beta \sim 5$ μm, and μ_s varies across the spectrum as $\bar{\nu}^{-1/2}$. At 2000 cm^{-1}, μ_s is approximately 6 μm; and at 500 cm^{-1}, μ_s has increased to about 12 μm. The thickest sample therefore represents the case where the strong bands are optically opaque and the

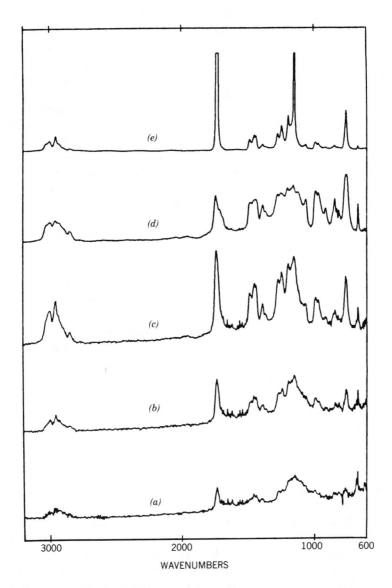

(e)

(d)

(c)

(b)

(a)

3000   2000   1000   600

WAVENUMBERS

Fig. 9.12. Spectra of Plexiglas films. Photoacoustic spectra of films of thickness (a) 0.3 μm, (b) 0.7 μm, (c) 3.7 μm, and (d) 13.8 μm. (e) Transmission spectrum of the 13.8-μm film plotted in absorbance. (Reproduced from [21], by permission of the Society for Applied Spectroscopy; copyright © 1981.)

sample is thermally thick over most of the spectrum. Thus, the more intense bands have become photoacoustically saturated, whereas the weaker bands continue to become more intense as the sample thickness is increased. If band intensity is plotted against concentration or path-length for isotropic samples, a nonlinear plot results when each band begins to become saturated. Because of these conclusions, Krishnan has warned that depth profiling studies can give ambiguous results. Ideally, if photoacoustic spectra are generated using a rapid-scanning interfero-meter, as the velocity of the mirror (and hence ω) is varied, a plot of band intensity against ω^{-1} should be linear for optically transparent isotropic materials. If the sample starts to become opaque for any band, the plot will become nonlinear at low ω. Thus, if depth profiling studies are to be performed by varying the velocity of the moving mirror of a rapid-scan-ning interferometer or the modulation frequency of a phase modulation interferometer, both μ_s and μ_β should be known if the data are to be accurately interpreted.

It should also be noted that the theory of Rosencwaig and Gersho assumes that the sample is isotropic. Fernelius [32] and Bennett and For-man [33] have introduced revised theories to take surface absorption into account. Because of these complexities, we believe that photoacoustic spectrometry is not as useful for depth profiling studies as, for example, ATR spectrometry (see Chapter 5, Section III.b). A recent report by Gardella et al. [34] appears to confirm this view.

Teng and Royce [35] have proposed a technique to permit quantita-tively accurate absorptivity spectra to be measured using a rapid-scanning interferometer, provided that data can be acquired at at least two scan speeds. It should be noted, however, that their method is not applicable for samples whose thickness is less than about six times the thermal dif-fusion depth. Thus, samples must be at least 50 μm thick for their pro-cedure to work, and photoacoustic saturation has often occurred with intense bands for samples measured at this thickness.

With powdered samples the situation can become even more compli-cated. Vidrine [20] has shown that although the relative band intensities in the photoacoustic spectra of polymeric samples do not depend strongly on the morphology of the sample, the absolute band intensities in the spectra of powdered samples can be several times greater than in the spectra of isotropic samples, presumably because of more efficient heat transfer to the contact gas, see Fig. 9.13. Rockley et al. [36] have shown that photoacoustic spectra of organic powders (for which μ_β is typically quite large, even for the stronger bands in the spectrum) are to a first approximation independent of particle size. Minor variations are observed because of front-surface reflection and changes in the relative magnitudes

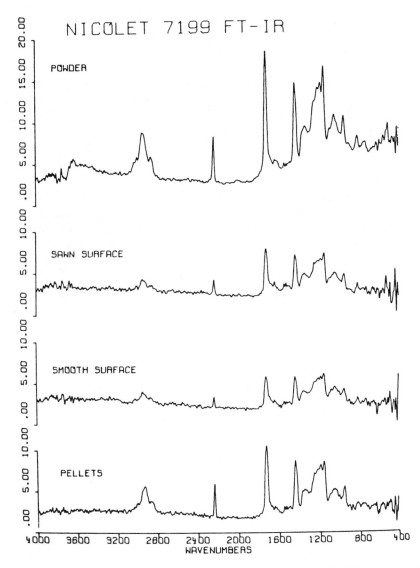

Fig. 9.13. Photoacoustic spectra of samples of a nitrile polymer with different morphologies. Note that the relative band intensities vary only slightly between the extreme cases of the isotropic sample with a smooth surface and the power (in contrast to the case of reflectance spectrometry) but the absolute intensities vary by about a factor of 4. (Reproduced from [20], by permission of the Society for Applied Spectroscopy; copyright © 1980.)

329

of l, μ_s, and μ_β as the particle diameter is increased. For mineral samples, the absorptivities of the strongest bands may be as much as an order of magnitude greater than those of organic samples. Thus, not only is μ_β very small but front surface reflection losses can also give anomolous results [37].

Tilgner [38] has suggested that photoacoustic spectra of powders cannot be readily interpreted in terms of the Rosencwaig–Gersho theory but are more adequately understood in terms of diffuse reflectance spectra treated using the Kubelka–Munk [39,40] theory. Obviously, much more work needs to be done in this area, especially in the infrared region of the spectrum where a further complication may be found in the fact that the average particle size may often be approximately the same as the wavelength of the radiation and also of the same order of magnitude as μ_β (for the more strongly absorbing bands in the spectrum) and μ_s.

A few quantitative measurements on powdered samples have been described. For example, Rockley et al. [41] reported that the composition of mixtures of $^{14}NO_3$ and $^{15}NO_3$ could be determined by photoacoustic FT–IR spectrometry, albeit with rather low precision, see Fig. 9.14. Yeboah et al. [42] demonstrated that when the intensity of strongly absorbing bands in the spectrum of caffeine–KCl mixtures is plotted against the

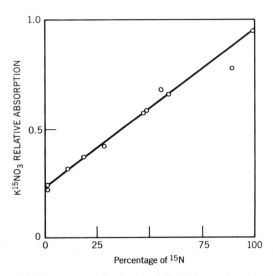

Fig. 9.14. Plot of relative photoacoustic intensity of the bands at 800 cm^{-1} (due to K $^{15}NO_3$) and at 825 cm^{-1} (due to K $^{14}NO_3$) against the percentage of K $^{15}NO_3$ in a mixture of the two salts. (Reproduced from [41], by permission of the Society for Applied Spectroscopy; copyright © 1981.)

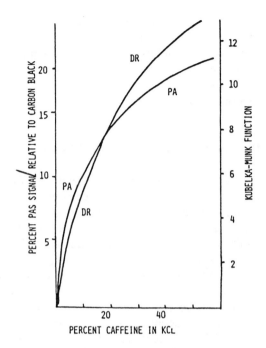

Fig. 9.15. (PA) Ratio of the intensity of the 1701-cm^{-1} band in the photoacoustic spectrum of caffeine: KCl mixtures to the intensity of the PA spectrum of carbon black at the same wavenumber and (DR) the value of the Kubelka–Munk function at 1701 cm^{-1} in the diffuse reflectance spectra of the same caffeine–KCl mixtures, each plotted against the percentage of caffeine in the mixture. (Reproduced from [42], by permission of Pergamon Press; copyright © 1981.)

concentration of caffeine, a nonlinear plot is observed, see Fig. 9.15. It is interesting to note that when diffuse reflectance spectra of the same samples were measured, the plot of the Kubelka–Munk function $f(R_\infty)$ against concentration was slightly *less* nonlinear than the corresponding plot of photoacoustic band intensity versus concentration. Presumably, at high concentration photoacoustic saturation has started to take place. This may be expected in view of the fact that the average particle size of the particles (10–20 μm) was greater than μ_β for several of the stronger bands in the caffeine spectrum.

For certain samples, photoacoustic FT–IR spectrometry has been shown to be the optimum sampling technique for quantitative analysis. For example, routine quality control assays of polymer pellets were achieved with a measurement time of 1 min per sample. No sample preparation was required. The success of this procedure is partially attributable to the reproducibility in the preparation of the pellets in the manufacturing process. The measurements were performed on a Digilab Qualimatic spectrometer, which has a very high optical throughput. Certainly photoacoustic spectrometry is one area where the greater the throughput of the instrument, the higher is the SNR of the measurement (see Chapter 5, Section I.e).

V. COMPARISON WITH OTHER SAMPLING TECHNIQUES

As a general rule, it can be stated that if the spectrum of a given sample can be measured using either photothermal spectrometry or an alternative "conventional" sampling technique, it is very likely that spectra measured using the alternative technique will have a higher SNR. Photoacoustic spectrometry can exhibit secondary advantages, however. For example, transmission spectra of powdered samples prepared as mulls, slurries, or pellets can exhibit severe distortions due to the Christiansen effect; these distortions are absent in photoacoustic spectra [43]. The Christiansen effect is also absent in diffuse reflectance spectra, but for strongly absorbing bands, especially in the spectra of minerals, distortions may be introduced in diffuse reflectance spectra because of specular reflection (*reststrahlen* bands). The magnitude of these distortions is significantly reduced in photoacoustic spectra.

KBr discs, mineral oil mulls, or diffuse reflectance spectra of very hard samples may be difficult to measure because of the difficulty of grinding the sample. Since photoacoustic spectrometry does not require a powdered sample, the increased measurement time of this technique may be more than offset by the decreased sample preparation time. One other possible advantage of photoacoustic spectrometry over diffuse reflectance measurements for powdered samples is the greater repeatability of band intensities in photoacoustic spectra. The reproducibility of band intensities in diffuse reflectance spectra is quite poor, although it can be improved by compressing the sample into a pellet before the spectrum is measured [44]. In this case, the band intensities in the diffuse reflectance spectra can be increased by up to a factor of 4 when compared to the corresponding values of $f(R_\infty)$ in the spectrum of the loose powder. In photoacoustic spectra, band intensities *decrease* on compacting the sample, but by a lesser amount than the increase in diffuse reflectance measurements [37].

Another type of sample that is very difficult to measure by transmittance or the attenuated total reflectance technique is a carbon-filled polymer, such as tire rubber. Photoacoustic FT–IR spectra of these samples are relatively easily measured. Coals can also be said to fall into this category. Several workers have compared the diffuse reflectance and photoacoustic spectra of coals and shown them to yield very similar information [20,21,45]. The SNR of the diffuse reflectance spectrum is usually superior to that of the photoacoustic spectrum of the same sample, however. If the sample can be ground (even at liquid nitrogen temperature), diffuse reflectance would therefore seem to be the technique of choice. If it cannot, then photoacoustic spectrometry allows the spectrum to be measured, albeit at lower SNR.

Another type of sample for which photoacoustic spectrometry has proved to be appropriate is conducting polymers. These samples can change their morphology on grinding, so that the preparation of KBr disks, mineral oil mulls, or even powders for diffuse reflection measurements does not permit the spectrum of the original polymer to be measured. As the degree of doping is increased, conducting polymers become highly opaque. In spite of these problems, acceptable photoacoustic spectra of conducting polymers have been reported [46–48].

One of the applications for which photoacoustic spectrometry has been claimed to be particularly appropriate is the characterization of surface species. For samples where the adsorbate is nonvolatile, several reports of surface species have been given. For example, Rockley and Devlin [49] have followed the oxidation of the surface of a freshly cleaved coal. Lowry et al. [50] have used photoacoustic spectrometry to investigate the bonding of a formulated pesticide to clay carriers. Gardella et al. [34] have compared ATR and photoacoustic spectrometry for characterizing the surface of polymer mixtures, whereas Porter et al. [51] have compared the detection capabilities of external reflection and photoacoustic spectrometry for the study of electrode-supported films. Only for the study of coal oxidation at a freshly cleaved surface does it appear difficult to obtain equivalent spectrometric data using an alternative sampling technique. If the adsorbate is only loosely bound to the surface so that there is an appreciable concentration in the vapor phase, photoacoustic measurements become even more difficult as the photoacoustic effect is about two orders of magnitude more efficient for molecules in the vapor phase than for solids.

An interesting potential application of a phase-modulated interferometer for photoacoustic measurements of surface species has been suggested by Débarre et al. [29]. Since all wavenumbers are modulated at the same frequency for phase modulation interferometry and a lock-in amplifier is used to process the signal, both the in-phase and in-quadrature interferograms may be measured. For the visible Fourier transform photoacoustic spectrum of a single crystal of $CaWO_4/Nd^{3+}$, adjusting the phase of LIA to give the maximum signal at zero retardation produces a broadband spectrum due to surface spoiling, which has taken place during the polishing procedure. The in-quadrature signal gave maximum discrimination against the surface signal, as illustrated in Fig. 9.16.

Some of the most impressive measurements of adsorbed species are starting to be reported using beam deflection photothermal spectroscopy. The adsorption of carbon monoxide onto a plethora of adsorbents has been investigated by many workers using a variety of sampling techniques. Excellent transmission and diffuse reflection data have been reported for CO adsorbed on metals dispersed on supports such as SiO_2, Al_2O_3, or TiO_2, which are fairly transparent near 2000 cm^{-1}. No data have been reported, however, using carbon as the support until very re-

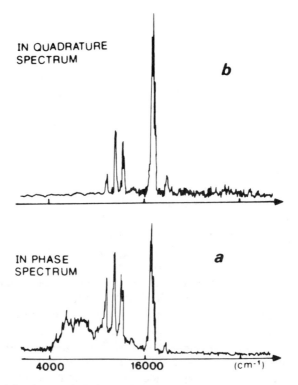

Fig. 9.16. (*a*) In-phase and (*b*) in-quadrature photoacoustic spectra of a CaWO$_4$/Nd^{3+} single crystal. The phase of spectrum (*a*) was chosen to maximize the broadband signal, which is due to surface spoiling during the polishing procedure, while the phase of spectrum (*b*) was chosen for maximum discrimination against the surface signal. (Reproduced from [28], by permission of the Optical Society of America; copyright © 1981.)

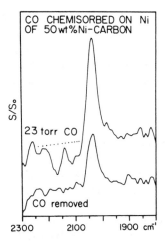

Fig. 9.17. Beam deflection photothermal FT–IR spectrum of carbon monoxide adsorbed on a 50 wt% nickel–carbon catalyst in the presence of 23 torr of CO (*above*) and under vacuum (*below*). (Reproduced from [52], by permission of Marcel Dekker Publishing Co.; copyright © 1982.)

cently because of the intense absorption by the carbon. Recently, Low et al. [52,53] have reported the spectra of CO adsorbed on a nickel–carbon catalyst (see Fig. 9.17). Although these spectra do not have a particularly high SNR, we do not believe that it is possible to obtain the vibrational spectrum of CO adsorbed on this catalyst in any other way.

In summary, it can be seen that photothermal spectrometry represents a significant tool that can be added to the FT–IR spectroscopist's arsenal. Further fundamental studies into this technique are needed before all photothermal spectra can be interpreted unequivocally. Nevertheless, we believe that no analytical laboratory in which a wide range of samples is to be characterized using an FT–IR spectrometer can afford to be without a photoacoustic accessory. The distinction between beam deflection and microphonic detection is more difficult to make at this time. During the next year or two, there will doubtless be several comparisons between these two techniques reported in the scientific literature.

REFERENCES

1. A. G. Bell, *Phil. Mag.*, **11**, 510 (1881).
2. J. Tyndall, *Proc. Roy. Soc.* (*Lond.*), **31**, 307 (1881).
3. W. C. Rontgen, *Phil. Mag.*, **11**, 308 (1881).

4. L. B. Kreuzer and C. K. N. Patel, *Science,* **173,** 45 (1971).
5. W. R. Harshbarger and M. B. Robin, *Chem. Phys. Lett.,* **21,** 462 (1973).
6. W. R. Harshbarger and M. B. Robin, *Acc. Chem. Res.,* **6,** 329 (1973).
7. C. K. N. Patel and A. C. Tam, *Rev. Mod. Phys.,* **53,** 517 (1981).
8. E. Voightman, A. Jurgensen, and J. Winefordner, *Anal. Chem.,* **53,** 1442 (1981).
9. M. J. D. Low and G. A. Parodi, *Appl. Spectrosc.,* **34,** 76 (1980).
10. G. Busse and B. Bullemer, *Infrared Phys.,* **18,** 631 (1978).
11. K. Krishnan, unpublished results (1982).
12. M. G. Rockley, *Chem. Phys. Lett.,* **68,** 455 (1979).
13. M. G. Rockley, *Appl. Spectrosc.,* **34,** 405 (1980).
14. P. Mahmoodi, R. W. Duerst, and R. A. Meiklejohn, *Appl. Spectrosc.,* **38,** 437 (1984).
15. S. M. Riseman and E. M. Eyring, *Spectrosc. Lett.,* **14**(3), 163 (1981).
16. S. A. C. Boccara, D. Fournier, and J. Badoz, *J. Appl. Phys. Lett.,* **36,** 130 (1980).
17. M. J. D. Low and G. A. Parodi, *Chem. Biomed. Environm. Instrum.* **11,** 265 (1982).
18. M. J. D. Low and M. Lacroix, *Infrared Phys.,* **22,** 139 (1982).
19. M. J. D. Low, M. Lacroix, and C. Morterra, *Appl. Spectrosc.,* **36,** 582 (1982).
20. D. W. Vidrine, *Appl. Spectrosc.,* **34,** 314 (1980).
21. K. Krishnan, *Appl. Spectrosc.,* **35,** 549 (1981).
22. K. Krishnan, S. Hill, J. P. Hobbs, and C. S. P. Sung, *Appl. Spectrosc.,* **36,** 257 (1982).
23. J. F. McClelland, *Anal. Chem.,* **55,** 89A (1983).
24. M. J. D. Low, M. Lacroix, and C. Morterra, *Spectrosc. Lett.,* **15**(1), 57 (1982).
25. M. M. Garrow, R. K. Burnham, and E. M. Eyring, *Appl. Phys. Lett.,* **33,** 735 (1978).
26. L. B. Lloyd, R. K. Burnham, W. L. Chandler, E. M. Eyring, and M. M. Farrow, *Anal. Chem.,* **52,** 1595 (1980).
27. L. B. Lloyd, S. M. Riseman, R. K. Burnham, E. M. Eyring, and M. M. Farrow, *Rev. Sci. Instrum.,* **51,** 1488 (1980).
28. D. Débarre, A. C. Boccara, and D. Fournier, *Appl. Optics,* **20,** 4281 (1981).
29. M. M. Farrow, R. K. Burnham, M. Auzanneau, S. L. Olsen, N. Purdie, and E. M. Eyring, *Appl. Optics,* **17,** 1093 (1978).
30. L. B. Lloyd, R. C. Yeates, and E. M. Eyring, *Anal. Chem.,* **54,** 549 (1982).
31. A. Rosencwaig and A. Gersho, *J. Appl. Phys.,* **47,** 64 (1976).
32. N. C. Fernelius, *Appl. Optics,* **21,** 481 (1982).
33. H. S. Bennett and R. A. Forman, *J. Appl. Phys.,* **49,** 2313 (1978).
34. J. A. Gardella, G. L. Grobe, W. L. Hopson, and E. M. Eyring, *Anal. Chem.,* **56,** 1169 (1984).
35. Y. C. Teng and B. S. H. Royce, *Appl. Optics,* **21,** 77 (1982).

36. N. L. Rockley, M. K. Woodard, and M. G. Rockley, *Appl. Spectrosc.*, **38**, 329 (1984).

37. S. A. Yeboah, Ph.D. dissertation, Ohio University, Athens, Ohio (1982).

38. R. Tilgner, *Appl. Optics*, **20**, 3780 (1981).

39. P. Kubelka and F. Munk, *Z. Tech. Phys.*, **12**, 593 (1931).

40. P. Kubelka, *J. Opt. Soc. Am.*, **38**, 448 (1948).

41. M. G. Rockley, D. M. Davis, and H. H. Richardson, *Appl. Spectrosc.*, **35**, 185 (1981).

42. S. A. Yeboah, W-J. Yang, and P. R. Griffiths, *Proc. Soc. Photo-Opt. Instrum. Eng.*, **289**, 118 (1981).

43. G. Laufer, J. T. Huenke, B. S. H. Royce, and Y. C. Teng, *Appl. Phys. Lett.*, **37**, 517 (1980).

44. S. A. Yeboah, S-H. Wang, and P. R. Griffiths, *Appl. Spectrosc.*, **38**, 259 (1984).

45. P. R. Solomon and R. M. Carangelo, *Fuel*, **61**, 663 (1982).

46. S. M. Riseman, S. I. Yaniger, E. M. Eyring, D. A. MacInnes, A. G. MacDiarmid, and A. J. Heeger, *Appl. Spectrosc.*, **35**, 557 (1981).

47. S. I. Yaniger, S. M. Riseman, T. Frigo, and E. M. Eyring, *J. Chem. Phys.*, **76**, 4298 (1982).

48. S. I. Yaniger, D. J. Rose, W. P. McKenna, and E. M. Eyring, *Appl. Spectrosc.*, **38**, 7 (1984).

49. M. G. Rockley and J. P. Devlin, *Appl. Spectrosc.*, **34**, 407 (1982).

50. S. R. Lowry, D. G. Mead, and D. W. Vidrine, *Anal. Chem.*, **54**, 546 (1982).

51. M. D. Porter, D. H. Karweik, T. Kuwana, W. B. Theis, G. B. Norris, and T. O. Tiernan, *Appl. Spectrosc.*, **38**, 11 (1984).

52. M. J. D. Low, C. Morterra, and A. B. Severdia, *Spectrosc. Lett.*, **15**, 415 (1982).

53. M. J. D. Low, C. Morterra, A. G. Severdia, and M. Lacroix, *Appl. Surface Sci.*, **13**, 429 (1982).

CHAPTER

10

QUANTITATIVE ANALYSIS

I. BEER'S LAW

Many of the more important data processing algorithms used in FT–IR spectrometry, such as spectral subtraction, Fourier self-deconvolution, and factor analysis, rely on the intensities of spectral bands being linearly proportional to the concentration of each component in the sample. For absorption spectrometry, the law relating band intensity to concentration is the well-known Bouguer–Beer–Lambert law, which we will simply abbreviate to *Beer's law*. For a single solute in a nonabsorbing solvent, Beer's law gives the absorbance at any wavenumber $\bar{v}$ as

$$A(\bar{v}) = -\log_{10}\tau(\bar{v}) = a(\bar{v})bc \tag{10.1}$$

where $\tau(\bar{v})$ and $a(\bar{v})$ are the transmittance and absorptivity at $\bar{v}$, respectively, b is the pathlength, and c is the concentration of the sample. For a mixture of N components, the total absorbance at $\bar{v}$ is given by

$$A(\bar{v}) = \sum_{i=1}^{N} a_i(\bar{v})bc_i \tag{10.2}$$

Deviations from Beer's law are know to occur for many reasons, which have been extensively covered in nearly all introductory texts on instrumental analysis. Suffice it to say that deviations from linearity for plots of $A(\bar{v})$ against c_i can be caused by the effect of stray radiation, insufficient resolution, and chemical effects. The stray light levels in FT–IR spectra are very small, provided that the instrument is operating correctly, the scan speed is constant, the effects of folding are negligible, and phase correction is performed correctly. These criteria are generally fulfilled fairly well on most (but not all) commercial FT–IR spectrometers. Chemical interaction and reactions are independent of the nature of the instrument being used to make the measurement. For FT–IR spectrometers, the principal factor leading to deviations from Beer's law is the effect of insufficient resolution.

The shape of each band in the absorption spectrum of materials in the condensed phase is given, to a good approximation, by the Lorentzian (or Cauchy) function

$$A(\bar{\nu}) = A^t_{peak} \frac{\gamma^2}{\gamma^2 + (\bar{\nu} - \bar{\nu}_0)^2} \qquad (10.3)$$

where A^t_{peak} is the true peak absorbance at the wavenumber of maximum absorption, $\bar{\nu}_0$, and γ is the half-width at half-height. (Note that the FWHH = 2γ.)

The effect of measuring a spectrum at a resolution that is too low is best discussed in terms of the *resolution parameter* ρ, where ρ is the ratio of the nominal resolution to the FWHH of the absorption band. For an FT–IR spectrometer, the nominal resolution is defined as the reciprocal of the maximum retardation. It should be recognized that as the extent of apodization is increased, the FWHH of the ILS function increases even though the *nominal* resolution remains constant.

The measured spectrum is the convolution of the true spectrum and the ILS function. The true spectrum is, of course, the single-beam spectrum and not the absorbance spectrum. Anderson and Griffiths [1] have shown the way in which the measured, or apparent, peak absorbance, A^a_{peak}, varies as a function of A^t_{peak} and ρ. The variation of A^a_{peak} with A^t_{peak} was shown (on a logarithmic scale) for different values of ρ in Fig. 1.11 for boxcar truncation and in Fig. 1.12 for triangular apodization.

For bands of weak and medium intensity ($A^t_{peak} < 0.7$), the gradient of these plots is approximately unity for both boxcar truncation and triangular apodization. In this case, A^a_{peak} is linearly proportional to A^t_{peak} (and hence to sample concentration) even for spectra measured at very low resolution. For any given value of ρ, A^a_{peak} is much closer to A^t_{peak} if interferograms are unweighted than if a triangular apodization function is applied. When $A^t_{peak} < 0.5$ and $\rho < 0.5$ for all bands in the spectrum, the FT–IR spectrum computed without apodization is a very close approximation to the true spectrum.

For intense peaks ($A^t_{peak} > 1$), the behavior of absorbance spectra computed with and without apodization may be quite different. When ρ is large (i.e., very low resolution), A^a_{peak} is always much less than A^t_{peak}. Indeed, for triangularly apodized spectra, A^a_{peak} is always less than A^t_{peak}, no matter at what resolution the spectrum was measured. For unapodized spectra when A^t_{peak} is large and the width of the band is approximately equal to the nominal resolution ($\rho \sim 1$), the measured transmittance may take negative values. Obviously, under these

LINEARITY PERFORMANCE
OVER A 0-3 ABSORBANCE RANGE

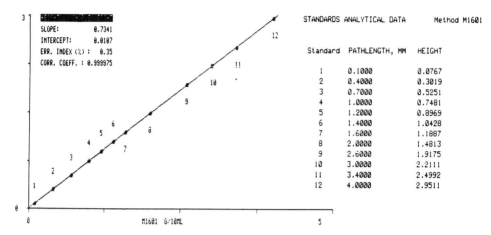

SLOPE: 0.7341
INTERCEPT: 0.0107
ERR. INDEX (%) : 0.35
CORR. COEFF. : 0.999975

STANDARDS ANALYTICAL DATA   Method M1601

Standard	PATHLENGTH, MM	HEIGHT
1	0.1000	0.0767
2	0.4000	0.3019
3	0.7000	0.5251
4	1.0000	0.7481
5	1.2000	0.8969
6	1.4000	1.0428
7	1.6000	1.1887
8	2.0000	1.4813
9	2.6000	1.9175
10	3.0000	2.2111
11	3.4000	2.4992
12	4.0000	2.9511

Fig. 10.1. Linear Beer's law plot over three absorbance units. (Courtesy of the Perkin-Elmer Corporation.)

circumstances, quantitative analysis through the application of Beer's law is not possible.

In summary, quantitative analysis is best performed on bands where peak absorbance is less than 0.7. However, for broad bands ($\rho \leq 0.1$), linear plots of A_{peak}^a against sample concentration can be obtained to absorbances as great as 3.5, as shown, for example, in Fig. 10.1.

II. SPECTRAL SUBTRACTION

a.   General Method

In order to determine the concentration of components in a mixture, it is often necessary to isolate spectral bands due to each component so that measurement of their peak absorbances is possible. Similarly, for the qualitative analysis of components, it may be desirable to eliminate all bands due to the rest of the mixture. One method whereby this was accomplished traditionally in dispersive spectrophotometry has been to match the sample with a suitable reference. For solution spectra, the reference cell usually contains the same solvent as found in the sample cell and the cells are matched for optical thickness. In this way, solvent bands are compensated, and only the absorption bands of the solute are measured. For the accurate determination of absorbances of trace solutes (<0.1%), precision in the optical thicknesses of the two cells becomes critical. An adjustable pathlength cell may be used for reference, but the thickness would have to be controlled to better than 100 nm for many applications. This is beyond the mechanical limits of all variable-pathlength cells. If there is an additional interferent besides the solvent, not only the optical thickness but also the exact composition of the reference solution must be controlled. Usually this is not possible.

Gas phase samples can be handled in a similar way to liquids. In many respects, this is simpler because the partial pressure of gases can often be easily varied to control their absorbances, and hence pathlength becomes less of a problem. Judicious mixing of several gases in the reference cell can eliminate the spectrum of each gas from the spectrum of a mixture.

Unfortunately, these methods are not usually amenable to solid samples. It is often impossible to construct solid samples and references of identical optical thicknesses. For example, polymer films that may have an additive whose concentration is to be determined are difficult to match with another polymer film. One notable exception is semiconductor wafer materials, which can be polished to precise thicknesses, but other methods are more easily implemented for this type of sample. (see Chapter 15, Section III.)

Computerized spectral subtraction permits an interferent or interferents to be removed without a reference of precisely the same optical thickness. Usually two single-beam spectra are collected, one of the sample of interest and the other of the pure solvent or matrix whose spectrum is to be compensated. Both spectra are ratioed against a blank (no cell) reference and absorbance spectra are calculated. Different effective pathlengths can be compensated for by multiplying the reference spectrum by a factor k and subtracting the scaled reference spectrum from the sample spectrum. The scaling factor k is calculated such that

$$(A_1 - kA_2) = 0 \qquad (10.4)$$

for interferent bands, where A_1 is the absorbance of the sample spectrum interferent band and A_2 is the absorbance of the band of the interferent in the reference spectrum. Equation 10.4 states that a given band in the sample and reference spectra is forced to have the same optical thickness.

This is known as *scaled absorbance subtraction*. The methods of calculation of the scaling factor may include the visual comparison of peak absorbances or integrated peak areas, or a least-squares curve-fitting method. In several commercial data systems, the integrated area of selected peaks is calculated, and the scaling factor is estimated from the ratio of these areas. There are several instances where this procedure may not be appropriate, especially when the matrix and analyte have overlapping bands. The best method by which spectral subtraction is executed is to isolate a single band of the component to be removed in both spectra and then to calculate k based on this single band. This factor should be used for the entire spectrum to be subtracted when the difference spectrum is calculated. The only conclusive test for the accuracy of the operation is a careful examination of the difference spectrum. If the interferent or component bands to be removed still exist or are negative in the difference spectrum, the scaling factor must be corrected.

Spectral subtraction is not only used to remove interferents. Koenig [2] has established the principal uses of absorbance subtraction (see Fig. 10.2). The first of these uses is a simple confirmation of the identity of a sample. If a reference spectrum is subtracted from a sample spectrum and the difference is zero, it is clear that the two spectra are of the same compound. This method is not explicitly implemented in spectral search systems because it is too cumbersome, but it is viable for authentication

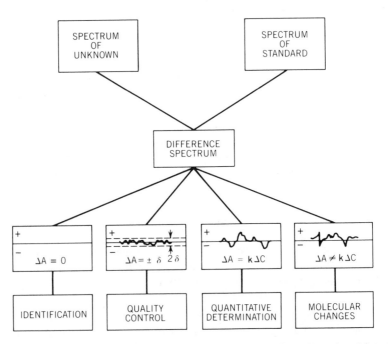

Fig. 10.2. Principal uses of scaled absorbance spectral subtraction. (Reproduced from [2], by permission of the Society for Applied Spectroscopy; copyright © 1975.)

of a sample. In fact, this is probably the best test for unequivocally matching a sample and a reference.

Quality control can be performed using absorbance subtraction spectrometry. If a standard spectrum for a product of known composition is recorded, it can be used as a reference for products produced subsequently. In this case, the scaling factor is set to 1, and difference spectra will show positive or negative bands, indicating an excess or deficiency of particular components. Tolerance limits can be set on these deviations from the standard, and the product can be rejected or accepted. The absorbance subtraction operation forms the basis of many commercial quantitative infrared systems.

As discussed above, scaled absorbance subtraction can be used to remove impurities, solvents, or major components. In this way, bands of constituents of interest may be isolated so that accurate peak measurements can be made. In this application, unexpected contaminants may be discovered as part of the normal procedure.

The final principal use of scaled absorbance subtraction is the identification of molecular changes. In this case, a spectrum of a sample is measured and the sample subjected to a reaction, such as photolysis, adsorption of a gas, temperature change, and so on; then a spectrum of the sample is again measured. A difference spectrum will indicate any molecular changes as any constant features will be removed by the subtraction operation. Small spectral differences are easy to determine by this method. New bands appear as positive absorbances, and bands that have been removed appear as negative absorbances. Derivative-like bands appear when slight band shifts occur due to the change in environment of an infrared chromophore.

It should be noted that scaled absorbance subtraction, or difference spectrometry, is not a panacea for all problems. Scaled absorbance subtraction is not possible if the absorbances are too high. It has been shown [1] that a maximum absorbance for accurate difference spectrometry is 0.7 absorbance units unless the absorbances of the component to be removed in the sample and reference spectra are very similar, so that $k \sim 1$. As discussed in the preceding section, the apparent peak absorbance varies linearly with concentration for even low resolutions (ρ large) when $A < 0.5$. In other words, to ensure that Beer's law is obeyed, a good rule of thumb for absorbance subtraction is to be certain the spectra under study have absorbances less than 0.5 absorbance units. This is a fairly stringent criterion, but if it is ignored, erroneous conclusions may be drawn from difference spectra. It is possible to subtract a series of spectra of minor components from the spectrum of a highly absorbing sample and see no negative bands appear. The temptation is to conclude that all the minor components are present in the sample. Conversely, two peaks of high absorbance may be subtracted and a positive residual remains. This may be assigned to a minor component. In fact, all one may be seeing is a failure in Beer's law.

b. Instrumental Effects

Fourier transform infrared spectrometry exhibits effects that may lead to errors in quantitative analysis. Some of these effects are inherent to FT–IR spectrometry and others pertain to IR spectrometry in general. One of the effects that is restricted to FT–IR spectrometer performance is the stability of the interferometer. Instability in the interferometer may cause frequency shifts in spectra, which may be caused by noncoherence of coadded interferograms. A difference spectrum from two wavenumber-offset spectra will lead to photometric errors.

Instabilities in an interferometer may be classified into three broad types. First, there are short-term instabilities that cause errors in the interferogram during a single scan. These errors include mirror velocity fluctuations and the effect of mirror tilt, which have been covered in Chapter 1. The effect of mirror velocity error is to produce ghost peaks in the spectrum, and a poor mirror drive (tilt) degrades the resolution. Both these effects can be recognized by comparing spectra produced by the instrument with standard spectra. In quantitative analysis, these errors will produce anomalies from which incorrect conclusions may be drawn.

Medium-term fluctuations occur on a scan-to-scan basis or have an effect within a short time, that is, a few minutes. These fluctuations have been studied in detail by de Haseth [3], but they are not particularly important for quantitative work when long signal averaging is used. This is because these fluctuations are either cyclical or random in nature, and their effects are largely canceled by signal averaging. If small numbers of scans are taken, or if single-scan spectra are used for quantitative work, their effects are similar to the long-term fluctuations.

Long-term fluctuations can be exhibited by drifts in the system that occur over several hours. These effects often are not noticeable over a short time of, say, a few minutes. The long-term fluctuations have been studied by van Kasteren [4]. These fluctuations are important when two signal-averaged spectra are collected at different times and a drift or error has been introduced into the interferograms. The resulting spectra may have the wavenumber positions of peaks slightly offset, which can intro-duce photometric errors in difference spectra. Van Kasteren collected 500 scan signal-averaged interferograms every 25 min over a period of

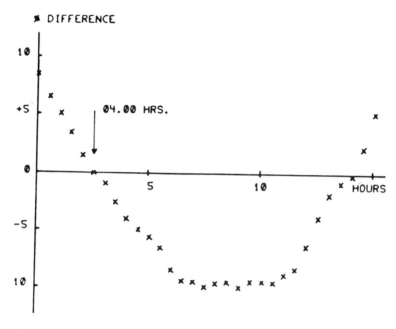

Fig. 10.3. Deviation of interferogram centerburst height over a 15-hr period. Interferograms of 500 scans were collected approximately every 25 min. The interferogram collected at 4 hr was used as a reference for these data. (Reproduced from [4], by permission of D. Reidel Publishing Company; copyright © 1980.)

about 15 hr. One interferogram was chosen as a standard, and it was subtracted from all the other interferograms in the series. A wide variation was found in the difference interferograms in terms of the residual centerburst height. Differences of almost 10% with respect to the original standard interferogram centerburst height were found. The deviation from the standard was found to be cyclic, and the data are shown in Fig. 10.3. Van Kasteren showed that these deviations arose from a sampling error in the interferograms, as discussed in Chapter 1 (see Eq. 1.28). The interferometer used in this study had a separate white-light and He–Ne interferometer, and the coupling between this and the infrared interferometer may have led to the error. A slow thermal or material elasticity change could account for this behavior. Another explanation is that the purge gas was not at a uniform temperature or composition in each arm of the interferometer. This leads to path differences between the arms and offsets in the data. Similar behavior has been seen in the short-term stability study [3]. Instabilities and photometric errors in general are most noticeable on the subtraction of bands of high absorbances because as

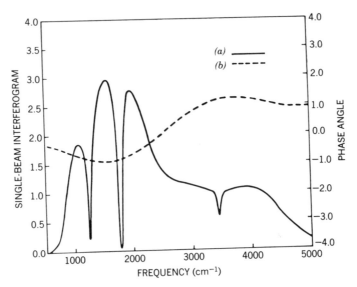

Fig. 10.4. (*a*) Single-beam synthetic spectrum. (*b*) Synthetic phase error spectrum. (Reproduced from [5], by permission of the Society for Applied Spectroscopy; copyright © 1982.)

the samples become more opaque, greater SNRs are required to ensure that noise does not become more significant than the extremely low difference signal. Of course, subtraction of bands with absorbances of greater than 0.5 absorbance units is not recommended, so the tolerances on interferometer instability can be more relaxed.

Sampling error in the interferogram should not be a serious consideration because it should be possible to compensate for any constant sampling error by phase correction. Chase has examined phase correction in detail with respect to photometric errors in the spectrum [5]. The object of the study was to determine if the phase correction algorithm introduced any errors into the spectrum. Inherent with the phase correction is the apodization function, and this must be considered as part of the process. Chase generated a synthetic spectrum for the study that had three absorption bands of 1, 10, and 40% transmittance. The synthetic spectrum and the actual phase error spectrum computed from a Nicolet 7199 are shown in Fig. 10.4. The phase-error function was used to generate an asymmetric interferogram from the synthetic spectrum. The interferogram contained 4096 points on the long side of the centerburst, and the number of points used on the short side varied with the phase correction parameters. The complete interferogram was zero filled to 16-k points.

To calculate the precision of the phase correction algorithms, Chase defined the function percent transmittance error as

$$PTE(\bar{\nu}) = \frac{I_{calc}(\bar{\nu}) - I_{true}(\bar{\nu})}{I_{bkgd}(\bar{\nu})}$$

where $I_{calc}(\bar{\nu})$ is the calculated phase-corrected spectrum, $I_{true}(\bar{\nu})$ is the true spectrum, and $I_{bkdg}(\bar{\nu})$ is the true background spectrum (single-beam spectrum without the absorption bands).

In Chapter 3, the Mertz and Forman phase correction algorithms were described, and these were the two algorithms tested by Chase. Many commercial spectrometers use the Mertz method, where the phase function is calculated from a short double-sided interferogram of 256 points or fewer. The PTE curves for using 256, 512, and 1024 data points in the phase calculation interferogram are shown in Fig. 10.5. If fewer than 256

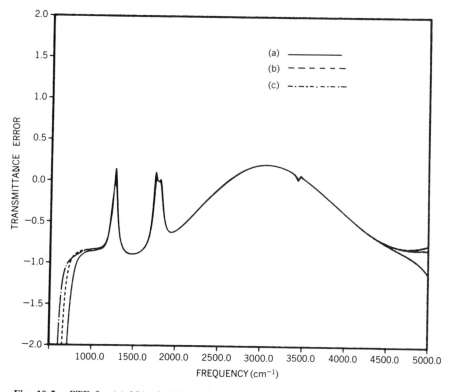

Fig. 10.5. PTE for (a) 256, (b) 512, and (c) 1024 data points in the double-sided phase correction interferogram using the Mertz method. (Reproduced from [5], by permission of the Society for Applied Spectroscopy; copyright © 1982.)

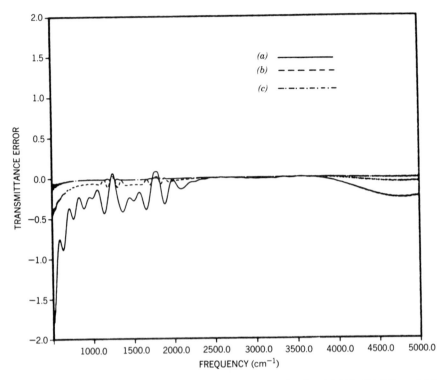

Fig. 10.6. PTE for the Forman phase correction method using (*a*) 255, (*b*) 511, and (*c*) 1023 points in the convolving interferogram. (Reproduced from [5], by permission of the Society for Applied Spectroscopy; copyright © 1982.)

points are used, the PTE is more severe. A few manufacturers employ the Forman phase correction method, which uses convolution in the interferogram domain to remove phase error. PTE curves for the Forman method are shown in Fig. 10.6, and it is clear that this method introduces smaller errors than the Mertz method for a corresponding number of points. Interestingly, the Forman method introduces worse errors than the Mertz method when fewer than 256 points are used for the phase spectrum. Chase also showed that when both the sample and reference single-beam spectra are phase corrected using the same algorithm, the PTE is reduced by a factor of 3–4. The apodization function used to calculate the phase function also has some effect on the PTE, and Chase concluded that triangular apodization was adequate for the Mertz method and a cosine apodization function, $A(x)$,

$$A(x) = \frac{1 + [\cos(\pi \mid x \mid / x_0)]}{2} \qquad (10.5)$$

was best for the Forman method. In $A(x)$, x_0 is the endpoint of the double-sided interferogram with the data centered at $x = 0$. This function is similar in form to the function employed by Forman et al. [6] (see Eq. 3.45).

The source of photometric errors in the Mertz method is the ramp function used to correct for the asymmetrically sampled interferogram (see Chapter 3). The ramp function artificially introduces an extra imaginary component to the complex transform, which manifests itself as photometric error. For absolute photometric accuracy, either the Forman method using the appropriate apodization function or the Mertz method with a symmetrically sampled interferogram, no ramp function, and triangular apodization should be used. A few commercial FT–IR spectrometer manufacturers offer double-sided Mertz computations, and this could be a consideration if absolute accuracy is required. The Mertz phase correction algorithm will probably continue to be favored over the Forman method because it is more computationally efficient, even when array processors are used. Faster, larger computer memories allow double-sided interferograms to be processed, and more manufacturers may incorporate this option in the future.

Another potential source of photometric errors in infrared spectrometry arises from digitization of the signal. Many dispersive as well as FT–IR spectrometers are computer interfaced, but the problems that can be encountered with signal digitization are more severe in FT–IR spectrometry. This is due to the dynamic range problems in FT–IR spectrometry and the required word length for coadded interferograms. In dispersive infrared spectrometry, the dynamic range is smaller and the computer word can contain fewer bits. It is feasible to coadd spectra in FT–IR spectrometry, and this is done in at least one commercially manufactured system. Nevertheless, transform round-off or round-up error can become greater than the errors associated with data storage word size for coadded interferograms. Foskett examined the computer word size requirements for the coaddition of interferograms [7] and showed that words of 24–25 bits are necessary to preserve photometric accuracy for extensively signal averaged interferograms measured using a 15-bit ADC. This is purely a noise consideration, and higher SNRs are achievable with longer word lengths. The number of bits in the computer word is also important for postprocessing of the data. In order to preserve the SNR obtained after the transform, word lengths of 25–26 bits are required for the accurate computation of logarithms in the calculation of the absorbance spectrum.

At this point, it may seem that FT–IR spectrometers contain several sources of photometric inaccuracies. Undoubtedly this is true, but one must realize that an FT–IR spectrometer is capable of outstanding spectral SNRs and is hence capable of demonstrating the sources of photometric inaccuracies described above. Ultimately, all these errors can be recognized and corrected, allowing a very high degree of photometric accuracy in spectral measurements. Clearly, the advantages outweigh the potential disadvantages. The interested reader who would like a more detailed review of photometric inaccuracy or problems in FT–IR spectrometry is referred to an excellent article by Hirschfeld [8].

c. Sample Effects

The sample can have an effect on the photometric accuracy of all infrared spectrometers, but again this effect may be more noticeable in FT–IR spectrometry because of the high SNRs that can be achieved. One problem that develops in infrared spectrometry is the measurement of spectra of thin samples. If a sample is thin and the sides are parallel, interference fringes (also known as channel spectra) will appear in the spectrum. These fringes can obscure small features or at least alter the absorbances. Interference fringes are often encountered in polymer samples, and the fringes are clearly recognizable in the spectrum of a polystyrene film, the ubiquitous standard of infrared spectrometry.

Interference fringes arise from constructive and destructive interference of internally reflected waves inside the sample. This phenomenon is not restricted to polymers but can be observed in virtually any sample. The number of interference fringes, N, over the wavenumber range $\bar{v}_1 - \bar{v}_2$ in a spectrum can be related to the thickness of the sample, b, and its refractive index n:

$$b = \frac{1}{2n} \frac{N}{(\bar{v}_1 - \bar{v}_2)} \tag{10.6}$$

Interference fringes are sinusoidal in nature and are hence seen as a spike (alternately called a secondary burst or signature) in the interferogram. Removal of the secondary burst removes the interference fringes in the spectrum, and two general methods have been developed for their removal [9–12].

One method is to replace the region in the interferogram that contains the spike with zeroes. Upon Fourier transformation, the interference fringes are gone, but of course there is another error introduced. Zero filling an interferogram within the data introduces a discontinuity in the interferogram (actually a discontinuous apodization function). This effect is distributed over the entire spectrum and effectively increases the noise slightly. Its effect, in most cases, is less severe than the interference fringes unless the sample is very thin, where the spike is located very close to the centerburst.

The second method involves taking two interferograms of the same sample, but each one with the sample tilted slightly in the beam path with respect to the other. This has the effect of changing the pathlength inside the sample and hence the number of fringes within a given wavenumber range (Eq. 10.6). The spike in the interferogram is also displaced. Only one interferogram is necessary for calculating the spectrum so its spike is effectively removed by grafting corresponding interferogram points from the other interferogram into the first interferogram. This procedure produces fairly accurate, high SNR results. The two-interferogram method has a second advantage: it can be used to locate the secondary centerburst. It is not uncommon for this feature to occur near the centerburst of the interferogram where it cannot be visually detected. Subtraction of the two interferograms makes the location of the interference signatures evident.

There are two other methods by which interference fringes can be reduced in amplitude or compensated completely. In the first of these, the sinusoidal fringe is simulated and subtracted [13]. In the second, the sample is held at Brewster's angle to the beam [14]. Each of these techniques exhibits the same deficiency for accurate quantitative analysis since each involves the implicit assumption that the refractive index does not vary with wavenumber. In practice, there is always a variation of refractive index across an absorption band. Thus, the largest errors are found in exactly the regions where the effect of interference fringes should be compensated best for accurate quantitative measurements.

There is yet another manner in which interference fringes may be removed, that is, by sample wedging. In this case, the sample does not have parallel sides, and interference fringes cannot occur. Hirschfeld has studied the effects of sample wedging and has shown that this too can lead to photometric error [15]. The effects of sample wedging are small but certainly observable. Figure 10.7 shows a true Lorentzian absorption band (A) and the absorption band of the same sample with a thickness variation of $\pm 25\%$ (B). Hirschfeld then subtracted the two spectra ($B - A$) using three different criteria for choosing the scaling factor. When no negative value was produced by the subtraction, curve C was produced (shown expanded 25 times). Curve D resulted from nulling the peak integral, also shown expanded 25 times. The final difference spectrum, curve E, resulted from nulling the peak centers (again expanded 25 times). In all cases, there was a sizeable differrence. The difference is a Taylor series function in the true absorbance that of course decreases as the true absorbance decreases. If the true absorbance is reduced from 1.0 absorbance

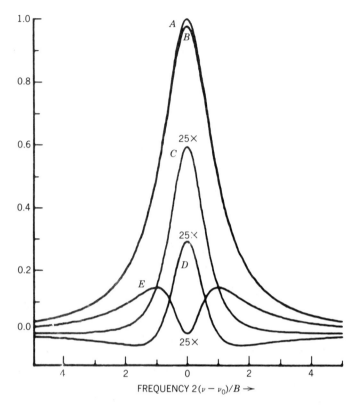

Fig. 10.7. Absorbance subtract spectra for the difference between a true Lorentzian absorbance peak (*A*), and the absorbance peak of a wedged sample (*B*) that has a thickness range of ±25% (*C*) difference spectrum, *B − A*, resulting from no negative value criterion; (*D*) difference spectrum from peak integral nulling criterion; and (*E*) residual spectrum using peak center nulling criterion. (Reproduced from [15], by permission of the American Chemical Society; copyright © 1979.)

unit, as shown in Fig. 10.7, to 0.5 absorbance units, as recommended for absorbance subtraction, the sample wedging error is reduced by at least a factor of 4. Hirschfeld has also shown that in some cases sample wedging absorbance errors can be corrected algorithmically [15].

Other interferences and sources of error may be found in FT–IR spectra. Second-order spectra arise where radiation from the interferometer is reflected back into the interferometer where it is modulated again. The twice-modulated radiation can distort absorption bands and baselines. The source of the back-reflected radiation is often cell windows normal to the interferometer beam.

Gain ranging, that is, applying a gain factor to the wings of an interferogram, can be a useful technique for increasing the dynamic range of an interferogram and improving the SNR of a spectrum. For gain ranging to work, the amplifier gain step must be identical to the nominal gain factor to a precision dictated by the precision of the ADC. Before transformation, the amplifier gain ranging factor must be removed from the interferogram by division with the nominal gain factor. If the two gain factors are not identical, a discontinuity exists in the interferogram and noise is introduced.

Each of these anomalies can be readily recognized by experts, and correction for their effects is usually possible [8]. It should be stressed, however, that FT–IR spectrometry is only *apparently* very susceptible to photometric errors. This is because the SNR of FT–IR spectra can be extremely high and small anomolies can then be detected. Most of these problems can be avoided with reasonable care, observation, and "constructive paranoia" [8].

III. SINGLE-COMPONENT ANALYSIS

Single-component analysis in FT–IR spectrometry is straightforward and largely follows classical methods [16,17]. Traditional spectrometric quantitative analyses were usually carried out in two modes, cell-in-cell-out and baseline. The former was used for fixed-wavenumber analyses where the wavenumber was not scanned and is hence not applicable to FT–IR spectrometry. The latter method requires that a baseline is established from constant spectral features, such as baseline regions between peaks. As long as the determination of the baseline is consistent, the absorbances will be consistent provided there are no deviations from Beer's law.

In classical spectrophotometry, any solvent effects are nullified by using matched sample cells and having the appropriate pathlength of solvent in the reference cell. Computerized spectrometry, both dispersive and Fourier transform, can provide better solvent or reference compensation by scaled absorbance subtraction, as explained in Section II above.

The potential linearity of Beer's law plots for quantitive analysis in FT–IR spectrometry is high. Figure 10.1 shows linearity over more than 3 absorbance units, but this was generated by increasing the pathlength rather than the concentration of the analyte. Linearity may be achieved over more than two absorbance units, as shown in Fig. 10.8a, by increasing the analyte concentration. The original data that produced this plot are given in Table 10.1, and the fact that the last four data were recorded sometime after the initial data attests to the high SNR and photometric

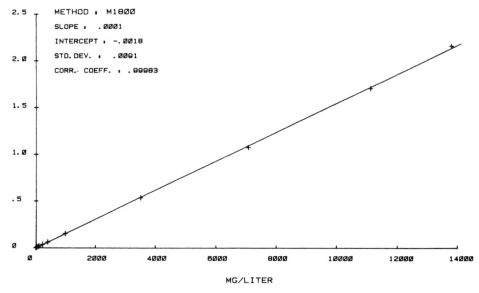

(a)

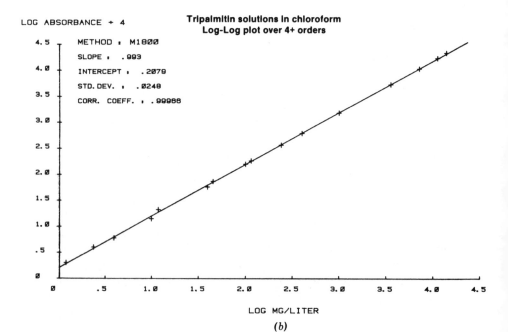

(b)

Table 10.1.   Model 1800 Ordinate Linearity Data: Tripalmitin in Chloroform, Carbonyl at 1737 cm^{-1}

Concentration mg/Liter	Peak Height Absorbance
1.2	0.0002
2.4	0.0004
3.9	0.0006
10.0	0.0014
11.9	0.0021
39.6	0.0057
45.1	0.0073
100.0	0.0158
114.4	0.0182
239.5	0.0366
401	0.0615
1000	0.1516
3526*	0.5354
7080*	1.0731
11115*	1.7032
13804*	2.1550

Data from four sets of samples. Samples marked (*) were analyzed one month after the other samples.

accuracy attainable with FT–IR spectrometers. Figure 10.8b is a log-log plot of the data in Fig. 10.8a to show that the data near zero absorbance are indeed linear.

Once peak absorbances are measured, the concentration of an unknown may be calculated using standard procedures such as least-squares approximations [16].

IV.   MULTICOMPONENT ANALYSIS

The quantitative analysis of single-component systems in infrared spectrometry is important, but this is not the usual situation. It is relatively rare that a single pure solute is present; in practice, one usually finds several components in a sample. For this reason, multicomponent quantitative analysis is becoming more important, and infrared spectrometry is capable of very precise results. Many methods of multicomponent quantitative analysis have been proposed. The major methods are reviewed here.

←

Fig. 10.8.   (a) Linear Beer's law plot of tripalmitin in chloroform as the concentration of the analyte was changed. (b) Log-log plot of the data in (a). (Courtesy of the Perkin-Elmer Corporation.)

a. Factor Analysis

Factor analysis is a matrix algebraic technique that can identify the number of components in a mixture, determine the concentration, and reproduce the spectrum of each component. This may seem to be a supernatural technique, but all these tasks are realizable within the constraints imposed by the spectral noise, the adherence to Beer's law for each component, and, to a certain extent, the ability of the analyst.

Factor analysis is a general technique and has been applied to a wide variety of problems [18], but it was first used for infrared multicomponent analyses by Antoon et al. [19]. It was shown previously that Beer's law for a series of components at a single absorbance could be written as

$$A(\bar{\nu}) = \sum_{i=1}^{N} a_i(\bar{\nu})bc_i \tag{10.2}$$

where the subscript i refers to the ith component of the mixture. A spectrum of a mixture may be represented as

$$\mathbf{A}_j = \sum_{i=1}^{N} \mathbf{a}_i c_{ji} \tag{10.7}$$

where $\mathbf{A}$, $\mathbf{a}$, and $\mathbf{c}$ are vectors, indicating they have many components. The subscript j indicates this is the jth spectrum, and the pathlength term b has been dropped on the assumption that it is a constant scalar and can be included in the absorptivity vector $\mathbf{a}_i$. If a series of absorbance spectra is collected in which the concentrations of the components are linearly independent, Eq. 10.7 becomes a matrix equation:

$$\mathbf{A} = \mathbf{EC} \tag{10.8}$$

where $\mathbf{E}$ is the matrix equivalent of the vector $\mathbf{a}_i$. The matrix $\mathbf{A}$ is rectangular, having the number of columns equal to the number of absorption wavenumbers and as many rows as these are spectra. By the conventions of matrix algebra, $\mathbf{E}$ has as many columns as $\mathbf{A}$ but has as many rows as components. The $\mathbf{C}$ matrix has a column for every component and a row for every spectrum. Unfortunately, no information is known about $\mathbf{E}$ and $\mathbf{C}$. If the number of components and their molar absorptivities were known, the problem would reduce to a collection of simultaneous equations. All that is known is $\mathbf{A}$, yet $\mathbf{E}$ and $\mathbf{C}$ can be calculated if the additional condition is imposed that no component concentration is a constant ratio of the concentration of another component.

Factor analysis reduces the matrix A to a diagonal matrix (only the diagonal terms are nonzero). The nonzero diagonal terms are called the eigenvalues. The number of nonzero eigenvalues is often called the rank of the matrix. It is a practical circumstance of spectrometry that data contain noise, and the noise prevents the occurrence of zero eigenvalues. In other words, some discretion must be used in the interpretation of results.

The method by which the rank of A is determined is to transform it linearly to a new matrix D:

$$D = TA + B \qquad (10.9)$$

where T is a unit diagonal matrix (all 1's along the diagonal), and each row of B is equal to $-x_j$, the mean absorbance in the jth spectrum. Matrix D is used to calculate the mean absorbance covariance matrix Z:

$$Z = DD^T \qquad (10.10)$$

where D^T is the transpose of D, that is, the rows and columns have been interchanged. Here Z is a real symmetric matrix; that is, the diagonal should contain all ones and the data are reflected about the diagonal.

The covariance matrix can be reduced to the diagonal matrix that contains all the appropriate eigenvalues by decomposing it with the appropriate eigenvector matrix X:

$$X^T Z X = \Lambda \qquad (10.11)$$

where Λ is the diagonal matrix containing the eigenvalues λ_i. In an ideal case the λ_i eigenvalues would have nonzero values for the number of components in the mixture, but in the presence of noise, every eigenvalue will probably be nonzero. Indicator functions can be calculated that test the eigenvalues to determine which components should not be attributed to noise. One such indicator function goes through a minimum as the correct number of eigenvalues is found [20]. The form of the indicator function, IND, is

$$\text{IND} = \frac{\text{RSD}}{(p - n)^2} \qquad (10.12)$$

where RSD is the residual standard deviation calculated from the eigenvalues, p is the number of mixtures, and n is the number of eigenvalues. The residual standard deviation is calculated from the eigenvalues:

$$
\text{RSD} = \left(\frac{\displaystyle\sum_{j=n+1}^{p} \lambda_j^0}{r(p - n)} \right)^{1/2}
\tag{10.13}
$$

where r is the number of absorbances and λ^0 corresponds to the eigenvalues due to noise.

An example of factor analysis has been given for mixtures of compatible and incompatible polymers [21]. A compatible polymer mixture will give a spectrum of a "complex" that is a result of interaction of the components. An incompatible mixture forms no complex, so only the original components are present in a mixture. The two mixtures formed for the

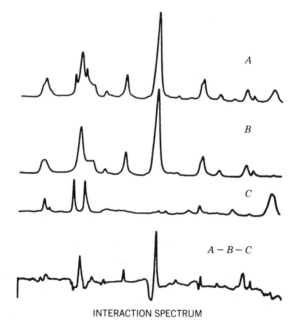

INTERACTION SPECTRUM

Fig. 10.9. (*A*) Spectrum of PPO/PS compatible polymer blend, (*B*) pure PPO spectrum, (*C*) pure PS spectrum, (*D*) complex spectrum resulting from $A - B - C$. Spectra are recorded over the range $1700-700\ \text{cm}^{-1}$. (Reproduced from [21], by permission of the Society for Applied Spectroscopy; copyright © 1981.)

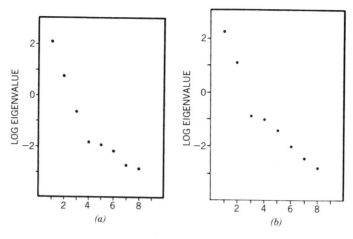

Fig. 10.10. (*a*) Plot of log eigenvalues versus component number for the PPO/PS blend. (*b*) Plot of log eigenvalues versus component number for the P4ClS/PPO blend. (Reproduced from [21], by permission of the Society for Applied Spectroscopy; copyright © 1981.)

test were poly(phenylene oxide), PPO, with polystyrene, PS, for the compatible mixture and PPO with poly(4-chlorostyrene), P4ClS, for the incompatible mixture. Therefore, the PPO–PS mixture should produce at least three components and the P4ClS–PPO mixture only two. Spectra of the PPO–PS mixture and its components are shown in Fig. 10.9. The difference of the mixture, *A*, minus both pure components, PPO(*B*) and PS(*C*), yields a nonzero difference. This difference spectrum is the mixture complex spectrum. The mixture complex spectrum is sometimes called an action spectrum.

Nine different component combinations were made of the PPO–PS mixture and eight of the P4ClS–PPO system. Spectra were collected of each sample and factor analysis applied to the data for both mixture systems. Eigenvalues were calculated, and plots of the log eigenvalues versus component number for the PPO–PS and P4ClS–PPO mixtures are shown in Figs. 10.10*a,b*, respectively. As can be seen in the figures, there is no clear indication as to how many components are in each mixture. There is a clear break in each curve, but it is unclear whether the breakpoint belongs to the component or the noise set. The indicator functions for these data are presented in Fig. 10.11, and it is clear there is a minimum of three components for the PPO–PS mixture (lower curve) and two components for the incompatible system. At this point, the number of components in a mixture has been established.

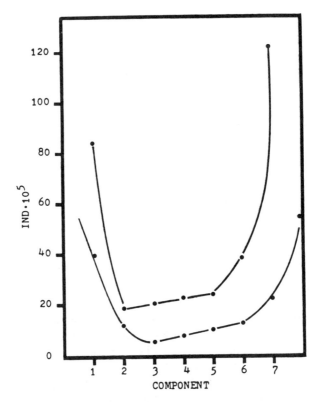

Fig. 10.11. Plot of the indicator functions (IND) for the PPO/PS blend (lower curve) and the P4ClS/PPO blend (upper curve). (Reproduced from [21], by permission of the Society for Applied Spectroscopy; copyright © 1981.)

In Eq. 10.11, the eigenvalue matrix Λ was found by multiplying the covariance matrix Z with the eigenvector matrix X and its transpose. It can be shown that

$$X^T = C \qquad (10.14)$$

and

$$DX = E \qquad (10.15)$$

Equations 10.14 and 10.15 do not have the same physical significance as one might expect; they are simply representations of concentration and molar absorptivities. Matrix X^T must be rotated by a nonorthogonal rotation matrix **R** to bring **C** into the proper hyperspace orientation. That is,

$$C = X^T R \tag{10.16}$$

and it follows that

$$E = DC^T(CC^T)^{-1} \tag{10.17}$$

Once the proper orientations of **C** and **E** have been found, the component spectra can be generated. The generation of the nonorthogonal rotation matrix **R** is based on a method of finding unique wavenumbers for each component of the system. This method was originally developed by Knorr and Futrell for mass spectra [22].

An example of the rotation method can be demonstrated with a hexane–chloroform mixture [23]. Six mixtures were prepared, and their spectra are shown in Fig. 10.12. The factor analysis results are shown in Fig. 10.13, and indeed, spectra of the pure components have been generated. The anomaly in the hexane spectrum at 1220 cm^{-1} is probably due to nonlinearity in Beer's law.

Factor analysis is not a multicomponent quantitative analytical tech-

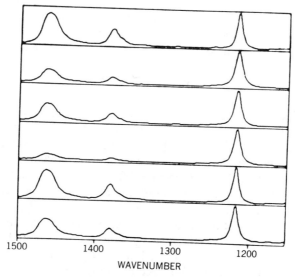

Fig. 10.12. Mixture spectra of hexane and chloroform. (Reproduced from [23], by permission of the American Chemical Society; copyright © 1983.)

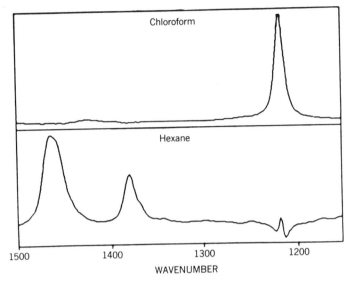

Fig. 10.13. Pure component spectra from mixtures in Fig. 10.12 using factor analysis. (Reproduced from [23], by permission of the American Chemical Society; copyright © 1983.)

nique per se, but it is a method by which the number of components in a mixture may be identified and their pure component spectra generated. Once the pure components are known, difference spectrometry can be used to determine the concentration of the components in mixture spectra.

b. Least-Squares Methods

One of the simple methods by which multicomponent quantitative analysis may be accomplished is by scaled absorbance subtraction as long as Beer's law is obeyed. However, spectral subtraction requires that an isolated feature common to two spectra can be removed and used as a point of reference. If this is done, the difference spectra can be used for single-component concentration determinations and the total composition determined. If no isolated band, or point of reference, can be found, scaled spectral subtraction is not feasible, or at least not very accurate.

Least-squares regression analysis methods have been applied to quantitative analysis of infrared spectra. Least-squares approximations have been in use for some time [16], but recently more sophisticated approaches have been presented. It must be noted, however, that these techniques are not specific to FT-IR spectrometry but are applicable to all forms of computerized infrared spectrometry.

Early least-squares methods fit Gaussian or Lorentzian peak shapes to the spectra, but these band shapes are not general and hence not as accurate as other criteria. One procedure of least-squares curve fitting was presented by Antoon et al. [24]. In this method, no prescribed band shapes were fit to spectra; rather standard spectra were fit to the mixture. The general form of the equation is

$$\sum_{i=1}^{N} R_{i,k} = \sum_{j=1}^{N} \left(\sum_{i=1}^{M} W_i R_{i,j} R_{i,k} \right) x_j \qquad (10.18)$$

where N is the number of spectral data points; M is the number of standard spectra; $R_{i,k}$ is the absorbance at the ith data point in the kth standard spectrum; W_i is the weighting factor, which is the reciprocal of the absorbance at the ith data point in the mixture spectrum; and x_j is the scaling factor for each of the standard spectra at each data point to fit the mixture spectrum.

Error analysis was completed on the least-squares analysis and standard deviations of the scaling values, x_j, were obtained for model and practical systems [24]. Antoon et al. were able to show that these regression analysis techniques were able to determine the components in a mixture of the three xylene isomers to within 1% (absolute) of the true values. Similar results were found for polymer blends.

The method presented by Antoon et al. [24] made two general assumptions. First, it assumed that the baseline was zero. Second, the weighting factor W_i was equal to $1/A_i$, where A_i is the absorbance. Haaland and Easterling [25] used the procedure of Antoon et al. [24] but modified the two assumptions mentioned above. They applied four different criteria for the baseline and included these in the least-squares fit. The four criteria are: (1) the baseline is zero, (identical to the method of Antoon et al. [24]); (2) there is a linear baseline across the spectral region of interest; (3) the baseline is linear over each individual peak in the region of interest; and (4) the baseline is nonlinear but continuous. The second assumption involved the weighting factor x_j. The original work had related the scaling factor to $1/A_i$, which assumes the noise is a function of the absorbance. (This is true for gamma ray spectrometric detectors, for which the technique was originally developed [26].) For transmittance spectra, the noise is constant across the spectral range, and hence W_i should be equal to T_i^2, where T_i is the transmittance.

By fitting the spectra and the baseline, Haaland and Easterling [25] were able to improve on the accuracy by which each component in a mixture could be determined. For a mixture of the three xylene isomers, the relative error was kept below 0.25% for all three isomers for high-SNR (250:1) spectra. Even for very low SNR (2.5:1) spectra, the relative error was kept below 20%. The three nonzero baseline criteria yielded different results depending on the SNR of the mixture spectrum. The baseline fit could be chosen on this basis. Regardless, inclusion of the baseline in the least-squares regression fit significantly improves the accuracy of the method.

Least-squares regression analyses need not use pure components as the standards and calculate scaling factors. Another method involves using mixtures of known composition as the standards and calculating the concentrations of an unknown mixture directly [27,28]. It is possible to write Beer's law as

$$A = kc \qquad (10.19)$$

where k is equivalent to the product of the absorptivity and the pathlength. For a multicomponent system, absorbances are additive, so the total absorbance at a single wavenumber is

$$A = \sum_{i=1}^{N} k_i c_i \qquad (10.20)$$

where k_i is the proportionality constant for the ith component, c_i is its concentration, and N is the number of components. A series of such equations may be written for each wavelength in a spectrum:

$$A_j = \sum_{i=1}^{N} k_{ji} c_i \qquad j = 1, 2, \ldots, N \qquad (10.21)$$

where A_j is now the absorbance at the jth wavenumber and k_{ji} is the proportionality constant at the jth wavenumber for the ith component. Equation 10.21 can be reduced to matrix notation:

$$\mathbf{A} = \mathbf{KC} \qquad (10.22)$$

where $\mathbf{A}$ and $\mathbf{C}$ represent one-dimensional matrices, or vectors. The $\mathbf{K}$ matrix can be determined from a series of pure standards, and the concentration C of the components of an unknown can be calculated if the absorbances A are measured:

$$\mathbf{C} = \mathbf{K}^{-1}\mathbf{A} \qquad (10.23)$$

This is largely equivalent to the least-squares fit [24,25] and does not take interactions of components into accounts. If mixtures of standards are used, A becomes **A** where now there is a column of absorbances for each mixture and C becomes **C** and there is a column of concentrations for each mixture:

$$\mathbf{A} = \mathbf{KC} \tag{10.24}$$

Both the number of data points and number of components can be over-determined. In other words, to form Eq. 10.24, one needs an absorbance and standard mixture for each component in the mixture. If extra absorbances are measured and extra standard mixtures used, redundancy is added and random errors are reduced.

Unfortunately, overdetermination of the data forces the matrices to be no longer square, and the inverse cannot be calculated. However, other operations are available so that **K** and **C** can be calculated. Once the absorbances of a set of mixture standards have been measured, the **K** matrix must be calculated, so

$$\mathbf{AC^T} = \mathbf{KCC^T} \tag{10.25}$$

where $\mathbf{C^T}$ is the transpose of **C**, and the product of any matrix with its transpose produces a square matrix, of which the inverse can be calculated:

$$\mathbf{K} = \mathbf{AC^T(CC^T)^{-1}} \tag{10.26}$$

Once **K** has been calculated and the absorbances of an unknown system have been measured, **C** can be calculated using Eq. 10.23. If **K** is not square, **C** can still be determined:

$$\mathbf{C} = \mathbf{(K^TK)^{-1}K^TA} \tag{10.27}$$

This procedure has been called the K-matrix representation of Beer's law, and it is quite versatile. If Beer's law is nonlinear, the law can be approximated as being linear over a finite range of concentrations. This then gives the linear portion of the curve a nonzero intercept at zero concentration. This effect can be incorporated into the K-matrix representation by putting an extra column of k terms (for the intercept) into the **K** matrix and an extra row of 1's into the **C** matrix. The **A** matrix is unaffected.

The problem with the K-matrix representation is that the concentration is not calculated directly. The concentration matrix must be inverted to calculate the proportionality matrix, and hence the determinant of the concentration matrix must be non-zero. If the determinant is zero, the inversion produces an indeterminate result. The K-matrix representation also requires that the standards must not have impurities. If impurities are present, they must be included in the K-matrix representation. In other words, their presence must be known. The potential problems with the K-matrix approach can be overcome using the P-matrix representation [27,28].

In the P-matrix representation, concentration is written as a function of absorbance:

$$\mathbf{C} = \mathbf{PA} \qquad (10.28)$$

where $\mathbf{A}$ and $\mathbf{C}$ have the same meanings as before, but now the proportionality matrix is called $\mathbf{P}$ to distinguish it from $\mathbf{K}$. Nonlinear Beer's law conditions can be imposed on Eq. 10.28 where an extra column is added to the $\mathbf{P}$ matrix for the intercept and a row of 1's is added to the $\mathbf{A}$ matrix. The $\mathbf{P}$ matrix can be solved by the general least-squares approach:

$$\mathbf{P} = \mathbf{CA}^T(\mathbf{AA}^T)^{-1} \qquad (10.29)$$

Once $\mathbf{P}$ is determined, $\mathbf{C}$ can be calculated for an unknown mixture using Eq. 10.28.

Brown [27,28] examined the P-matrix representation for several model systems under conditions of linear and nonlinear adherence to Beer's law. For mixtures of compounds with very similar spectra, p-xylene and 1,2,4-trimethylbenzene, the P-matrix approach gave correct concentrations within 1% (absolute) of unknowns over a wide range of concentrations provided the appropriate sampling procedures were observed [27]. That is, the spectral regions from which $\mathbf{A}$ was measured had to have significant spectral differences; otherwise, the method could not distinguish the spectra. Brown also measured the concentrations of lipids in mixtures using the P-matrix representation [29]. In this study, the concentrations of the lipids showed a deviation from Beer's law; yet the errors were low in the determinations if the absorbances did not become too high ($A^t \leq 1.0$).

The P-matrix approach has its shortcomings too. The primary problem is that many mixtures must be prepared and measured to ensure that enough terms are present in the P matrix. The main limitation to this approach is, of course, that it can be very time consuming.

V. CONCLUSIONS

Infrared quantitative analysis has reached the point where it is an extremely reliable tool for single and multicomponent analyses. Many manufacturers of infrared spectrometers, both dispersive and Fourier transform, are beginning to address the quality assurance/quality control market with low-cost instrumentation specifically for quantitative analysis. As spectrometers, software, and computer hardware become more reliable and sophisticated, undoubtedly we shall see more general acceptance and higher performance of infrared quantitative analysis.

REFERENCES

1. R. J. Anderson and P. R. Griffiths, *Anal. Chem.*, **47**, 2339 (1975).
2. J. L. Koenig, *Appl. Spectrosc.*, **29**, 293 (1975).
3. J. A. de Haseth, *Appl. Spectrosc.*, **36**, 544 (1982).
4. P. H. G. van Kasteren, Quantitative Aspects of FT-IR in Industrial Applications, in *Analytical Applications of FT-IR to Molecular and Biological Systems*, (J. R. Durig, ed.), D. Reidel, Dordsect, Holland, 1980.
5. D. B. Chase, *Appl. Spectrosc.*, **36**, 240 (1982).
6. M. L. Forman, W. H. Steel, and G. A. Vanasse, *J. Opt. Soc. Am.*, **56**, 59 (1966).
7. C. Foskett, *Appl. Spectrosc.*, **30**, 531 (1976).
8. T. Hirschfeld, Quantitative FT-IR: A Detailed Look at the Problems Involved, in *Fourier Transform Infrared Spectroscopy: Applications to Chemical Systems*, Vol. 2 (J. R. Ferraro and L. J. Basile, eds.), Academic Press, New York (1979).
9. T. Hirschfeld and A. W. Mantz, *Appl. Spectrosc.*, **30**, 552 (1976).
10. F. R. S. Clark and D. J. Moffatt, *Appl. Spectrosc.*, **32**, 547 (1978).
11. A. Baghdadi and R. A. Forman, *Appl. Spectrosc.*, **35**, 473 (1981).
12. P. R. Griffiths, *Appl. Spectrosc.*, **36**, 319 (1982).
13. T. Hirschfeld, *Appl. Opt.*, **17**, 1400 (1978).
14. Harrick Scientific Corporation, Box 1288, Ossining, NY 10562, sample attachment BPH-S1G, Polarizer and Brewster's Angle Film Sample Holder.
15. T. Hirschfeld, *Anal. Chem.*, **51**, 495 (1979).
16. R. P. Bauman, *Absorption Spectroscopy*, Wiley, New York (1962).
17. W. J. Potts, Jr., *Chemical Infrared Spectroscopy*, Vol. I, Wiley, New York (1963).
18. E. R. Malinowski and D. G. Howery, *Factor Analysis in Chemistry*, Wiley-Interscience, New York (1980).
19. M. K. Antoon, L. D'Esposito, and J. L. Koenig, *Appl. Spectrosc.*, **33**, 351 (1979).
20. J. L. Koenig, *Adv. Polym. Sci.*, **54**, 87 (1983).

21. J. L. Koenig and J. M. Tovar Rodriquez, *Appl. Spectrosc.*, **35**, 543 (1981).
22. F. J. Knorr and J. H. Futrell, *Anal. Chem.*, **51**, 1236 (1979).
23. P. C. Gillette, J. B. Lando, and J. L. Koenig, *Anal. Chem.*, **55**, 630 (1983).
24. M. K. Antoon, J. H. Koenig, and J. L. Koenig, *Appl. Spectrosc.*, **31**, 518 (1977).
25. D. M. Haaland and R. G. Easterling, *Appl. Spectrosc.*, **36**, 665 (1982).
26. J. A. Blackburn, *Anal. Chem.*, **37**, 1000 (1965).
27. C. W. Brown, P. F. Lynch, R. J. Obremski, and D. S. Lavery, *Anal. Chem.*, **54**, 1472 (1982).
28. M. A. Maris, C. W. Brown, and D. S. Lavery, *Anal. Chem.*, **55**, 1694 (1983).
29. H. J. Kisner, C. W. Brown, and G. J. Kavarnos, *Anal. Chem.*, **55**, 1703 (1983).

DISPERSIVE FOURIER TRANSFORM
SPECTROMETRY

I. INTRODUCTION

For most spectrometric measurements made using a Michelson interferometer, the sample is held between the interferometer and the detector. Therefore, both partial waves from the interferometer experience the same attenuation and phase shift when they interact with the sample. Because the phases of both waves are shifted by the same amount, this shift has no effect on the resultant phase of the measured interferogram. If a sample of thickness d is placed in *one arm* of the interferometer, the optical pathlength in that arm is increased by an amount equal to $2d\{n(\bar{v}) - 1\}$, where $n(\bar{v})$ is the refractive index of the sample at wavenumber $\bar{v}$ and the phase can shift dramatically.

This effect has been nicely illustrated by Bell [1] for measurements of polyethylene and Mylar made in the far infrared. Figure 11.1 shows an interferogram measured in the conventional fashion using a slow-scanning interferometer (which generates a very symmetrical interferogram). Below this is shown an interferogram measured when a sheet of polyethylene is inserted in one arm of the interferometer. The refractive index of polyethylene is essentially constant across the spectral range, and consequently the interferogram is merely shifted with respect to the background interferogram but is not distorted. Measurement of the refractive index of polyethylene would be trivial in this case if the sample thickness d were accurately known. When a sheet of Mylar is inserted in the same arm, however, the resultant interferogram is asymmetrical, demonstrating that the refractive index of Mylar in the far infrared varies with wavenumber. Calculation of the refractive index (more precisely, the *complex* refractive index, since the absorption coefficient can also be calculated) is no longer trivial, but it can be achieved. It is the purpose of this chapter to discuss briefly the sampling and computational techniques used in dispersive Fourier transform spectrometry (DFTS), where the sample is placed in one arm of the interferometer.

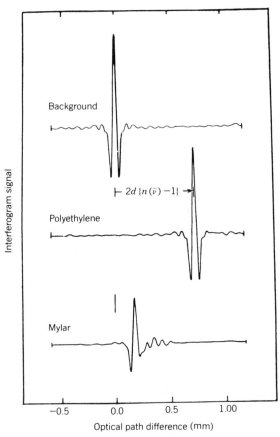

Fig. 11.1. Far-infrared interferograms plotted in the region of the centerburst. (*Upper trace*) Background interferogram with no sample in either arm of the interferometer; (*middle*) interferogram measured with a 1.51-mm-thick polyethylene film in one arm of the interferometer; (*bottom*) interferogram measured with 0.25-mm-thick Mylar film in one arm of the interferometer. (Reproduced from [1].)

To understand the DFTS exponent, one must consider the *complex insertion loss* $\hat{L}(\bar{\nu})$, which is the complex factor by which the insertion of a sample into the path of an electromagnetic wave changes the complex electric field at some point further along the path. It may be written as

$$\hat{L}(\bar{\nu}) = L \exp(i\phi) \qquad (11.1)$$

where ^ indicates a complex quantity and L and ϕ are the modulus and phase, respectively; both L and ϕ are wavenumber dependent. Typical schematic optical configurations for conventional and dispersive measurements are shown in Fig. 11.2 for transmissive and reflective samples [2].

For transmission measurements of samples with a Fresnel transmission coefficient of $\hat{t}(\bar{v})$, the complex insertion loss for one pass through the sample is

$$\hat{L}(\bar{v}) = \hat{t}(\bar{v}) \cdot \exp(-2\pi i \bar{v} d) \tag{11.2}$$

and it can be shown that in this case

$$\hat{L}(\bar{v}) = \frac{4\hat{n}}{(1 + \hat{n})^2} \cdot \exp[2\pi i(n - 1)d] \cdot \exp\left(\frac{\alpha d}{2}\right) \tag{11.3}$$

where $\hat{n}$ is the complex refractive index given by

$$\hat{n} = n + ik \tag{11.4}$$

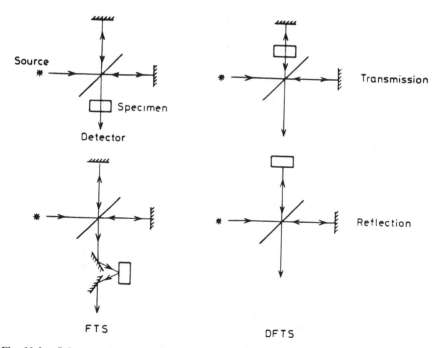

Fig. 11.2. Schematic representations of the conventional and dispersive FT–IR interferometer configurations used for transmission and reflection measurements on solid samples. The DFTS configurations for liquids and gases are similar to those shown for solids but require the use of windows to contain the sample and define its shape. (Reproduced from [2], by permission of the author.)

where k is the absorption index and α is the power absorption coefficient given by $4\pi\bar{\nu}k$.

The insertion loss is also related to the complex spectra of the sample, $\hat{S}_S(\bar{\nu})$, and of the reference with no sample present, $\hat{S}_R(\bar{\nu})$, by

$$\hat{L}(\bar{\nu}) = \frac{\hat{S}_S(\bar{\nu})}{\hat{S}_R(\bar{\nu})} \tag{11.5}$$

In transmission DFTS measurements, the beam usually passes through the sample twice, so that

$$\{\hat{L}(\bar{\nu})\}^2 = \frac{\hat{S}_S(\bar{\nu})}{\hat{S}_R(\bar{\nu})} \tag{11.6}$$

Reflection measurements in which the sample exactly replaces a plane mirror may be treated in an analogous fashion. If the Fresnel reflection coefficient of the sample is $\hat{r}(\bar{\nu})$, the complex insertion loss is given by

$$\hat{L}(\bar{\nu}) = \hat{r}(\bar{\nu})\cdot\exp(-i\pi) \tag{11.7}$$

assuming that the mirror that was replaced was perfect [with a reflectivity of $1.0\exp(i\pi)$]. In this case, the phase term $-\pi$ represents the loss of

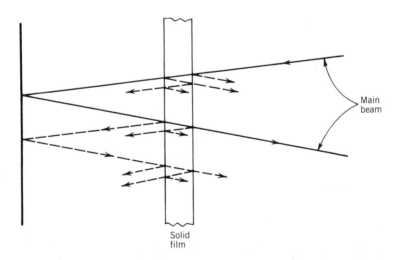

Fig. 11.3. The beams that can reach the detector caused by reflections from various surfaces of a sample inserted in one arm of the interferometer. Only the beam indicated by the solid line is of interest.

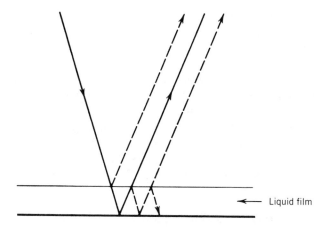

Fig. 11.4. The number of unwanted beams is reduced for liquid samples if the liquid is held on the fixed mirror of the interferometer, provided that this mirror is held in a horizontal plane.

the mirror phase when it is replaced by the sample. Thus,

$$\hat{L}(\bar{v}) = \frac{1 - \hat{n}}{1 + \hat{n}} \cdot \exp(-i\pi) \qquad (11.8)$$

If contributions to the interferogram from internally reflected rays are resolved, $\hat{t}(\bar{v})$ or $\hat{r}(\bar{v})$ may be expressed as the summation of the amplitudes of these rays. Thin films of the type used for transmission measurements of samples for which k is large often generate several rays because of reflections from the front and back surfaces, see Fig. 11.3. For reflection measurements, these rays can also be measured, as shown in Fig. 11.4, but the sample is usually much thicker than for transmission measurements so that the path difference imposed by the sample (or the window, if one is present) can often be made greater than the maximum optical retardation of the interferometer. In practice, reflection measurements are usually made when the absorption of the sample is so high that transmission measurements become unfeasible, largely because of the difficulty of accurately measuring the phase of the insertion loss in a reflection measurement.

The phase of the insertion loss can be obtained by the expression

$$\phi = \text{ph}\{\hat{S}_S(\bar{v})\} - \text{ph}\{\hat{S}_R(\bar{v})\} \qquad (11.9)$$

where ph{ } refers to the phase of the complex spectrum referred to inside the brackets. For this equation to be useful for the determination of refractive index, the phases of the two complex spectra must be referred to exactly the same mirror position. Originally [3] this statement was interpreted to mean that the position of the zero retardation points in the sample and reference interferograms must be known very accurately, and elaborate interpolation procedures were developed to refer the two-phase spectra to exactly the same point.

More recently [4,5] it has been pointed out that these procedures are not only unnecessary but are also susceptible to systematic errors. It was shown that by application of the Fourier shift theorem the phase of the complex insertion loss can be determined from the measured spectra by using the expression

$$\phi = ph\{\hat{S}_S(\bar{\nu})\} - ph\{\hat{S}_R(\bar{\nu})\} + 2\pi\bar{\nu}\bar{\delta} \qquad (11.10)$$

where $\bar{\delta}$ is the optical retardation between the two points used as the origins of computation for the two complex transforms used to calculate $\hat{S}_S(\bar{\nu})$ and $\hat{S}_R(\bar{\nu})$. Since $\bar{\delta}$ is an integral number of sampling intervals, it is known exactly, and therefore application of Eq. 11.10 removes the systematic error.

In reflection measurements, the phase of the insertion loss may be only a few tens of milliradians, so that even a small phase error could constitute a major part of the measured phase. Extremely stable interferometers are therefore needed for this work. In transmission measurements, phase shifts of many radians can be generated. Therefore, in this respect (contrary to the conclusion that was drawn earlier), transmission measurements are to be preferred over reflection measurements.

A very large number of far-infrared DFTS measurements has been described in the research literature. A bibliography with over 130 references in this field was published by Birch and Parker in 1979 [6], and the level of effort in DFTS does not appear to have slowed down over the following years. Most of the applications fall into the realm of chemical physics and therefore are somewhat outside the scope of this monograph. Nevertheless, a relatively brief summary of some of the published work in far-infrared DFTS will be attempted in the remainder of this chapter. It should be noted that very little work in the mid or near infrared has been described.

II. GASES AND VAPORS

An accurate knowledge of the refractive index spectrum of atmospheric species in the far infrared is starting to become necessary in view of the interest in telecommunications, radar, and remote sensing using submil-

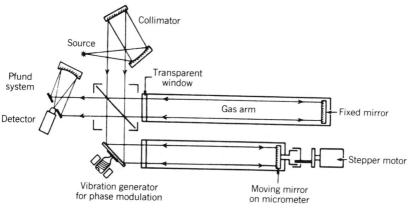

Fig. 11.5. Schematic representation of a dispersive phase-modulated interferometer for measuring the refractive index spectra of gases. (Reproduced from [7], by permission of Pergamon Press; copyright © 1978.)

limeter and near-millimeter wavelengths, especially for calculations of electromagnetic propagation through turbulent atmospheres. It is also possibly easier to calculate the integrated absorption of a line from the refractive index dispersion across it for this purpose because of the need to measure the contributions from the wings of the line if conventional absorption spectrometry were to be used [2].

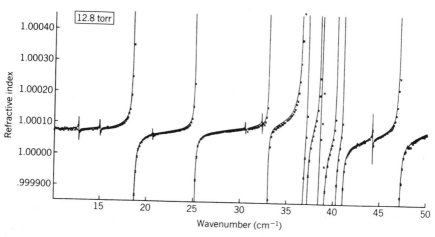

Fig. 11.6. Refractive index spectrum of water vapor at a pressure of 12.8 torr and a temperature of 293 K. ×, experimental data points; solid line, computed spectrum. (Reproduced from [8], by permission of Pergamon Press; copyright © 1978.)

It is difficult to measure DFTS spectra of gases at long pathlengths using most of the designs for interferometers shown in Chapter 4. A long-arm interferometer has been described by Birch [7] that allows a pathlength of 90 cm, see Fig. 11.5. In the spectral region below 50 cm^{-1}, refractive indices of gases could be measured with an uncertainty of about 4×10^{-7} using this instrument. The refractive index spectrum of water vapor between 10 and 50 cm^{-1} was measured by Kemp et al. [8], and this spectrum together with the spectrum computed using the theory of Gross [9] is shown in Fig. 11.6. The refractive index spectra of other important atmospheric species are reported elsewhere [10,11], as are the spectra of several other molecules in the vapor phase [12–14].

III. LIQUID SAMPLES

DFTS measurements of liquid samples can be made either in the transmission or reflection mode. For relatively weakly absorbing molecules,

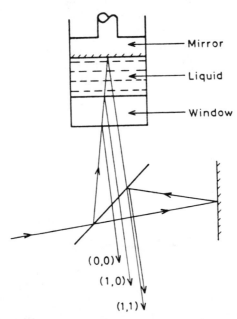

Fig. 11.7. Beams encountered in DFTS studies of liquid samples using a closed, variable-thickness cell. The first rays reflected from the three interfaces of the cell (interferometer–window, window–liquid, and liquid–mirror) are shown. (Reproduced from [2], by permission of the author.)

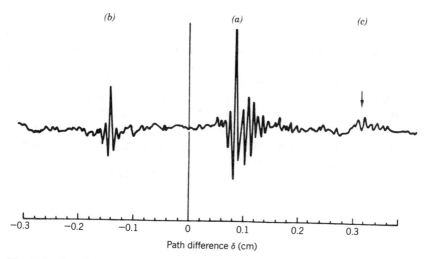

Fig. 11.8. Interferograms for a free-standing, 0.68-mm-thick film of *sym*-tetrabromoethane on the surface of the fixed mirror of a far-infrared interferometer. The signature (*a*) is due to the partial wave which was reflected once from the lower surface of the sample, corresponding to a sample thickness of 1.36-mm. The signature (*b*) is due to the partial wave reflected from the top surface of the liquid, while (*c*) is due to the partial wave that has been internally reflected once from the upper surface of the liquid. (Reproduced from [16], by permission of Pergamon Press; copyright © 1967.)

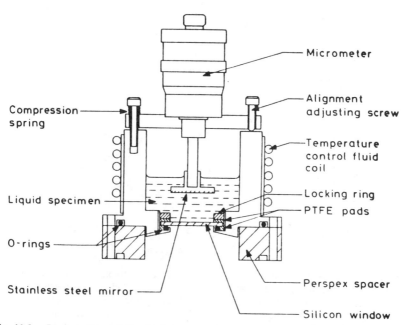

Fig. 11.9. Design of a variable-thickness cell for DFTS measurements of liquids. (Reproduced from [2], by permission of the author.)

377

the former technique must be used, whereas for strongly absorbing samples, it is necessary to use reflection measurements. A variable-thickness cell suitable for transmission measurements by DFTS is shown schematically in Fig. 11.7, with the rays labeled according to the notation of Honijk et al. [15]. In this measurement, the only ray of interest is (1, 1), but it can be seen that the (1, 0) and (1, 1) rays give rise to an infinite set of rays that have undergone multiple internal reflections before reentering the interferometer. This behavior leads to an interferogram that has localized interference signatures, as illustrated in Fig. 11.8 [16]. In principle,

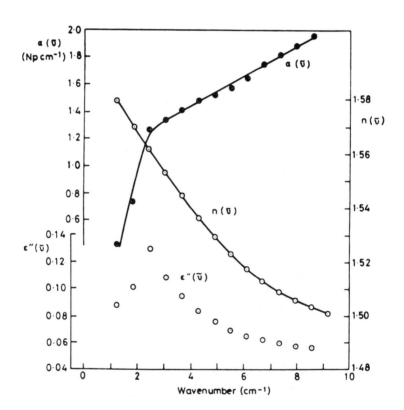

Fig. 11.10. The measured absorption spectrum, $\alpha(\bar{\nu})$, and refractive index spectrum, $n(\bar{\nu})$, of a 0.018-mol fraction solution of acetonitrile in CCl_4 at millimeter wavelengths. The imaginary part of the complex permittivity, $\epsilon''(\bar{\nu})$, calculated from these spectra is also shown. The maximum in this curve indicates a value of 2.1 psec for the rotational correlation time. (Reproduced from [21], by permission of Pergamon Press; copyright © 1981.)

there are many methods of reducing the results of liquid cell measurements to complex refraction spectra, each depending on the Fourier transforms of different combinations of signatures [15,17–19]. Most recent measurements are simplified by configuring the cell so that only the three rays (0, 0), (1, 0), and (1, 1) are measured. A more practical version of a cell suitable for transmission DFTS studies was described by Birch et al. [20] and is shown in Fig. 11.9.

Measurements of the DFTS spectra of pure liquids and solutions in the very far infrared can give significant information on the molecular dynamics of these systems. The rotational correlation time is directly calculable from the *dielectric loss* spectrum, where the dielectric loss ϵ'' is given by

$$\epsilon'' = \frac{n\alpha}{2\pi\bar{\nu}} = 2nk \qquad (11.11)$$

Figure 11.10 shows the variation of α, n, and ϵ'' for a 0.018 mol fraction solution of acetonitrile in CCl_4 in the very far infrared. The maximum value of ϵ'' is seen at 2.5 cm^{-1}, from which one can calculate that the rotational correlation time is 2.1 psec [21].

As noted above, for strongly absorbing liquids, the pathlength required for transmission measurements can become so small that large errors in the measured values of n and k can result. In this case, reflection DFTS techniques must be used. A schematic cell for reflection measurements is shown in Fig. 11.11 [2]. The basis of all reflection DFTS methods for liquids is the determination of the complex Fresnel reflectivity of the interface between a plane parallel window and the liquid. To obtain this parameter, the measured spectrum has to be normalized by some other spectrum in order to remove the frequency-dependent instrumental response. When a TPX (poly-4-methylpentene-1) window is used, it is necessary to replace the liquid sample with mercury, which is assumed to have a reflectance of unity [22]. If silicon is used instead of TPX, its refractive index is large enough that the reflection of the window–empty cell interface can be used for the reference measurement, and absolute complex reflectivities can be determined [23].

In reflection DFTS, the measured absorption spectrum is largely determined by the phase of the interface reflectivity. The measurement of phase can be strongly affected by the scan-to-scan repeatability of slow-scanning interferometers of the type typically used for these measurements. Interferometers that employ stepper-motor-driven micrometer drives for the moving mirror can exhibit sufficiently large backlash errors from one scan to the next that the accuracy of these measurements can be degraded. One technique that has been used to align the sample and reference interferograms [24,25] was to make the measurements at a high enough resolution that the signature due to the (0, 0) reflection from the lower window surface was observed. An alternative technique [26] only requires a full cell measurement, with the complex reflectivity of the interface between the window and the liquid determined by comparing the complex spectra corresponding to the (1, 0) and (0, 0) rays, as these contain all the necessary information.

One of the most strongly absorbing liquids in the far infrared is water, and the refractive index and absorption spectra of water in this region have been measured by reflection DFTS [24–27]. These spectra can provide insights on the molecular and intermolecular motions of water and solute ions. It is noteworthy that the cell shown in Fig. 11.9 can be used for reflection as well as transmission DFTS measurements. Refraction and absorption spectra of liquid water measured by reflection DFTS using this cell are shown in Fig. 11.12.

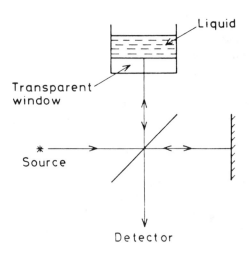

Reflection method

Fig. 11.11. Basic features of an interferometer used for the investigation of strongly absorbing liquids by the reflection method. (Reproduced from [2], by permission of the author.)

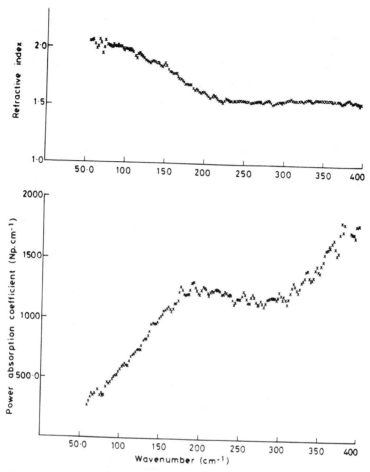

Fig. 11.12. Far-infrared refraction and absorption spectra of liquid water determined by reflection DFTS using the cell shown in Fig. 11.11. (Reproduced from [2], by permission of the author.)

IV. SOLID SAMPLES

From a computational viewpoint, measurement of the DFTS spectra of solids is easier than that of liquids since there is no need to allow for window effects. However, sample preparation can be a significant problem since it is often difficult to prepare samples with a large flat surface. Samples may not be isotropic, so that polarization studies may be necessary, or they may not be homogeneous, so that radiation is scattered. Absorption bands of solids are typically much sharper than those of liquids, and samples may go from being transparent to virtually opaque in a very short interval, so that both transmission and reflection techniques may have to be used. Finally, data are often required over a very large temperature range, often from liquid helium to above ambient temperatures.

In spite of the experimental difficulties, several different designs for DFTS measurements of solid samples have been described [28–32]. Parker et al. [28] described an interferometer suitable for measuring single-pass spectra of solid samples. This instrument, which incorporates corner retroreflectors, is shown in Fig. 11.13. The sample could be either held at ambient temperature or mounted in a liquid nitrogen dewar. Single-pass measurements are much more useful than double-pass measure-

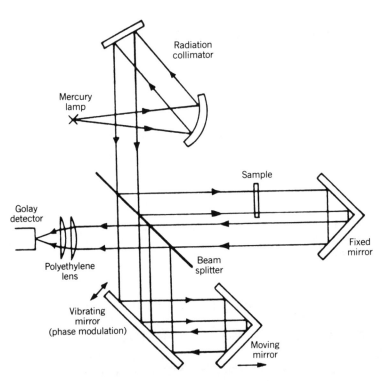

Fig. 11.13. Optical schematic of a Fourier spectrometer used for single-pass dispersive transmission measurements. (Reproduced from [28], by permission of Pergamon Press; copyright © 1978.)

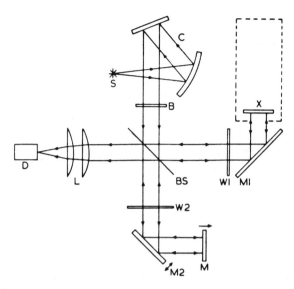

Fig. 11.14. Schematic diagram of a dispersive far-infrared Fourier transform spectrometer and cryostat. S, mercury lamp source; C, collimator; B, black polyethylene filter; BS, beamsplitter; W1, white polyethylene vacuum window; W2, compensating window; M1, plane mirror; X, sample; M2, phase modulation mirror; M, moving mirror; L, polyethylene focusing lenses; D, detector. The sample is mounted on the base plate of the liquid helium can of the cryostat indicated by the dotted lines. (Reproduced from [36], by permission of Pergamon Press; copyright © 1978.)

ments, since in the latter type of measurement the complex insertion loss is squared (see Eq. 11.6), rendering many samples essentially opaque.

Most of the recent developments in solid-state DFTS have involved reflection measurements. The instrumentation falls into two types depending on how the reference spectrum is measured. In an alternative approach illustrated in Fig. 11.14, part of the surface of the sample is coated with aluminum, and the field of view of the interferometer is divided by a set of screens such that either the aluminized or nonaluminized portion of the sample can be illuminated [33–35]. With the latter method, the SNR of the measurement is reduced since only a fraction of the beam is being observed, but the technique is less susceptible to systematic phase errors than is the replacement method. In view of the fact that the latter technique is also more easily applicable to samples held in a liquid helium dewar [36–39], it is probably the technique of choice.

A very large number of measurements of dispersive Fourier transform spectra of solid samples, including alkali halides [40,41], cadmium telluride [42], polymers [43], quartz [44], soda-lime glass [45], silicon [23], wire grids [46,47], and even hydrated lysozyme [48], have been reported. The interested reader is referred to several review articles [2,22,49] for further details.

REFERENCES

1. E. E. Bell, Aspen Int. Conf. on Fourier Spectrosc., 1970 (G. A. Vanasse, A. T. Stair, and D. J. Baker, eds.), AFCRL-71-0019, p. 71 (1971).
2. J. R. Birch, Proc. Soc. Photo-Opt. Instrum. Eng., 289, 363 (1981).
3. J. Chamberlain, J. E. Gibbs, and H. A. Gebbie, Infrared Phys., 9, 185 (1969).
4. J. R. Birch and C. E. Bulleid, Infrared Phys., 17, 279 (1977).
5. J. R. Birch and T. J. Parker, Infrared Phys., 19, 103 (1979).
6. J. R. Birch and T. J. Parker, Infrared Phys., 19, 201 (1979).
7. J. R. Birch, Infrared Phys., 18, 275 (1978).
8. A. J. Kemp, J. R. Birch, and M. N. Afsar, Infrared Phys., 18, 827 (1978).
9. E. P. Gross, Phys. Rev., 97, 395 (1955).
10. J. R. Birch and C. E. Bulleid, Conf. on Precision Electromagnetic Measurement, Boulder, CO (June 1976), pp. 48–49, IEEE Cat. No. 76 CH1 099-1M.
11. J. R. Birch and M. N. Afsar, Conf. on Precise Electrical Measurement, Sussex, England (Sept. 1977), pp. 67–69. IEEE Conf. Publ. 152.
12. R. B. Sanderson, Appl. Opt., 6, 1527 (1967).
13. W. H. Robinette and R. B. Sanderson, Appl. Opt., 8, 711 (1969).
14. J. R. Birch and M. N. Afsar, Spectrochim. Acta, 35A, 669 (1979).
15. D. D. Honijk, W. F. Passchier, M. Mandel, and M. N. Afsar, Infrared Phys., 17, 9 (1977).
16. J. Chamberlain, A. E. Costley, and H. A. Gebbie, Spectrochim. Acta, 23A, 2255 (1967).
17. D. D. Honijk, W. F. Passchier, and M. Mandel, Infrared Phys., 16, 257 (1976).
18. W. F. Passchier, D. D. Honijk, and M. Mandel, Infrared Phys., 15, 95 (1975).
19. W. F. Passchier, D. D. Honijk, M. Mandel, and M. N. Afsar, Infrared Phys., 17, 381 (1977).
20. J. R. Birch, G. O'Neill, J. Yarwood, and M. Bennouna, J. Phys. E., 15, 684 (1982).
21. J. R. Birch, M. N. Afsar, J. Yarwood, and P. L. James, Infrared Phys., 21, 9 (1981).
22. N. W. B. Stone, Appl. Opt., 17, 1332 (1979).
23. J. R. Birch, Infrared Phys., 18, 613 (1978).
24. M. N. Afsar and J. B. Hasted, J. Opt. Soc. Am., 67, 902 (1977).
25. M. N. Afsar and J. B. Hasted, Infrared Phys., 18, 843 (1978).
26. J. R. Birch and M. Bennouna, Infrared Phys., 21, 224 (1981).
27. M. Bennouna, H. Cachet, J. C. Lestrade, and J. R. Birch, Proc. Soc. Photo-Opt. Instrum. Eng., 289, 266 (1981).

28. T. J. Parker, W. G. Chambers, J. E. Ford, and C. L. Mok, *Infrared Phys.*, **18**, 571 (1978).
29. E. E. Russell and E. E. Bell, *Infrared Phys.*, **6**, 75 (1966).
30. J. Gast and L. Genzel, *Opt. Comm.*, **8**, 26 (1973).
31. K. E. Gauss, H. Happ, and G. Rother, *Phys. Stat. Sol. B*, **72**, 623 (1975).
32. J. R. Birch and D. K. Murray, *Infrared Phys.*, **18**, 283 (1978).
33. T. J. Parker and W. G. Chambers, *IEEE Trans. Microwave Theory and Tech.*, Vol. MTT-22, 1032 (1974).
34. T. J. Parker, W. G. Chambers, and J. F. Angress, *Infrared Phys.*, **14**, 207 (1974).
35. P. R. Staal and J. E. Eldridge, *Infrared Phys.*, **17**, 299 (1977).
36. D. G. Mead, *Infrared Phys.*, **18**, 257 (1978).
37. D. G. Mead and L. Genzel, *Opt. Comm.*, **27**, 95 (1978).
38. T. J. Parker, R. P. Lowndes, and C. L. Mok, *Infrared Phys.*, **18**, 565 (1978).
39. T. J. Parker and R. P. Lowndes, *J. Phys. E*, **12**, 495 (1979).
40. P. R. Staal and J. E. Eldridge, *Infrared Phys.*, **19**, 625 (1979).
41. A. Memon and T. J. Parker, *Intl. J. Infrared and Millimeter Waves*, **2**, 839 (1981).
42. T. J. Parker, J. R. Birch, and C. L. Mok, *Sol. St. Comm.*, **36**, 581 (1980).
43. J. R. Birch, J. D. Dromey, and J. Lesurf, *Infrared Phys.*, **21**, 225 (1981).
44. T. J. Parker, J. E. Ford, and W. G. Chambers, *Infrared Phys.*, **18**, 215 (1978).
45. J. R. Birch, R. J. Cook, and G. W. F. Pardoe, *Sol. St. Comm.*, **30**, 693 (1979).
46. C. L. Mok, W. G. Chambers, T. J. Parker, and A. E. Costley, *Infrared Phys.*, **19**, 437 (1979).
47. J. A. Beunen, A. E. Costley, G. F. Neill, C. L. Mok, T. J. Parker, and G. Tait, *J. Opt. Soc. Am.*, **71**, 184 (1981).
48. I. Golton, Ph.D. thesis; University of London (1980).
49. J. F. Angress and T. J. Parker, Chapter 7 in *Neutron Interferometry*, (H. Bonse and U. Rauch, eds.), Oxford University Press (1981).

CHAPTER

12

FT–IR SPECTROMETRY OF TIME-DEPENDENT PHENOMENA

I. INTRODUCTION

Many of the applications discussed in the remainder of this book deal with stable samples. Nevertheless, systems that exhibit a time dependence are also of interest. The rate of change of composition of reacting systems can vary greatly. The most easily studied systems have a relatively slow rate of change with respect to the scanning time of the interferometer. In these cases, a signal-averaged spectrum of the initial reactants can be recorded; the reaction is then allowed to progress for some time, and another signal-averaged spectrum is recorded. The process can, of course, be repeated at several intervals as the reaction proceeds. An example of such a slow reaction is the photolytic production of smog in atmospheric chemistry. In many instances, dispersive spectrometers have been used to monitor the composition of these dynamic systems. Regardless, the use of FT–IR spectrometers is usually preferable when the throughput is low or a high spectral resolution is required.

A second type of dynamic system is one in which the time constant approaches the scan time of the spectrometer. As the scan speed of the spectrometer increases, the temporal resolution of the experiment can be increased. Examples of these systems include gas chromatography/Fourier transform infrared (GC/FT–IR) spectrometry and liquid chromatography/Fourier transform infrared (LC/FT–IR) spectrometry. These particular topics have sufficiently broad scopes that they are treated individually in Chapters 18 and 19, respectively. However, other systems exist that also fall into this category.

The final broad category includes those systems that have component changes or transient lifetimes shorter than the scan time of the spectrometer. The spectra of transients can be measured by forcing the reaction into a steady state where the transient can be observed for a long time period or in some instances by time resolution of the spectrometric data.

II. SPECTROMETRY FOR DYNAMIC SYSTEMS

FT–IR spectrometry is not the sole method available to chemists to record the infrared spectra of time-dependent systems. As was stated above, in many instances conventional dispersive spectrometry can record the spectra adequately if the samples do not require high SNR or if the resolution is beyond the capabilities of available spectrometers. Thus, it is instructive to examine several of the methods available for recording the infrared spectra of time-dependent systems.

The use of dispersive (or filter) infrared spectrometers has been quite common in the past for investigating steady-state systems. Chemiluminescent reactions are good examples of systems falling into this category. Chemiluminescent species constantly emit radiation by the establishment of a steady-state or continuous reaction. A dispersive spectrometer can scan the desired bandwidth slowly and record a high-SNR spectrum. If the system under study is not in a steady state or is slowly varying, rapid scanning of the spectrometer or gated sampling of a fixed wavenumber becomes necessary. Such techniques have been successful in the ultraviolet and visible regions of the electromagnetic spectrum [e.g., 1–4] but are often inadequate in the relatively energy-starved mid infrared.

The primary difference between ultraviolet–visible and mid-infrared spectrometry lies in the detection systems. Optical (i.e., ultraviolet–visible) detectors are, for the most part, shot-noise limited. This means that these detectors can record discrete events such as single photon emissions from the source and the error of the measurement of spectral intensities is limited by the counting system. Shot noise increases with the photon flux, however, and this factor largely offsets the Fellgett (or multiplex) advantage of a Fourier transform spectrometer. Because there is little multiplex advantage, a dispersive optical spectrometer may be almost as efficient as a Fourier transform optical spectrometer, and hence there is little reason to use the interferometer. (This statement is not completely accurate as some advantage can be gained by the increased throughput, but this does not always warrant using an FT spectrometer.)

Rapid-scanning dispersive infrared spectrometers have been used in limited cases, but they suffer from poor temporal resolution, poor spectral resolution, or both. Because mid-infrared detectors are not shot-noise limited but detector-noise limited, the error in the measurement is independent of the photon flux. As a result, the SNR of rapid-scanning dispersive mid-infrared spectrometers tends to be quite low. It is primarily for this reason that FT–IR spectrometers are more successful in the study of dynamic systems than are infrared dispersive spectrometers. The multiplex advantage reduces the effect of detector noise, and the SNR is enhanced by $M^{1/2}$, where M is the number of resolution elements, assuming the collection time remains constant (See Chapter 7, Section IV).

The majority of commercially available mid-infrared FT–IR spectrometers incorporate rapid-scanning interferometers. To obtain the greatest sensitivity, quantum detectors, rather than thermal detectors, should be used. The sensitivity (D^*) of photoconductive detectors increases with modulation frequency up to about 1 kHz, and D^* remains constant between about 1 and 100 kHz (see Chapter 5). Under these circumstances, high mirror velocities can be used, and spectra with good SNR can be collected rapidly. Indeed, it is possible to collect a full-bandwidth mid-infrared spectrum with an FT–IR spectrometer at 8 cm^{-1} resolution in approximately 0.02 sec. At this collection rate, it should be possible to acquire spectra down to a temporal resolution of about 40 msec. It will be shown later (Section VI) that temporal resolution can be extended to less than 100 μsec by utilizing the discrete sampling nature of FT–IR spectra and by the repetition of dynamic experiments. This technique, called time-resolved spectrometry (TRS) or stroboscopic spectrometry, is somewhat limited in terms of spectral resolution, but nevertheless spectra of fairly high SNR have been reported (see this chapter, Section VI).

Although FT–IR spectrometry has proved to be extremely useful in the study of dynamic processes, it does not exhibit the highest spectral and temporal resolutions of all available techniques. Bethune and co-workers [5,6] have developed the technique of infrared spectral photography, which upconverts infrared radiation to visible frequencies by laser techniques. The visible radiation is then dispersed and recorded photographically. The temporal resolution of this laser upconversion technique is on the order of 5 nsec but produces only a narrow-bandwidth infrared spectrum.

III. TRANSIENT SPECIES IN A STEADY STATE

It is often possible to obtain infrared spectra of short-lived transient species by generating a steady-state concentration of the species and recording the spectrum over a long period of time. Even in steady-state applications, the concentrations of the transient species may be very low, and their absorption spectra may be obscured by the spectra of reactants and products of the system under study. For this reason, the most widely studied steady-state reactions have been chemiluminescent reactions where emission spectra of transients can be recorded easily, often without spectral interference from other species.

An early example of chemiluminescence studies was reported by Stair [7] where the spectrum of NO_2^* was measured. Stair designed a flowing afterglow experiment where NO and atomic oxygen were mixed in an integrating sphere. The integrating sphere was used to gather the chemiluminescent radiation and direct it to the spectrometer due to the low intensity for this reaction. An illustration of the apparatus is given in Fig. 12.1.

McDonald and co-workers took a slightly different approach to chemiluminescence studies of weakly emitting systems. These workers lowered the temperature of the spectrometer and its surroundings, thereby reducing the intensity of the background radiation to improve the sensitivity of the detector [8]. This was necessary because they used a helium-cooled copper-doped germanium detector, which is shot-noise limited. This is in contrast to most infrared detectors of lower D^*. At a temperature of 300 K, the photon flux from the spectrometer is about a factor of 10^6 greater than the photon flux at 77 K, which was shown to give an improvement in the SNR of $\sim 10^3$ when the spectrometer was cooled. The spectrometer is a modified commercial instrument, the details of which are given in Chapter 15. McDonald and co-workers recorded the chemiluminescent spectra of halogen atom–olefin reactions [8,9]. Infrared emission from

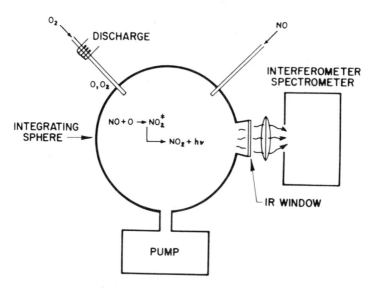

Fig. 12.1. Schematic drawing of apparatus for flowing afterglow chemiluminescence experiment using an internally gold-coated integrating sphere (Reproduced from [7], by permission of the author.)

these systems was weak due to the long radiative lifetimes of the species under study. A flow system was used for the collision reaction, and a large volume was necessary to improve the residence time of the chemiluminescent species. The radiative lifetimes were on the order of 0.1–1 sec, and the residence time of each species was only about 1 msec. As a result, less than 1% of the transients can emit during the residence time. To simplify spectral identification, the pressure in the cell was kept low to minimize secondary reactions by collision. Total quantities of each reactant in the cell were on the order of 10–50 nmol, thus leaving very low emissive populations. It was necessary to signal-average the reaction systems for up to 1600 sec in order to obtain an adequate SNR. More details of the experiments and results are presented in Chapter 15.

The chemiluminescence studies presented thus far provide spectrometric information about the nature of the transient species, but no information about lifetimes, reaction mechanisms, and kinetics has been gained directly. Sung and Setser pursued the study of chemiluminescent reactions and were able to extract kinetic data [10]. Their approach was to use a flowing afterglow apparatus that consisted of a flow tube with five NaCl windows, each separated by 4.6 cm, along the wall of the tube. By flowing the reactants, and consequently the products, through the tube at 85 m sec^{-1}, spectra of different temporal events in the reactions could be recorded at each of the windows. The changes in the spectra were used to deduce the kinetic data. These workers studied the reactions of fluorine atoms with hydrogen bromide and hydrogen iodide. Using an InSb detector (at 77 K) and an interferometer with a Fe_2O_3/CaF_2 beamsplitter, 2 cm^{-1} resolution emission spectra of HF* were collected and near-infrared emission bands due to I* and Br* were observed. With this system, operating at ambient temperature, between 500 and 2000 scans were required per spectrum.

IV. SLOWLY VARYING SYSTEMS

Slowly varying systems can be considered to be those where the rate of change of reaction is slow compared to the time necessary to collect a single interferogram. Many of these reactions are sufficiently slow that the use of dispersive spectrometry would be feasible in terms of spectral recording time. Nevertheless, FT–IR spectrometry is often necessary to obtain sufficiently high signal-to-noise spectra for accurate characterization. Some of the earliest work in long-term kinetic studies by FT–IR spectrometry has been accomplished in the characterization of atmospheric reactions, especially in the formation of chemical smog in the troposphere. The kinetics or rates of reactions in atmospheric chemistry are of interest here, not the mechanisms.

Most reactions in the production of chemical smog are promoted by ultraviolet–visible radiation and are thus induced in the troposphere by solar radiation. Calvert and co-workers use a 170-m modified White cell to conduct their studies [11,12]. A schematic of the apparatus is presented in Fig. 12.2. The output beam of the spectrometer is passed into a 32-pass, 6.3-m White cell that is surrounded by black-light fluorescent lamps. The reactants of interest are placed in the cell with synthetic air at the desired pressure (to mimic altitude) and interferograms are recorded, both before the lights around the cell have been turned on and during illumination. These particular lamps were chosen to simulate solar radiation in the atmosphere near ground level. Although the rate of reactions in the atmosphere may be considered slow, it is often necessary to record spectra at intervals of approximately 1 min. In the study of the rates of reaction for the formation of HOCl, Calvert and co-workers [12] photolyzed mixtures of Cl_2, H_2, O_3, O_2, and N_2 and recorded data at discrete intervals throughout the photolysis and for some time after illumination was halted. All spectra were of 1 cm^{-1} resolution, and the sampling interval was slightly in excess of 1 min. Figure 12.3 shows the variation of the v_2 band of HOCl during the experiment. The second through the seventh spectra were recorded with the illumination on, and the remaining seven were

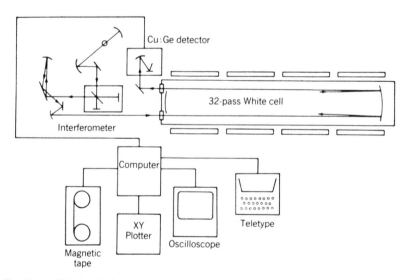

Fig. 12.2. Schematic diagram of apparatus for long-path atmospheric studies. (Reproduced from [12], by permission of the American Chemical Society; copyright © 1979.)

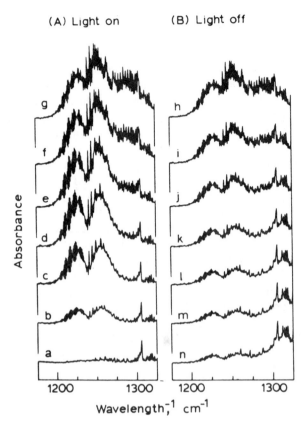

Fig. 12.3. Absorbance spectra of two bands of HOCl as it grows and decays with UV radiation on (*a*) and off (*b*). Reaction results from a $Cl_2–H_2–O_3–O_2–N_2$ mixture and spectra *a* through *n* were taken at times 0, 0.61, 1.74, 2.87, 3.99, 5.12, 6.25, 7.43, 8.55, 9.68, 11.93, 14.18, 16.43 and 20.48 min, respectively. (Reproduced from [12], by permission of the American Chemical Society; copyright © 1979.)

recorded after the light was turned off so that HOCl decay could be monitored. Difference spectra were used to measure product concentrations from estimates of molar absorptivities. From studies such as these, Calvert and co-workers have been able to calculate relative rates of reactions for the various species involved in atmospheric reactions. Niki et al. have performed similar experiments on different systems to study, for example, the rates of reaction for a hydroxyl radical with aldehydes [13] and the kinetics of chlorine atoms with formaldehyde [14]. The procedures are quite similar to those of Calvert, except that much higher resolution spectra (0.06 cm^{-1}) were collected and usually 16 scans were signal averaged per spectrum.

Fourier transform infrared spectrometry has been used in accelerated aging tests, for example, to study the photodecomposition of pigment in paint. Chase et al. [15] exposed paint panels to carbon-arc radiation that accelerates natural sunlight weathering effects. Here it was necessary to use a diffuse reflectance technique (see Chapters 5 and 17) to record spectra from the surface of the panel. Once these spectra were recorded, the diffuse reflectance spectrum of the pure paint binder was subtracted to yield the spectrum of the pigment. Figure 12.4 shows the diffuse reflectance spectrum of a paint panel after the binder spectrum has been subtracted and a diffuse reflectance spectrum of the pure paint pigment. As can be seen readily, the two spectra are quite similar and are of the pigment dioxabiphthalide. Figure 12.5 shows the spectra of the panel before and after exposure to the carbon-arc source. The photodecomposition mechanism was believed to be the dimerization of the pigment, and this was shown to be the case when the exposed panel spectrum was compared with the diffuse reflectance spectrum of the pure dioxabiphthalide dimer,

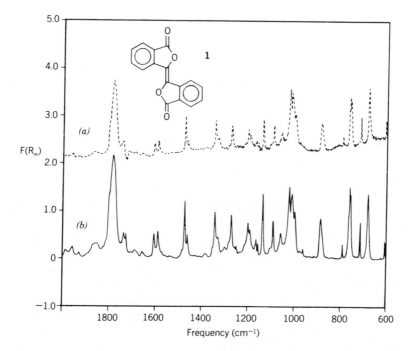

Fig. 12.4. (a) Diffuse reflectance spectrum of paint panel after the binder spectrum has been subtracted. (b) Diffuse reflectance spectrum of pure dioxabiphthalide, the pigment. (Reproduced from [15], by permission of the Society for Applied Spectroscopy; copyright © 1982.)

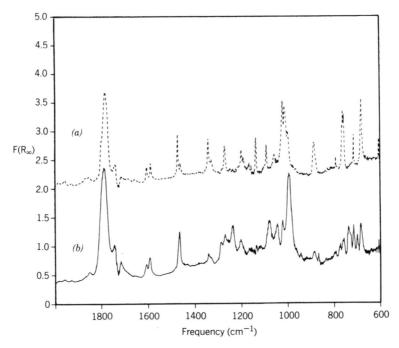

Fig. 12.5.   Diffuse reflectance spectra of (*a*) unexposed panel and (*b*) exposed panel. (Reproduced from [15], by permission of the Society for Applied Spectroscopy; copyright © 1982.)

Fig. 12.6. Although no kinetic data for this study were determined and the photodecomposition process was apparently slow (exposure was for 44 hr), FT–IR spectrometry proved useful for the measurement of the diffuse reflectance spectra of a time-dependent process.

Another application of FT–IR spectrometry to time-dependent phenomena has been presented by Kinney et al. in the photochemical study of a surface-confined organometallic [16]. In this study, a cobalt carbonyl was surface confined on silica and then the organometallic was exposed to reactive gas under photochemical conditions. These studies were done *in situ* by employing photoacoustic spectrometry (FT–IR/PAS). (FT–IR/PAS is covered in detail in Chapter 9.) Here the surface-confined sample was placed in the photoacoustic cell and the spectrum recorded. In order to record photoacoustic spectra, the cell must be filled with an inert gas. Argon was used in this study. The reactive gas (CO, O_2, or PF_3) was introduced to the cell and the sample plus reactive gas were irradiated with a 150-W xenon lamp for 20 min. The reactive gas was removed and replaced with argon so that a new spectrum could be recorded. Differences between the spectra yielded the photochemical changes. Figure 12.7*a* illustrates the photochemical reaction between the surface-confined cobalt–carbonyl and phosphorus trifluoride, where curve 1 is the unreacted surface-confined cobalt carbonyl, denoted $[S] \gtrsim SiCo(CO)_4$, curve 2 is the spectrum of the $[S] \gtrsim SiCo(CO)_4$ after reaction with PF_3 and ultraviolet irradiation to give $[S] \gtrsim SiCo(CO)_n(PF_3)_{4-n}$. Curve 3 of Fig. 12.7*a* is the spectrum of regenerated $[S] \gtrsim SiCo(CO)_4$ by irradiating the product in curve 2 in the presence of CO. Figure 12.7 shows the spectral changes of $[S] \gtrsim SiCo(CO)_4$ at approximately $-50°C$ (curve 1) and after irradiation of the sample under 100 torr of 2-methylpropene for 10 min to produce $[S] \gtrsim SiCo(CO)_3(2\text{-methylpropene})$. This technique illustrates the ability to examine dynamic changes in reactants with a minimum of sample preparation when the sample can be left *in situ*.

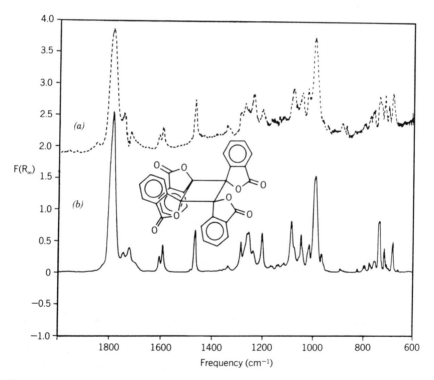

Fig. 12.6. Diffuse reflectance spectra of (*a*) exposed panel and (*b*) pure dioxabiphtahalide dimer. (Reproduced from [15], by permission of the Society for Applied Spectroscopy; copyright © 1982.)

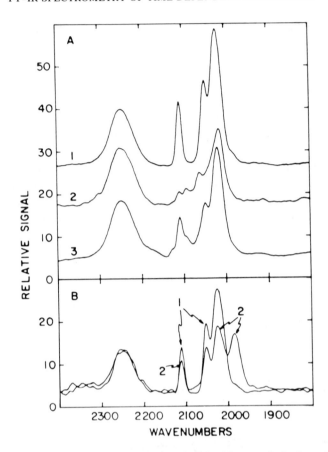

Fig. 12.7. (*a*) Initial FT–IR/PA spectrum of [S]≧SiCo(CO)₄ sample is shown in curve 1, and curve 2 is the spectrum obtained after the sample is irradiated and put under 1 atm of PF₃ for 10 min. The product is [S]≧SiCo(CO)ₙ(PF₃)₄₋ₙ. Curve 3 is regenerated [S]≧SiCo(CO)₄ after the product in curve 2 is irradiated under 1 atm of CO. (*b*)[S]≧SiCo(CO)₄, curve 1, is irradiated at −50°C under 100 torr for 10 min to produce [S]≧SiCo(CO)₃ (methylpropene) (curve 2). (Reproduced from [16], by permission of the American Chemical Society, copyright © 1981.)

Photocatalyzed olefin isomerism by pentacarbonyliron(0) has been presented by Chase and Weigert [17]. The photocatalysis reactions were monitored in real time by placing a solution of the olefin (in a solvent such as benzene) in an infrared cell with the iron catalyst. Spectra of the system were recorded in the dark and then as the system was irradiated with a 2000-W xenon lamp filtered to pass the near-ultraviolet–visible region. Interferograms were collected at a resolution of 4 cm^{-1}, and two interferograms were coadded for each spectrum. In the rapid-scanning mode, the authors were able to record a spectrum every 0.5 sec. During these studies, four carbonyl bands were monitored, at 1800, 1969, 1995, and 2020 cm^{-1}, as well as two olefin bands at 1641 and 994 cm^{-1}. Changes in the band intensities were measured in terms of changes in the absorbance (Δ OD) by subtracting a spectrum of the system before irradiation from the spectra recorded during the experiment. Figure 12.8 shows the change in absorbance for the six bands of interest for a 20% solution of 1-pentene in benzene with an olefin–iron ratio of 300:1. Although detailed analysis of the quantum yields or catalyst cycles cannot be made from these studies due to the construction of the experiment, Chase and Weigert have shown an *in situ* analysis with a temporal resolution of less than 1 sec.

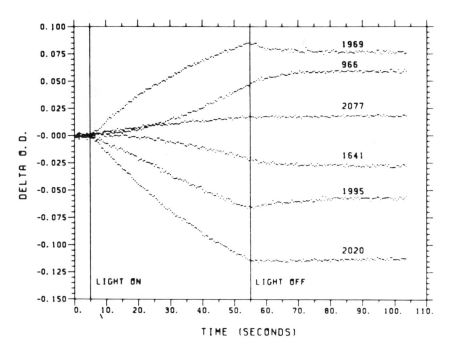

Fig. 12.8. Rate of change of six infrared bands before, during, and after irradiation of Fe(CO)$_5$ and 1-pentene. (Reproduced from [17], by permission of the American Chemical Society; copyright © 1981.)

V. RAPIDLY VARYING SYSTEMS

In the study of rapid time-dependent systems, we should consider various types of chromatographs interfaced to FT–IR spectrometers. For GC/FT–IR and LC/FT–IR spectrometry, the effluent of the chromatograph

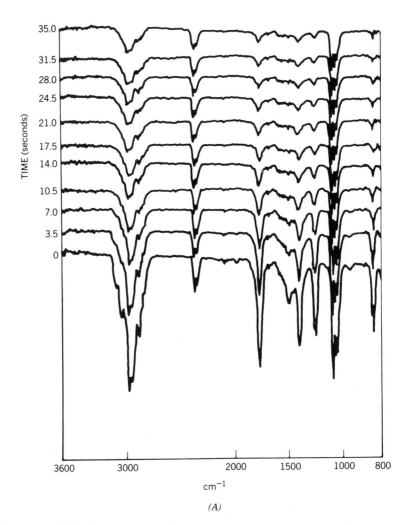

(A)

Fig. 12.9. (*A*) Spectra of a mixture of acetone, benzene, and methanol over 0.5 g activated charcoal at 3.5-sec intervals. (*B*) Absorbance profiles of data in Fig. *A* for a, benzene; b, acetone; and c, methanol. (Reproduced from [18],.by permission of the Society for Applied Spectroscopy; copyright © 1975.)

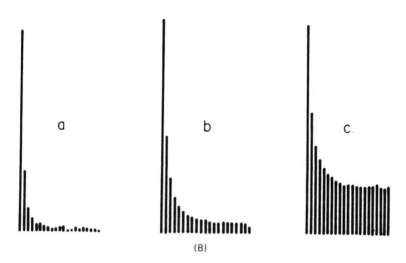

(B)

Fig. 12.9. (continued)

is often passed through a flow cell where the composition of the effluent is monitored constantly. The spectral resolutions required are usually 8–4 cm^{-1} as the nature of the samples limit the resolution that can be achieved. Temporal resolutions of up to 1 sec are desirable, especially for capillary column GC/FT–IR measurements. The subject matter of GC/FT–IR and LC/FT–IR measurements is sufficiently broad to be discussed separately in Chapters 18 and 19, respectively.

Rapidly varying systems are by no means restricted to the chromatographic separations. An early interesting kinetic study was reported by Lephardt and Vilcins [18] for the reaction of butadiene with NO$_2$ and the adsorption of various organics on activated carbon. A standard 10-cm gas cell with 49.5-mm windows was used as the reaction chamber. Lephardt and Vilcins placed a 0.5-g sample of activated carbon in the cell and then introduced a 10-μL sample of an equal-volume solution of acetone, methanol, and benzene. Interferograms at a resolution of 8 cm^{-1} were collected at 3.5-sec intervals with a pre-A series Digilab FTS-14 spectrometer. Figures 12.9A and B show the spectra collected and the absorption profiles with time for the absorbed gases. In addition to this system, Lephardt and Vilcins also studied the reaction of butadiene with NO$_2$. Spectra were collected at 6.5-sec intervals and indicated the consumption of NO$_2$, N$_2$O$_4$, and butadiene as well as the production of a nitroproduct (see Figs. 12.10A and B).

One of the more interesting instrumental aspects of this paper was the development of a frequency-dependent apodization model for kinetic

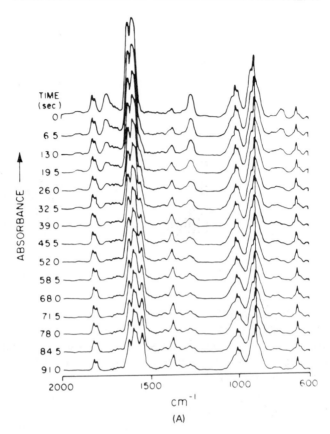

Fig. 12.10. (*A*) Absorbance spectra of a mixture of butadiene and NO_2–N_2O_4 at intervals of 6.5 sec. (*B*) Absorbance profiles of data in Fig. 12.10 *A* for a, NO_2 (1628 cm^{-1}); b, N_2O_4 (1260 cm^{-1}); c, butadiene (910 cm^{-1}); and d, a nitro product (1550 cm^{-1}). (Reproduced from [18], by permission of the Society for Applied Spectroscopy; copyright © 1975.)

data. This model was used to account for the spectral effects produced by systems that change composition rapidly with respect to the scan time of the interferometer. The spectral effects distort the absorption profiles based on the rate of sample concentration change. Lephardt and Vilcins explained how to deconvolve the spectra to recover the true band shape, but to the best knowledge of these authors, this analysis has never been employed. The interested reader is referred to the original article [18].

In a separate study, Vilcins and Lephardt [19] studied the effects of cigarette smoke aging. It had been reported in the literature that cigarette smoke contained methyl nitrite, a known carcinogen, and Vilcins and

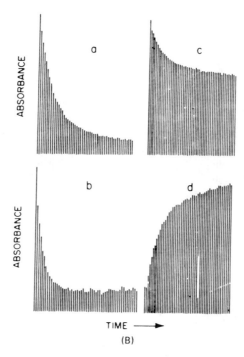

Fig. 12.10. (continued)

Lephardt sought to study the kinetics of the formation reaction. In a static cell experiment, NO, a known component of cigarette smoke was mixed with methanol, another known component. No formation of methyl nitrite was observed in the absence of oxygen. It was found that methyl nitrite formed only in the presence of air, which indicated that the carcinogen was not formed appreciably until the smoke was exhaled by the smoker. Figure 12.11 shows the absorbance profiles of the reactants and products of this reaction.

An early flow system was developed by Liebman, Ahlstrom, and Griffiths [20] for the on-line analysis of combustion and pyrolysis products. In this study, a gas chromatographic pyrolysis unit was interfaced directly to an early GC/FT–IR light-pipe cell. This light pipe was 4 × 4 mm in cross section and 30 cm in length. The infrared beam from a Digilab FTS-14 spectrometer was focused onto the entrance window of the cell, and the beam, as it exited from the other end of the cell, was focused onto a TGS detector. This cell transmitted about 20% of the radiation from the spectrometer. Three basic methods of sample introduction were used. An adsorbent was used to concentrate dilute volatile products, and then these

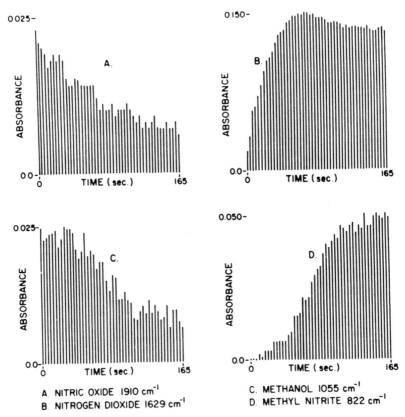

Fig. 12.11.   Absorbance profiles of the aging process of cigarette smoke from nitrate-enriched burley tobacco. (Reproduced from [19], by permission of the Society of Industrial Chemistry; copyright © 1975.)

volatiles were desorbed into the transfer line to the light-pipe cell. The second method involved the pyrolytic, or combustive, reaction of the sample with a rapid heat pulse to take the final sample temperature to between 100 and 1000°C. Finally, the volatiles were produced by programmed heat treatments at rates of up to 40°C min^{-1}. In all instances, the volatiles were introduced to the gas cell at a 10-mL min^{-1} flowrate with helium as the carrier gas.

Liebman and co-workers collected spectra at 4 cm^{-1} resolution. For the programmed heat treatment, they studied the pyrolysis of polyvinyl chloride at 20°C min^{-1}. A 2–3-mg sample was pyrolyzed in a helium atmosphere starting at 100°C and ending at 600°C. The temperature rate change was sufficiently slow that spectra could be averaged for periods of 2 min. The evolution of products such as HCl, benzene, and ethylene were readily detected. Some of these compounds, particularly HCl, could not be detected by the flame ionization detector (FID) attached to a pyrolysis gas chromatograph. Liebman et al. also demonstrated a much higher temporal resolution than 2 min. When a pulse pyrolysis or combustion was performed, the product evolved more rapidly and it was necessary to signal-average only four scans to give a temporal resolution of 8 sec. The resulting spectra contained many identifiable compounds for which relative concentrations could be determined. Unfortunately, with so few scans signal averaged, the SNR at high wavenumbers (>2600 cm^{-1}) was too low for the small quantities of components that absorb in that region to be observed. The authors noted that this technique was advantageous for the identification of species due to its specificity, but also pointed out that the gas chromatograph was more sensitive to many of the organic components.

The interest in passing an inert or reactive gas over a heated solid and continuously monitoring the products was not great for several years. Recently, however, there appears to be a growing realization of the potential of such techniques. Saperstein [21] devised a system whereby a reactor was placed inside a GC oven and gases evolved during the reactions were transported to an internally gold-coated light pipe with a 2 mm i.d. and 42 cm in length. Spectra were measured using an MCT detector. The reactor was a stainless steel tube in which was placed a catalyst or reactants. The GC oven controlled the temperature of the reactor and the GC plumbing and controls were used to meter reactants and carrier gases. (The interested reader is urged to consult the original article for a detailed description.) Saperstein demonstrated the ability to follow fairly rapid catalytic reactions, including the Fischer–Tropsch synthesis of methane from H_2 and CO using chromium cobalt molybdate as a catalyst as well as the conversion of methanol to methane with the same catalyst. A series of studies was completed showing the selectivity of the catalyst as a function of temperature.

The logical extension of the pyrolysis, combustion, and thermal reaction experiments is to interface thermal analyzers with an FT–IR spectrometer. Cody et al. [22] interfaced a DuPont thermogravimetric analyzer (TGA) with a Nicolet 7199 FT–IR spectrometer. The spectrometric interface itself was a GC/FT–IR light pipe as in some of the previously described studies. These workers utilized the standard Nicolet GC/FT–IR software to detect when gases were evolved from the samples and to determine which interferograms to save on disk. If every interferogram measured when the interferometer is employed in a rapid-scanning mode is stored, disk space is rapidly exhausted (see Chapter 18). Using the

TGA/FT–IR technique, Cody et al. were able to analyze components of compounds associated with weight loss. The decomposition of several simple inorganic compounds such as calcium oxalate monohydrate was studied, and acceptable detection limits were calculated. For example, approximately 5 μg of the strong absorber CO_2 present in the cell could be detected at carrier gas flow rates of 15 mL min^{-1}.

Lephardt and Fenner [23–26] have developed a slightly different evolved gas analysis system (FT–IR/EGA). The main feature of this apparatus is that the temperature of the sample, rather than that of the oven, is carefully controlled. In this way, precise temperature adjustment of the sample can be provided, which can overcome the exothermic or endothermic transitions in the sample that could affect the analysis. In most gaseous evolution reactions, the temperature of the gas will vary as the oven temperature varies. To overcome this effect, Lephardt and Fenner employed a quartz wool/stainless steel filter heat sink to equilibrate all the gases to 180°C. A Digilab FTS-14 with a GC/FT–IR light-pipe interface was used for these measurements. Lephardt and Fenner took pains to ensure that the carrier gas flow rate for evolved gases was sufficiently high to separate the products at different temperatures but low enough that the residence time of compounds in the light pipe was long enough that spectra of adequate SNR could be measured. At the end of the experiment, absorption spectra and temperature data were transferred to a Xerox Sigma 9 computer, and the absorption spectra were analyzed and the results plotted. Rather than each spectrum being plotted, the absorbance of specific bands was monitored and plotted versus temperature. The temporal resolution is only limited by the scan rate of the spectrometer. This system is very reproducible as evidenced by data in Fig. 12.12. Figure 12.12 shows two curves each for the evolution of CO_2 and CH_4 from burley tobacco heated in nitrogen with a flowrate of 60 mL min^{-1} and a temperature ramp of 6°C/min. By changing the carrier gas, the rate of evolution of certain gases changes, as evidenced by the evolution of CO_2 from a bright tobacco sample with different carrier gas compositions (Fig. 12.13). The spectra produced by this technique, of course, may represent several compounds; however, these compounds can be isolated by spectral subtraction (Chapter 10). In the FT–IR/EGA investigation of Kraft pine lignin, formaldehyde and water are evolved simultaneously. Figure 12.14 shows a spectrum of the evolved species H_2O and H_2CO before and after the spectral subtraction of the water. The SNR is quite high, and the formaldehyde can easily be identified. In a recent review [26], it was indicated that Lephardt has updated his spectrometer system to a Digilab FTS-20/E system and that he has made improvements in the FT–IR/EGA interface. We eagerly await the results of these studies of what we consider to be an important technique.

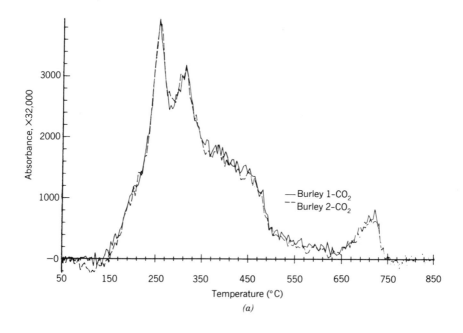

(a)

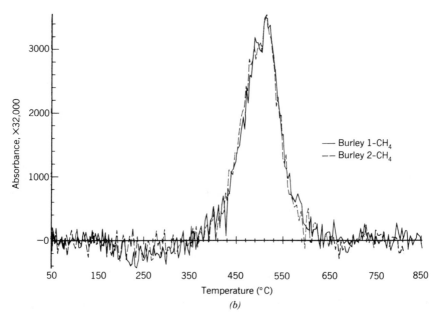

(b)

Fig. 12.12. (*a*) Absorbance profiles of CO_2 evolving from two samples of burley tobacco heated in nitrogen at 6°C/min. (*b*) Absorbance profiles of CH_4 evolving from two samples of burley tobacco, same conditions as (*a*). (Reproduced from [23], by permission of the Society for Applied Spectroscopy; copyright © 1980.)

405

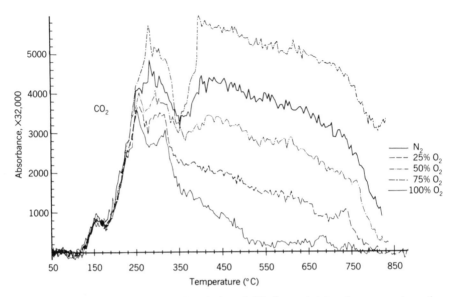

Fig. 12.13. Absorbance profiles of evolution of CO_2 from a bright tobacco sample as the oxygen content of the carrier gas is changed. (Reproduced from [23], by permission of the Society for Applied Spectroscopy; copyright © 1980.)

A very different flow system has been developed by Gendreau [27,28] to study the adsorption of flowing blood onto surfaces. These studies involved the use of an attenuated total reflectance (ATR) cell to study the surface absorption of adsorbed proteins. To study the thrombogenic nature of a polymer that could potentially be used for an artificial implant, various blood samples or blood component solutions were caused to flow across the surface of an internal reflection element that had been coated with a thin polymer film. In the latter paper [28], a temporal resolution of 0.8 sec was demonstrated by signal averaging four consecutive scans of 0.2 sec duration. The SNR of the spectra was sufficiently high that even at this high temporal resolution the rate of deposition of different proteins could be followed. These experiments are discussed in greater detail in Chapter 14.

An instrument that appears to be particularly well suited to the study of rapidly varying systems is the Nicolet 60-SX. The manufacturer's specifications indicate that the spectrometer can collect 50 scans per second at 16 cm^{-1} resolution. Data can be acquired as the moving mirror of the interferometer travels in both directions. Interferograms from each direction are stored in separate files because they have different phase er-

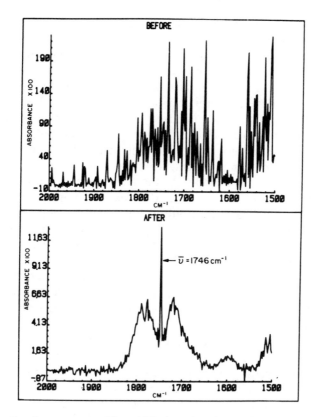

Fig. 12.14. Absorbance spectra of formaldehyde carbonyl stretching region before and after subtraction of coevolved water. (Reproduced from [25], by permission of the American Chemical Society; copyright © 1981.)

rors. (Once phase correction has been accomplished and the interferograms transformed, the spectra can be added to improve the SNR.) Results from time-dependent experiments, such as stopped-flow reactions, have been presented, but at the time of this writing none has been published. Demonstrated temporal resolutions are on the order of 0.02 sec. Publication of these results is anticipated in the near future.

VI. TIME-RESOLVED SPECTROMETRY

Time-resolved spectrometry (TRS) refers to those situations in FT–IR spectrometry where the lifetime of the species being studied is shorter than the scan rate of the interferometer. Two basic methods have been devised for "time-resolving" time-dependent phenomena. In actuality, both techniques are equivalent, but one was developed for step-scanning FT–IR and the other for rapid-scanning FT–IR measurements.

A method whereby the time-resolved spectra of emission reactions could be recorded to a temporal resolution on the order of tens of microseconds was developed for step-scan interferometry by Murphy and Sakai [29,30]. The technique is based on the criterion that the reaction under study is repetitive and reproducible. Murphy and Sakai investigated various gas phase reactions that were initiated by either microwave discharge or electron bombardment. The reactions were carried out in a continuous reactant flow mode. As the reactants flow into the cell, a 2-liter gold-coated integrating sphere, the system initiator (microwave discharge or electron beam) was pulsed for a short period. For example, during the reaction of N_2 and O_2 [30], chemiluminescent radiation was produced by the product species NO, N_2O, and NO_2. The initiator was pulsed for a relatively short period with respect to the entire reaction period. The reaction period was the time for all products to form and all emissions to cease. After the reaction ended, the cycle was repeated because by this time new reactants were in the discharge or beam and no interference from prior products is likely to result. As stated above, this reaction mechanism may be considerably shorter than the single scan time of the interferometer, particularly for a step-scan system.

Murphy and Sakai produced time-resolved data by repeating the reaction many times at each step retardation of the interferometer. An analog tape recorder was interfaced to the spectrometer, detector, and reaction initiator trigger. Data from the entire reaction were recorded on the tape and signal-averaged for as many times as the reaction was repeated and at each retardation of the interferometer. The reaction itself was time resolved because the data on the tape were collected at short discrete intervals of only a few tens of microseconds. Therefore, signal-averaged data of the reaction of short time intervals were collected for each retardation of the interferometer. Figure 12.15 represents the data as collected on tape. It must be remembered that the tape is a serial device and therefore the data are stored sequentially. For simplicity, only one experiment cycle has been collected at each retardation δ and each ordinate value represents a sampled datum at equal time intervals, Δt. The tape format is represented in tabular form to explain the sorting process more easily.

When the data tape is played back, a trigger signal (denoted *) is read, indicating the start of data collection. The data recorded at the same time as the trigger pulse, t_0, represent the detector response at experiment initiation at the particular retardation. That datum is read, processed by

Retardation

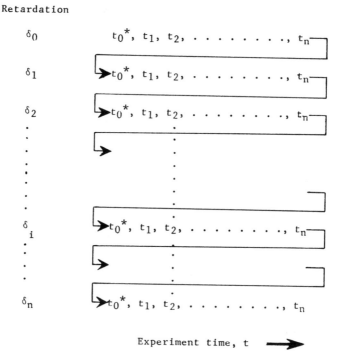

Fig. 12.15. Schematic drawing of tape format for data collected for stepped-scan FT–IR/ TRS experiment. Only a single experiment cycle is shown at each retardation for clarity.

an ADC, and stored in the computer. All other data are ignored until the next trigger pulse is read. If the t_0 datum corresponds to the experiment at the same retardation, then the processed datum is added to the datum in memory to improve the SNR. On the other hand, if the datum corresponds to reaction initiation at the subsequent retardation step, it is processed and stored as the subsequent datum in the interferogram. Once the entire tape has been read following the prescribed algorithm, an interferogram results that is constant in time. A second reading of the tape can be used to process all t_1 data to give another interferogram constant in time but at Δt after reaction initiation. Subsequent readings of the tape produce other constant-in-time interferograms. The result of this sorting procedure is shown in Fig. 12.16. The temporal resolution Δt is only limited by the response time of the detector and how rapidly the recording system can store these data.

The constant-in-time interferograms are transformed to produce the time-resolved spectra. Figure 12.17 is a series of time-resolved spectra

Retardation

$\delta_0 \quad \delta_1 \quad \delta_2 \quad \delta_3 \quad \cdots \cdots \cdots \cdots \cdots \quad n$

Experiment
Time

$* \quad = 0 \qquad t_0 \quad t_0 \quad t_0 \quad t_0 \quad \cdots \cdots \cdots \cdots \cdots \cdots \quad t_0$

$* + \Delta t = t_1 \qquad t_1 \quad t_1 \quad t_1 \quad t_1 \quad \cdots \cdots \cdots \cdots \cdots \quad t_1$

$* + 2\Delta t = t_2 \qquad t_2 \quad t_2 \quad t_2 \quad t_2 \quad \cdots \cdots \cdots \cdots \cdots \quad t_2$

$* + j\Delta t = t_j \qquad t_j \quad t_j \quad t_j \quad t_j \quad \cdots \cdots \cdots \cdots \cdots \quad t_j$

$* + n\Delta t = t_n \qquad t_n \quad t_n \quad t_n \quad t_n \quad \cdots \cdots \cdots \cdots \cdots \quad t_n$

Fig. 12.16. Schematic drawing of sorted data from Fig. 12.15. Each interferogram is now constant in time.

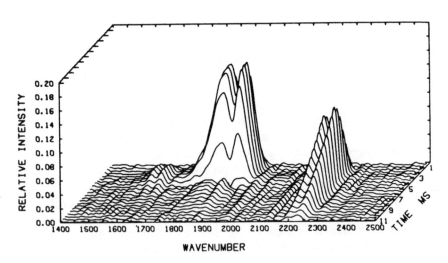

Fig. 12.17. Time-resolved emission spectra for the production of NO, N_2O, and NO_2 from a mixture of N_2/O_2 at 500 μsec temporal resolution over a period of 11 msec. (Reproduced from [30], by permission of the Optical Society of America; copyright © 1975.)

410

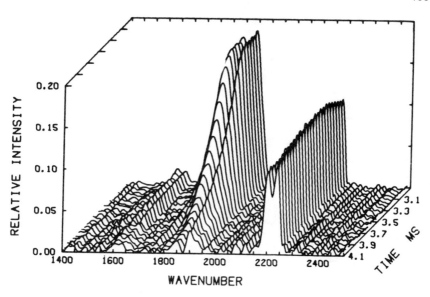

Fig. 12.18. Time-resolved emission spectra of the system in Fig. 12.17, but at 50 μsec temporal resolution. (Reproduced from [30], by permission of the Optical Society of America; copyright © 1975.)

for the N_2-O_2 system where NO, N_2O, and NO_2 are produced. The temporal resolution of this figure is 500 μsec, and the entire production and relaxation of the products can be observed. It should be noted that the time scale is in such a direction that the relaxation phenomena can be observed most clearly. Of course, if the time scale were reversed, production would be observed. The same data are presented in Fig. 12.18 at a temporal resolution of 50 μsec for a 1.2-msec portion of the reaction. The temporal resolution is quite good; however, the response time for the system as a whole is about 60 μsec so the spectra are not temporally independent. In a later paper, Sakai and Murphy [31] improved the data collection electronics and were able to match the speed of data acquisition to the time response of the spectrometer detector and increase the spectral bandwidth. With this technique, the spectral resolution should be limited only by the spectral resolution of the spectrometer. Figure 12.19 shows three different time-resolved spectra of the relaxation of CO_2 and CO in the presence of N_2, He, and Ar. Electron beam bombardment was used to initiate these reactions, and in all cases the pressure of the CO_2 reactant was 0.5 torr, with 50 torr of N_2, He, or Ar. The spectral resolution of these data was about 2 cm^{-1}, and the temporal resolution was 100 μsec. Sakai and Murphy believe it would be possible to demonstrate temporal

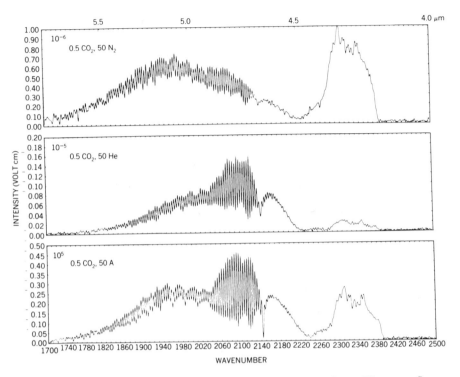

Fig. 12.19. Time-resolved relaxation spectra of CO_2 with three different filler gases. Spectral resolution is about 2 cm^{-1}. (Reproduced from [31], by permission of the Optical Society of America; copyright © 1978.)

resolutions of 1 μsec and spectral resolutions of 0.1 cm^{-1}, but very long measurement times would be needed.

A similar method for recording time-resolved spectra for rapid-scanning FT–IR spectrometry has been developed by Mantz [32,33]. The procedure is quite similar to that of Murphy and Sakai [29,30] except that all the data cannot be recorded at a single retardation and the interferogram stepped to the next retardation, as the mirror position is changing constantly. As for the stepped-scan system, the experiment must be reproducible, but in this case a signal from the interferometer must trigger the reaction initiator. The retardation of most commercial rapid-scanning interferometers is maintained by a 632.8-nm He–Ne laser. The laser radiation is modulated by the interferometer, and data are collected at the zero crossings. The time between crossings is used to regulate the velocity of the moving mirror (see Chapters 2 and 4). In an interferometer that has a moving mirror traveling at 1.58 mm sec^{-1}, the He–Ne laser is modulated at 5 kHz. In other words, every second zero crossing (one wavelength) occurs at 200 μsec intervals, and for mid-infrared measurements, data are sampled at this interval. The same sampling clock (that is, a datum sampled every 200 μsec) that is used to trigger data acquisition can likewise be used to clock an FT–IR/TRS experiment.

Let us assume a reaction can be initiated by some stimulus (microwave discharge, electron beam, electrical spark, ultraviolet radiation pulse, etc.) and be complete in a relatively short time, say 20 msec. Now let us connect the sampling clock to the reaction initiator so that the clock can be used as a trigger. As the first scan of the interferometer is begun, the initiator is triggered simultaneously to the collection of the first datum of the interferogram. This corresponds to t_0 in our previous example of stepped-scan interferometry. Two hundred microseconds later, the moving mirror is at the next sampling position and the reaction has proceeded in time by one sampling interval, Δt, which is datum t_1 of the previously defined convention. This process continues such that t_2 is collected at the next sampling step, t_3 at the next, and so on. After 100 sampling steps, the reaction is complete because 20 msec has elapsed. On the 101st zero crossing, the initiator is triggered once more and the process repeated. The initiator is triggered again on the 201st zero crossing, the 301st zero crossing, and so on. If the interferogram has 8192 data points, the experiment would be triggered 82 times. This exact process can be repeated as many times as desired to achieve the appropriate SNR. At this point, we have a single interferogram in which sets of 100 data represent different times in relation to the initiation of our experiment.

Now the whole process of collecting a signal-averaged interferogram is repeated, except that the reaction is triggered simultaneously to the second datum, that is, the second retardation position. Datum t_0 is collected where t_1 was collected on the previous interferogram. Similarly, a third interferogram is recorded where the reaction is triggered simultaneously to the collection of the datum at the third retardation position. One hundred such interferograms are collected in this example, with each successive interferogram having the reaction trigger offset by one datum to the previous interferogram. The results of such a data collection scheme are presented in Fig. 12.20, where $n = 100$. These data can be sorted to produce interferograms that are constant in time. If we select the 1st datum from the 1st interferogram, the 2nd from the 2nd, the 3rd from the 3rd, until we select the 100th from the 100th interferogram (or the nth from the nth), we have selected the first 100 (or n) data of a constant-in-time interferogram. The subsequent 100 data are selected by taking the 101st datum [or $(n + 1)$st] from the 1st interferogram, the 102nd from the 2nd, the 103rd from the 3rd, until we extract the 200th from the 100th

Interferogram Number	δ_1	δ_2	δ_3	δ_4	$\cdots$	δ_n	δ_{n+1}	δ_{n+2}	δ_{n+3}	$n+4$	$\cdots$	δ_{2n}	δ_{2n+1}	δ_{2n+2}	δ_{2n+3}	δ_{2n+4}	$\cdots$
1	t_1^*	t_2	t_3	t_4	$\cdots$	t_n	t_1^*	t_2	t_3	t_4	$\cdots$	t_n	t_1^*	t_2	t_3	t_4	$\cdots$
2		t_1^*	t_2	t_3	$\cdots$	t_{n-1}	t_n	t_1^*	t_2	t_3	$\cdots$	t_{n-1}	t_n	t_1^*	t_2	t_3	$\cdots$
3			t_1^*	t_2	$\cdots$	t_{n-2}	t_{n-1}	t_n	t_1^*	t_2	$\cdots$	t_{n-2}	t_{n-1}	t_n	t_1^*	t_2	$\cdots$
4				t_1^*	$\cdots$	t_{n-3}	t_{n-2}	t_{n-1}	t_n	t_1^*	$\cdots$	t_{n-3}	t_{n-2}	t_{n-1}	t_n	t_1^*	$\cdots$
$\cdot\cdot\cdot\cdot\cdot\cdot\cdot$																	
$n-1$						t_2	t_3	t_4	t_5	t_6	$\cdots$	t_2	t_3	t_4	t_5	t_6	$\cdots$
n						t_1^*	t_2	t_3	t_4	t_5	$\cdots$	t_1	t_2	t_3	t_4	t_5	$\cdots$

interferogram. As can be seen from Fig. 12.20, if this cyclic pattern is repeated, a complete constant-in-time interferogram can be constructed for t_0 in this case. To construct the constant-in-time interferogram for a time t_1 after each reaction was initiated, the 2nd datum from the 1st interferogram is collected as the 2nd datum in a new interferogram. Next comes the 3rd datum of the 2nd interferogram, the 4th of the 3rd, and so on. An examination of Fig. 12.20 provides the pattern for sorting. Care must be taken to ensure that each datum is preserved in its retardation position; that is, care must be taken not to shift data. A representation of the constant-in-time interferogram set is given in Fig. 12.21.

There is one striking difference between Figs. 12.16 and 12.21; that is, in Fig. 12.16, data are available for all retardations, whereas in Fig. 12.21, successively larger portions of the data are missing at the beginning of the constant-in-time interferograms. These data can be zero filled (Chapter 3, Section V) without any measurable effects because the number of zero-filled data is small.

It should be noted that temporal resolution is regulated by the modulation rate of the interferometer. The temporal interval can be increased, by undersampling the data (Chapter 2, Section I) so that the bandwidth is decreased or, by selectively sorting the data. The temporal resolution can be increased (i.e., the interval decreased) by two methods, both of which greatly increase the number of data collected and consequently prolong the total time of the experiment. One could simply increase the data modulation rate, but this would increase the number of interferograms to be collected. Doubling the modulation rate doubles the number of data in a reaction time period, which doubles the number of interferograms to be collected. The second alternative is to collect one set of time-resolved data triggered normally and a subsequent set (all n interferograms) with the trigger offset to actual data collection. Thus, data in the two sets are temporally offset. For example, if in the first set the initiator is triggered simultaneously with the laser interferogram zero crossing, the temporal resolution is equivalent to the time between successive zero crossings. If in the other set we trigger the initiator one-half a sampling interval late, the temporal resolution is the same as the first set but offset by one-half the time of the sampling interval. If the two sets are interleaved, a new set results after sorting that has an effective temporal resolution double that of either original sets. To double the temporal resolution again requires twice the number of interferograms. A compro-

---

Fig. 12.20. Schematic drawing of data collection procedure for rapid-scanning FT–IR/TRS. Times t_1^* correspond to experiment cycle initiations.

Retardation

Sorted Interferogram Number	δ_1	δ_2	δ_3	δ_4	…	δ_n	δ_{n+1}	δ_{n+3}	δ_{n+4}	…	δ_{2n}	δ_{2n+1}	δ_{2n+2}	δ_{2n+3}	δ_{2n+4}	…
1	t_1^*	t_1^*	t_1^*	t_1^*	…	t_1^*	t_1^*	t_1^*	t_1^*	…	t_1^*	t_1^*	t_1^*	t_1^*	t_1^*	…
2		t_2	t_2	t_2	…	t_2	t_2	t_2	t_2	…	t_2	t_2	t_2	t_2	t_2	…
3			t_3	t_3	…	t_3	t_3	t_3	t_3	…	t_3	t_3	t_3	t_3	t_3	…
4				t_4	…	t_4	t_4	t_4	t_4	…	t_4	t_4	t_4	t_4	t_4	…
⋯																
n−1						t_{n-1}	t_{n-1}	t_{n-1}	t_{n-1}	…	t_{n-1}	t_{n-1}	t_{n-1}	t_{n-1}	t_{n-1}	…
n							t_n	t_n	t_n	…	t_n	t_n	t_n	t_n	t_n	…

mise can be made by running the experiment at low temporal and spectral resolutions to determine the time of maximum interest. Once that time has been selected, the experiment can be repeated at the desired resolution, but only the data pertaining to the time most relevant to the reaction are stored. This procedure does not reduce the data collection time, but the times for sorting and transformation are greatly reduced. As with the stepped-scan process, the spectral resolution is limited only by the spectrometer, and the temporal resolution is only limited (in theory, at least) by the detector response time.

The first TRS data acquired using a rapid-scanning interferometer were reported by Mantz, who performed two basic experiments, the photolysis of acetone [32] and the photolysis of acetaldehyde [33]. In both experiments, absorption spectra were measured, as opposed to the measurement of emission spectra by Murphy and Sakai [29–31]. By performing the measurements in this way, it was hoped that transient species could be seen in both emission and absorption as the sample cell was placed before the interferometer. For the photolysis of acetone, the well-known reaction

$$(CH_3)_2CO \xrightarrow{h\nu} 2CH_3 + CO$$

was expected. The spectra from this photolysis are presented in Fig. 12.22. No CH_3 bands can be observed in the spectra, and the CO band at 2149 cm^{-1} is superimposed on the acetone overtone. The only distinct feature is at 3138 cm^{-1}, which was identified tentatively as being due to ketene ($CH_2{=}C{=}O$), a previously unknown reaction product. These spectra have a temporal resolution of 400 μsec and a spectral resolution of 2 cm^{-1}. The photolysis of acetaldehyde [31] produced methane, carbon monoxide, and HCO, as expected. Figure 12.23 is a 0.25-cm^{-1}-resolution spectrum at the peak of CH_4 absorption.

It was shown later by Garrison et al. [34] that the results obtained by Mantz were not spectral absorptions or emissions but only artifacts. This resulted from the fact that the concentrations in the reaction cell were not constant throughout the entire experiment. As the reactions proceeded, the concentration of the products changed as they accumulated in the cell due to the inadequate flow rates of the gaseous reactants. Upon sorting, the concentration changes added a second modulation to the constant-in-time interferograms. Transformation resulted in ghost peaks displaced from the actual absorptions. It was shown that the "ketene" peak

---

Fig. 12.21. Schematic drawing of sorted, constant-in-experiment-time, rapid-scanning FT–IR/TRS data.

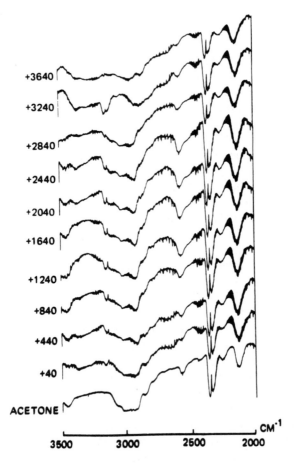

Fig. 12.22. Time-resolved spectra from photolysis of acetone with a temporal resolution of 400 μsec. (Reproduced from [32], by permission of the Society for Applied Spectroscopy; copyright © 1976.)

in the photolysis of acetone was displaced CO_2 from an inadequate purge and the HCO band in the photolysis of acetaldehyde was displaced CO. These phenomena have been computer modeled by de Haseth [35], who showed that the positions of these displaced bands could be predicted in a straightforward manner.

Unfortunately, the discovery of these experimental errors all but stopped work in the area of FT–IR/TRS. To the best knowledge of these authors, no additional papers on flow systems have been published. Fateley and Koenig [36] have used FT–IR/TRS to study the physical deformation of polypropylene films. In this study, the reaction initiator was a mechanical device that elongated a film by approximately 1%. The entire cycle involved stretching and relaxation of the film, which lasted about 100 msec. Figures 12.24 a–c are time-resolved spectra of polypropylene before stretching, at the stretch midpoint, and at the extreme stretch point, respectively. No bands are seen to shift, but the intensities of the bands at 974 and 995 cm^{-1} change with elongation. The ratio of intensities of these two bands has been correlated to polypropylene crystallinity. The results of this experiment are consistent with crystalline changes due to stress in this polymer.

In a later paper, Graham, Grim, and Fateley [37] presented additional data for the stretching of isotactic polypropylene. The film was pre-stretched 5% to reduce buckling upon relaxation and an additional 5% stretching was accomplished at a rate of 17.9 Hz. Single-scan spectra were collected to avoid spectral artifacts that may be caused by instrumental drifts. A total of 50 data points was collected per cycle, and a temporal resolution of approximately 1.12 msec was achieved. The greatest changes in the time-resolved spectra were observed for the CH$_2$ and

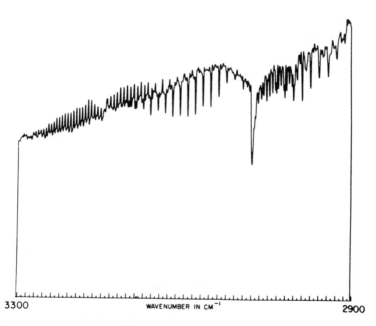

3300   WAVENUMBER IN CM^{-1} 2900

Fig. 12.23. Time-resolved spectrum of photolysis products of acetaldehyde. (Reproduced from [33], by permission of the Optical Society of America; copyright © 1978.)

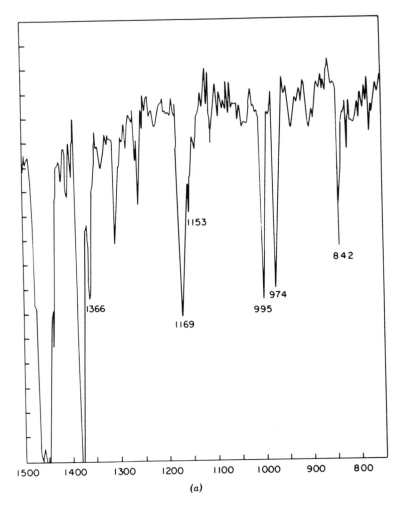

(a)

Fig. 12.24. Time-resolved polypropylene spectra (a) before stretching; (b) midway through stretch cycle; and (c) at extreme of stretching cycle. (Reproduced from [36], by permission of John Wiley and Sons, Inc.; copyright © 1982.)

CH_3 deformation bands at 1457 and 1378 cm^{-1}. The absorption spectra from the two extremes of the stretch cycle are presented in Fig. 12.25. Clearly, the ratio of the two bands changes relative to the intensity between the most stretched (Fig. 12.25a) and least stretched (Fig. 12.25b) points in the deformation cycle. The ratio between the two absorbances, A_{1378}/A_{1457}, ranges between 0.93 (least stretched) and 1.34 (most stretched).

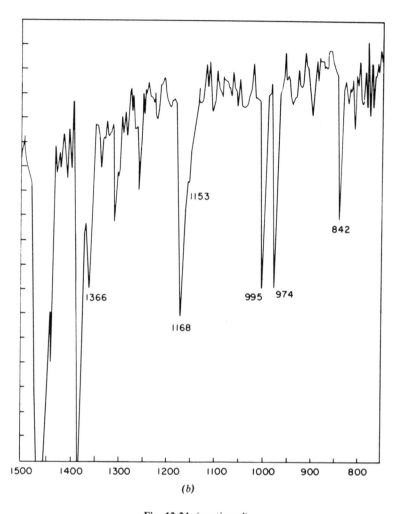

1153

842

1366   995   974

1168

1500   1400   1300   1200   1100   1000   900   800

(b)

Fig. 12.24. (continued)

Another set of FT–IR/TRS experiments has been directed to the study of polymer deformations. Hsu [38,39] has modified the rapid-scanning TRS experiment to operate with slow experiment cycles. It is a requirement of conventional rapid-scanning TRS that the experiment cycle is shorter than the interferometer scan initiation to centerburst time. If it is longer than this time, the centerburst will not be sampled at large cycle offsets, see Fig. 12.21. Hsu and co-workers [38] devised a modified TRS procedure where the stretching cycle of the polymer ran independently

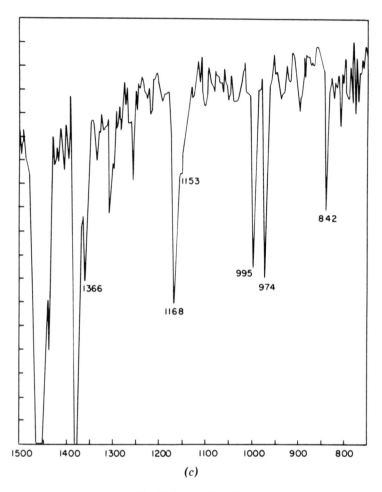

Fig. 12.24. (continued)

of the interferometer. Interferograms of the polymer were collected and flagged in order to correlate the interferograms to the stretching cycle. Single-scan interferograms were collected, in essentially a random order; then interferograms were coadded on the basis of the flag and hence the offset in the stretching cycle. This method obviated the problem of a cycle that was too long, and the full number of data points was always collected at the beginning of the interferogram.

A modification of the above procedure was reported by Hsu and coworkers [39]. The polymer stretcher was synchronized to the interfero-

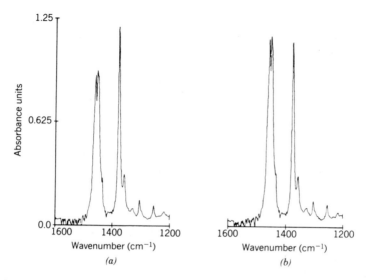

Fig. 12.25. Time-resolved absorbance spectra of polypropylene (*a*) at extreme of stretch cycle and (*b*) least stretched. (Reproduced from [37], by permission of Elsevier Scientific Publishing Co.; copyright © 1984.)

meter scan via a microprocessor. The interferometer does not initiate the stretch cycle, as in the conventional experiment. Instead, a pulse is sent to indicate with the interferometer scan is initiated. Using this method, coaddition of scans can be completed sequentially rather than randomly. Results of these two papers [38,39] are presented in Chapter 13.

Time-resolved infrared spectrometry clearly has a great potential in the understanding of dynamic systems. Although FT–IR/TRS has not yet been demonstrated for rapid-scanning interferometers for short experiment cycle times (<20 msec), experiments under these conditions undoubtedly will be performed in the near future. Flow systems have inherent problems (primarily a difficulty to reproduce the experimental conditions exactly), but these too should be solved.

REFERENCES

1. J. G. Calvert and J. N. Pitts, *Photochemistry,* Wiley, New York (1966).
2. G. Herzberg, *The Spectra and Structure of Simple Free Radicals,* Cornell University Press, Ithaca, N.Y. (1971).
3. N. J. Turro, *Modern Molecular Photochemistry,* Benjamin/Cummings, Menlo Park, Ca. (1978).

4. H. Okabe, *Photochemistry of Small Molecules*, Wiley, New York (1978).

5. D. S. Bethune, J. R. Lankard, and P. O. Sorokin, *Opt. Lett.*, **4**, 103 (1979).

6. D. S. Bethune, J. R. Lankard, M. M. T. Loy, and P. O. Sorokin, *IBM J. Res. Develop*, **23**, 556 (1979).

7. A. T. Stair, Jr., Aspen Int. Conf. on Fourier Spectrosc., 1970 (G. A. Vanasse, A. T. Stair, and D. J. Baker, eds.), AFCRL-71-0019, p. 127 (1971).

8. J. G. Moehlmann, J. T. Gleaves, J. W. Hudgens, and J. D. MacDonald, *J. Chem. Phys.*, **60**, 4790 (1974).

9. J. F. Durana and J. D. MacDonald, *J. Chem. Phys.*, **64**, 2518 (1976).

10. J. P. Sung and D. W. Setser, *Chem. Phys. Lett.*, **48**, 413 (1977).

11. W. H. Chan, W. M. Uselman, J. G. Calvert, and J. H. Shaw, *Chem. Phys. Lett.*, **45**, 240 (1977).

12. F. Su, J. G. Calvert, C. R. Lindley, W. M. Uselman, and J. H. Shaw, *J. Phys. Chem.*, **83**, 912 (1979).

13. H. Niki, P. D. Maker, C. M. Savage, and L. P. Breitenbach, *J. Phys. Chem.*, **82**, 132 (1978).

14. H. Niki, P. D. Maker, L. P. Breitenbach, and C. M. Savage, *Chem. Phys. Lett.*, **57**, 596 (1978).

15. D. B. Chase, R. L. Amey, and W. G. Holtje, *Appl. Spectrosc.*, **36**, 155 (1982).

16. J. B. Kinney, R. H. Staley, C. L. Reichel, and M. S. Wrighton, *J. Am. Chem. Soc.*, **103**, 4273 (1981).

17. D. B. Chase and F. J. Weigert, *J. Am. Chem. Soc.*, **103**, 977 (1981).

18. J. O. Lephardt and G. Vilcins, *Appl. Spectrosc.*, **29**, 221 (1975).

19. G. Vilcins and J. O. Lephardt, *Chem. Ind.*, **22**, 974 (1975).

20. S. A. Liebman, D. H. Ahlstrom, and P. R. Griffiths, *Appl. Spectrosc.*, **30**, 355 (1976).

21. D. D. Saperstein, *Anal. Chem.*, **52**, 1565 (1980).

22. C. A. Cody, L. DiCarlo, and B. K. Faulseit, *Am. Lab.*, **13**(1), 93 (1981).

23. J. O. Lephardt and R. A. Fenner, *Appl. Spectrosc.*, **34**, 174 (1980).

24. J. O. Lephardt and R. A. Fenner, *Appl. Spectrosc.*, **35**, 95 (1981).

25. R. A. Fenner and J. O. Lephardt, *J. Agric. Food Chem.*, **29**, 846 (1981).

26. J. O. Lephardt, *Appl. Spectrosc. Rev.*, **18**, 265 (1982–83).

27. R. M. Gendreau and R. J. Jakobsen, *J. Biomed. Mater. Res.*, **13**, 893 (1979).

28. R. M. Gendreau, *Appl. Spectrosc.*, **36**, 47 (1982).

29. R. E. Murphy and H. Sakai, Aspen Int. Conf. on Fourier Spectrosc., 1970 (G. A. Vanasse, A. T. Stair, and D. J. Baker, eds.), AFCRL-71-0019, p. 301 (1971).

30. R. E. Murphy, F. M. Cook, and H. Sakai, *J. Opt. Soc. Am.*, **65**, 600 (1975).

31. H. Sakai and R. E. Murphy, *Appl. Opt.*, **17**, 1342 (1978).

32. A. W. Mantz, *Appl. Spectrosc.*, **30**, 459 (1976).

33. A. W. Mantz, *Appl. Opt.*, **17**, 1347 (1978).

34. A. A. Garrison, R. A. Crocombe, G. Mamantov, and J. A. de Haseth, *Appl. Spectrosc.*, **34**, 399 (1980).

35. J. A. de Haseth, *Proc. Soc. Photo-Opt. Instrum. Eng.*, **289**, 34 (1981).

36. W. G. Fateley and J. L. Koenig, *J. Polym. Sci., Polym. Lett. Ed.*, **20**, 445 (1982).

37. J. A. Graham, W. M. Grim, III, and W. G. Fateley, *J. Mol. Struct.*, **113**, 311 (1984).

38. J. E. Lasch, D. J. Burchell, T. Masoaka, and S. L. Hsu, *Appl. Spectrosc.*, **38**, 351 (1984).

39. S. E. Molis, W. J. MacKnight, and S. L. Hsu, *Appl. Spectrosc.*, **38**, 529 (1984).

FT–IR STUDIES OF POLYMERS

I. INTRODUCTION

Virtually every chemist involved with infrared spectrometry is familiar, at least at a cursory level, with the spectra of polymers. Infrared spectrometers are often supplied with a sample of polystyrene film; this is used to check wavenumber accuracy in dispersive spectrophotometers and can be used to demonstrate the performance of FT–IR instruments. Although the ubiquitous polystyrene film has become a standard in infrared spectrometry, the study of macromolecular structures goes far beyond this well-known example.

Vibrational spectrometry has been applied to the identification of polymeric materials for both qualitative and quantitative determination of the chemical composition. Many parameters can be investigated, including polymer end groups, chain branching, configuration and conformation, as well as steric and geometric isomerism. Infrared spectrometry has also been used to identify and determine the concentration of impurities, antioxidants, emulsifiers, additives, plasticizers, fillers, and residual monomers in polymeric materials. Many reaction mechanisms of polymers are relatively slow so that processes such as vulcanization, polymerization, and degradation can be followed using rapid-scanning spectrometers. The effects of external conditions on polymers have been studied as well. The changes of state with temperature and pressure as well as the effects of irradiation, deformation, fatiguing, and weathering are of interest.

The techniques used for the measurement of FT–IR spectra of polymers are for the most part identical to those applied to conventional samples. Polymers may be handled as solutions, solids, powders, and films. Standard instrumental techniques include transmission, attenuated total reflectance, diffuse reflectance, reflection–absorption, emission, and photoacoustic spectrometry. These techniques are covered in other chapters, and additional examples will not be presented here.

The analysis of polymers using FT–IR spectrometry relies heavily on the isolation of spectral features. Spectral subtraction is frequently employed to isolate spectral features of a component in a polymer blend, to remove solvent bands, or to isolate spectral features after a chemical reaction or physical change in the sample. Many polymers can be compressed or solvent-cast into thin films. Preparation of samples as thin films is often the most desirable sampling technique because there is no solvent interference; however, most characteristics imposed by the processing and thermal activity of the sample may be lost. Thin films do present another problem in that interference fringes are often present and may mask spectral features. Interference fringes may be removed by employing various computer methods (Chapter 10); hence their presence is not an insurmountable problem.

Thin films of polymer samples are easily handled and are often measured by conventional transmission spectrometry. They are also suitable for ATR measurements and, when present on the surface of a metal, may also be studied by reflection–absorption spectrometry. Other sampling methods for transmission spectrometry include compressing polymer powders with KBr to form pellets or preparation as a mull. Polymer solutions in nonaqueous solvents are handled in standard liquid cells, and of course aqueous solutions can be studied using FT–IR spectrometry by transmission or ATR spectrometry (see Chapter 14).

Microanalysis can be applied to polymer samples. Minute samples can be located under an aperture of a suitable diameter and placed in a $4\times$ or $6\times$ beam condenser. This technique is often used to examine relatively large impurities or occlusions in polymer samples [1,2]. Diamond anvil cells can be used to examine small areas of more opaque samples. The sample is compressed between two diamond windows so that the pathlength is reduced and transmission can be achieved. The diamond anvil cell is then placed in a beam condenser because the anvil windows are rarely much greater than 2 or 3 mm in diameter. It should be noted that although this method can be used to examine difficult samples, the diamond anvil cell deforms the sample and may hence alter its characteristics. In addition, the region from approximately 1900 to 2300 cm^{-1} is opaque due to absorption by the diamond.

Transmission ultramicrosampling has been demonstrated for polymer films and fibers by Curry et al. [3]. Curry and co-workers obtained transmission spectra of polymer films through apertures of 35 μm and as small as 12.5 μm. Figure 13.1a is the spectrum of a 12-μm-thick film of poly(ethylene terephthalate) (PET) through a 35-μm aperture measured using an MCT detector. This 4-cm^{-1}-resolution spectrum was collected using an $8\times$ beam condenser and was acquired in 55 min and represents approximately 16 ng of the sample. Comparison with a standard reference spectrum of PET (Fig. 13.1b) shows that the spectrum through the ap-

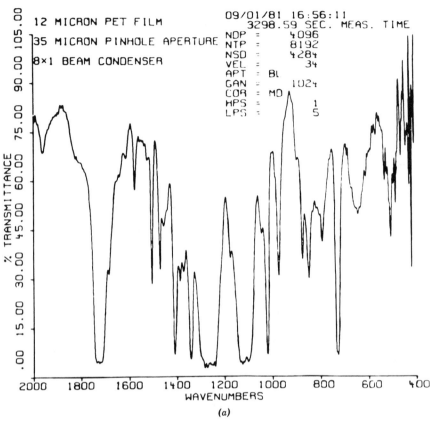

(a)

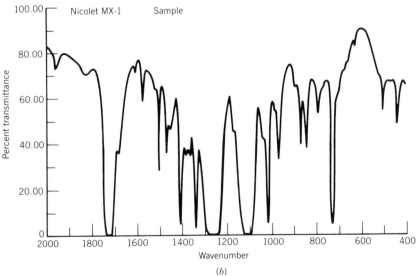

(b)

428

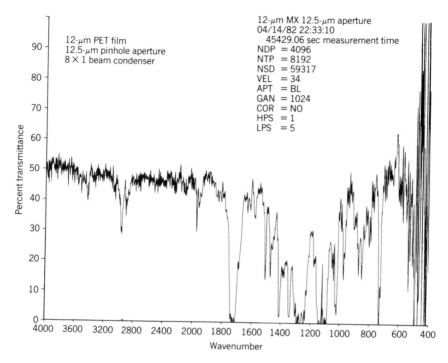

Fig. 13.2. A 12-μm-thick PET film through a 12.5-μm aperture using an 8× beam condenser. (Reproduced from [3], by permission of the Society for Applied Spectroscopy; copyright © 1985.)

erture is readily recognizable as PET. Reduction in the aperture size to 12.5 μm decreases the SNR, as shown in Fig. 13.2. The spectrum is still recognizable as PET, but the collection time increased to 12.5 hr, and spectral information in the range 400–600 cm^{-1} was lost due to diffraction effects. The authors also show spectra of fiber samples using slit apertures [3].

A recent development in FT–IR spectrometry is the interfacing of a transmission or reflection microscope. These microscopes permit the investigation of samples that are as small as 10–15 μm in diameter. The size of the sample is determined by diffraction effects, not by the reso-

←———————————————————————————————

Fig. 13.1. (*a*) A 12-μm-thick PET film through a 35-μm aperture using an 8× beam condenser. (*b*) Reference transmission spectrum of PET. (Reproduced from [3], by permission of the Society for Applied Spectroscopy; copyright © 1985.)

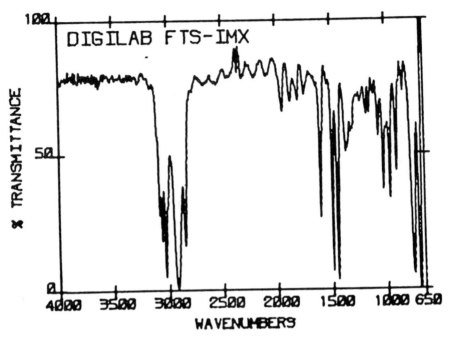

Fig. 13.3. Spectrum of polystyrene measured with an IR microscope through a 10-μm aperture. (Reproduced from [4], by permission of the American Chemical Society; copyright © 1984.)

lution of the microscope. It is anticipated that infrared microscopes will be very useful in the nondestructive analysis of defects or occlusions in polymers. Figure 13.3 shows the spectrum of polystyrene measured with an infrared microscope through a 10-μm aperture [4].

Of much interest in polymer analysis is the change in chemical composition due to factors such as reaction conditions, oxidation, thermal decomposition, and so on. Often much of the effect of the chemical history of a sample is altered by preparing a sample in the form of a pellet, mull, thin film, or solution for transmission spectrometry. For this reason, more emphasis is being placed on nondestructive analytical methods. One such method that is starting to enjoy a great deal of success for the study of powdered polymers is diffuse reflectance (DR) spectrometry. Diffusely reflected radiation is collected and directed to the detector in the manner discussed in Chapter 5. The spectrum is obtained by ratioing the diffuse reflectance spectrum of the sample to the spectrum of a nonabsorbing

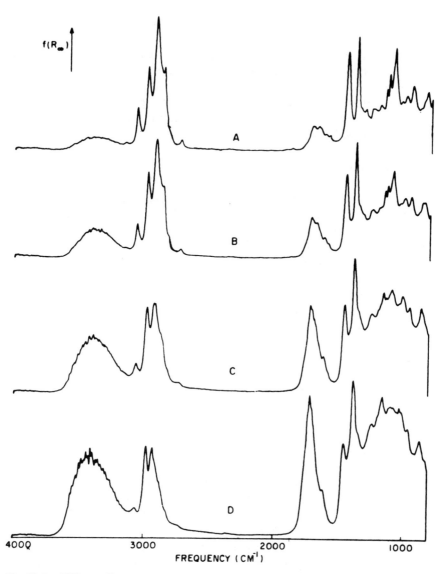

Fig. 13.4. Diffuse reflectance spectra of poly(dimethylfulvene) as a function of time after sample preparation: (A) 15 min, (B) 70 min, (C) 255 min, and (D) 1285 min. The oxidation rate appears to be diffusion controlled, at least in the latter stages of oxidation. (Reproduced from [5], by permission of the American Chemical Society; copyright © 1978.)

subtrate, usually powdered KCl or KBr, and then converting to the Ku-belka–Munk function. Diffuse reflectance spectrometry is very useful for powdered polymer samples where virtually no sample preparation is needed, except possibly for dilution in the KCl or KBr matrix. Ther-mosetting polymers may usually be ground at room temperature whereas thermoplastic polymers may need to be cooled to 77 K before grinding. Provided that the grinding time is short, the morphology of the sample is usually not altered. Examples of DR spectra are shown in Fig. 13.4 in which powdered poly-(dimethylfulvene) is shown as a function of time after sample preparation [5]. This polymer is unstable and susceptible to air oxidation. Had the sample been compressed into a disk, the rate of oxidation would have been far slower.

Diffuse reflectance spectrometry of nonpowdered solid samples, such as pellets, is almost invariably unsuccessful. This is because the specular component of the reflection can become large especially in the region of strong absorption where band inversion can sometimes be noticed unless the polymer powder has been diluted in KCl or KBr. Quantitative analysis under these conditions may not be possible.

Photoacoustic spectrometry (PAS) is another technique that has held promise for polymer samples, since sample preparation is negligible. PAS has the additional advantage that the form of the sample (e.g., pellet, powder, film) does not effect the relative intensity of most bands in the spectrum [6]. Photoacoustic spectra of a nitrile plastic with four different surface morphologies are shown in Fig. 9.13. The photoacoustic spectrum of a piece of insulating foam, prepared simply by cutting a small sample from a larger piece and mounting it in the cup, is shown in Fig. 13.5. PAS is now starting to be used for quantitative quality control measurements of polymer pellets, primarily because of the minimal sample preparation required. Nevertheless, the depth of penetration of the incident radiation is highly dependent on the modulation frequency (see Chapter 9), so that the velocity of the moving mirror must never be altered once the desired value has been determined. PAS is also very susceptible to environmental interferences, such as noise in the laboratory. Considerable effort is being directed to the solution of these problems, and it is anticipated that PAS will start to manifest its true potential in the near future.

The entire field of polymer studies by FT–IR spectrometry is extremely broad, and a comprehensive review is beyond the scope of this text. Extensive reviews of infrared polymer spectrometry are available [7–11]. The detailed review by Koenig [11] is particularly recommended. Much of the remainder of this chapter is devoted to analytical methods in pol-ymer analysis that are not generally covered elsewhere. These methods

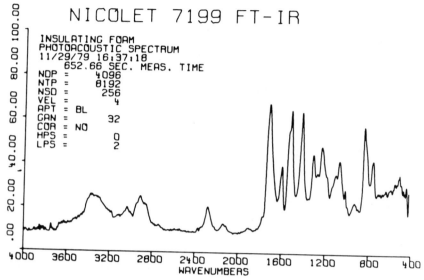

Fig. 13.5. Photoacoustic spectrum of a piece of insulating foam. Measurement time was 11 min. (Reproduced from [6], by permission of the Society for Applied Spectroscopy; copyright © 1980.)

include polarization spectrometry of polymers and stress studies using time-resolved spectrometry.

II. SUBTRACTION STUDIES

The theory of spectral subtraction has been presented in detail in Chapter 10, and examples were given for the quantitative analysis of mixture components. Absorbance subtraction spectrometry can also be used to isolate components, for identification, and for quality control. One aspect of spectral subtraction that was not discussed was its use as a spectrometric separation technique. This is particularly germane to some problems in polymer chemistry.

It is possible to perform a spectrometric separation of a heterophase system that exhibits a composite spectrum. Specifically, it is possible to obtain spectra of pure crystalline or pure amorphous phase materials when a pure sample of either does not exist [12]. Because the crystalline phase often has a different, and more ordered, orientation than the amorphous phase, these differences can lead to different spectral characteristics. If samples can be obtained in which the relative concentrations of crystalline and amorphous phases change, then spectra of the pure components can be obtained.

The spectrum of a heterophase system can be considered as being a composite of all the absorbances of all the phases. These absorbances are the crystalline phase absorbance, $A_c(\bar{\nu})$; the amorphous phase absorbance, $A_a(\bar{\nu})$; and the absorbance that is phase independent, $A_i(\bar{\nu})$. Hence the total absorbance $A_t(\bar{\nu})$ is

$$A_t(\bar{\nu}) = A_c(\bar{\nu}) + A_a(\bar{\nu}) + A_i(\bar{\nu}) \qquad (13.1)$$

If a sample is obtained that contains a higher crystallinity, its total absorbance may be represented as

$$A_t'(\bar{\nu}) = A_c'(\bar{\nu}) + A_a'(\bar{\nu}) + A_i'(\bar{\nu}) \qquad (13.2)$$

The total absorbance spectrum of a sample with lower crystallinity can be represented by

$$A_t''(\bar{\nu}) = A_c''(\bar{\nu}) + A_a''(\bar{\nu}) + A_i''(\bar{\nu}) \qquad (13.3)$$

Careful annealing studies can provide information as to which bands are due to the amorphous phase and which are due to the crystalline phase. To obtain the pure crystalline spectrum of the sample, a subtraction is performed such that

$$[A_a'(\bar{\nu}) - kA_a''(\bar{\nu})] = 0 \qquad (13.4)$$

where k is the scale factor to bring all the amorphous phase absorbance bands to the baseline.

The resultant crystalline phase absorbance spectrum is then

$$[A_t'(\bar{\nu}) - kA_t''(\bar{\nu})] = [A_c'(\bar{\nu}) - kA_c''(\bar{\nu})] + [A_i'(\bar{\nu}) - kA_i''(\bar{\nu})] \qquad (13.5)$$

A similar set of equations can be written to obtain the absorbance spectrum of the pure amorphous phase. Of course, this technique is not limited to heterophase crystalline/amorphous polymer samples, but it can be applied to a wide variety of similar problems.

Isotactic polypropylene exhibits both crystalline and amorphous phases, and these can be separated spectrometrically [13]. An isotactic polymer has highly ordered macromolecular chains in which each monomer of the chain has an identical stereochemical orientation. In this study, thin films were prepared by pressing small sample amounts at 200°C. Quenching the thin films in baths of ice water, liquid nitrogen, or dry ice–acetone produced samples with high amorphous phase content. Samples with high crystalline phase content were produced by annealing the quenched samples in a vacuum oven at 130°C.

Figure 13.6 shows the difference spectrum (*C*) of annealed (*A*) and quenched (*B*) isotactic polypropylene samples. In this study [13], the amorphous and crystalline features had not been previously identified. The methyl deformation mode at 1376 cm^{-1}, however, had a slightly different absorption wavenumber in the amorphous phase, at 1378 cm^{-1}. Once this difference was noted, the correct subtraction factor (*k* from Eq. 13.4) was chosen such that the absorption at 1378 cm^{-1} was at the baseline and not negative. The spectrum in Fig. 13.6*c*, is consistent with

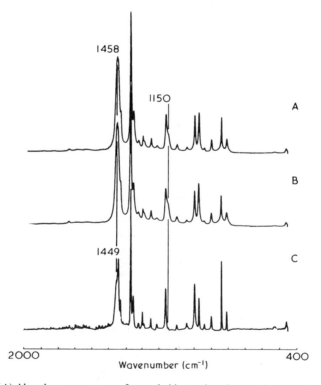

Fig. 13.6. (*A*) Absorbance spectrum of annealed isotactic polypropylene sample; (*B*) spectrum of quenched sample; and (*C*), difference spectrum (*A*–*B*), which shows the spectrum of the pure crystalline phase of the sample. (Reproduced from Painter, P. C., Watzek, M., and Koenig, J. L. *POLYMER*, 1977, **18**, 1169–1172, by permission of the publishers, Butterworth & Co. (Publishers) Ltd. ©)

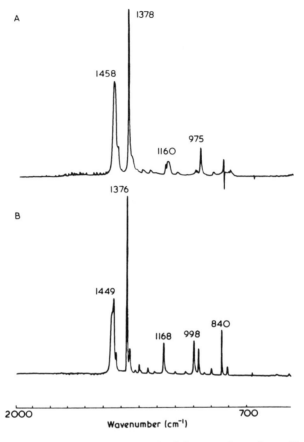

Fig. 13.7. (A) Difference absorbance spectrum of the amorphous phase of isotactic poly-propylene produced by subtracting the annealed sample spectrum from that of the quenched sample. (B) Difference spectrum characteristic of the crystalline phase of the sample (see Fig. 13.6c). (Reproduced from Painter, P. C., Watzek, M., and Koenig, J. L. *POLYMER*, 1977, **18**, 1169–1172, by permission of the publishers, Butterworth & Co. (Publishers) Ltd. ©)

crystalline phase spectra as it has the characteristic sharp absorption bands.

In Fig. 13.7 are difference spectra for both the amorphous (A) and crystalline (B) phases of isotactic polyproylene. The amorphous phase spectrum exhibits a derivative-like peak at 841 cm^{-1}. This is because the CH$_2$ rocking mode at this wavenumber has different absorption wavenumbers for the ordered crystalline phase and the 3$_1$ helical phase in the amorphous phase. The wavenumber difference is small, but the subtraction method accentuates this difference.

Another characteristic of polymers that can often be detected by absorbance subtraction is compatible and incompatible blends. If the assumption is made that the preferred state of a polymer is the crystalline phase, then polymer blend compatibility can be explained easily. In an incompatible mixture, two or more polymers are mixed and the crystalline chains of one polymer do not interact with the others. Therefore, the spectrum of the blend is equivalent to the sum of the spectra of the pure polymers in the appropriate concentrations. On the other hand, a compatible blend has an interaction between the crystalline chains such that significant changes in the chain conformations occur and there is a chemical interaction that affects the hydrogen or dipolar bonding sufficiently to perturb the normal vibrations of one or both polymer chains. Either change in a compatible blend can alter the spectra of the pure components so that the spectrum of the blend is not a linear combination of the pure component spectra.

The effects of compatible and incompatible blends were discussed in Chapter 10, and an absorbance subtraction of pure component spectra from a two-component compatible blend is shown in Fig. 10.9. After the subtraction, a residual spectrum remains that corresponds to the interaction of the compatible components. A great deal of study of compatible and incompatible polymer blends has been completed by M. M. Coleman. The list of papers in these studies is too numerous to be referenced here, but the manuscript by Coleman et al. [14] on this subject is noteworthy.

III. POLARIZATION STUDIES

The molecular orientation of chains in polymers determines their mechanical properties, such as strength, ductility, and glass transition temperature. The procedures used to process a polymer can alter the orientation of the macromolecular chains and hence alter the mechanical properties of the polymer. Various methods have been employed to measure the chain orientation, for example, bifringence, sonic modulus, X-ray diffraction, nuclear magnetic resonance spectrometry, polarized Raman and fluorescence spectrometries, as well as ultraviolet and infrared dichroism [15,16]. It is useful to be able to investigate the orientation of crystalline and amorphous polymers, and vibrational spectrometry using polarized radiation can often be used to examine specific modes that are associated with the different structures. Fourier transform infrared spectrometry is well suited to dichroic measurements due to the high SNR and photometric accuracy that can be achieved in reasonably short measurement times.

A series of studies on uniaxially oriented polymers using infrared dichroism provides a good example of macromolecular chain orientation [17–19]. The first of these studies [17] involves isotactic polystyrene. Isotactic configurations are chosen because each monomer in the macromolecular chain has an identical stereochemical orientation. Thus, specific absorption bands can be correlated to the orientation. In syndiotactic or atactic structures, the spectral features are not usually as well defined and analysis is more difficult.

In a uniaxially oriented polymer, there is only one axis and analysis is rather straightforward. Infrared dichroism measurements can be used to measure the orientation of a single unit of the chain, which has its own coordinate system, with respect to the orientation of the polymer, which too has its coordinate system. The rotation of the unit Cartesian coordinate system into the polymer reference coordinate system describes the orientation of the macromolecular unit. The rotation is defined by the three Eulerian angles θ, ϕ, and ψ. In uniaxially oriented polymers, structural units are randomly distributed about the axis and are hence random with respect to ϕ and ψ, so only θ can be determined. The relation between the orientation of a single unit of the chain and the polymer is called the orientation distribution function.

In this case, the orientation distribution function is a function of θ only and this can be written as [17–20]

$$f(\theta) = \sum_{n=0}^{\infty} (n + \tfrac{1}{2}) \langle P_n(\cos \theta) \rangle_{av} P_n(\cos \theta) \qquad (13.6)$$

where the functions $P_n(\cos \theta)$ are the spherical harmonic functions

$$P_2(\cos \theta) = \tfrac{1}{2}(3 \cos^2 \theta - 1)$$
$$P_4(\cos \theta) = \tfrac{1}{8}(35 \cos^4 \theta - \cos^2 \theta + 3) \qquad (13.7)$$
$$\vdots$$

The orientation angle θ can be determined by placing the polymer under stress and measuring the dichroic ratio of an isotactic vibration. The dichroic ratio R is defined as

$$R = \frac{A_{\parallel}}{A_{\perp}} \qquad (13.8)$$

where $A_{\parallel}$ is the measured absorption maximum for an isotactic vibration when the electric vector of the infrared radiation is parallel to the polymer draw direction whereas for $A_\perp$ the electric vector is perpendicular to the draw or stress direction. The dichroic ratio can be related to the harmonic orientation function $\langle P_2(\cos \theta)\rangle_{av}$ by

$$\langle P_2(\cos \theta)\rangle_{av} = \frac{3\langle\cos^2 \theta\rangle_{av} - 1}{2} = \frac{(R - 1)}{(R + 2)}\frac{(R_0 + 2)}{(R_0 - 1)} \quad (13.9)$$

where α is the angle between the chain axis and the dipole moment vector of the vibration and $R_0 = 2 \cot^2 \alpha$. The angle α can often be determined from theoretical considerations [21], and θ is the angle between the chain axis and the stress direction.

One procedure by which infrared linear dichroism spectra are obtained is to place a polarizer of high efficiency (to obviate the necessity of correcting for polarization) between the sample and the interferometer; a wire grid polarizer is often used in this measurement. The polarizer should be placed just before the sample because the interferometer and occa-

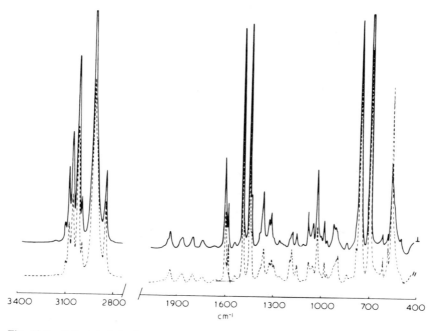

Fig. 13.8. Infrared dichroic spectra of isotactic polystyrene. (Solid trace) perpendicular polarization, (broken line) parallel polarization. (Reproduced from Jasse, B. and Koenig, J. L. *POLYMER*, 1981, **22**, 1040–1044, by permission of the publishers, Butterworth & Co. (Publishers) Ltd. ©)

sionally some aspherical mirrors can distort the polarization. Often the beamsplitter efficiency may be strongly dependent on the polarization (see Chapter 4) so that the intensity is not equal in all directions. For this reason, the polarizer is rotated until maximum transmission is attained. The sample, rather than the polarizer, is rotated 90° for each orientation, so that the single-beam background spectrum against which the polymer spectrum is ratioed is not altered. The dichroic ratio, $R = A_{\parallel}/A_{\perp}$, is calculated using the absorbances at the maximum absorptions for $A_{\parallel}$ and $A_{\perp}$. An example of the measurement of a dichroic ratio spectrum, in this case of isotactic polystyrene, is shown in Fig. 13.8.

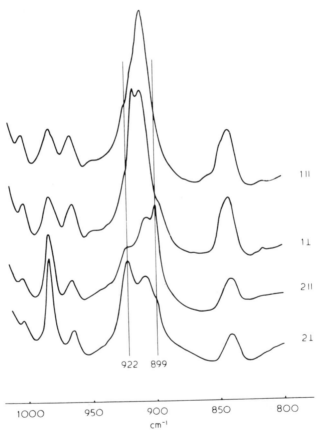

Fig. 13.9. Spectral comparison between oriented (1) amorphous and (2) crystalline isotactic polystyrene around the 906-cm^{-1} atactic absorption. (Reproduced from Jasse, B. and Koenig, J. L. *POLYMER*, 1981, **22**, 1040–1044, by permission of the publishers, Butterworth & Co. (Publishers) Ltd. ©)

Two regions are of interest in the dichroism spectra of oriented isotactic polystyrene, 850–950 cm^{-1} and 1000–1100 cm^{-1}. Not only can differences be observed between the parallel and perpendicular orientations, differences exist between amorphous and crystalline isotactic polystyrene [17]. As can be seen from Fig. 13.9, the spectra for the amorphous sample in its orientations (1) and the crystalline sample (2) are quite different. In this figure, the range from 800 to 1000 cm^{-1} is shown. (It should be noted that atactic polystyrene exhibits only one band in this region, the ν_{17b} out-of-plane mode of the benzene ring, at 906 cm^{-1}.) The oriented isotactic polystyrene spectra exhibit four bands in the region 850–950 cm^{-1}. In the amorphous isotactic sample, the atactic band at 906 cm^{-1} is split into two bands at 908 cm^{-1} (parallel) and 915 cm^{-1} (perpendicular). The crystalline isotactic sample shows two bands at 899 cm^{-1} (parallel) and 922 cm^{-1} (perpendicular), although there is a large amorphous component at 908 cm^{-1}. These bands were interpreted to indicate that the amorphous sample has a 3_1 helix structure of the chain and that these are not organized in a crystal lattice. The existence of the 3_1 helical structure (trans–gauche configuration) was confirmed by an absorption band at ~560 cm^{-1}, although gauche–trans–trans–gauche conformational defects were indicated by a band at ~540 cm^{-1}. The 899- and 922-cm^{-1} bands in the crystalline isotactic polystyrene arise from interactions in the crystal lattice.

Atactic polystyrene exhibits an absorption at 1069 cm^{-1} that splits into four bands in the isotactic form. These four bands are shown in Fig. 13.10, where amorphous and crystalline sample spectra are presented. The four bands have absorptions at 1083 (perpendicular), 1075 (parallel), 1064 (parallel), and 1052 (perpendicular) cm^{-1}. These bands are associated with ring and chain vibrations. The bands have been assigned to A and E modes of vibrations in the 3_1 helix in the amorphous form. From the spectra shown in Fig. 13.9 and 13.10 and from other spectral features in the dichroic ratio spectrum, it was determined that the unit chain axis is as shown in Fig. 13.11 and not along the helix axis. This is primarily because of the trans–trans conformational defects in the helix. Figure 13.11 clearly illustrates the orientation rotation angle θ and the dipole moment vector angle α.

Various strains were placed on a series of samples in the form of a draw ratio. The draw ratio is defined as the stretched length of the polymer, l, versus the relaxed length, l_0. The relaxed length was initially 1 cm, and the samples (films) were marked accordingly. The film was then stretched and the draw ratio determined. For the amorphous samples, draw ratios ranged from 1.20 to 1.87. The dichroic ratios were determined for four bands, 1602, 1585, 1028, and 565 cm^{-1}. All calculations were

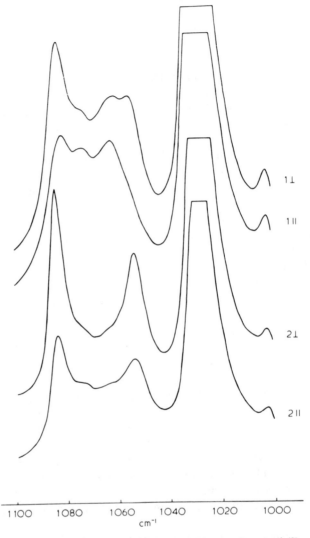

Fig. 13.10. Spectral comparison between oriented (1) amorphous and (2) crystalline isotactic polystyrene around the 1069-cm^{-1} atactic absorption. (Reproduced from Jasse, B. and Koenig, J. L. *POLYMER*, 1981, **22**, 1040–1044, by permission of the publishers, Butterworth & Co. (Publishers) Ltd. ©)

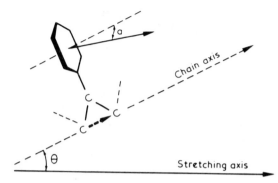

Fig. 13.11. Chain axis and orientation angles for the polystyrene structural unit. (Reproduced from Jasse, B. and Koenig, J. L. *POLYMER*, 1981, **22**, 1040–1044, by permission of the publishers, Butterworth & Co. (Publishers) Ltd. ©)

based on the 1028-cm^{-1} band as it is unaffected by combination bands. It was determined that the orientation of the benzene ring is at an angle of about 35° to the chain axis. This is in close agreement with theoretical calculations and birefringence measurements [17]. The crystalline samples exhibited no orientation until the draw ratio was sufficiently high that the sample necked. After "necking," a tilt angle of 31° for the benzene ring was observed.

Similar studies have been performed on uniaxially oriented poly(2,6-dimethyl-1,4-phenylene oxide)–atactic polystyrene blends [18] and on uniaxially oriented polytetrafluoroethylene (Teflon by one trade name) [19].

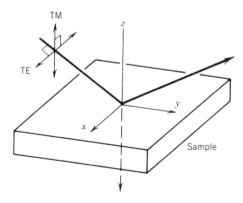

Fig. 13.12. Coordinate system for ATR infrared dichroism. The x axis is the draw axis. (Reproduced from [22], by permission of the American Chemical Society; copyright © 1981.)

The studies of macromolecular chain orientation cited above were based on transmission studies that examined the bulk properties of the sample. The surface characteristics of polymers can be quite different, and methods have been devised for studying the surface of a polymer sample [22,23]. By using ATR spectrometry, it is possible to examine only the surface of a sample. Three optical constants for a polymer film, k_x, k_y, and k_z, are related to the reflectivities of the transverse electric (TE) and transverse magnetic (TM) waves. The axes x, y, and z are in reference to a sample film, where x is the draw axis, y is parallel to the film plane and perpendicular to x, and z is orthogonal to x and y. The coordinate system is illustrated in Fig. 13.12. The film and polarizer can assume a total of four different orientations, that is, the TE and TM waves can both be set perpendicular and parallel to the x and y axes. This combination of orientations can be described by the equations

$$\ln R_{TE_x} = -ak_x \tag{13.10}$$

$$\ln R_{TM_x} = -bk_y - ck_z \tag{13.11}$$

$$\ln R_{TE_y} = -ak_y \tag{13.12}$$

$$\ln R_{TM_y} = -bk_x - ck_z \tag{13.13}$$

where R is the ratio of the sample and reference spectra for this orientation, and a, b, and c are functions of the refractive indices of the sample, the ATR crystal, and the angle of incidence. Measurements of values for R_{TE_x} and R_{TM_x} are obtained by rotating the polarizer 90°. Values for R_{TE_y} and R_{TM_y} are obtained in the same manner when the sample is first rotated by 90°. All three optical constants can be calculated from any three of the measurements; however, when surface orientations are of interest, these can be measured by the dichroic ratio k_x/k_y. The dichroic ratio can be calculated from Eqs. 13.10 and 13.12 by leaving the polarizer set and rotating the sample 90°.

The problem with rotating a sample in an ATR accessory is that it is difficult to reproduce the experimental conditions when the sample is removed from the IRE, rotated 90°, and remounted. To avoid this problem, Sung [22] devised an ATR sample holder in which it is unnecessary to dismantle the sample and IRE. Rather than use a standard 50 × 20 × 2-mm IRE, a parallelogram KRS-5 element was designed. This element is 25 × 25 × 3 mm in dimension and cut so that all four edges have 45° entrance planes. Two identical samples are drawn and placed on both sides of the IRE and clamped in place with pressure plates and a C-clamp, see Fig. 13.13. One of the pressure plates has two orthogonal sets of

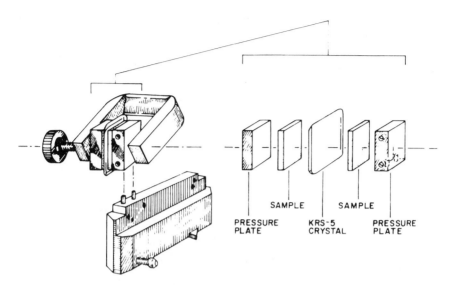

Fig. 13.13. Rotatable ATR sample holder. (Reproduced from [22], by permission of the American Chemical Society; copyright © 1981.)

register holes that fit on a set of register pins on the sample holder. The pins are positioned so that the IRE and C-clamp are in identical positions in each rotation position.

In one study [22], Sung examined both drawn and injection molded polypropylenes. Comparisons were made of surface (ATR) and bulk (transmission) dichroic ratios. The drawn polypropylene exhibited similar orientations on the surface and in the bulk sample, up to draw ratios of 8 (l/l_0, see above). The dichroic ratios of surface and bulk injection molded polypropylene differed greatly, with values of 2.0 and 1.2, respectively. All dichroic ratios were calculated from the 1168-cm^{-1} absorption band, which is a complex mode of C–C chain stretching, CH$_3$ rocking, and CH bending. Sung stated that much of the contribution of this band was from the crystalline phase [22], but others maintain that this band is more indicative of configuration [24]. The results were confirmed by birefringence measurements.

In a later paper [23], Sung and co-workers addressed penetration depth using ATR spectrometry as it related to three-dimensional surface structural analyses. This study involved not only uniaxially drawn films but biaxially drawn films as well. For these three-dimensional studies, all optical constants (or spatial absorbances) k_x, k_y, and k_z need to be calculated. Unfortunately, as the polarizer is rotated 90° to change between the orientation for TE and TM wave measurement, the effective depth of penetration is altered. This depth of penetration can be described by the following equations:

$$\frac{d_e(\text{TE})}{\lambda_1} = \frac{n_{21} \cos \theta}{\pi(1 - n_{21}^2)(\sin^2 \theta - n_{21}^2)^{1/2}} \qquad (13.14)$$

$$\frac{d_e(\text{TM})}{\lambda_1} = \frac{n_{21} \cos \theta (2 \sin^2 \theta - n_{21}^2)}{\pi(1 - n_{21}^2)[(1 + n_{21}^2)\sin^2 \theta - n_{21}^2](\sin^2 \theta - n_{21}^2)^{1/2}} \qquad (13.15)$$

where d_e/λ_1 is the effective penetration depth, λ_1 is the wavelength of the radiation divided by the refractive index of the IRE, n_{21} is the refractive index ratio of the polymer to the IRE, and θ is the incidence angle. The two effective depths of penetration can be made equal if the TE wave measurements are made at an incidence angle of 45° and the TM wave measurements are made at 51.2°. At these angles, the effective depth of penetration is about 3 μm at 1000 cm^{-1} using a KRS-5 IRE.

In order to adjust for the incidence angle, the rotatable ATR sample holder, as described above, was mounted on rotating stage. The rotation axis was designed so that it coincided with the entrance plane to the IRE. The angle of incidence could be adjusted by a gear and worm-screw to precise angles. A drawing of the cell is presented in Fig. 13.14.

The four possible dichroic infrared ATR spectra for uniaxially drawn polypropylene are shown in Fig. 13.15. From these spectra, all the necessary optical constants (k_x, k_y, and k_z) were measured. These were used to calculate the surface crystallinity of the sample. Similar studies were done on uniaxially drawn PET and biaxially and undrawn polypropylene. The ability to alter the angle of incidence on the ATR apparatus allowed depth profiling to be pursued. Uniaxially drawn polypropylene film was profiled from about 1 to 15 μm. This study is an elegant method of determining surface characteristics of polymeric materials.†

Infrared dichroism is a well-established technique, and the above examples only touch on the number of studies in the literature. One technique that holds a great deal of promise for infrared dichroism is polarization modulation linear dichroism as described by Marcott [25]. Linear dichroism is measured in an equivalent way as the dichroic ratio, but the

---

† As mentioned above, there is a differing opinion on the results obtainable from dichroic ratios of polyporopylene. Polypropylene does not exhibit pure crystalline and amorphous bands [24]. This in no way detracts from the usefulness of the technique for other polymers.

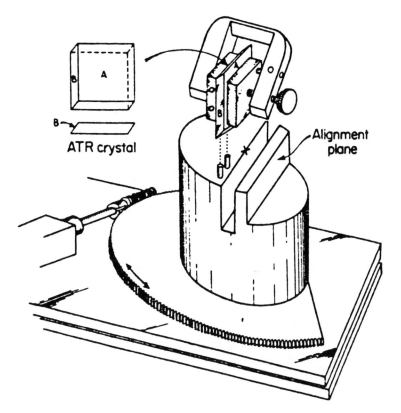

ATR crystal

Alignment plane

Fig. 13.14. Drawing of adjustable incidence angle, rotatable ATR sample holder, as seen from the detector side. (Reproduced from [23], by permission of the American Chemical Society; copyright © 1983.)

dichroic difference, ΔA,

$$\Delta A = A_\parallel - A_\perp \qquad (13.16)$$

is measured directly, as opposed to the dichroic ratio $R = A_\parallel/A_\perp$. The dichroic ratio and dichroic difference are related [25]:

$$R = 1 + \Delta A/A_\perp. \qquad (13.17)$$

The dichroic difference is usually more sensitive than the dichroic ratio, especially when $R \simeq 1$. The polarization modulation linear dichroism experiment is the same as the dichroic ratio experiment except that a photoelastic modulator (PEM) is placed between the polarizer and the sample. (PEMs are discussed in Chapter 8, Section II.b). This experiment is vir-

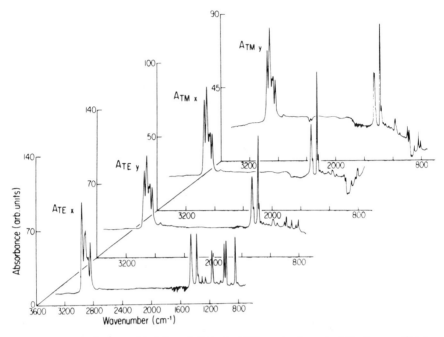

Fig. 13.15. The four possible infrared dichroic ATR spectra for uniaxially drawn polypropylene. (Reproduced from [23], by permission of the American Chemical Society; copyright © 1983.)

tually identical to the vibrational circular dichroism (VCD) experiment (also described in Chapter 8) except that the modulation amplitude of the PEM is doubled and the amplifier is locked in to the first harmonic of the modulation frequency of the PEM, $2\omega_m$. As the photoelastic modulator oscillates, the plane of polarization is rotated through 90° twice with each cycle. The frequency of the PEM should be at least 10 times the modulation frequency imposed by the interferometer. For example, Marcott [25] used a PEM frequency of 168 kHz whereas his interferometer modulated the He–Ne laser radiation at only 5 kHz and the infrared wavelengths at considerably lower frequency.

A polarization modulation linear dichroic transmission spectrum of oriented isotactic polypropylene is shown in Fig. 13.16a. The region between 1500 and 1375 cm^{-1} has been omitted because the absorption bands in this region exceed two absorbance units. The negative-going bands at 1103, 1220, 1297, and 1363 cm^{-1} are consistent with other studies involving oriented isotactic polypropylene [25]. Figure 13.16b shows the normal linear dichroic difference spectrum for the same sample collected on the same spectrometer. A spectrum was collected with the polarizer parallel to the orientation of the polymer. This spectrum was ratioed to a polarizer reference spectrum, and the absorbance spectrum was calculated. From this "parallel" absorbance spectrum, a "perpendicular" absorbance spectrum was subtracted. The perpendicular spectrum was obtained by rotating the sample 90°. As can be seen, the polarization modulation linear dichroic difference spectrum has a much higher SNR. Of course, the SNR of the normal linear dichroic difference spectrum can be increased by more signal averaging, but the polarization modulation

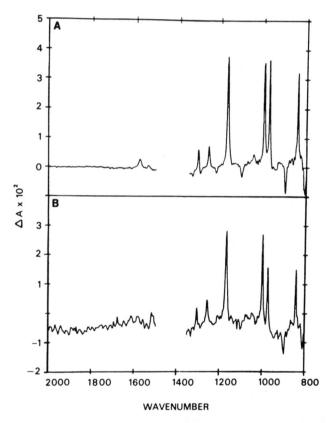

Fig. 13.16. (*a*) Polarization modulation linear dichroic difference spectrum of a 135-μm-thick film of oriented isotactic polypropylene, 100 scans coadded at 4 cm⁻¹ resolution. (*b*) Normal linear dichroic difference spectrum obtained by rotating the sample. Each orientation of the sample was the average of 100 scans. (Reproduced from [25], by permission of the Society for Applied Spectroscopy; copyright © 1984.)

technique obviates the need for rotating the sample, and fewer scans need to be collected.

Polarization modulation linear dichroic difference spectrometry has met with less acceptance than the dichroic ratio method. In large part, this can be attributed to the fact that the polarization modulation technique is a far more recent development. Another factor is the added expense of the PEM, lock-in amplifier, filter networks, and the necessary modification to the detector preamplifier. Nevertheless, these authors believe this technique will attain greater acceptance in time and may eventually rival dichroic ratio methods for FT–IR studies of polymer orientation.

IV.   TIME-RESOLVED STUDIES

The theory of FT–IR time-resolved spectrometry (TRS) is described in detail in Chapter 12. Some examples of polymer deformation studies are also presented in that chapter. There are, however, a couple of modifications to the more standard TRS experiment that in some ways simplify the experiment and extend the time duration of the experiment from milliseconds to seconds.

Hsu and co-workers [26,27] have studied polymer deformations by slowly stretching oriented films while spectrometrically investigating changes in the macromolecular structure. When using conventional rapid-scanning FT–IR/TRS (as described in Chapter 12), it is possible to have an experiment with a cycle time longer than the time it takes to collect the leading side of the interferogram up to the centerburst. If the experiment time is too long, this means that the TRS procedure cannot be completed because some interferograms will be collected without a centerburst.

In the first study [26], Lasch et al. chose not to synchronize the stretcher with the interferometer, as with more conventional experiments [28,29]. Instead, they chose to allow the stretcher to run continuously and have the interferometer collect data in random synchronization. The phase of the stretcher, that is, the point in the deformation cycle, was stored in the interferogram header record. After all the data were collected, all interferograms with identical phases were coadded. One other change was also implemented: the time resolution was degraded from the more usual interferogram sampling interval to several data points. Hsu's technique allows several points to be arranged into a single temporal element, giving a resolution of a few milliseconds. This process is legitimate because each temporal element is still small with respect to the experiment cycle and structural changes occurring in this interval are minimal. Because the temporal resolution windows were quite wide, the number of scans that had to be collected was relatively small. Because the collection was done in a random fashion, not all files had the same number of scans in them, but the number was within a factor of ~2.5 from the minimum to maximum number of scans.

An example of this modified TRS experiment is the deformation of crystalline-oriented poly(butylene terephthalate) (PBT) [26]. A PBT film was prestrained by stretching 2% and then placed in an oscillatory strain apparatus that stretched it by 3% with a frequency of 1 Hz. In this experiment, the cycle time was rather slow and a 4-cm^{-1}-resolution spectrum could be collected in 116 msec, so that each interferogram was considered to be an entire block. In this case, the interferogram was synchronized to the stretcher cycle, and the entire interferogram was collected within a short time compared to the cycle; hence there was no need to sort the data. The results of the time-resolved study on PBT are illustrated in Fig. 13.17.

The spectrum in Fig. 13.17a covers the methylene rocking region of PBT. Three bands are quite distinct: the amorphous band at 938 cm^{-1}, the α crystalline band at 917 cm^{-1}, and the β crystalline band at 960 cm^{-1}. Figure 13.17b shows the stress cycle as a function of time, the cycle taking 1000 msec to complete. A total of eight spectra was collected for the time-resolved study, giving a temporal resolution of about 125 msec. Because the strain cycle and interferometer were not synchronized, not all files had the same number of coadded scans, the file with the fewest scans had 21, the one with the most scans had 45.

The differences in the bands with respect to the first spectrum in the cycle are shown in Fig. 13.17c. As can be seen, the α and β crystalline absorption bands change with time in the cycle. These changes are plotted versus the cycle time in Figs. 13.17d,e. This technique is not true time-resolved spectrometry as discussed in Chapter 12, nor is it a conventional slowly varying technique; rather it is a hybrid of the two.

In a later paper from Hsu's group [27], a further modification in the time-resolved experiment was described. Data were still averaged in blocks, but the random synchronization between the experiment and the interferometer was removed. The experiment was designed so that the cycle time for the study is exactly the same as the cycle time for the interferometer (scan plus retrace). By this method, the experiment is synchronized to the interferometer cycle, and the desired number of scans is coadded. The synchronization between the experiment and the interferometer is offset, and a new set of scans is collected. This is directly akin to the more conventional TRS experiment (Chapter 12); however, the offset is several sampling points instead of just one. As in the previous

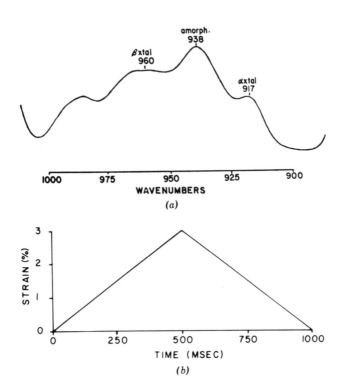

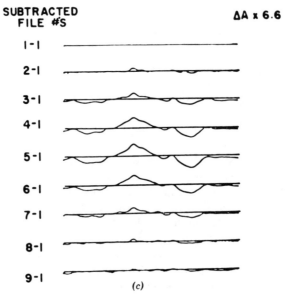

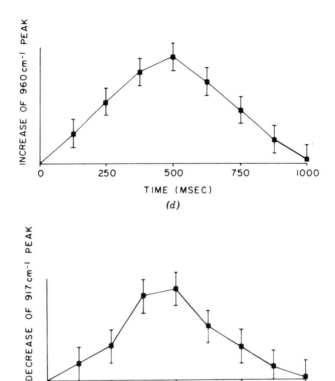

Fig. 13.17. (continued)

study [26], the rate of change of the spectrum is slow, so a block of data may be considered as being collected at a single time. Selecting experiments that have exactly the same cycle time as the interferometer is not a large drawback because the interferometer scan time can be varied over a fairly broad range.

An example of the feasibility of this experiment involves rotating an

Fig. 13.17. (a) Spectrum of the methylene rocking region of poly(butylene terephthalate) showing α-crystalline, β-crystalline, and amorphous bands. (b) Strain cycle ramp function. (c) Difference spectra for the methylene rocking bands in PBT with respect to the first spectrum in the cycle. (d) Increase in absorption peak height for the α-crystalline band. (e) Decrease in the absorption peak height for the β-crystalline band. (Reproduced from [26], by permission of the Society for Applied Spectroscopy; copyright © 1984.)

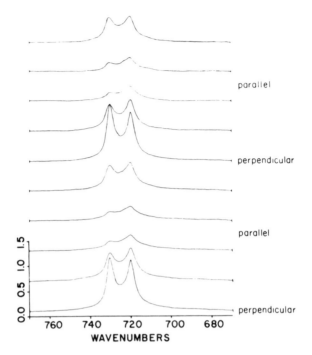

Fig. 13.18. CH$_2$ rocking region from oriented ethylene – vinyl acetate copolymer as a function of orientation to polarizer. (Reproduced from [27], by permission of the Society for Applied Spectroscopy; copyright © 1984.)

ethylene–vinyl acetate copolymer and measuring interferograms using polarized radiation. In this study, a copolymer film was rotated in synchronization with the interferometer scan cycle. That is, one complete revolution of the film took as much time as the interferometer scan cycle. A polarizer was placed in the infrared beam. The CH$_2$ rocking region, 670 cm^{-1}, is shown in Fig. 13.18. When this highly oriented film is perpendicularly polarized, a doublet at 720–730 cm^{-1} is clearly evident. This doublet is due to crystal-field splitting [27].

There is one problem with this method. The interferograms change amplitude during the cycle because the transmission of the oriented sample changes with respect to its orientation to the polarizer. This effect causes the sorted interferograms to be discontinuous between sample blocks. Mathematical corrections must be made to the sorted interferograms to remove these discontinuities, otherwise spectral artifacts occur [27].

V. CONCLUSIONS

Infrared spectrometric analysis of polymeric materials is an extremely large disipline. Many standard sampling methods are applicable to polymer studies for the identification and analysis of the polymers themselves. There is a growing interest in probing the macromolecular structure and thermomechanical properties of polymeric materials in detail, and techniques involving polarization and deformation are contributing to the effort. Undoubtedly, these techniques will find wider application, and new methods will continue to be developed in the future.

REFERENCES

1. S. S. T. King, *J. Agric. Food Chem.*, **21**, 256 (1973).
2. D. H. Anderson and T. E. Wilson, *Anal. Chem.*, **47**, 2482 (1975).
3. C. J. Curry, M. J. Whitehouse, and J. M. Chalmers, *Appl. Spectrosc.*, **39**, 174 (1985).
4. K. Krishnan, *Polymer Preprints*, **25**, 182 (1984).
5. M. P. Fuller and P. R. Griffiths, *Anal. Chem.*, **50**, 1906 (1978).
6. D. W. Vidrine, *Appl. Spectrosc.*, **34**, 314 (1980).
7. J. L. Koenig, *Chemical Microstructure of Polymer Chains*, Wiley, New York (1980).
8. L. D'Esposito and J. L. Koenig, Applications of Fourier Transform Infrared to Synthetic Polymers and Biological Macromolecules, in *Fourier Transform Infrared Spectroscopy: Applications to Chemical Systems*, Vol. 1 (J. R. Ferraro and L. J. Basile, eds.), Academic Press, New York (1978).
9. J. L. Koenig, Applications of Fourier Transform Infrared to Polymers, in *Analytical Applications of FT-IR to Molecular and Biological Systems* (J. R. Durig, ed.), D. Reidel, Dordrecht, Holland (1980).
10. H. W. Siesler, *J. Mol. Struct.*, **59**, 15 (1980).
11. J. L. Koenig, *Adv. Polym. Sci.*, **54**, 87 (1983).
12. M. M. Coleman, P. C. Painter, D. L. Tabb, and J. L. Koenig, *J. Polym. Sci., Polym. Lett. Ed.*, **12**, 577 (1974).
13. P. C. Painter, M. Watzek, and J. L. Koenig, *Polymer*, **18**, 1169 (1977).
14. M. M. Coleman, D. F. Varnell, and J. P. Runt, *Polym. Sci. Technol.*, **20**, 59 (1983).
15. I. M. Ward, *Structure and Properties of Oriented Polymers*, Applied Science Publications, London (1975).
16. R. J. Samuels, *Structural Polymer Properties*, Wiley, New York (1974).
17. B. Jasse and J. L. Koenig, *Polymer*, **22**, 1040 (1981).
18. D. Lefebvre, B. Jasse, and L. Monnerie, *Polymer*, **22**, 1616 (1981).
19. C. K. Young and B. Jasse, *J. Appl. Polym. Sci.*, **27**, 4587 (1982).
20. B. Jasse and J. L. Koenig, *J. Macromol. Sci. (C)*, **17**, 61 (1979).

21. R. Zbinden, *Infrared Spectroscopy of High Polymers*, Academic Press, New York (1964).

22. C. S. P. Sung, *Macromolecules*, **14**, 591 (1981).

23. J. P. Hobbs, C. S. P. Sung, K. Krishnan, and S. Hill, *Macromolecules*, **16**, 183 (1983).

24. R. W. Hannah, H. A. Willis, personal communication.

25. C. Marcott, *Appl Spectrosc.*, **38**, 442 (1984).

26. J. E. Lasch, D. J. Burchell, T. Masoaka, and S. L. Hsu, *Appl. Spectrosc.*, **38**, 351 (1984).

27. S. E. Molis, W. J. MacKnight, and S. L. Hsu, *Appl. Spectrosc.*, **38**, 529 (1984).

28. W. G. Fateley and J. K. Koenig, *J. Polym. Sci., Polym. Lett. Ed.*, **20**, 445 (1982).

29. J. A. Graham, W. G. Grim, III, and W. G. Fateley, *J. Mol. Struct.*, **113**, 311 (1984).

BIOCHEMICAL AND BIOMEDICAL APPLICATIONS

I. INTRODUCTION

Until the commerical introduction of mid-infrared Fourier spectrometers, infrared spectroscopy had not been widely used in the fields of biochemistry or biomedicine. The principal reason for this occurrence was that most measurements of biochemically important samples require these samples to be present in an aqueous environment. Although water does indeed absorb infrared radiation fairly strongly in several regions, FT–IR spectra of species dissolved in water are not particularly difficult to measure provided the pathlength of the cell is chosen correctly. Of course, it should be remembered that the common alkali halides cannot be used as window materials because of their high solubility in water, but there are many alternative materials available.

The spectrum of water in a 6-μm-thick barium fluoride cell is shown in Fig. 14.1. Several features may be observed from this spectrum [1]. First, water is essentially opaque between 3700 and 3050 cm^{-1} at this pathlength. If any data are to be measured in this region, a much shorter pathlength than 6 μm is needed, and the only experimentally feasible technique is to use a single-reflection ATR accessory. It may also be noted that the C–H stretching region may be observed, but the absorption due to water increases very rapidly above 3000 cm^{-1}.

The H–O–H bending mode of water absorbs at approximately 1640 cm^{-1}. This is an important region for measurements of protein spectra since the amide I band, which gives important information on the conformation of proteins, absorbs between 1660 and 1630 cm^{-1}. The frequency and intensity of the 1640-cm^{-1} water band are very sensitive to changes in temperature. Spectra of water recorded at different temperatures and subtracted can show remarkably large residual features, as shown in Fig. 14.1; inexperienced spectroscopists should be careful not to confuse these features with the amide I band of proteins. The 1640-cm^{-1} band is also sensitive to dissolved species (e.g., buffering agents), so that it is necessary to ensure that the reference is identical to the sample in terms

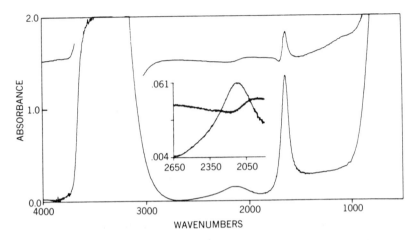

Fig. 14.1. Absorbance spectrum of water in a 6-μm BaF$_2$ cell at 33°C (lower curve) and difference spectrum (upper curve) obtained by subtracting the above spectrum from one recorded under identical conditions at 43°C. (*Inset*) Corresponding ordinate-expanded water and difference spectra of the water association band in the region in which the C–D stretching modes absorb. (Reproduced from [1], by permission of Elsevier Publishing Company; copyright © 1979.)

of pH, ionic strength, temperature, and so on, if useful data are to be obtained in this region.

The O–H stretching and H–O–H bending modes of water can of course be shifted to a lower wavenumber by using D$_2$O as the solvent instead of H$_2$O. The spectrum of D$_2$O in a BaF$_2$ cell is shown in Fig. 14.2. It should be noted that the low-wavenumber cutoff can be extended considerably through the use of cadmium telluride instead of barium fluoride. Deuterium oxide is essentially transparent in the C–H stretching region, and the sensitivity of spectra in this region can be increased by using cells of much greater pathlength than 6 μm.

An alternative approach, which is proving to be very useful for the study of lipids, is to deuterate the solute totally or selectively, so that the C–D stretching bands absorb in a region where H$_2$O does not absorb particularly strongly. The C–D stretch absorbs at about 2200 cm^{-1}, and it should be noted that the so-called *association band* of water absorbs in this region. Like the 1640-cm^{-1} band, the water association band is also quite temperature sensitive, as shown in Fig. 14.1.

It is apparent from the above discussion that most biochemical FT–IR studies should preferably be performed at a constant temperature. Conversely, some measurements require the temperature to be changed so that phase transitions can be investigated. A thermostated cell mount described by Cameron and Jones [2] is shown in Fig. 14.3. The body of the cell is machined from aluminum so that a preheated or cooled fluid can be passed around it to control the temperature. The front face is designed to provide a location for a standard infrared liquid cell, and springs press the cell against the mount for good thermal contact. A Bakelite plate is bolted to the back plane to provide some insulation between the cell and its heaters and the spectrometer. Temperature control to ±0.05°C has been claimed for this cell. Unfortunately, the temperature can only be changed slowly with this cell design. More rapid changes can be achieved by mounting the cell against one face of a Peltier plate [3].

David Cameron, formerly of the National Research Council of Canada, who pioneered the use of variable-temperature FT–IR measurements of lipids and membranes (*vide infra*, next section), generated a file of water reference spectra measured at 4°C temperature intervals and has written programs to interpolate between these spectra in order to calculate a reference spectrum at any desired temperature. Hardware to actuate changes in the thermostated temperature of the cell described above has

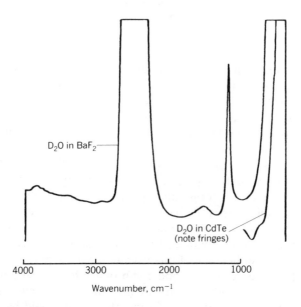

D$_2$O in BaF$_2$

D$_2$O in CdTe
(note fringes)

4000   3000   2000   1000

Wavenumber, cm^{-1}

Fig. 14.2. Absorbance spectra of D$_2$O in a 6-μm cell with BaF$_2$ windows (entire spectrum) and with CdTe windows (partial spectrum). Although the use of CdTe windows expands the spectral range to a lower wavenumber, its high refractive index leads to the appearance of interference fringes. (Reproduced from [1], by permission of Elsevier Publishing Company; copyright © 1979.)

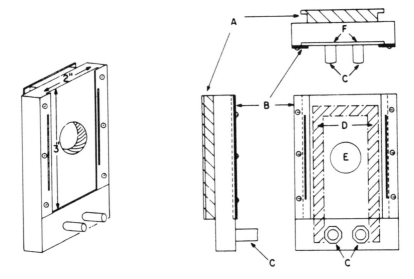

Fig. 14.3. Thermostated mount for liquid cells. A, mating plate; B, stainless steel pressure springs; C, inlet and outlet ports; D, 0.25-in.-diameter flow path for coolant; E, aperture for infrared beam; F, slot for mounting nonthermostated cells and pellet holders. (Reproduced from [2], by permission of the Society for Applied Spectroscopy; copyright © 1981.)

also been written so that a complete temperature profile experiment may be performed overnight with no operator intervention required [4].

For the sensitivity of infrared spectra of species in aqueous solution to be optimized, it is vital to select the correct cell pathlength, detector, filters, and optical retardation. It is well known from most introductory textbooks on spectroscopy that in order to obtain the highest possible SNR, the transmittance of the solvent should be $1/e$, that is, its absorbance should be approximately 0.43. For relatively strong, narrow bands, it is impossible to select an "optimum" pathlength since the transmittance changes too rapidly across the band. For example, a good pathlength for measurements of solutes in aqueous solution at exactly 1640 cm^{-1} could be about 2.5 μm, but this pathlength is much too short for optimum sensitivity 20 cm^{-1} on either side of this wavenumber. In practice, the use of a 6-μm-thick cell represents a good compromise for measurements made in this region. On the other hand, for measurements between 1500 and 1000 cm^{-1}, a 12-μm-, or even a 25-μm-, thick cell would be better if a TGS or DTGS detector were to be used, whereas good spectra around 2200 cm^{-1} can be measured using even thicker cells.

The choice of pathlength is made more difficult if cooled detectors are used. In Chapter 2, it was shown that digitization noise could limit the sensitivity of measurements made when the SNR of the interferogram exceeds the dynamic range of the ADC. The SNR of measurements made on an interferometer operating at close to the limiting optical throughput (~0.06 cm^2 sr, see Chapter 7) will be digitization noise limited if a cooled photodetector is used unless the sample has a low transmittance across the entire spectrum. Thus, for example, the spectrum of an aqueous solution held in a 12-μm-thick cell measured using a narrow-range MCT detector will probably be digitization noise limited because of the high transmittance of the solvent between 1800 and 3000 cm^{-1}. It is therefore necessary to insert an optical filter to eliminate radiation above 1800 cm^{-1} if the optimum sensitivity is to be achieved below this wavenumber.

Similarly, the highest sensitivity for measurements in the C–D stretching region (~2200 cm^{-1}) can be achieved using a 50-μ or 100-μm cell and an InSb detector [5]. However, the very high transmittance of water around 2700 cm^{-1} combined with the high D^* of the InSb detector make it necessary to use a bandpass filter to remove radiation in regions other than the one of interest, see Fig. 14.4. Obviously, when filters are used in this fashion, much of the multiplex advantage may be lost, but the benefit of measuring the spectra of aqueous solutes at very high SNR can more than compensate for this loss.

Finally, it is important to note that it is rarely necessary to measure a spectrum at high resolution since most solutes have rather broad absorption bands, especially when they are in aqueous solution. A good guide is that the resolution should be no better than one-half the full width at half-height of the narrowest band in the spectrum. If measurements are taken at higher resolution, the resultant spectra will have a poorer SNR and/or take a longer time to measure than if the above criterion were applied, without any significant gain in the information content. For cosmetic reasons only, it may be preferable to interpolate between the independent data points in order to produce a smooth-*looking* spectrum.

The choice of window materials is limited both by the solubility of infrared transparent materials in water and by their low-wavenumber cutoff. Some commonly used window materials for transmission spectrometery are listed in Table 14.1 together with their low-wavenumber cutoff and other desirable characteristics. In view of the rather low refractive index of water (~1.3), special care should be exercised when window materials of high refractive index are used for studies of aqueous solutions. In this case, interference fringes (channel spectra) of large amplitude can be produced and constitute an annoying source of error. Because of its low refractive index and fairly low wavenumber cutoff, barium fluoride is commonly used by many workers in the field of biochemical FT–IR spectrometry.

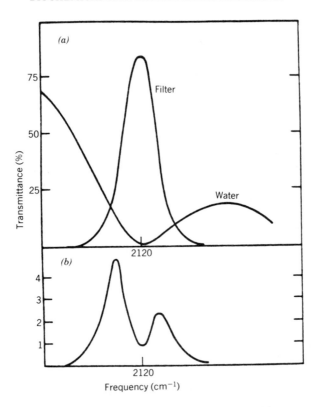

Fig. 14.4. (*a*) Spectra of a 100-μm film of water around 2120 cm^{-1} showing a minimum transmittance of about 1% and a narrow-bandpass filter centered at about the same wavenumber. The high transmittance on either side of the water band makes the dynamic range of the ADC the limiting source of noise when a high-sensitivity detector such as InSb is used for the measurement. By filtering out information from unwanted spectral regions with the filter, the overall transmittance has the profile shown in (*b*), and the spectrum again becomes detector noise limited.

Several of the problems described above can be avoided through the use of ATR spectroscopy provided that a sufficiently high incidence angle is used to avoid major spectral distortions. Because of the long pathlength through the IRE, the cutoff wavenumbers of materials used for ATR are usually higher than is shown in Table 14.1. Thus, silicon ($n = 3.6$) can only be used to about 1500 cm^{-1}, and even germanium ($n = 4.0$) is not very suitable for measurements below 900 cm^{-1}. For longer wavelengths, zinc selenide can be used, although its refractive index ($n = 2.4$) is less than optimum.

The major advantages of Fourier spectrometry are, of course, high

Table 14.1. Window Materials for Transmission Spectrometry of Aqueous Solutions

Material	Cutoff	Comments
Crystal quartz	2000 cm^{-1}	Only good for short wavelengths
Sapphire	2000 cm^{-1}	Only good for short wavelengths
Lithium fluoride	1400 cm^{-1}	Eliminates much of the fingerprint region
Calcium fluoride	1000 cm^{-1}	Slightly soluble in acid
Barium fluoride	700 cm^{-1}	Slightly soluble in base
Zinc selenide	600 cm^{-1}	Expensive, fairly high refractive index
Cadmium telluride	400 cm^{-1}	Expensive, fairly high refractive index
Silver chloride	400 cm^{-1}	Reacts with metals, very soft
Silicon	Far infrared	High refractive index, brittle
Germanium	Far infrared	High refractive index, brittle

sensitivity, the capability of measuring spectra in a very short period of time, and the high repeatability of the wavenumber scale. The benefits of each of these advantages have been reaped in various biochemical or biomedical applications. The high sensitivity is required because of the almost universal presence of water and the generally low concentrations of solutes. Several kinetic studies have been made, sometimes necessitating the resolution of processes occurring in less than 10 sec [6]. Finally, phase changes in lipids and membranes can be studied through the measurement of very small shifts in the position or width of a band. The instrumental requirements for optimizing the SNR of biochemical measurements has been discussed above. Wavenumber precision and bandwidth have been discussed in Chapter 6, Section II.b. These techniques are particularly applicable to biochemical FT–IR.

II. THERMOTROPIC BEHAVIOR OF LIPIDS AND BIOMEMBRANES

The work of Cameron and Mantsch illustrates the enormous amount of information on the structure of lipids that can be obtained by FT–IR spectrometry using the techniques described in the first section of this chapter and in Chapter 6, such as accurate determination of band maxima and widths, total and selective deuteration of the lipids, and the use of D_2O as a solvent.

The use of D_2O is illustrated in a study of the thermotropic phase behavior of the alkali metal salts of ascorbyl palmitate [7]. The potassium salt (APK) has the structure shown below:

$$
\begin{array}{l}
\text{O}\\
\|\\
\text{H} \quad \text{C}_1 \\
\text{O}-\text{C}_2 \quad \text{O}^H \quad \text{O} \quad \overset{2'}{\text{CH}_2} \quad \overset{4'}{\text{CH}_2} \quad \overset{6'}{\text{CH}_2}\\
\text{C}^3\text{-C}-\overset{5}{\text{C}}-\text{CH}_2 \quad \text{C}_{1'} \quad \underset{3'}{\text{CH}_2} \quad \underset{5'}{\text{CH}_2} \quad \underset{7'}{\text{CH}_2}\\
\text{K}^+\text{ }^-\text{O} \quad \overset{}{\text{O}} \quad \text{O} \quad \overset{8'}{\text{CH}_2} \quad \overset{10'}{\text{CH}_2} \quad \overset{12'}{\text{CH}_2} \quad \overset{14'}{\text{CH}_2} \quad \overset{16'}{\text{CH}_3}\\
\text{H} \qquad\qquad \underset{9'}{\text{CH}_2} \quad \underset{11'}{\text{CH}_2} \quad \underset{13'}{\text{CH}_2} \quad \underset{15'}{\text{CH}_2}
\end{array}
$$

This derivative of ascorbic acid is a lipophilic analog of vitamin C, which is as biologically active as its hydrophilic counterpart and may be used as a preservative for oils and fats. Its possible biological role as a lipophilic vehicle of vitamin C was therefore of interest. APK is structurally analogous to surfactants such as the alkyl sulfonates. The solubility in water of most surfactants increases dramatically above a certain temperature known as the Krafft point, and the solubility of APK exhibits the same behavior.

For Fourier spectrometry, 0.1 M solutions of APK in D_2O were prepared above the Krafft point. On cooling, the clear solution turns into a uniform curd. This curd is a semicrystalline mesophase in which the surfactant molecules form lamellar structures. According to the degree of hydration and ordering, the curd may be a *gel* (well hydrated and unilamellar) or a *coagel* (poorly hydrated and multilamellar). The Krafft point may be thought of as the temperature at which the transition from a coagel or gel phase to a micellar phase occurs.

Spectra of APK at several different temperatures are shown in Fig. 14.5. In the carbonyl stretching region, the changes are indicative of increased hydrogen bonding at the higher temperatures. Dramatic changes in the C–C stretching bands around 1580 cm^{-1} are also evident, and this can be seen even more clearly when the wavenumber of the maximum absorption is plotted against temperature (see Fig. 14.6), with the maximum gradient occurring at the Krafft point.

The changes in the C–H stretching region are shown in Fig. 14.7. Below 40°C, these bands are narrow, indicating relatively low acyl chain mobility and the absolute peak wavenumbers are characteristic of fully extended all-trans chains. As the temperature is raised, the bands broaden and shift, as shown in Fig. 14.8. The plot of peak wavenumber versus temperature

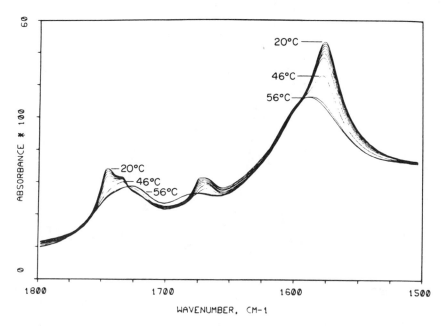

Fig. 14.5. Spectra in the region between 1800 and 1500 cm⁻¹ of 0.1 *M* APK in D₂O as a function of temperature. Spectra were measured from 20 to 56°C at intervals of 3.6°C. (Reproduced from [7], by permission of the author.)

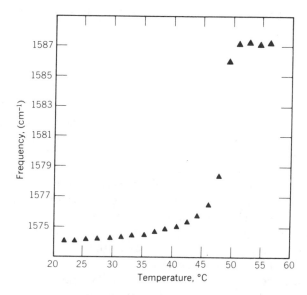

Fig. 14.6. Temperature dependence of the peak wavenumber of the C–C stretching vibration of APK in D₂O. (Reproduced from [7], by permission of the author.)

465

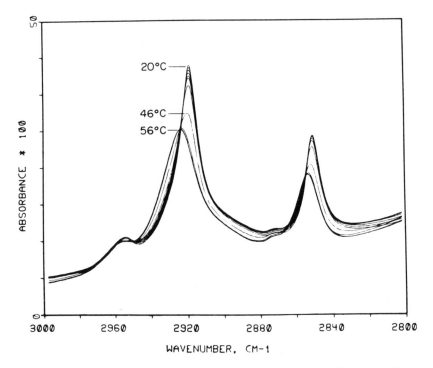

Fig. 14.7.   Infrared spectra of 0.1 *M* APK in D₂O in the C–H stretching region. For experimental details, see caption to Fig. 14.5. (Reproduced from [7], by permission of the author.)

is almost invariant below 40°C, but it may be noted that the bandwidth changes somewhat between 20 and 40°C, indicating an increase in acyl chain mobility in this temperature range. This behavior has also been observed prior to the melting temperature of phospholipids [8–12].

One of these studies [8] will also be described in some detail since it illustrates the use of specifically deuterated phospholipids; in this case, dipalmitoyl phosphatidyl choline (DPPC) bilayers were studied. DPPC was deuterated at the 2′, 3′, 7′, 8′, 13′, and 16′ positions of the *sn*-2-palmitoyl chain (see structure of APK above for numbering), so that changes occurring in different regions of the hydrophobic chain of the bilayer could be studied through changes in the C–D stretching bands occurring as a function of temperature.

Spectra were measured with the sample in 25- and 50-μm-thick CaF₂ cells using an InSb detector. A 2500-cm⁻¹ low-pass filter was employed to avoid detector saturation and the occurrence of digitization noise. For these studies, samples were dissolved in H₂O, since the O–D stretching band of D_2O would mask the C–D stretching band of the lipids. The effects of the water association band at 2150 cm^{-1} were removed by spectral subtraction.

Previous work had appeared to indicate two phase transitions in DPPC: a large transition at 41.5°C (T_m) from a gel phase in which the acyl chains are fully extended to a liquid crystalline phase in which there is a large population of gauche conformers and a pretransition event at about 36°C (T_{pre}) where there is an abrupt change in the unit cell in which the acyl chains are packed. In addition, there appears to be a continuous change in the gel phase packing as the temperature is reduced below T_{pre}.

The C–D stretching bands in the spectrum of DPPC-2'-d_2 exhibit quite different properties from those of the other specifically labeled DPPC isomers. Not only are the peak frequencies shifted with respect to the corresponding bands of the other isomers, but sharp changes in either the wavenumber or bandwidth were not apparent, see Fig. 14.9. The plots of half-width and $\Delta A/\Delta T$ (which monitors changes in peak absorbance) versus temperature possibly indicate a very small effect occurring around T_m, but the changes are not sharp. These data indicate that there is not a large change in the inductive interaction of the ester linkage with the acyl chain.

The behavior of the C–D stretching bands of the other labeled DPPC

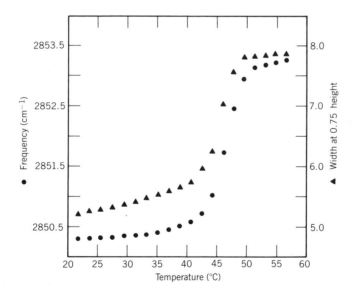

Fig. 14.8. Temperature dependence of the frequency (●) and the width at three-fourths the height (▲) of the symmetric CH$_2$ stretching band of APK in D$_2$O. (Reproduced from [7], by permission of the author.)

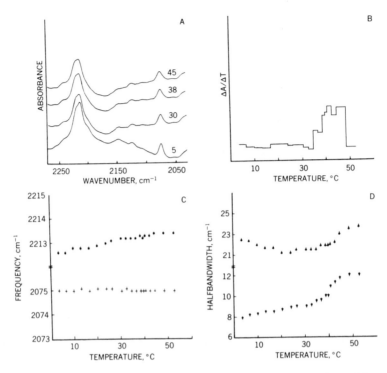

Fig. 14.9. Temperature dependence of the CD_2 stretching bands of hydrated bilayers of DPPC-2'-d_2. (*A*) Spectra recorded at 5, 30, 38, and 45°C, with all spectra plotted on the same scale. (*B*) Plot of $\Delta A/\Delta T$ versus temperature derived from the antisymmetric CD_2 stretching band near 2213 cm^{-1}. (*C*) Wavenumber of the antisymmetric (∗) and symmetric (+) CD_2 stretching band maxima versus temperature. (*D*) Half-bandwidth of the antisymmetric (▲) and symmetric (◆) CD_2 stretching bands versus temperature. (Reproduced from [8], by permission of the Biophysical Society; copyright © 1981.)

isomers is shown in Fig. 14.10. Sharp changes may be observed in all plots at T_m, but it is more difficult to observe changes around T_{pre}. Spectral changes near T_{pre} are negligible for DPPC-3'-d_2. On the other hand, the plots of bandwidth versus temperature for DPPC-7',8'-d_4 and DPPC-13'-d_2 show definite changes around 36°C. Without concomitant frequency shifts, such changes are indicative of increased rates of motion of the acyl group as the temperature is increased. Since the chains are all trans, this must result from librational and torsional motion about the long axes of the acyl chains. Changes in the CD_3 stretching modes of DPPC-16'-d_3 with temperature were found to be small, even around T_m, apparently resulting from an increase in the rate of rotation of the methyl group and the overall increased mobility in the liquid crystalline phase.

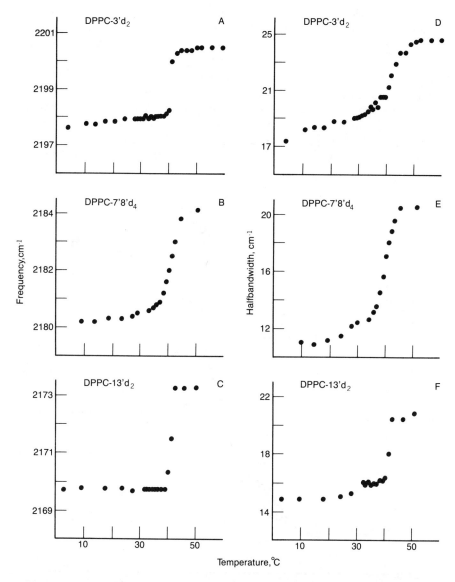

Fig. 14.10. (*A–C*) Temperature dependence of the peak wavenumbers of the antisymmetric CD_2 stretching bands of (*A*) DPPC-3'-d_2, (*B*) DPPC-7',8'-d_4, (*C*) DPPC-13-d_2. (*D–F*) Temperature dependence of the bandwidths of the antisymmetric CD_2 stretching bands of (*D*) DPPC-3'-d_2, (*E*) DPPC-7',8'-d_4, (*F*) DPPC-13-d_2. (Reproduced from [8], by permission of the Biophysical Society; copyright © 1981.)

469

The same group has performed many other studies of the thermo-tropic behavior of lipids, and their work is now starting to elucidate the structure of lipids in live cell membranes. Most of the early studies of biomembranes have been performed using samples isolated prior to the spectroscopic characterization, although the results of one previous investigation of intact cells by ^{2}H NMR spectrometry appeared to justify the assumption that the behavior of isolated membranes is similar to that of membranes in intact cells [13]. To test this assumption more accurately, Cameron et al. [14] grew deuterated samples of *Acholeplasma laidlawii B*. at 30°C in a fatty-acid-depleted medium supplanted with perdeutero-

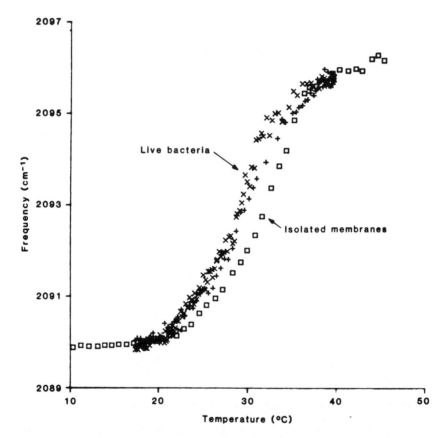

Fig. 14.11. Variation of the peak wavenumber of the symmetric CD_2 stretching bands with temperature for live *Acholeplasma laidlawii B*. bacteria and membranes isolated from these bacteria; see text for full experimental details. (Reproduced from [14], by permission of the American Association for the Advancement of Science; copyright © 1983.)

myristic acid and avidin. Avidin is a potent inhibitor of fatty acid synthesis, and its presence in the growth medium results in incorporation of exogenous fatty acids (perdeuteromyristic acid in this case) to high levels of homogeneity. The cells were harvested and washed and placed in a 50-μm BaF$_2$ cell. Data acquisition was automated so that one spectrum was collected every 10 sec while the temperature was raised from 20 to 39°C and then lowered to 16°C; 200 spectra were collected in 33 min. After the measurement, the cells were cultured and found to be 98–99% intact with no evidence of contamination.

The plasma membranes were then isolated and studied spectrometrically over the temperature range 5–50°C with a data acquisition time of 3 min per spectrum. Although the spectrum of the live cells was much more complex than that of the isolated membrane, the C–D stretching bands are not overlapped by any other bands, and band maxima could be accurately measured by computing $\bar{\nu}_{CG}$ from the topmost 2 cm^{-1} of the band after interpolation to one datum per reciprocal centimeter.

Figure 14.11 shows the variation of the wavenumber of the C–D symmetric stretching band with temperature for the live bacteria and the isolated membranes. The transition from gel to liquid crystal is seen to occur at a lower temperature for the live bacteria. At the growth temperature (30°C), the liquid crystalline phase content of the isolated membrane is only 20% as compared to the 50% content of live cell membranes. Under growth conditions, therefore, data on conformational order obtained from isolated membranes are not always applicable to the live microorganism. This important conclusion suggests either that the process of extraction changes the intrinsic properties of membranes or that part of the mechanism for regulating membrane fluidity is extrinsic to the membrane itself and hence is eliminated by membrane isolation. From further detailed studies of this type, the current body of knowledge on biochemical systems should be extended considerably.

III. KINETIC AND MECHANISTIC STUDIES OF ENZYME REACTIONS

The measurements of live cells described in the previous section illustrated how high-quality FT–IR spectra of biomolecules can be measured very rapidly, even when the molecules are in an aqueous environment. Other studies on biochemical systems have also illustrated the capability of obtaining kinetic and mechanistic information. One such study by Fisher et al. [15] involved the hydrolysis of β-lactam antibiotics by the β-lactamase enzyme. Until this work, little was known about the action of these hydrolytic enzymes. It had not even been ascertained whether the enzyme acts as a general acid–base catalyst to facilitate the direct attack of a water molecule on the β-lactam carbonyl group or instead provides a nucleophile to attack the β-lactam, leading to the formation of an acyl–enzyme intermediate.

Fisher et al. selected the β-lactam cefoxitin for this study in view of its relative stability to β-lactamases. They showed by ultraviolet–visible spectrometry that a reaction intermediate is probably formed during the hydrolysis of cefoxitin by β-lactamase and that its rate of formation is greater than its rate of breakdown. Experiments using radiolabeled [14C]cefoxitin showed that it was possible to isolate the intermediate. After isolation of this intermediate, the FT–IR spectrum could be measured at high SNR in the region of the carbonyl stretching modes provided that D_2O was used as the solvent. Good sensitivity was required since a single ester group on a protein molecule of molecular weight 29,000 had to be observed.

When the hydrolysis of cefoxitin by a stoichiometric amount of β-lactamase was investigated, absorption bands at 1767 cm^{-1} (due to the β-lactam carbonyl group) and 1753 cm^{-1} were observed. The latter band is in the region expected for an α-methoxy ester. After disappearance of the 1767-cm^{-1} band, the 1753-cm^{-1} band decayed with first-order kinetics. The spectra are shown in Fig. 14.12; the fact that the right-hand series of spectra is plotted with 0.02 a.u. full-scale indicates the need for high

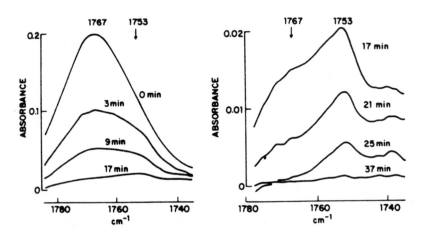

Fig. 14.12. Spectra measured during the hydrolysis of cefoxitin by wild-type β-lactamase. The top curve of the right-hand group of spectra is equivalent to the bottom curve of the left-hand group of spectra; the change in the ordinate scale should be noted. The initial cefoxitin concentration was 20.2 mM and the enzyme concentration was 8.2 mM. (Reproduced from [15], by permission of the American Chemical Society; copyright © 1980.)

sensitivity. The behavior of the 1753-cm^{-1} band is consistent with the formation of an acyl–enzyme intermediate.

Belasco and Knowles [16] have also measured the infrared spectrum of dihydroxyacetone phosphate bound to the enzyme triosephosphate isomerase by FT–IR spectrometry. Two carbonyl bands that appeared to be attributable to the bound substrate were observed with an intensity ratio of about 3:1. Relative to the carbonyl band of free dihydroxyacetone phosphate in H_2O solution, the more intense band is shifted by 19 cm^{-1}, providing evidence of enzyme-induced distortion of the substrate. This strain was believed to be due to an enzymic electrophile that polarizes the carbonyl group of the substrate and therefore promotes catalysis.

An enzymatic reaction that has been extensively studied by many workers using a variety of instrumental techniques is the photolytic reduction of dioxygen through cytochrome c oxidase. This enzyme contains two iron atoms, each of which is located in a heme group, and two copper atoms. The function of the metal atoms is only beginning to be understood, but it is certain that each is different. For example, in the oxidized form of the enzyme, EPR signals are observed for only one heme iron and one copper atom; these are designated as components of cytochrome a. The other heme and copper are designated as components of cytochrome a_3. In the fully reduced enzyme, iron and copper have even numbers of electrons and yield no EPR signal. A good understanding of the mechanism by which this enzyme acts and a knowledge of the structure of the oxygen binding site should lead to a more complete comprehension of the reaction mechanisms of other metalloproteins.

Fourier transform infrared measurements of the carbon monoxide complex with cytochrome c oxidase and other heme proteins reported by Alben's group at Ohio State University over the past decade are helping to elucidate the structure and function of this enzyme. Carbon monoxide has long been known to inhibit reduction of O_2 by cytochrome c oxidase. Carbon monoxide only binds to iron and copper in their lower oxidation states, Fe(II) and Cu(I), respectively. The wavenumber, half-width, absorptivity, and even the profile of the C–O stretching vibration are very sensitive to the nature of the metal complex. For example, the FWHH of bands due to the heme–CO complex of cytochrome c oxidase can be as low as 2.4 cm^{-1}, indicating the CO to be in a nonpolar, ordered environment. The complex can be photolyzed, after which the CO stretching band shifts by about 100 cm^{-1} and its FWHH more than doubles [17], indicating that the CO has moved to a less ordered and/or more polar environment. Structural information can also be obtained through measurements of the integrated intensity of this band. Through a series of very careful FT–IR measurements, Alben and his co-workers showed that

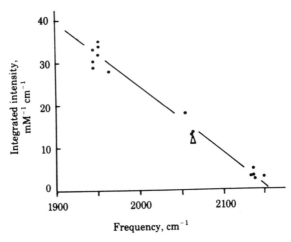

Fig. 14.13. Correlation of vibrational absorption frequencies with integrated absorptivities for CO–protein complexes at low temperatures. The point indicated by the triangle is calculated for cytochrome oxidase. (Reproduced from [17], by permission of the author.)

the integrated absorbance and peak wavenumbers of the C–O stretching mode of a series of CO–metalloprotein complexes are linearly correlated, see Fig. 14.13. These data were used subsequently to infer the nature of the metal to which CO is bound before and after photolysis [17,18].

Cytochrome c oxidase actually consists of seven subunits sequestered in the inner mitochondrial membrane. To prepare samples suitable for spectrometric investigation, mitochondria prepared from beef heart were saturated with CO to form the endogenous substrate-reduced CO complex. This complex was then dehydrated with CO-saturated glycerol to yield a waxlike glass whose infrared and visible spectrum could be measured in the regions of interest. For FT–IR spectrometry, 200-μm-thick samples were held between CaF_2 windows in a closed-cycle refrigerator, and interferograms were measured using an InSb detector. In view of the narrowness of the CO bands, spectra were measured at 1 cm^{-1} resolution. Since the absorbance of most of the features of interest was 0.01 or less and they had to be observed at high SNR, it was necessary to average over 4000 scans for most measurements. The samples could be photolyzed at low temperature using a tungsten lamp. Replicate experiments could be performed after relaxation of the complex in the dark at 210 K.

The infrared absorption spectrum of the CO complex of cytochrome c oxidase shows a relatively strong band at 1963 cm^{-1} and a weaker band at 1952 cm^{-1}, both of which are assigned to a nonbridging CO in the a_3 heme–CO complex. On photolysis, the CO absorption is shifted to 2062 cm^{-1} with a weaker band at 2043 cm^{-1}, see Fig. 14.14. The integrated absorbance and peak wavenumber of the new bands lay close to the corresponding values for other hemocyanin–CO complexes in which the CO is bound to copper, as indicated by the triangle in Fig. 14.13. Thus on photolysis, the CO is believed to be transferred to the a_3 copper atom (Cu$_B$). The increased width of the band due to Cu–CO (6 cm^{-1}) compared to the very narrow (2.4-cm^{-1}) Fe–CO band suggests that the Cu$_B$–CO complex is in a less ordered environment, that is, that the Cu–protein complex is more flexible than the heme protein, see Fig. 14.15.

The data appear to indicate that although the environments of the a_3 iron and copper atoms are quite different, the peak activation enthalpy of the reformation (measured as 40.3 kJ mol^{-1}) is large enough that the CO is not transferred spontaneously at low temperatures. The reformation of a_3 Fe–CO after photolysis was shown to be an apparent first-order reaction [18]. Although no FT–IR measurements on cytochrome c oxidase

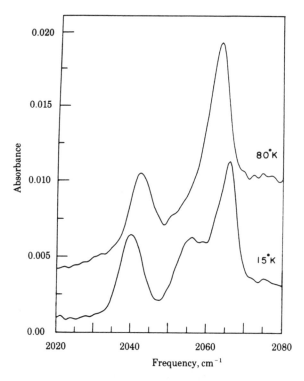

Fig. 14.14. Absorbance difference spectra (light minus dark) of the CO complex of cytochrome c oxidase at 80 and 15 K. (Reproduced from [17], by permission of the author.)

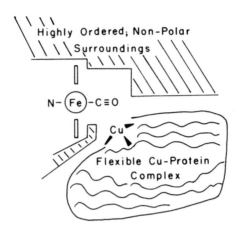

Fig. 14.15. Representation of the local interactions of CO in cytochrome c oxidase. (Reproduced from [17], by permission of the author.)

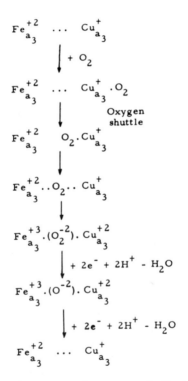

Fig. 14.16. Proposed reaction sequence for oxygen reduction by cytochrome a_3. The last two reactions represent reduction, proton shuttle, and water shuttle. (Reproduced from [17], by permission of the author.)

and O_2 have been reported, a mechanism for this reduction was proposed by Alben et al. [17] by extrapolating the interpretation of the results with carbon monoxide. This reaction sequence is shown in Fig. 14.16.

Aside from the biochemical significance of this work, it is noteworthy from an instrumental viewpoint that the intensities of the bands being studied were all rather low (typically with a peak absorbance of less than 0.01), yet the spectra were measured at a sufficiently high resolution and SNR and in a short enough time that accurate band intensities, half-widths, and kinetic data could be derived. Even band profiles could be investigated; for example, the observation that the C–O stretching bands of metal–CO complexes in frozen mitochondria are better fit by a Gaussian, rather than by a Lorentzian, profile is consistent with a heterogeneously broadened population and supports the kinetic evidence for a distribution of conformational states in the biological matrix at low temperatures. As Alben et al. [19] note: "The goodness of fit exemplifies the power of FT–IR spectroscopy to obtain information about biological systems that has been unattainable by other methods. We have only begun to make use of these capabilities. The future of Infrared Biospectroscopy looks bright indeed."

IV. ADSORPTION OF PROTEINS FROM BLOOD

Few applications of FT–IR spectrometry to biomedicine (as opposed to biochemistry) have yet been described. Nevertheless, the results of one type of measurement that falls into this category, namely the adsorption of blood proteins onto polymer surfaces, illustrate the potential of FT–IR spectrometry to the field of biomedical research. The work that will now be described requires all the speed and sensitivity of Fourier spectrometry, and even then, every phase of the procedure must be optimized before truly useful data can be obtained.

An understanding of the mechanism by which proteins are adsorbed from blood onto polymer surfaces is important because of the widespread use of synthetic polymers in heart valves, indwelling catheters, dialysis membranes, and artificial organs implanted in the body. The final stage of the adsorption process is the formation of a blood clot (thrombus), which can be fatal. Polymers that can be used with the lowest danger of thrombosis are called biocompatible, and Jakobsen, Gendreau, and their colleagues at Battelle's Columbus Laboratories have pioneered the application of FT–IR spectrometry to monitor the complex sequence of events that occur when a foreign surface is placed in contact with flowing blood.

The sampling technique selected for this investigation was ATR spectrometry, in which a thin layer of the polymer being studied is deposited on the surface of the IRE and blood is passed over its outer surface. Germanium is used as the IRE since its refractive index is so much higher than that of water or the polymer layer. Specially designed ATR cells have been designed with the following goals in mind:

1. laminar flow with no areas of the surface not reached by the flowing liquid,
2. a large number of reflections to maximize the signal due to species adsorbed on the surface and,
3. a high refractive index to minimize spectral distortions and to yield small penetration depths so that absorption by the bulk blood is also minimized.

Two designs of ATR accessories suitable for internal reflection spectrometry of flowing systems were discussed in Chapter 5 (see Figs. 5.19 and 5.20). The cross section of the rectangular cell used by the Battelle group is shown in Fig. 14.17. The parts of the element over which the flow is not uniform are masked by depositing a fairly thick metallic overlayer. The polymers were deposited as layers of 20–50 nm thickness, that is, much less than the penetration depth of the evanescent wave, which is 600 nm at 1650 cm^{-1}.

In the preliminary experiments, solutions of proteins in saline [20] or freshly drawn "bag blood" [21] were passed over the surface of the IRE

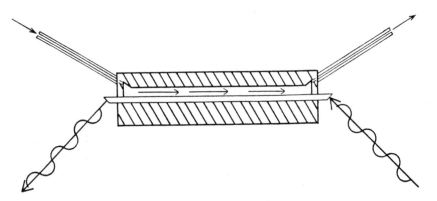

Fig. 14.17. Exploded view of the ATR cell used for biocompatibility studies of protein adsorption. Two long narrow internal reflection elements can be installed in the cell to minimize the number of times the entire cell need be removed from the ATR accessory. (Reproduced by permission of R. J. Jakobsen, Battelle's Columbus Laboratories.)

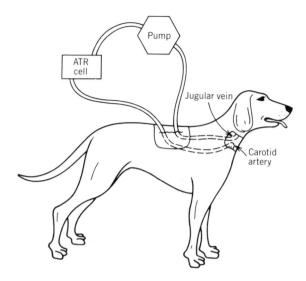

Fig. 14.18. Schematic diagram of carotid–jugular shunt placed in female beagle dog. Note that the pump is downstream from the ATR cell. (Reproduced from [24], by permission of the Society for Applied Spectroscopy; copyright © 1982.)

using tubing whose internal surface had been treated with the anticoagulant heparin. In later *ex vivo* studies, blood was passed directly from live animals to the ATR cell; both dogs [22,23] and sheep [24] have been used so far. Figure 14.18 shows a schematic diagram of the carotid–jugular shunt implanted in the animal prior to the experiment. When a study is not being performed, the shunt forms a harmless closed loop. During an experiment, the ATR cell and pump are connected so that the blood displaces the saline used to prime the system. The pump is downstream from the IRE and is used to regulate the flowrate through the circuit, with the actual energy being provided by the animal's heart. Flow rates of between 10 and 25 mL min^{-1} are used.

To begin the experiment, the shunt is opened and connected to the cell, displacing the saline. Immediately upon connection of the shunt, the spectrometer begins data acquisition using GC/FT–IR software. Spectra were typically collected at 5-sec intervals, with 25 scans per spectrum, but recently this time interval has been decreased [6]. In Fig. 14.19a, the spectra of the first five scan sets after blood enters the cell are shown. An uncoated IRE was used for this experiment; normally the polymer bands would also be seen in these spectra. The water spectrum is then subtracted (as would be the polymer spectrum), and the resulting spectra of the adsorbed protein are shown in Fig. 14.19b. The amide I and II

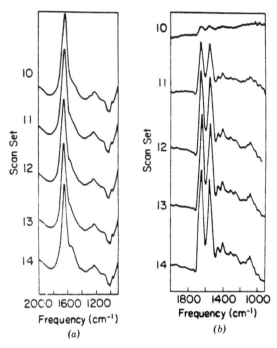

Fig. 14.19. (*a*) First five scan sets collected after blood entered the ATR cell before subtraction. (*b*) First five scan sets after blood entered the ATR cell after water subtraction plotted in the region between 2000 and 900 cm^{-1}. Note the appearance of the amide I and II bands at 1640 and 1550 cm^{-1}, respectively. (Reproduced from [24], by permission of the Society for Applied Spectroscopy; copyright © 1982.)

bands are dominant in these spectra, but several weaker bands can be seen at lower wavenumbers.

Previous studies have indicated that most of the qualitative information on the nature of the adsorbed proteins must be obtained from these weaker bands since all protein spectra show the amide I and II bands. In Fig. 14.20*a*, the spectra seen in Fig. 14.19*b* are replotted in the region between 1500 and 900 cm^{-1} using ordinate expansion. Even these spectra are quite similar, and in order to observe differences occurring in the time scale of this experiment, further subtractions of scan pairs must be performed, see Fig. 14.20*b*.

The center spectrum, 12-11, represents proteins adsorbed from blood in the first 5 sec of the experiment. From the position of certain bands in this spectrum, albumin is suspected to adsorb first, but the presence of carbohydrate bands at 1080 and 1040 cm^{-1} suggests the coadsorption of a glycoprotein. A different species adsorbs after the first few seconds, as seen from 23-12 in Fig. 14.20*b*. The data appear to suggest that albumin is replaced by large amounts of fibrinogen and possibly some gamma globulins. The carbohydrate bands continue to rise, indicating the initial glycoprotein is still being adsorbed. Even after a few minutes, the replacement of one protein by another was still observed to continue. After about 10 min, clot formation occurred and the experiment was terminated.

The amide II band at 1550 cm^{-1} and another band at 1400 cm^{-1} were present in all protein spectra with approximately the same absorptivity. Plots of the absorbance of these bands as a function of time, as illustrated in Fig. 14.21, indicate the kinetics of total protein adsorption throughout

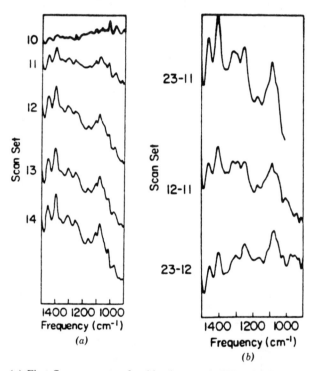

Fig. 14.20. (*a*) First five scan sets after blood entered ATR cell after water subtraction plotted in the region between 1500 and 900 cm^{-1}, shown expanded with respect to the spectra shown in Fig. 14.19 to illustrate detail of bands at wavenumbers below the amide II band. (*b*) Subtraction pairs. Top: subtraction of 5-sec spectrum from 2-min spectrum (difference is due to layer adsorbed from 5 sec to 2 min). Middle: subtraction of 5-sec spectrum from 10-sec spectrum. Bottom: Subtraction of 10-sec spectrum from 2-min spectrum. These spectra show that much of the protein deposits in the first 5 sec after blood makes contact with the ATR element. (Reproduced from [24], by permission of the Society for Applied Spectroscopy; copyright © 1982.)

the experiment. The initial adsorption is seen to be very fast for about 45 sec. The apparent break at about 2 min was not particularly reproducible and was at first believed to be due to the formation of a partial clot within the cell, which was subsequently removed. After about 4 min, the adsorption increases only slowly, with the rate being strongly dependent on the protein system being studied.

Subsequent work with blood from live sheep [24] shows a similar behavior to that in Fig. 14.21. This study indicates that the break in the plot of the intensity of the amide II band versus time formerly attributed to a temporary clot is almost certainly caused by the initiation of another type of process. From an instrumental viewpoint, it is noteworthy that improvements in sensitivity since the first report on this subject have enabled better time resolution to be attained [6] since spectra can be measured

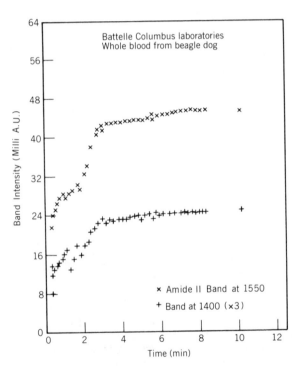

Fig. 14.21. Plot of the intensity of the amide II band at 1550 cm^{-1} and the band at 1400 cm^{-1} versus time: ×, Amide II band at 1550; +, band at 1400 (×3). By following baseline-corrected intensities of these bands, which are approximately invariant with the nature of the protein, the total amount of adsorbed protein can be followed in real time. (Reproduced from [24], by permission of the Society for Applied Spectroscopy; copyright © 1982.)

Live sheep run

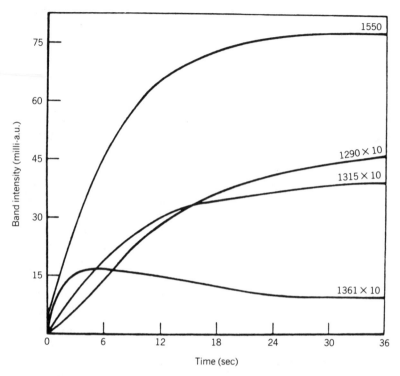

Fig. 14.22. Plot of the intensity of several bands of deposited proteins versus time for the first 36 sec of blood flow. Note that the maximum absorbance of the 1290-, 1315-, and 1361-cm^{-1} bands is less than 0.005 absorbance units. (Reproduced from [24], by permission of the Society for Applied Spectroscopy; copyright © 1982.)

every 0.8 sec. The variation of the absorbance of several bands in the spectra of the deposited proteins during the first half minute of the experiment is shown in Fig. 14.22. It should be noted that the *maximum* absorbance of the 1361-cm^{-1} band is only 0.0016 a.u., corresponding to a transmittance of 99.63%. If this band is to be measured with a SNR of 10, the noise level on the baseline must be about 0.03%, indicating that every phase of this measurement has been optimized.

    This investigation certainly illustrates the complex processes taking place during the early stage of blood clotting. A great deal more work is needed before the behavior of individual proteins can be followed accurately during deposition from whole blood, and it is noteworthy that results from other laboratories are now starting to appear in the literature [25]. It can certainly be said that investigations of this type are helping to provide a better insight into the molecular level basis for blood–polymer biocompatibility. Indeed, FT–IR spectrometry should prove to be an important tool for the investigation of many biomedical processes.

V. OTHER WORK

The studies described in some detail above were meant to give an idea of the potential of FT–IR spectrometry for biochemical and biomedical research and were not designed to be a comprehensive overview of this subject. The recent book by Parker [26] should be consulted for that purpose. Several other areas of biochemistry are susceptible for investigation by FT–IR spectrometry. Much of the work described since 1980 has made use of spectral subtraction techniques to focus on reactions or interactions taking place on specific sites of large molecules. For example, Theophanides [27] has investigated changes in the spectrum of calf thymus DNA after treatment with the anticancer drug Cisplatin. Liquier et al. [28] have studied changes in the secondary structure of DNA induced by various nucleohistone proteins.

Rothschild et al. [29–31] have made a detailed study of the membrane protein bacteriorhodopsin (bR), which is the light-activated proton pump from the purple membrane of *Halobacterium halobium*. This system has been extensively studied in view of its importance as a model for investigating the mechanism of energy transduction and ion transport in biological membranes. The related membrane protein rhodopsin is the primary light receptor for mammalian vision. Much previous work on these systems had been performed using resonance Raman spectrometry, but the laser beam almost certainly perturbs the sample during these measurements so that care has to be taken in their interpretation.

Using FT–IR difference spectrometry, Rothschild's group has shown evidence that alteration in the vibrations of both the retinylidene Schiff base chromophore and the protein groups of bR associated with the photocycle occur [29]. Using similar techniques, they have obtained spectroscopic evidence that the Schiff base is protonated [30]. More recently, Rothschild et al. [31] have studied intermediates in rhodopsin bleaching and claim that the techniques developed for their work should also be applicable to other membrane-based systems, such as calcium transport.

The exciting new data being obtained on biochemical systems causes us to believe that biochemical and biomedical spectrometry will represent one of the larger growth areas for the next few years. It is certainly true that the number of papers being published in this area in the 1980s is continuing to rise and shows no sign of slowing down.

REFERENCES

1. D. G. Cameron, H. L. Casal, and H. H. Mantsch, *J. Biochem. Biophys. Methods*, **1**, 21, (1979).
2. D. G. Cameron and R. N. Jones, Appl. Spectrosc., **35**, 448 (1981).
3. J. F. Blazyk, Ohio University, Athens, Ohio, unpublished work (1982).
4. D. G. Cameron and G. M. Charette, *Appl. Spectrosc.*, **35**, 224 (1981).
5. J. O. Alben and G. H. Bare, *Appl. Opt.*, **17**, 2985 (1978).
6. R. M. Gendreau, *Appl. Spectrosc.*, **36**, 47 (1982).
7. H. Sapper, D. G. Cameron, and H. H. Mantsch, *Can. J. Chem.*, **59**, 2543 (1981).
8. D. G. Cameron, H. L. Casal, H. H. Mantsch, Y. Boulanger, and I. C. P. Smith, *Biophys. J.*, **35**, 1 (1981).
9. J. Umemura, D. G. Cameron, and H. H. Mantsch, *Biochim. Biophys. Acta*, **602**, 32 (1980).
10. H. L. Casal, D. G. Cameron, I. C. P. Smith, and H. H. Mantsch, *Biochemistry*, **19**, 444 (1980).
11. D. G. Cameron, H. L. Casal, and H. H. Mantsch, *Biochemistry*, **19**, 3665 (1980).
12. D. G. Cameron, E. F. Gudgin, and H. H. Mantsch, *Biochemistry*, **20**, 4496 (1981).
13. H. C. Jarrell, K. W. Butler, R. A. Byrd, R. Deslauriers, I. Ekiel, and I. C. P. Smith, *Biochim. Biophys. Acta*, **688**, 622 (1982).
14. D. G. Cameron, A. Marin, and H. H. Mantsch, *Science*, **219**, 180 (1983).
15. J. Fisher, J. G. Belasco, S. Khosla, and J. R. Knowles, *Biochemistry*, **19**, 2895 (1980).
16. J. G. Belasco and J. R. Knowles, *Biochemistry*, **19**, 472 (1980).
17. J. O. Alben, P. P. Moh, F. G. Fiamingo, and R. A. Altschuld, *Proc. Natl. Acad. Sci. U.S.A.*, **78**, 234 (1981).
18. F. G. Fiamingo, R. A. Altschuld, P. P. Moh, and J. O. Alben, *J. Biol. Chem.*, **257**, 1639 (1982).
19. J. O. Alben, F. G. Fiamingo, and R. A. Altschuld, *Proc. Soc. Photo-Opt. Instrum. Eng.*, **289**, 175 (1981).
20. R. M. Gendreau, R. I. Leininger, S. Winters, and R. J. Jakobsen in *Biomaterials: Interfacial Phenomena and Applications* (S. L. Cooper, and N. A. Peppas, eds.), ACS Advances in Chemistry Series, Washington, D.C., 371 (1982).
21. R. J. Jakobsen, L. L. Brown, S. Winters, and R. M. Gendreau, *J. Biomaterials Res.*, **16**, 199 (1983).
22. R. J. Jakobsen, S. Winters, and R. M. Gendreau, *Proc. Soc Photo-Opt. Instrum. Eng.*, **289**, 469 (1981).
23. R. M. Gendreau, S. Winters, R. I. Leininger, D. Fink, C. R. Hassler, and R. J. Jakobsen, *Appl. Spectrosc.*, **35**, 353 (1981).
24. S. Winters, R. M. Gendreau, R. I. Leininger, and R. J. Jakobsen, *Appl. Spectrosc.*, **36**, 404 (1982).

25. R. Kellner and G. Gidaly, *Mikrochim. Acta,* **1,** 119 (1981).
26. F. S. Parker, *Application of Infrared Spectroscopy in Biochemistry, Biology, and Medicine,* 2nd ed., Plenum Publishing, New York (1984).
27. T. Theophanides, *Appl. Spectrosc.,* **35,** 461 (1981).
28. M. C. Liquier, *Nucleic Acids Res.,* **6,** 1479 (1979).
29. K. J. Rothschild, M. Zagaeski, and W. A. Cantore, *Biochem. Biophys. Res. Commun.,* **103,** 483 (1981).
30. K. J. Rothschild and H. Marrero, *Proc. Natl. Acad. Sci. USA,* **79,** 4045 (1982).
31. K. J. Rothschild, W. A. Cantore, and H. Marrero, *Science,* **219,** 1333 (1983).

CHAPTER

15

LOW-TEMPERATURE STUDIES

Temperature plays an important role in infrared spectrometry: the mean temperature of the sample determines the populations of the rotational and vibrational states, and temperature defines the emissive properties of the sample. Low-temperature studies have been undertaken for four basic reasons: to observe weak emissions such as chemiluminescence, to isolate chemically unstable or transient species, to depopulate excited rotational energy levels so as to confine all molecules to the ground state and observe the fundamental vibrations, or to enhance phonon bands. The first type of experiment often requires that the spectrometer be cooled; the second, third, and fourth require that only the sample be cooled.

I. LOW-TEMPERATURE SPECTROMETERS

The first low-temperature FT–IR spectrometer developed for chemical spectrometric studies was developed under the direction of McDonald [1–5]. McDonald was interested in polyatomic chemiluminescence for the study of reaction dynamics. In the first of these studies [1], a system was described for the observation of the substitution reactions of fluorine with monosubstituted ethylene compounds. The reactions of interest were of the general form

$$F + C_2H_3X \rightarrow X + C_2H_3F$$

$$(15.1)$$

where X is Br, Cl, CH_3, or hydrogen. These reactions released between 40 and 180 kJ mol^{-1}. Due to the rather poor noise characteristics of infrared detectors (see Chapter 5), these weak chemiluminescent emissions cannot be observed at room temperature. To prevent collisional relaxation of the luminescent states and to reduce the probability of secondary reactions of the products to less than 10%, the pressures under which these gas reactions take place must be on the order of 10^{-4} torr or lower. These conditions also decrease the emission flux and hence make the reactions difficult to measure.

Moehlmann et al. [1] calculated the photon fluxes for the substitution reactions at these low pressures. The total reactant flow rate was about 100 μmol sec^{-1}, and the total quantity of excited molecules was probably produced at the rate of about 1 mol sec^{-1}. The acceptance angle of the spectrometer was rather restricted with only 1% of the reaction chamber visible to the spectrometer, which translates to an observable concentration of about 5×10^{15} molecules/sec. Combining these data with an average molecular residence time of 3×10^{-4} sec, the typical radiative lifetime was 1 sec, or a radiative flux of 10^{12} photons/sec. Only 10^7 photons/sec could be detected due to the throughput of the spectrometer. The detector used was a mercury-doped germanium bolometer cooled to 4.2 K. When the sample and spectrometer are at room temperature, the detector has a noise equivalent photon flux of 5×10^8 sec^{-1}, which gives a SNR of the 10^7-photon/sec source of about 0.02. Obviously, this SNR is too low for practical experiments. By reducing the background temperature, including the sample cell and the interferometer, to 77 K, the noise-equivalent photon flux becomes 5×10^5 photons/sec, which yields a SNR of 20.

A schematic diagram of the apparatus is reproduced in Fig. 15.1. A Digilab Model 296 Michelson interferometer (pre-A series) was converted from a commercial interferometer that operates at 41°C to one that could be cryogenically cooled to 80 K. The interferometer was placed in a liquid nitrogen jacket that was isolated from the laboratory by a polyurethane foam heat shield. A few modifications were made to the interferometer itself. The gas used for the air bearing was changed to cooled helium to prevent condensation of oxygen (or nitrogen) inside the bearing. The cooling of the bearing gas also removed any residual water that might foul the system. The gas-bearing pressure line was changed from plastic to a stainless steel bellows to prevent fracture of the line. The He–Ne laser and white-light reference interferometer cube was mounted on Invar (a nickel steel alloy) because it has a similar expansion coefficient to quartz, the reference cube material. This prevented undue stress on the reference cube as it was cooled. The He–Ne laser was removed from its mounting on top of the interferometer and placed outside the polyurethane heat shield. This removed the heat produced by the laser.

Two wavelength ranges could be covered with this instrument. The range 1200–6000 cm^{-1} was covered by using a standard Fe_2O_3 beamsplitter with a calcium fluoride substrate. The other beamsplitter covered the range 720–2300 cm^{-1} and was made of germanium. One germanium plate had a single surface coated to prevent reflection, and the other surface

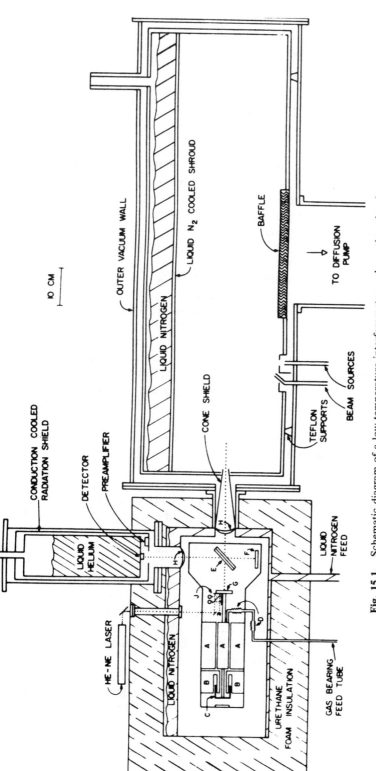

Fig. 15.1. Schematic diagram of a low-temperature interferometer and reaction chamber for chemiluminescence studies. Key: A, gas bearing; B, scanning drive magnet; C, scanning drive electromagnet; D, flexible steel bellows air-bearing gas delivery tube; E, beamsplitter; F, fixed mirror; G, moving mirror; H, germanium lenses; J, reference interferometer. (Reproduced from [1], by permission of the author and the American Institute of Physics; copyright © 1974.)

had a reflectivity of 30%, which acted as the actual beamsplitter interface. A compensator plate of germanium with both surfaces antireflection coated was employed. Provisions were made to align the fixed mirror during cooling because the alignment changed as the interferometer contracted.

The sample chamber and liquid helium bolometer were isolated thermally from the interferometer, and germanium lenses were used to collect the chemiluminescence radiation from a spot in the reduction chamber and to focus the radiation from the interferometer to the detector. The detector was modified for low-background radiative flux by reducing the thermal noise in the preamplifier. This was done by cooling a portion of the preamplifier to 5 K and other components to 80 K.

The sample chamber was an aluminum box cooled to 80 K with a liquid nitrogen shroud. The chamber was thermally isolated by placing it within a vacuum, the inner walls were painted flat black, and a large diffusion pump was used to maintain a low pressure within the sample chamber. The reactants were introduced through narrow tubes that were directed so that the reactant streams crossed about 2 cm into the chamber.

Data handling was not accomplished by standard coaddition for signal enhancement. Reaction and background scans were alternatively collected. A reaction scan was collected as the reactants were flowing and the interferogram was added into memory. The reactants were then shut off by use of a solenoid valve and a background interferogram collected. The background interferogram was then subtracted from the reactant interferogram to produce a difference interferogram. This process was repeated until an adequate SNR was achieved; usually between 500 and 10,000 scans were required. The Fourier transform of the difference interferogram produces a spectrum of the emission multiplied by the instrument function. The instrument function is the product of response function of the detector and the transmission of the optics. The instrument function was removed by employing the emission spectrum of a known source, in this case a small tungsten filament. Once the filament emission spectrum was recorded, it was divided into the difference spectrum, which removed the instrument function. Photometric correction of the product emission spectrum was achieved by multiplying the product emission spectrum by the relative photon emittance spectrum of the tungsten filament. Change in the instrument function by absorption of reaction products on the system optics was monitored by the collection of tunsten emission spectra periodically through the experiments.

The first set of experiments performed under the direction of McDonald [1] involved the reaction of atomic fluorine with ethene, vinyl chloride, propene, and vinyl bromine. A common product to all reactions was vinyl fluoride. The emission spectra from these four reactions are shown in Fig. 15.2. All four spectra exhibit three common bands that may be attributed to vinyl fluoride. The band at 1280 cm^{-1} from the fluorine plus propene reaction was identified as emission from the leaving group, a methyl radical. The band at 1250 cm^{-1} from the fluorine plus ethene reaction was identified as emission from C_2H_3 radicals produced by a hydrogen abstraction reaction where HF is the other product in addition to the radical.

Moehlmann and McDonald [2] completed a study on the substitution reactions between fluorine and various olefinic and aromatic compounds. The objective of these studies was to understand the molecular dynamics of the reactions. The substances studied were specifically the reactions of several chloroethenes (including the cis, trans, and gem isomers of dichloroethene) as well as the substitution reactions of fluorine with propenes, 2-butene, and some substituted benzenes. This study, and the preceding work [1], showed that the fluorine substitution reactions yielded products that exhibited a statistical distribution of vibrational energy. This, of course, was determined from the chemiluminescence spectra. It should be noted, however, that not all reactants, specifically three non-halogenated alkenes, ethene, propene, and 2-butene, exhibited nonstatistical product vibrational energy distributions. All the spectra shown in these two studies [1,2] were collected at low resolution, that is at 8 cm^{-1}. Because the chemiluminescence was measured some milliseconds after the reaction had ended and extensive rotational relaxation had taken place in that time, the rotational distribution was virtually identical within the emitting states for all the vibrational levels. In other words, no more information could be obtained by going to a higher spectrometric resolution.

As mentioned above, two types of reactions between atomic fluorine and the various substrates are possible, substitution and abstraction. Moehlmann and McDonald [3] studied some of the abstraction reactions between fluorine and unsaturated compounds. The abstraction reaction produces HF and a polyatomic radical. In these reactions, an emission spectrum can be recorded for HF but not for the polyatomic radical because the radical receives only a fragment of the available energy and because the substitution reaction is more favorable. Figure 15.3 shows an HF emission spectrum from the abstraction reaction between F and H_2 at 5×10^{-5} torr. The spectrum from the reaction between F and chlorobenzene is shown in Fig. 15.4. The bands at the lower wavenumbers, between 1200 and 1600 cm^{-1}, arise from C_6H_5F, which is produced in the substitution reaction. The abstraction reaction study provides complementary information to the substitution reactions.

Of course, these substitution and abstraction reactions need not be

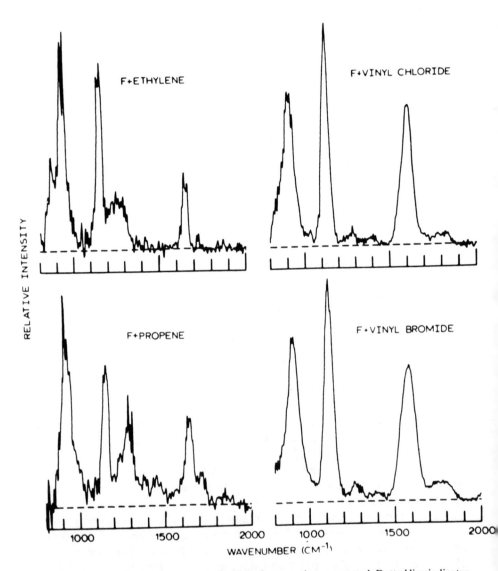

Fig. 15.2. Chemiluminescence spectra of the four reactions annotated. Dotted line indicates zero intensity for the normalized emission spectra. (Reproduced from [1], by permission of the author and the American Institute of Physics; copyright © 1974.)

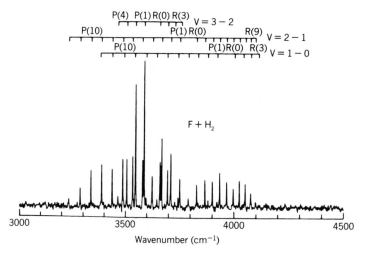

Fig. 15.3. Infrared emission spectrum of HF for the abstraction reaction H_2 + F. (Reproduced from [3], by permission of the author and the American Institute of Physics; copyright © 1975.)

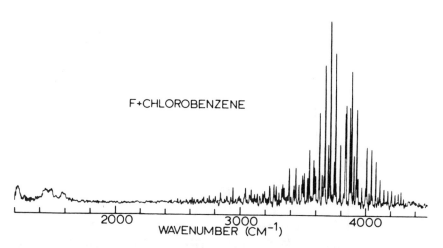

Fig. 15.4. Infrared emission spectrum of HF for the abstraction reaction F + C_6H_5Cl. The band between 1200 and 1600 cm^{-1} is due to C_6H_5F from the substitution reaction. (Reproduced from [3], by permission of the author and the American Institute of Physics; copyright © 1975.)

restricted to fluorine. Durana and McDonald [4] studied chlorine substitution reactions with brominated unsaturated hydrocarbons. Two changes were made in the optics to increase the emissive flux reaching the spectrometer. The two germanium lenses that were used to collect radiation from the reaction chamber and to focus the modulated radiation from the interferometer onto the detector were replaced with zinc selenide lenses. These lenses increased the overall transmittance by about 40%. The reaction chamber was made into a Welsh cell [5,6] by placing confocal rhodium-plated brass mirrors at the ends of the chamber. The collection efficiency was increased by up to a factor of 20 over the original design. The mirrors in the Welsh cell were aligned at room temperature and remained within tolerances after being cooled to 77 K. As in the previous studies [1–3], the background emission changed as products contaminated the reaction chamber, so calibration spectra were regularly taken throughout the experiment by using a tungsten filament source.

Figure 15.5 shows the emission spectra from the substitution reaction between atomic chlorine and vinyl bromide, 1-bromopropene and 2-bromopropene. These spectra were signal averaged with 600–1000 scans. When compared to the spectra in Fig. 15.2, the SNR is improved by at least a factor of 10 even though the total number of scans is probably less. McDonald has also applied this technique to the reactions of methyl radicals with oxygen and fluorine [7].

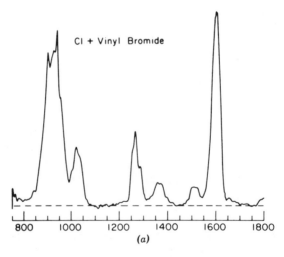

Fig. 15.5. Infrared emission spectra of the three reactions annotated. Dashed line indicates zero intensity for the normalized emission spectra. (Reproduced from [4], by permission of the author and the American Institute of Physics; copyright © 1976.)

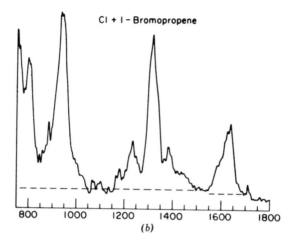

(b)

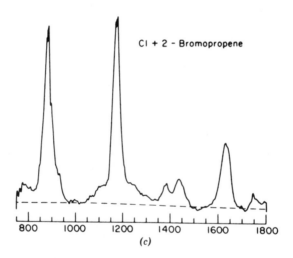

(c)

Fig. 15.5. (*continued*)

More recently, McDonald and co-workers have investigated laser-induced intramolecular vibrational relaxation (infrared fluorescence) using low-temperature FT–IR spectrometry [8–11]. The same interferometer was used as in the previous studies [1–3], except that the computer system was replaced. To measure infrared fluorescence, McDonald used a tunable infrared laser to excite species from a pulsed molecular beam. The laser radiation was from a 1-cm⁻¹-bandwidth Nd–YAG pumped optical parametric oscillator (OPO) that provided 2–4 mJ of energy per 14 nsec pulse. The laser was tunable over the range 2800–3200 cm⁻¹ and had a repetition rate of 7.5 Hz. The laser radiation was directed into a White cell where the radiation was multipassed and intersected the molecular beam at the focus of the field of view of the FT–IR detector. The White cell was baffled to reduce stray radiation.

Data collection was highly modified to accommodate the slow repetition rate (7.5 Hz) of the laser. Because infrared fluorescence is associated only with the excitation, pulse data were collected in synchronization with the laser. At a mirror velocity of 0.16 cm/sec, He–Ne zero crossings occur every 100 μsec. If the bandwidth of the spectrum is restricted to 0–3950 cm⁻¹, data are collected every 400 μsec (see Chapter 2). This means that when data are collected in synchronization with the 7.5-Hz excitation laser, data are available every 333rd possible sample point. To collect a single interferogram for this system, 333 scans have to be run, each with a different offset of the excitation laser pulsing with respect to interferometer retardation. Enough scans were collected so that the complete interferogram was effectively signal averaged for five scans.

McDonald and co-workers have primarily investigated the intramolecular vibrational relaxation in the C–H stretch region. Relaxation (fluorescence) spectra of dimethyl ether are shown in Fig. 15.6 [10] and of 1,4-dioxane in Fig. 15.7 [11]. The laser pump wavenumber (in reciprocal centimeters) is printed above each spectrum. Different pump wavenumbers produce different spectra because different rotational energy levels are excited. This work is particularly significant and has produced high-SNR spectra of a low-probability spectrometric phenomenon.

A low-temperature spectrometer has also been used to measure the emission spectrum of a molecular monolayer at 300 K. Allara, Teicher, and Durana [12] constructed a spectrometer based largely on the design of McDonald [1]. A schematic of the instrument for monolayer emission studies is shown in Fig. 15.8.

The samples were turned so that the angle between the collected beam and the sample was 35–40°, although Fig. 15.8 does not indicate this. This study presented the results for only one substance, *p*-nitrobenzoic acid (PNBA) chemisorbed on a copper–copper oxide substrate. The emission spectrum was corrected to remove the instrument function in a manner similar to that employed by Moehlmann and McDonald [1]. The major difference was that 2000 scans of the sample plus substrate and the substrate alone were collected separately prior to subtraction. The resulting PNBA chemisorbed spectrum is shown in Fig. 15.9*a*, and a transmission spectrum of bulk copper *p*-nitrobenzoate on NaCl is given in Fig. 15.9*b*. Clearly, there are differences between the two spectra, especially in the

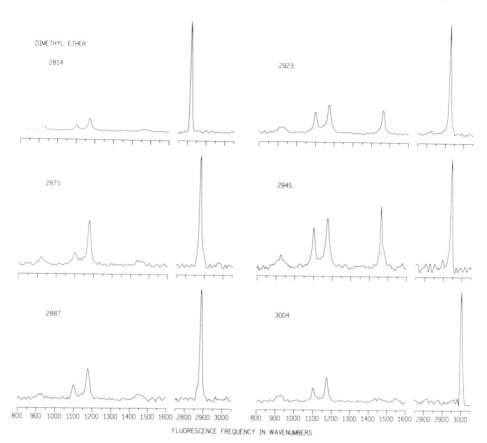

Fig. 15.6. Dimethyl ether fluorescence spectra. The spectral features show the CH_3 bending modes (1400–1500 cm^{-1}), the C-O-C symmetric stretch at 920 cm^{-1}, and the asymmetric C-O-C stretch at 1100 and 1175 cm^{-1}. The laser pump wavenumber (in cm^{-1}) is above each spectrum. The intensity of each spectrum is arbitrary and each spectrum is normalized to the height of the largest peak. (Reproduced from [10], by permission of the author and the American Institute of Physics; copyright © 1984.)

intensities and peak positions of the bands between 1500 and 1700 cm^{-1}. For example, the band at 1559 cm^{-1} in Fig. 15.9*b* is due to the asymmetric stretch of the carboxylate group. This band is shifted by nearly 30 cm^{-1} in the chemisorbed sample, although some of this shift (<10 cm^{-1}) can be attributed to optical factors. This study illustrated that cryogenic emission studies can be carried out to observe the properties of surfaces. However, a detailed study of the relative sensitivity of surface measurements made by emission spectrometry and reflection–absorption spectrometry (with or without polarization modulation) has not yet been reported.

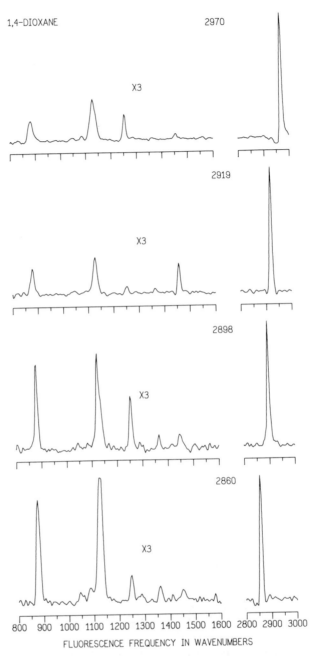

1,4-DIOXANE

2970

2919

2898

2860

X3

800 900 1000 1100 1200 1300 1400 1500 1600 2800 2900 3000

FLUORESCENCE FREQUENCY IN WAVENUMBERS

Fig. 15.7. 1,4-Dioxane fluorescence spectra from excitation in the C–H stretching region. The laser pump wavenumber (in cm^{-1}) is above each spectrum. The intensity of each spectrum is arbitrary and each spectrum is normalized to the height of the largest peak. (Reproduced from [11], by permission of the author and the American Institute of Physics; copyright © 1984.)

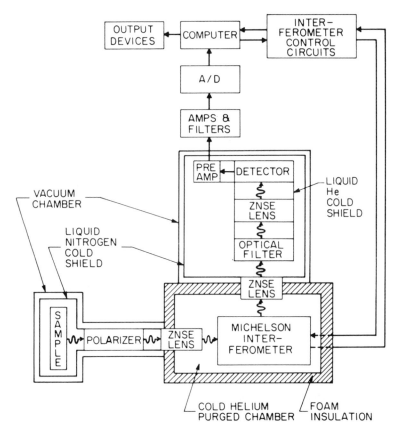

Fig. 15.8. Schematic of low-temperature spectrometer for surface studies. (Reproduced from [12], by permission of the North Holland Publishing Company; copyright © 1981.)

The number of chemical applications of using low-temperature inter-ferometers has been rather restricted thus far. The primary impediment to other workers joining this field has been the difficulty in obtaining or constructing a cryogenic spectrometer. As the designs for interferometers become more refined and construction becomes simpler, it may become more feasible to construct low-temperature systems. These authors an-ticipate more emission and kinetics studies with cryogenic spectrometers in the future.

II. MATRIX ISOLATION

Matrix isolation is a cryogenic technique in which a gaseous or vapor sample is mixed with a diluent gas and the mixture is frozen onto a window

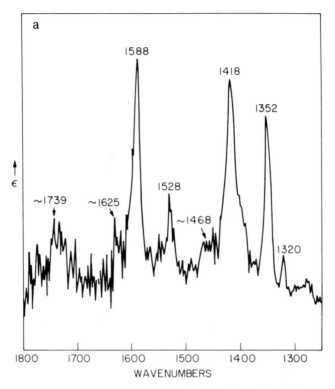

Fig. 15.9. (*a*) Normalized emission spectrum of a monolayer of *p*-nitrobenzoic acid. (*b*) Transmission spectrum of bulk copper *p*-nitrobenzoate on NaCl. (Reproduced from [12], by permission of the North Holland Publishing Company; copyright © 1981.)

or mirror for spectrometric investigation. The diluent (or matrix) gas is usually argon or nitrogen, but other matrix gases such as SF_6 or even *n*-alkanes can be employed. Several reviews on matrix isolation spectrometry are available [13,14], and at least two pertaining to FT–IR spectrometry have been written [15,16]. One advantage of matrix isolation (MI) spectrometry in the infrared is that individual molecules or radicals are isolated from others of the same compound within a transparent medium. Furthermore, the samples are held at temperatures approaching absolute zero (usually less than 20 K). Due to the structure of the inert matrix, fairly large molecules can be trapped in such a manner that all rotational movement is prohibited.

Matrix isolation is usually achieved by mixing a volatile substance at low pressure with the matrix gas. The matrix gas concentration exceeds that of the compound under study and is usually expressed as a ratio of

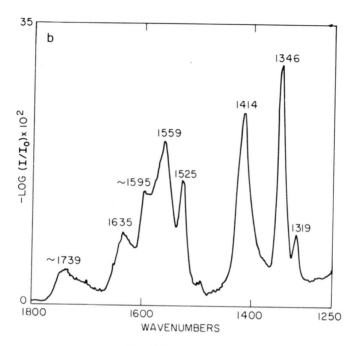

Fig. 15.9. (*continued*)

matrix to sample. Ratios range from about 500:1 up to about 10,000:1. When the concentration of the matrix falls below this range, intermolecular interactions are more likely to occur; higher concentrations than 10,000:1 effectively decrease the concentration of the sample (or solute) and decrease the detectability.

Once the sample is mixed with the matrix gas, the mixture is slowly deposited onto a cryogenic surface. Deposition rates are generally on the order of 1–15 mmol of the matrix per hour. These slow deposition rates allow the matrix to become evenly deposited without cracking. The cryogenic surface, either a mirror or window, is thermally in contact with a dewar or cryostat. Dewars are filled with liquid helium and have large heat capacities, yet the temperature is fixed. Closed-cycle two-stage cryostats based on the Stirling cycle lead to the liquefaction of helium, which gives an operating temperature of 12 K. These cryostats are relatively simple and inexpensive to operate. Heliplex three-stage cryostats can in theory operate as low as 4 K.

A schematic diagram of a window attached to a cryostat is shown in Fig. 15.10. In this case, the window, CsI, is connected to the cryostat cold finger, the actual surface where the cooling takes place. The con-

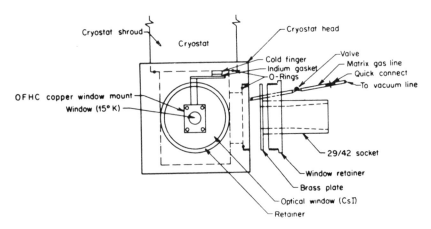

Fig. 15.10. Schematic of cryostat head. (Reproduced from [23], by permission of the Society for Applied Spectrosopy; copyright © 1979.)

nection to the cold finger is made such that there is maximum heat transfer to ensure that the cryogenic surface will be as cold as possible. A cover or shroud surrounds the cryogenic surface, and the interior of the shroud is evacuated to minimize heat transfer from the laboratory. It is common practice to be able to rotate the shroud so that deposition and spectral scanning can be made from the same angle.

The majority of the work done with MI spectrometry in the infrared has not involved analytical applications. This work mainly has been directed toward band assignments, force constants, the implementation of normal coordinate analysis studies, as well as the characterization of intermediates and unstable species. In fact, because the sample is trapped and is usually available in sufficient quantities, most infrared matrix isolation studies have been carried out using dispersive instruments.

One area where MI/FT–IR has found analytical applications is in the area of complex mixture analysis. An example of this problem is in the identification of a mixture of isomers that are not separated easily by chromatographic methods and that have virtually identical room temperature infrared spectra. Mamantov, Wehry, and co-workers have published a series of studies dealing with polycyclic aromatic hydrocarbons (PAHs) [16–20].

Because rotational modes are almost totally eliminated for large molecules such as PAHs by matrix isolation, identification can be based solely on vibrational band assignments. The infrared spectra of isomers can be very similar when the sample is in a vapor or condensed phase. Once the rotational broadening is removed, small shifts in some fundamental vi-

brational bands become more readily apparent. These shifts are reproducible and sufficient to identify isomers. To illustrate this fact, the MI/
FT–IR spectra of five PAHs are presented in Fig. 15.11. These spectra
were recorded at 2 cm^{-1} resolution, and clearly all the bands lack broadening due to rotation. (The peaks in the spectra in Fig. 15.11 at 1100 cm^{-1}
are caused by an experimental artifact, an impurity on the CsI substrate.)

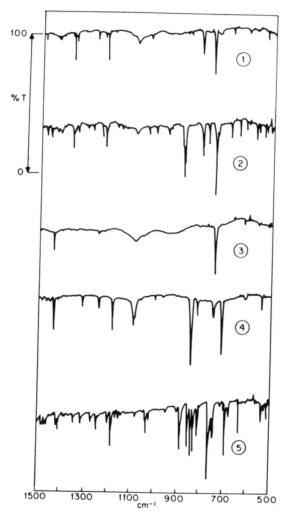

Fig. 15.11. Matrix isolation spectra of (1) chrysene; (2) benz[*a*]anthracene; (3) triphenylene;
(4) pyrene; and (5) benzo[*a*]pyrene, each in a nitrogen matrix, nitrogen to solute ratio 10^3.
(Reproduced from [18], by permission of the American Chemical Society; copyright © 1977.)

A high-resolution spectrum of matrix-isolated pyrene (spectrum 4 in Fig. 15.11) is shown in Fig. 15.12. This spectrum was collected at a spectrometer resolution of 0.5 cm⁻¹, and the band FWHH approaches the resolution. It has been found that when spectrometer resolution is increased to at least 0.125 cm⁻¹, the bandwidth for the fundamental bands of matrix-isolated large molecules is consistent with the instrument line shape [21].

A more dramatic demonstration of the capabilities of MI/FT–IR spectrometry is the spectrum of a matrix-isolated mixture of 50 μg each of 1,3-, 1,4-, 1,5-, 2,3,-, and 2,6-dimethylnaphthalene, shown in Fig. 15.13. This spectrum was recorded at 1 cm⁻¹ resolution. Clearly, each component of the mixture is identifiable, whereas if this sample is not rotationally restricted, there is severe band overlap, making component identification

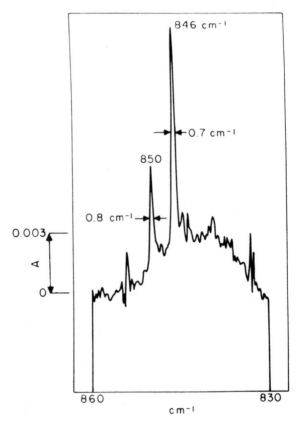

Fig. 15.12. Matrix isolation spectrum of 5 μg pyrene at a nominal spectrometer resolution of 0.5 cm⁻¹. (Reproduced from [18], by permission of the American Chemical Society; copyright © 1977.)

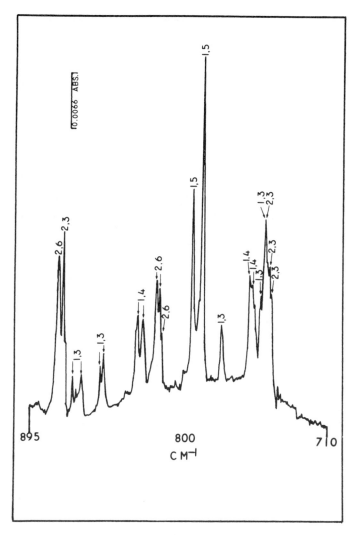

Fig. 15.13. Matrix isolation spectrum of 50 μg each of 1,3-, 1,4-, 1,5-, 2,3-, and 2,6-dimethylnaphthalene. (Reproduced from [19], by permission of Marcel Dekker, Inc.; copyright © 1979.)

impossible. A good example of the spectral differences between matrix-isolated samples and room temperature samples is shown in Fig. 15.14 [22]. In Fig. 15.14, the top spectrum is 10 μg of matrix-isolated naphthalene and the bottom is 150 μg of the same compound in the KBr disk. Clearly, the matrix-isolated sample has a more identifiable spectrum.

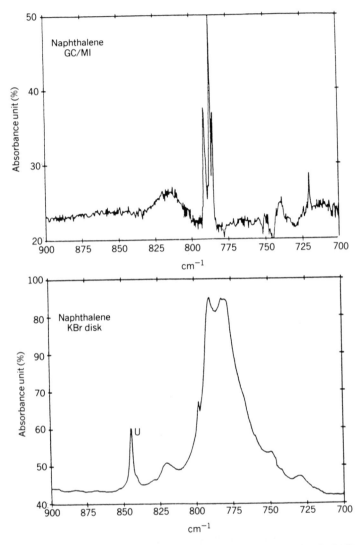

Fig. 15.14. Spectra of 10 μg of naphthalene by matrix isolation (*top*) and of 150 μg naphthalene using a KBr disk (*bottom*) as measured by Hembree [22]. The feature marked U is an unknown impurity.

One definite benefit to MI/FT–IR spectrometry is that quantitative analysis is performed rather easily. Linear Beer's law plots are obtained over approximately two decades of concentration for most PAHs, usually in the range 1–50 μg. It should be noted, however, that because the bands are very sharp with matrix-isolated samples, the spectra must be recorded at higher than usual resolutions to avoid deviations from Beer's law due to resolution errors. A more thorough treatment of quantitative analysis is given in Chapter 10.

The preceding samples, that is, PAHs, are not amenable to standard matrix isolation deposition methods. Many of these compounds are thermally labile and have low vapor pressures. As a consequence, Mamantov, Wehry, et al. have devised a deposition method based on the use of a Knudsen cell [23]. The PAH mixtures are placed inside a small glass cell with an injection port and an effusion orifice. Schematic diagrams of microsampling and ultramicrosampling Knudsen cells are shown in Fig. 15.15. The cells are placed inside heating coils that heat the cell to about 100°C. The cell is then placed inside the cryostat head beside the deposition window. Most PAHs can be vacuum sublimed below this temperature, and thus thermal degradation is not a problem. Inert matrix gas is supplied by a separate nozzle during the vacuum sublimation.

One very interesting application of MI/FT–IR spectrometry involves the identification of species eluting from a gas chromatograph (GC) [24–26]. GC/FT–IR spectrometry using light-pipe gas cells is covered in detail in Chapter 18; however, the application of MI techniques to GC/FT–IR

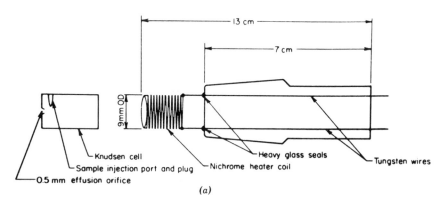

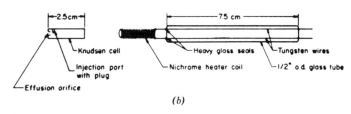

Fig. 15.15. (*A*) Microsampling and (*B*) ultramicrosampling Knudsen cell sampling assemblies. (Reproduced from [23], by permission of the Society for Applied Spectroscopy; copyright © 1979.)

is a special modification that is discussed in this section. Gas chromatography is usually a dynamic experiment in which the eluates pass through the detector for only a few seconds. If the standard GC detector is replaced by an FT–IR spectrometer, spectra must be measured during the elution period, which often does not allow sufficient time for long-term signal averaging. In the case of PAHs or coal derivatives, the vapor phase spectra are not suitable for accurate eluate identification because of the similarity between the spectra of isomers. If the eluates can be matrix isolated, the rotational structure is lost and full advantage can be taken of signal averaging because the eluates are trapped.

GC/MI–FT–IR can be accomplished by directing GC eluates into a cryostat chamber and depositing them onto a reflective surface. A multisided or cylindrical mirror can serve to collect a series of eluates. Matrix isolation is accomplished by using nitrogen as a carrier gas. The GC effluent can be split and sent to a flame ionization detector (FID) so that

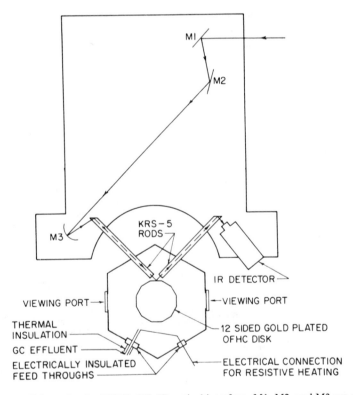

Fig. 15.16. Schematic of a GC/MI–FT–IR optical interface. M1, M2, and M3 are mirrors. (Reproduced from [26], by permission of the American Chemical Society; copyright © 1981.)

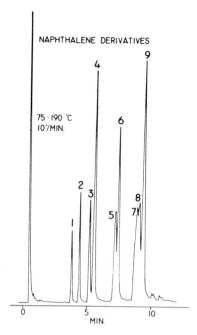

NAPHTHALENE DERIVATIVES

75 - 190 °C
10°/MIN.

Fig. 15.17. Gas chromatogram of a synthetic mixture of naphthalene derivatives using a 39-m SE-30 SCOT column, temperature programmed from 74–190°C at 10°C/min. The peaks correspond to the following components: (1) *cis*-decahydronaphthalene; (2) *trans*-decahydronaphthalene; (3) 1,2-dihydronaphthalene; (4) naphthalene and 1,4-dihydronaphthalene; (5) 2-methylnaphthalene; (6) 1-methylnaphthalene; (7)2,6-dimethylnaphthalene; (8) 1,3-dimethylnaphthalene; and (9) 2,3- and 1,5-dimethylnaphthalene. (Reproduced from [26], by permission of the American Chemical Society, copyright © 1981.)

eluates can be detected, and when an eluate is detected, the effluent is directed to the cryostat; otherwise, the effluent is vented.

A schematic diagram of the optical interface for a GC/MI–FT–IR system [26] is shown in Fig. 15.16. As can be seen from this figure, the modulated infrared beam is directed to one of the multisided mirror surfaces via an internal reflection KRS-5 rod. Another rod collects the reflected radiation and channels it to a detector. Figure 15.17 shows a chromatogram of a synthetic mixture of naphthalene derivatives. The mixture contained 11 components, but only nine peaks were resolved; naphthalene and 1,4-dihydronaphthalene coeluted as peak 4 and 2,3- and 1,5-dimethylnaphthalene coeluted as peak 9. The first five eluates of the mixture were deposited on one surface of the mirror, the next two on another surface, and the remaining four on a third surface. The resulting spectra

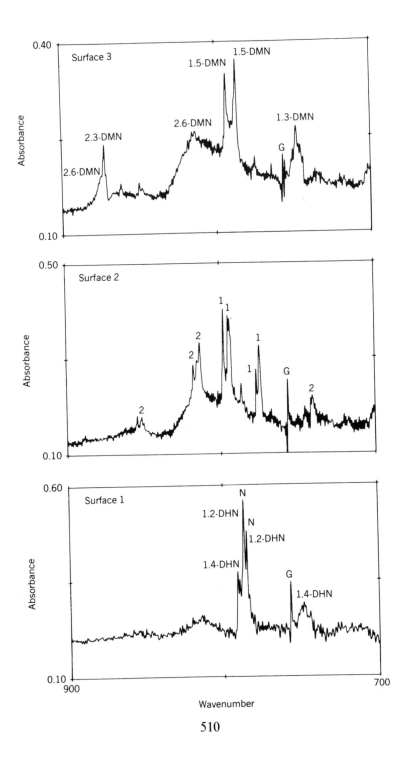

from the three surfaces are shown in Fig. 15.18, and all 11 components can be readily identified.

Mattson Instruments [27] has introduced a commercial GC/MI–FT–IR accessory for FT-IR spectrometers. The commercial system designed in conjunction with Cryolect Scientific is a self-contained unit that includes a cryostat, optics, and detector. Unlike the previous example, the modulated beam is focused onto the cryogenic deposits using off-axis paraboloids rather than KRS-5 rods. A cylindrical mirror is rotated continuously as deposition takes place. The GC carrier gas is a He–N_2 mixture, and additional N_2 is added during deposition. The helium gas is removed from the cryostat chamber by diffusion pumping. The apparatus marketed by Mattson Instruments can collect depositions for 5 hr, and the chromatographic resolution is approximately 6 sec. An optical diagram and a photograph of the accessory are shown in Fig. 15.19. The Cryolect GC/MI-FT-IR accessory is discussed further in Chapter 18, Section II.4, and representative spectra are shown in Fig. 18.27.

It should be noted that MI/FT–IR spectrometry is not a panacea for all difficult samples. Matrix isolation spectra are susceptible to various matrix effects that can shift absorption bands by a few reciprocal centimters. Extreme care must be taken in band assignment of unknown compounds, which usually involves the collection of spectra of pure compounds with assignment being accomplished by direct comparison of spectra. The matrix isolation experiment is also rather time consuming; it generally takes several hours to cool down the sample window and deposit the matrix. Another hour maybe spent signal averaging the spectrum, especially if a poor detector such as DTGS is used. It is worth noting that time may be saved in the matrix isolation experiment if a pulse deposition technique is employed. Such a technique was developed by Rochkind for the quantitative analysis of matrix-isolated mixtures by dispersive infrared spectrometry [28]. This technique is much faster than conventional slow continuous deposition and produces excellent matrices. To the knowledge of these authors, this technique has not been employed for analytical MI/FT–IR.

An interesting development in the field of rotationally cooled samples for infrared spectrometry has been reported by Colson [29–31]. Rather than isolate molecules in a solid matrix, Colson chose to rotationally cool the sample by supersonic jet expansion. This technique can, perhaps

←——————————————————————————————

Fig. 15.18. Spectra from eluates shown in Fig. 15.15 on three different surfaces. Key: N, Naphthalene; DHN, dihydronaphthalene; DMN, dimethylnaphthalane; 1, 1-methylnaphthalene; 2, 2-methylnaphthalene; G, spurious noise peak. (Reproduced from [26], by permission of the American Chemical Society; copyright © 1981.)

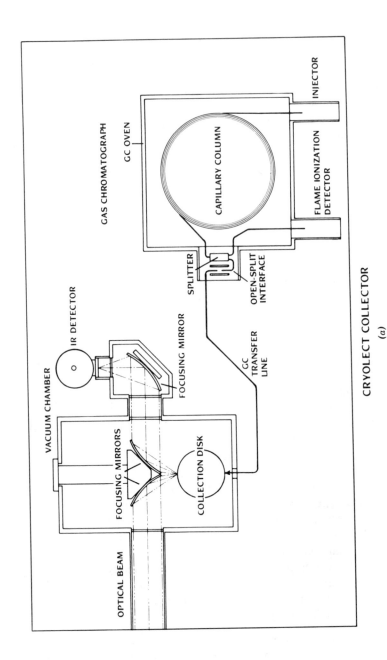

Fig. 15.19. (*a*) Optical layout of commercial GC/MI–FT–IR interface. (*b*) Photograph of cryostat optics. (Courtesy Mattson Instruments, Inc.)

512

Fig. 15.19 (*continued*)

somewhat crudely, be regarded as a form of vapor phase matrix isolation. Under certain conditions, the vibrational spectra of molecules seeded into a supersonic expansion are essentially identical with the spectra obtained in conventional low-temperature solids prepared under ordinary matrix isolation conditions [32]. To date, rotationally cooled ammonia, 112–116 K [29], methylene chloride, 150 K [30], and benzene, 80 K [31], have been studied. This is an area that should receive increased attention in the near future.

III. SEMICONDUCTOR MEASUREMENTS

The semiconductor industry is extremely interested in the quality assurance of semiconductor materials, primarily the purity of silicon. The most common contaminants in silicon ingots prepared by drawing from a melt, the Czochlarski method, are oxygen and carbon. The silicon is kept in its molten state in a quartz crucible while being heated with a graphite furnace. Although the Czochlarski process is performed under a vacuum, there is always residual oxygen present that can react with the graphite to produce carbon dioxide. Silicon dioxide can be extracted from the surface of the quartz crucible and finds its way into the melt. At the melting point of silicon, the solubility of oxygen is approximately 2×10^{18} atoms/cm^3 and the solubility of carbon is about 5×10^{17} atoms/cm^3. Oxygen is most often found in semiconductor quality silicon as occupying interstitial lattice sites, which means bonds of the type Si–O–Si. Carbon

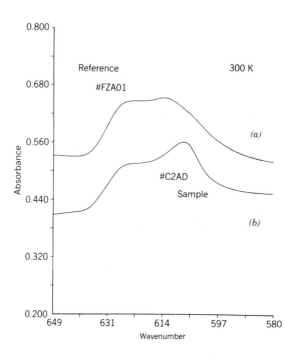

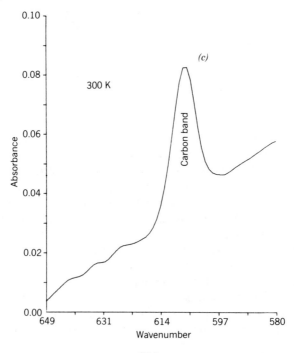

514

may substitute for silicon so that Si–C bonds are found. Oxygen adversely affects the electrical properties of the semiconductor, but carbon is electrically neutral. Using the Czochlarski method, impurity ranges of 10–20 ppm atomic (A) oxygen and 2–20 ppm A carbon can be expected. The Czochlarski method is the method of choice for the production of high-purity silicon even though other methods, such as vacuum float zone and float zone refining, may produce higher-purity silicon. The reason the Czochlarski method is chosen over the others is primarily economic. Because of the relatively high impurity levels in silicon produced by the Czochlarski method, good analytical methodology has had to be developed.

Nondestructive testing of semiconductor impurity levels can be completed by infrared spectrometry. The American Society for Testing and Materials (ASTM) [33,34] and Deutsche Industrie Nummer (DIN) [35,36] have adopted standard techniques using dispersive infrared spectrometry. The dispersive spectrometer standards call for double-beam operation with a pure silicon wafer used as the reference material. There are several shortcomings with this standard method, such as the need for the reference and sample wafers to have identical thicknesses. The method can be slow if a dispersive spectrophotometer is used. In addition, both the sample and reference wafers should be cooled for optimum results (*vide infra*, this section).

As implied above, dispersive infrared analysis of silicon wafers involves ratioing the sample to be tested against a reference that is impurity free. Early measurements taken with FT–IR spectrophotometers used identical methods [37]. A method that has greater flexibility has been developed by recording an absorbance spectrum of the reference silicon wafer versus the background spectrum (empty beam) and an absorbance spectrum of the sample wafer also versus the background. In this method, the reference wafer need not be impurity free (although it is preferable), but the impurity concentrations in the reference wafer must be known. It is not necessary for the thicknesses of the sample and reference spectra to be identical. Absorbance subtraction of the two spectra yields difference oxygen and carbon bands from which the respective impurity concentrations in the sample wafer can be calculated. Application of spectral subtraction techniques is the standard method when FT–IR spectrometry is used for silicon analysis [38–40]. Figure 15.20a shows the spectrum of

---

Fig. 15.20. (a) Reference silicon wafer and (b) a wafer that is carbon contaminated. (c) Difference spectrum of the contaminated wafer spectrum minus the reference wafer spectrum showing the substitutional carbon band at 607 cm^{-1}. (Reproduced from [38], by permission of the Society for Applied Spectroscopy; copyright © 1980.)

a silicon reference wafer at room temperature, and the spectrum of an unknown containing a higher carbon impurity is shown in Fig. 15.20b. The difference spectrum (Fig. 15.20c) clearly shows an absorption band at 607 cm^{-1} that corresponds to the absorption of substitutional carbon.

The absorption bands in the silicon wafer are due to phonon interactions and result from the combination of the longitudinal and transverse acoustic and optic modes. These modes show a strong temperature dependence that leads to an increase in band intensity with decreasing temperature. For this reason, it is preferable to measure impurity concentrations at liquid nitrogen temperature or below. At 20 K, the sensitivity of the determinations of substitutional carbon and interstitial oxygen are increased by a factor of 5 over the sensitivity for room temperature measurements [39]. A difference spectrum of a silicon wafer with an oxygen impurity is shown in Fig. 15.21. The bands at 1130 cm^{-1} indicate the various modes of phonon interaction. The modes have not been resolved as this spectrum was taken at a resolution of 6 cm^{-1} in order to avoid channel spectra. When the spectrum of a thin wafer is recorded, interference fringes will result if the two sides are parallel. Although these fringes (or channel spectral features) can be removed by a combination of experi-

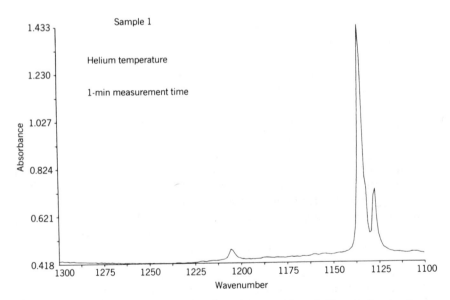

Fig. 15.21. Low-temperature spectrum of oxygen-contaminated silicon wafer showing interstitial oxygen band at 1130 cm^{-1}. (Reproduced from [39], by permission of the Society for Applied Spectroscopy; copyright © 1980.)

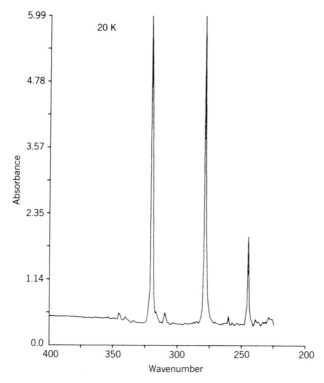

Fig. 15.22   Far-infrared low-temperature boron spectrum of boron-doped silicon wafer. (Reproduced from [39], by permission of the Society for Applied Spectroscopy; copyright © 1980.)

mental technique and software [41–44], low-resolution spectra suffice for quantification and identification of impurities.

Substitutional carbon and interstitial oxygen are not the only impurities that can be measured. Dopants such as phosphorus, boron, gallium, arsenic, indium, antimony, and aluminum are used as free carriers in the semiconductor substrates. These dopants usually have concentrations of less than 5×10^{16} atoms cm^{-3}, and their concentration can be measured directly from the absorption spectrum of the wafer. An absorbance spectrum (versus an empty cell background) of a cooled silicon wafer in which boron is present is shown in Fig. 15.22. Three major boron–silicon phonon bands can be clearly seen.

The use of cryogenic cooling to sharpen absorption features is not limited to the study of silicon or even semiconductors but can be applied to any crystalline material that is infrared transmitting. The bandpass of the spectrometer would have to be adjusted to accommodate the phonon interactions, but this is the only restriction. New applications involving the use of cryogenic cooling of materials can be anticipated in the future.

REFERENCES

1. J. G. Moehlmann, J. T. Gleaves, J. W. Hudgens, and J. D. McDonald, *J. Chem. Phys.*, **60**, 4790 (1974).
2. J. G. Moehlmann and J. D. McDonald, *J. Chem. Phys.*, **62**, 3052 (1975).
3. J. G. Moehlmann and J. D. McDonald, *J. Chem. Phys.*, **62**, 3061 (1975).
4. J. F. Durana and J. D. McDonald, *J. Chem. Phys.*, **64**, 2518 (1976).
5. H. L. Welsh, C. Cummings, and E. J. Stansbury, *J. Opt. Soc. Am.*, **41**, 712 (1951).
6. H. L. Welsh, E. J. Stansbury, J. Romanko, and T. Felshman, *J. Opt. Soc. Am.*, **45**, 338 (1955).
7. M. G. Moss, J. W. Hudgens, and J. D. McDonald, *J. Chem. Phys.*, **72**, 3486 (1980).
8. G. M. Stewart and J. D. McDonald, *J. Chem. Phys.*, **78**, 3907 (1983).
9. G. M. Stewart, M. D. Ensminger, T. J. Kulp, R. S. Ruoff, and J. D. McDonald, *J. Chem. Phys.*, **79**, 3190 (1983).
10. G. Stewart, R. Ruoff, T. Kulp, and J. D. McDonald, *J. Chem. Phys.*, **80**, 5353 (1984).
11. T. Kulp, R. Ruoff, G. Stewart, and J. D. McDonald, *J. Chem. Phys.*, **80**, 5359 (1984).
12. D. L. Allara, D. Teicher, and J. F. Durana, *Chem. Phys. Lett.*, **84**, 20 (1981).
13. H. E. Hallam, *Vibrational Spectroscopy of Trapped Species*, Wiley, London (1973).
14. B. Meyer, *Low Temperature Spectroscopy*, American Elsevier, New York (1971).
15. D. W. Green and G. T. Reedy, Matrix Isolation Studies with Fourier Transform Infrared in *Fourier Transform Infrared Spectroscopy: Applications to Chemical System* (J. R. Ferraro and L. J. Basile, eds.), Academic Press, New York (1978).
16. G. Mamantov, A. A. Garrison, and E. L. Wehry, *Appl. Spectrosc.*, **36**, 339 (1982).
17. E. L. Wehry, G. Mamantov, R. R. Kemmerer, H. O. Brotherton, and R. C. Stroupe, Low-temperature Fourier Transform Infrared Spectroscopy of Polynuclear Aromatic Hydrocarbons, in *Carcinogenesis, Vol. 1; Polynuclear Aromatic Hydrocarbons: Chemistry, Metabolism and Carcinogenesis* (R. I. Freudenthal and P. W. Jones, eds.), Raven Press, New York, p. 299, (1976).
18. G. Mamantov, E. L. Wehry, R. R. Kemmerer, and E. R. Hinton, *Anal. Chem.*, **49**, 86 (1977).

19. E. R. Hinton, Jr., G. Mamantov, and E. L. Wehry, *Anal. Lett.*, **12**, (A13), 1347 (1979).

20. A. A. Garrison, G. Mamanotov, and E. L. Wehry, *Appl. Spectrosc.*, **36**, 348 (1982).

21. G. Mamantov, personal communication.

22. D. M. Hembree, Ph.D. dissertation, The University of Tennessee, Knoxville, Tennessee (1980).

23. D. M. Hembree, E. R. Hinton, Jr., R. R. Kemmerer, G. Mamantov, and E. L. Wehry, *Appl. Spectrosc.*, **33**, 477 (1979).

24. G. T. Reedy, S. Bourne, and P. T. Cunningham, *Anal. Chem.*, **51**, 1535 (1979).

25. S. Bourne, G. T. Reedy, and P. T. Cunningham, *J. Chromatogr. Sci.*, **17**, 460 (1979).

26. D. M. Hembree, A. A. Garrison, R. A. Crocombe, R. A. Yokley, E. L. Wehry, and G. Mamantov, *Anal. Chem.*, **53**, 1783 (1981).

27. Mattson Instruments, Inc., 6333 Odana Road, Madison, WI 53719.

28. M. M. Rochkind, *Anal. Chem.*, **39**, 567 (1967).

29. D. L. Snavely, S. D. Colson, and K. B. Wiberg, *J. Chem. Phys.*, **74**, 6975 (1981).

30. D. L. Snavely, K. B. Wiberg, and D. L. Colson, *Chem. Phys. Lett.*, **96**, 319 (1983).

31. D. L. Snavely, V. A. Walters, S. D. Colson, and K. B. Wiberg, *Chem. Phys. Lett.*, **103**, 423 (1984).

32. T. G. Gough, D. G. Knight, and G. Scoles, *Chem. Phys. Lett.*, **97**, 155 (1983).

33. Annual Book of ASTM Standards, F121-43.

34. Annual Book of ASTM Standards, F123-43.

35. German Standard DIN 50 438, Parts 1 and 2.

36. German Standard FNH-AQ, 22-74.

37. K. L. Kizer and M. W. Scott, Digilab FTS/IR Notes No. 15, December 1974.

38. D. G. Mead and S. R. Lowry, *Appl. Spectrosc.*, **34**, 167 (1980).

39. D. G. Mead, *Appl. Spectrosc.*, **34**, 171 (1980).

40. K. Krishnan, Digilab FTS/IR Notes No. 39, October 1981.

41. T. Hirschfeld and A. W. Mantz, *Appl. Spectrosc.*, **30**, 552 (1976).

42. F. R. S. Clark and D. J. Moffatt, *Appl. Spectrosc.*, **32**, 547 (1978).

43. A. Baghdadi and R. A. Forman, *Appl. Spectrosc.*, **35**, 473 (1981).

44. P. R. Griffiths, *Appl. Spectrosc.*, **36**, 319 (1982).

CHAPTER

16

ATOMIC EMISSION SPECTROMETRY

I. INSTRUMENTAL CONSIDERATIONS

Although the subject matter of this book is in fact Fourier transform *infrared* spectrometry, several reports of atomic spectrometry in the visible and ultraviolet regions, as well as the near infrared, using Michelson interferometers have been published. In view of the importance of atomic spectrochemical measurements to analytical chemists and the interesting ramifications of this work to the theory of multiplex spectrometry, it seems worthwhile to summarize these results in this book. It can immediately be said that there appears to be no advantage in applying Fourier transform techniques to atomic *absorption* spectrometry using hollow cathode lamp sources since only a single wavelength is usually monitored. Conversely, in atomic *emission* spectrometry, many atomic transitions can be monitored. At first glance, therefore, it seems likely that a multiplex measurement would be beneficial.

The three most commonly used instruments for obtaining spectrochemical data by emission spectrometry at this time are

1. the spectrograph, in which emitted radiation is dispersed and recorded on a photographic plate;
2. the scanning monochromator with photomultiplier tube (PMT) detection; and
3. the direct reader, or polychromator, in which a separate exit slit/ PMT combination is required for each wavelength being monitored.

The most versatile of the three instruments is the scanning monochromator with PMT detection, since the PMT has good linearity, wide dynamic range, excellent sensitivity, and an electrical output proportional to the signal. Nevertheless, data acquisition on this instrument can be very slow if either high resolution is needed or a wide spectral range must be covered. Modern image sensor technology has also been applied to atomic emission spectrometry, and both vidicons and photodiodes have

520

been used as detectors. Neither of these devices has a particularly wide spectral range (typically only 5–50 nm), and they have neither the sensitivity nor the dynamic range of a PMT.

The spectrograph is a multiplex or, more accurately, a multichannel spectrometer capable of recording many hundreds of lines simultaneously on a photographic plate in a very short time. It is still the instrument of choice for arc/spark spectrometry. The limitations of the spectrograph are the same as the limitations of photographic plate detection, namely, nonlinear response to signal intensity (and hence a limited dynamic range) and the lack of a direct electronic readout. Because of these limitations, the spectrograph is used today primarily for rapid qualitative and semi-quantitative analysis.

The direct reader is an attempt to combine the strengths of the scanning monochromator and the spectrograph. This type of instrument is most useful for applications where between 30 and 100 wavelengths must be monitored on a routine basis. Direct readers do not have the versatility of either the scanning monochromator or the spectrograph. With a spectrograph, a decision must be made as to which wavelengths are to be observed and hence which elements may be monitored. Unexpected interferences and matrix effects are less easy to detect on a direct reader than for either of the other techniques, and changes in the analytical wavelengths are difficult to effect. An instrument technique giving both the wide coverage and multichannel advantage of a photographic plate as well as the sensitivity and dynamic range of a PMT is obviously needed, and a Fourier spectrometer is a likely candidate.

It may appear surprising that so few analytical chemists have studied the combination of a Michelson interferometer and a PMT for atomic spectrochemical analysis. One of the reasons for this less than overwhelming acceptance of the concepts of Fourier transform ultraviolet–visible spectrometry was discussed in Chapter 7 and relates to the properties of a photomultiplier tube. Unlike most detectors for the mid infrared, PMTs are shot noise (or photon noise) limited, that is, the noise increases with the square root of the signal level. If a continuous source is used, the shot noise *disadvantage* varies as $M^{-1/2}$ (where M is the number of resolution elements) and precisely offsets Fellgett's advantage, which varies as $M^{+1/2}$. Even Jacquinot's advantage is less, since instead of the SNR improvement being directly proportional to the ratio of the optical throughputs of an interferometer and a monochromator operating at the same resolution, the increased shot noise caused by the increased signal at the detector reduces this advantage only to the square root of the ratio of throughputs.

In addition to shot noise, scintillation noise (or *flicker noise*, as it is usually called in atomic spectrometry) can also seriously degrade the SNR of atomic emission measurements made using a Michelson interferometer. This type of noise is multiplicative rather than additive, so that it is particularly deleterious for Fourier spectrometry, as discussed in Chapter 7.

Despite these problems, there are still several reasons why the use of Fourier spectrometry for atomic emission spectrochemistry may be beneficial. For example, if a laser-referenced interferometer is being employed, the wavenumber precision is very high, and the calibration needed to ensure equally high accuracy is very simple. The procedure to calibrate an interferometer is as follows: an interferogram of a source emitting a single strong line, preferably at the high-wavenumber end of the spectrum, is measured. Spectra are computed with several different values of the laser wavenumber entered into the data system of the spectrometer. The spectrum must be interpolated in the region of each emission line, and the center wavenumber of each is measured by determining the center of gravity of the five or six topmost data points. (It should be remembered that the wavenumber with the largest ordinate value does not necessarily correspond to the line center, as discussed in Chapter 3.) The line centers found in this way are plotted against the value of the laser wavenumber entered into the data system for each computation. The laser wavenumber giving the accepted value for this line is then determined and used for all subsequent calculations. After this operation, the only times when recalibration of the instrument should be needed are

1. if measurements are performed using a different optical throughput (see Eq. 1.47),
2. if the position of the detector is changed, or
3. if a new laser is installed on the spectrometer or the present laser is realigned.

Any of the above conditions will probably lead to a spectral shift, but one that is still small when compared with the wavelength accuracy and precision of monochromators. With the above calibration technique, it should be possible to obtain wavenumbers for all lines (in a noise-free spectrum) accurate to about 0.005 cm^{-1}, or 1 part in 10^6–10^7.

The second advantage for a Fourier spectrometer concerns the resolution attainable. While most infrared spectrometrists might consider 8 cm^{-1} to be a rather low resolution, since it corresponds to a resolving power of 500 or less across the mid-infrared spectrum, the same resolution at 250 nm (40,000 cm^{-1}) corresponds to an order of magnitude greater resolving power. At the same wavelength, a retardation of 1 cm will yield a resolving power of 4×10^4. This resolution is routinely available in the visible region of the spectrum on most top-of-the-line commercial FT–

IR spectrometers. Although many of these instruments have a maximum mirror travel that should be enough to generate a resolution of 0.1 cm^{-1} (corresponding to a further increase in resolving power by a factor of 10), the drive on several of these instruments may often not be good enough to give this resolution in practice at ultraviolet wavelengths (see Chapters 1 and 4). Nevertheless, interferometers with cats-eye retroreflectors have been used for atomic emission spectrometry at resolutions even higher than 0.1 cm^{-1} [1–5] (see Figs. 16.1 and 16.2).

One other factor pertaining to resolution should also be raised. The Mertz method of phase correction, which is the method chosen by most manufacturers of FT–IR spectrometers, is applicable to interferograms of broadband sources and not to interferograms of sources emitting discrete lines. Horlick, who with his co-workers published more on the subject of atomic spectrometry using a Michelson interferometer than any

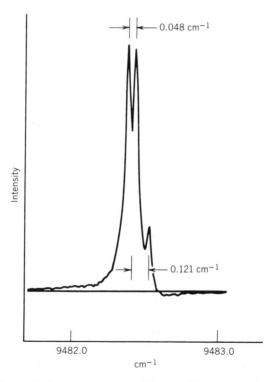

Fig. 16.1. Curium emission spectrum near 9482.4 cm^{-1} illustrating self-reversal and the isotope shift of ^{246}Cm. This is a small region of the emission spectrum of a hollow cathode lamp showing 1743 lines assignable to curium, measured using a high-resolution cats-eye interferometer and computed from an 800,000-point interferogram. (Reproduced from [3], by permission of Pergamon Press; copyright © 1976.)

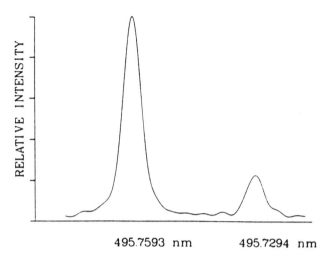

Fig. 16.2. Representative section of iron spectrum in zone 4 of an ICP measured using the Kitt Peak Observatory cats-eye interferometer. The two iron 1 lines shown here have a wavelength separation of 0.03 nm and are completely resolved. (Reproduced from [4], by permission of Pergamon Press; copyright © 1984.)

other worker in this field [6–8], recommends that a symmetrically double-sided interferogram be measured so that the magnitude spectrum can be calculated. Because the mirror is required to scan equal retardations on either side of the centerburst, the maximum resolution obtainable from a symmetrically double-sided interferogram is only about one-half of that which could be attained from a single-sided interferogram measured using the same total mirror travel.

For most measurements in the visible and near ultraviolet, Horlick's group have used a standard photomultiplier (1P28 from RCA). This PMT has its peak response at 350 nm (28,520 cm^{-1}), and its response falls to 10% of peak at 185 nm (54,050 cm^{-1}) and 650 nm (15,550 cm^{-1}). For measurements at shorter wavelengths, a "solar blind" PMT (R166 from Hamamatsu) was used, which peaks at 220 nm (45,450 cm^{-1}) and falls to 10% of its peak response at 160 nm (62,500 cm^{-1}) and 300 nm (33,330 cm^{-1}). Since PMTs are usually configured to detect dc signals, a relatively small change was made in the electronics of the dynode chain. The final three dynodes were capacitively coupled to permit ac detection and their gain was also increased [8]. For measurements in the near infrared, a silicon photodiode was used.

Both flame and inductively coupled plasma (ICP) sources have been studied. For flame measurements, the normal slot burners were found to give excessively large flicker noise [8]. Cylindrical burner heads constructed after the design of Aldous et al. [9] and Winefordner and Har-

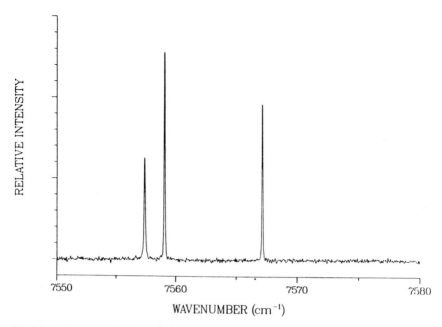

Fig. 16.3. Segment of ICP emission spectrum of argon measured at 0.04 cm^{-1} resolution using 6×10^5 data points; the lines are located at 7555.951, 7557.591, and 7565.668 cm^{-1}. (Reproduced from [5], by permission of Pergamon Press; copyright © 1985.)

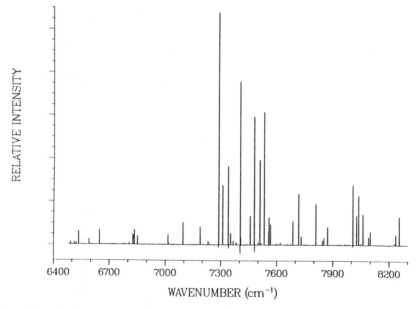

Fig. 16.4. Background spectrum from an inductively coupled argon plasma measured at high resolution using a silicon photodiode detector. (Reproduced from [5], by permission of Pergamon Press; copyright © 1985.)

525

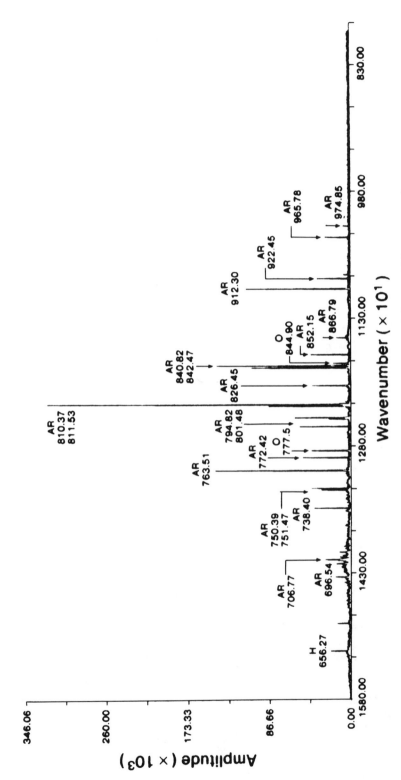

Fig. 16.5. Background spectrum from an inductively coupled plasma at higher wavenumber but lower resolution than the spectrum shown in Fig. 16.4. (Reproduced from [8], by permission of Academic Press; copyright © 1982.)

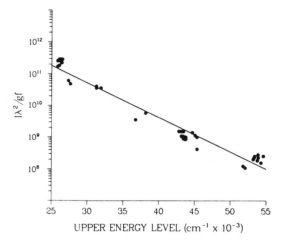

Fig. 16.6. Boltzman plot of $\log(I\lambda^2/gf)$ versus E_{upper} for iron I lines in zone 2 of an ICP. The slope of this graph is $(-1/kT)$. (Reproduced from [4], permission of Pergamon Press; copyright © 1984.)

aguchi [10] were found to give substantially reduced flicker noise. The radiation emitted from an ICP was found to be more stable, but it was less easy to collect light from this source and pass it into the interfero-meter. Another disadvantage of ICPs is the rather rich background emis-sion spectrum in the near infrared (see Figs. 16.3–16.5).

Faires et al. [4] recently reported how the cats-eye interferometer de-veloped for astronomical measurements at Kitt Peak National Observa-tory could be used for temperature determination in the ICP. Eighty-six lines of Fe(I) were identified in a particular zone of the plasma, and the intensity of each line (I) at wavelength λ was measured. Using literature values for the oscillator strength (gf) of each line, a plot of $\ln(I\lambda^2/gf)$ against the energy level of the upper excited state should be linear with a slope of $1/kT$. That this is indeed the case can be seen from Fig. 16.6. The high resolution of this measurement (0.08 cm^{-1}) permitted many lines to be resolved and interferences to be minimized. The same criteria could be applied to emphasize why Fourier spectrometry should be applicable to multielement analysis by ICP emission despite the shot noise limitation of PMT detectors.

II. SAMPLING FREQUENCY AND ALIASING

Ideally, one would like to be able to cover the entire near-infrared–vis-ible–ultraviolet spectrum in one measurement, although the spectral range of PMT detectors precludes this possibility. Nevertheless, the 1P28 pho-

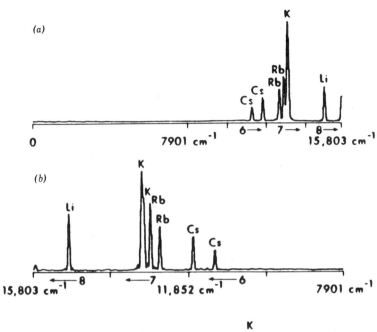

(a)

0   7901 cm^{-1} 15,803 cm^{-1}

(b)

15,803 cm^{-1} 11,852 cm^{-1} 7901 cm^{-1}

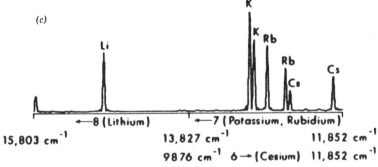

(c)

15,803 cm^{-1} 13,827 cm^{-1} 11,852 cm^{-1}

9876 cm^{-1} 6→(Cesium) 11,852 cm^{-1}

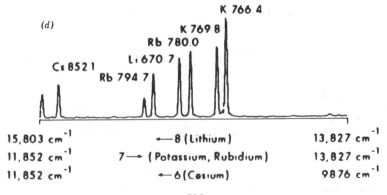

(d)

Cs 852 1

Rb 794 7

Li 670 7

Rb 780.0

K 769 8

K 766.4

15,803 cm^{-1} ←8 (Lithium)   13,827 cm^{-1}

11,852 cm^{-1} 7→ (Potassium, Rubidium)   13,827 cm^{-1}

11,852 cm^{-1} ←6(Cesium)   9876 cm^{-1}

tomultiplier does permit the entire visible region to be measured along with the near ultraviolet to 54,050 cm^{-1} (185 nm). Coverage of this entire region would necessitate a sampling interval of $\frac{1}{2} \times$ 185 nm, or 92.5 nm. To achieve a sampling interval this small, the frequency of the sinusoidal reference interferogram generated using a He–Ne laser would have to be doubled three times or else a much shorter wavelength (and much more expensive) reference laser would have to be installed. In view of the decreased sampling interval, the amount of memory required to store a double-sided interferogram for an 8-cm^{-1}-resolution atomic emission spectrum with a $\bar{v}_{max}$ of 63211.2 cm^{-1} is 32-k words. This may be compared with the 2-k words needed to store a single-sided interferogram yielding an 8-cm^{-1}-resolution mid-infrared spectrum with a $\bar{v}_{max}$ of 7901.4 cm^{-1}.

To get around this problem, which is of course accentuated for higher-resolution measurements, Horlick et al. [8] recommend that a longer sampling interval be employed. If, for example, a sampling interval of $\frac{1}{2} \times$ 0.6328 μm is employed, $\bar{v}_{max}$ becomes the wavenumber of the laser, that is, 15,802.8 cm^{-1}. The entire region covered by the 1P28 PMT can now be divided into four 15.802.8-cm^{-1}-wide regions, 15,802.8–31,605.6 cm^{-1} (region I), 31,605.6–47,408.4 cm^{-1} (region II), and 47,408.4–63,211.2 cm^{-1} (region III), all of which are folded back into the region between 0 and 15802.8 cm^{-1} (region 0). Note that region 0 falls almost entirely below the lowest wavenumber able to be detected with the 1P28 PMT. Therefore, a strong audio-frequency high-pass filter should be installed to eliminate all noise with frequencies between dc and $2v \times$ 15,802 Hz, where v is the velocity of the moving mirror of the Michelson interferometer.

Under these circumstances, any line from region I with a wavenumber of 15,802.8 + x would appear in region 0 with an apparent wavenumber of 15,802.8 − x. A line in region II with a true wavenumber of 31,605.6 + x would appear in region 0 at x. Similarly, a line in region III with a true wavenumber of 47,408.4 + x would be folded into region 0 with an apparent wavenumber of 15,802.8 − x.

Provided that the spectrum is fairly sparse, so that a line from one region is not folded to exactly the same wavenumber of a line from another region, this technique is a good way of reducing the data acquisition rate and memory requirements for atomic spectrometry in the visible and ultraviolet. Even longer sampling intervals can be used in the near infrared. To illustrate these concepts, the principal emission lines of the alkali metals are shown schematically in Fig. 16.7a. If folding the sodium doublet

Fig. 16.7. Flame emission spectra of Li, K, Rb, and Cs using a sampling interval of 0.6328x micrometers, where x is (a) 0.5; (b) 1; (c) 2; (d) 4. (Reproduced from [8], by permission of Academic Press; copyright © 1982.)

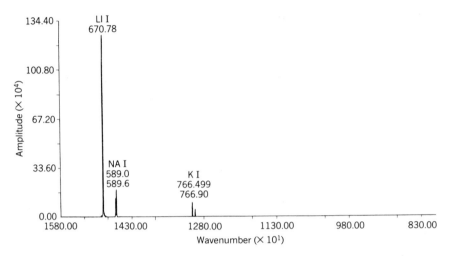

Fig. 16.8. Flame emission spectrum of Li (20 ppm), Na (35 ppm), and K (2 ppm). (Reproduced from [8], by permission of Academic Press; copyright © 1982.)

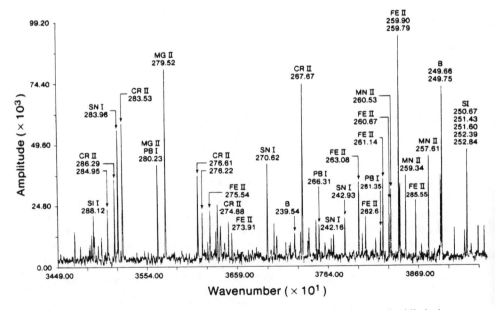

Fig. 16.9. Multielement plasma emission spectrum measured using a solar blind photomultiplier detector. (Reproduced from [8], by permission of Academic Press; copyright © 1982.)

530

at 16,978 and 16,961 cm^{-1} is to be avoided, a very short sampling interval ($\frac{1}{4}$ × 0.6328 μm if a He–Ne reference laser is used) must be employed. Doubling that sampling interval folds the sodium doublet into the window between the lithium and potassium lines, Fig. 16.7b. Quadrupling the sampling interval merely folds this spectrum into the empty region below 7901.4 cm^{-1}, Fig. 16.7c. The actual flame emission spectrum of these alkali metals excited in an air–acetylene flame and measured with a silicon photodiode is shown in Fig. 16.8. The wavenumber scale on this spectrum is correct for all lines except the sodium doublet.

For very rich spectra, such as the multielement plasma emission spectrum measured using a solar blind PMT detector shown in Fig. 16.9, careful filtering (both optical and electrical) is needed to ensure that noise is not folded into a neighboring spectral region. If noise is uniformly distributed across the spectrum, the SNR will be degraded by a factor of $2^{1/2}$ each time the spectrum is folded. In this case, the measurement time must be doubled each time the spectrum is folded in order to recover the original SNR of the unfolded spectrum. However, it should be noted that the memory of contemporary mini- or microcomputer data systems is getting larger in capacity, lower in cost, and faster than in the recent past. Therefore, one can forecast that the need to apply the folding techniques developed by Horlick is smaller now than it was even in 1982.

III. NOISE IN ATOMIC SPECTROMETRY

The behavior of noise is obviously critical to the success or failure of atomic spectrochemical measurements on a Michelson interferometer. Several workers have suggested that shot noise due to a single line will be distributed uniformly across the spectrum [3,11–16]. Hirschfeld [15] has called this the *distributive advantage* of Fourier spectrometry. In the case of an emission spectrum with many lines of approximately equal intensity, the distributive advantage should give measurements made on a Fourier spectrometer an advantage over measurements of the same spectrum sample performed on a monochromator. On the other hand, if there is a large range of line intensities, the noise due to the more intense lines distributed across the spectrum could bury weaker lines that might have been detectable using a monochromator.

The above arguments are based on the hypothesis that shot noise is distributed evenly across the spectrum, but there is a strong doubt that this is actually the case for atomic emission spectrometry. Horlick et al. [8] investigated this effect using an ICP source with 100 ppm of magnesium (which emits in the ultraviolet at about 280 nm) and 0.1 ppm of calcium

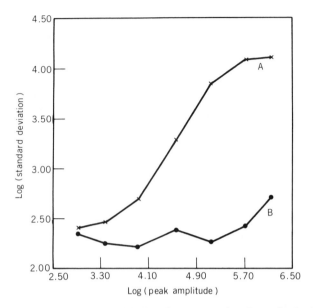

Fig. 16.10. (*a*) Standard deviation of lithium line versus Li peak amplitude. (*b*) Standard deviation of baseline in the same series of spectra. (Reproduced from [8], by permission of Academic Press; copyright © 1982.)

(which emits in the visible at around 395 nm). Under these conditions, the magnesium lines were about 10 times more intense than the calcium lines. The variance of the intensity of the calcium lines was measured both in the presence of the magnesium lines and also with these lines filtered out optically. The relative standard deviation of the measured intensities of the calcium lines did not change significantly.

The same effect has been observed for flame emission spectrometry. For example, the standard deviation of the peak height of a lithium line and that of the baseline is shown as a function of lithium concentration in Fig. 16.10. This behavior indicates that the noise is localized at the line and is *not* distributed across the baseline. An explanation for the sloping portion of this curve was offered in Chapter 7. What was not explained is the fact that the gradient of this curve becomes small again at high analyte concentrations. Detector saturation is not indicated since the analytical curve is still linear at the highest concentration. Horlick et al. [8] report this behavior to be reproducible over several experiments, and it would appear that further investigations into the effects of noise are still necessary.

IV. SUMMARY AND PROGNOSTICATION

It is apparent that there are still several experiments that must be performed to determine the source of noise in Fourier transform atomic spectrometry. Once the origins of noise in emission spectra measured using this technique are well understood, we will be able to say with more certainty whether it will be useful to many analytical chemists. Only if atomic emission spectra can be measured rapidly at high SNR is it likely that an instrument would be introduced commercially. This spectrometer would have to include a good high-resolution interferometer and a fairly large data system in view of the size of the transforms involved. It is

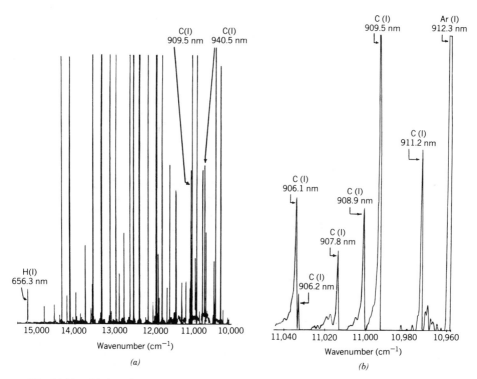

Fig. 16.11. (*a*) Atomic carbon and hydrogen emissions in the spectrum of an ICP for a CH_4 sample measured at a resolution of 2 cm^{-1}; data acquisition time, 1 sec (*b*) Expanded region of the C(I) emission for a CH_4 sample measured at a resolution of 0.25 cm^{-1}; data acquisition time, 10 sec. The asymmetry of these lines testifies to the difficulty in phase correcting interferograms of line sources using standard algorithms. (Reproduced from [22], by permission of the American Chemical Society; copyright © 1984.)

unlikely that the cost of interferometer whose mirror travels more than 5 cm will go down much in the foreseeable future. However, microcomputer-based data systems with 1 Mbyte of semiconductor memory and a Winchester-type disk are starting to become available in the $10,000 price range. Therefore, it is possible that the cost of a Fourier transform atomic emission spectrometer with an ICP source could become competitive with that of other ICP spectrometers that give lower performance.

It is also noteworthy that the rapid-scanning capability of Michelson interferometers could permit on-line elemental analysis of peaks eluting from a gas chromatograph. Brown et al. [17,18] have shown how near-infrared (NIR) emissions can be used for this purpose. Hughes and Fry [19,20] have reported the near-infrared and red emission of carbon, hydrogen, nitrogen, oxygen, sulfur, chlorine, and bromine in an argon inductively coupled plasma and have constructed a photodiode array spectrometer to measure ICP emission spectra rapidly in this region [21]. Recently, Schleisman et al. [22] reported that a commercial medium-resolution Fourier transform spectrometer could be used for the determination of nonmetals such as carbon, hydrogen, and sulfur using an ICP (see Fig. 16.11) and suggested that C–H ratios could be measured with good accuracy. It is not unreasonable to suggest that FT–NIR spectrometry could be used for the on-line elemental analysis of GC peaks. At this point, one may even be allowed to speculate that a special-purpose GC analyzer could be constructed in which the infrared spectrum and the elemental composition of each peak are measured simultaneously using two (or possibly even one) interferometers. Such an instrument must remain in the realm of speculation until further detailed investigations into the sources of noise in interferometrically measured ICP emission spectra are performed.

REFERENCES

1. S. Gerstenkorn, P. Luc, A. Perrin, and J. Chaurille, *Astron. Astrophys.*, **58**, 255 (1977).

2. P. Luc and S. Gerstenkorn, *Appl. Opt.*, **17**, 1327 (1978).

3. J. G. Conway, J. Blaise, and J. Vergés, *Spectrochim. Acta*, **31B**, 31 (1976).

4. L. M. Faires, B. A. Palmer, R. Engleman, and T. M. Niemczyk, *Spectrochim. Acta*, **39B**, 819 (1984).

5. L. M. Faires, B. A. Palmer, R. Engleman, and T. M. Niemczyk, *Spectrochim. Acta*, **40B**, 545 (1985).

6. G. Horlick and W. K. Yuen, *Anal. Chem.*, **47** 775A (1975).

7. W. K. Yuen and G. Horlick, *Anal. Chem.*, **49**, 1446 (1977).

8. G. Horlick, R. H. Hall, and W. K. Yuen, Chapter 2 in *Fourier Transform Infrared Spectroscopy: Applications to Chemical Systems*, Vol. 3 (J. R. Ferraro and L. J. Basile, eds.), Academic Press, New York, pp. 37–81 (1982).

9. K. M. Aldous, R. F. Browner, R. M. Dagnall, and T. S. West, *Anal. Chem.*, **42**, 939 (1970).

10. J. D. Winefordner and H. Haraguchi, *Appl. Spectrosc.*, **31**, 195 (1977).

11. L. Mertz, *Transformations in Optics*, Wiley, New York, (1965).

12. A. S. Filler, *J. Opt. Soc. Am.*, **63**, 589 (1973).

13. T. L. Chester, J. J. Fitzgerald, and J. D. Winefordner, *Anal. Chem.*, **48**, 779 (1976).

14. J. D. Winefordner, R. Avni, T. L. Chester, J. J. Fitzgerald, L. P. Hart, D. J. Johnson, and F. W. Plankey, *Spectrochim. Acta*, **31B**, 1 (1976).

15. T. Hirschfeld, *Appl. Spectrosc.*, **30**, 68 (1976).

16. R. F. Knacke, *Appl. Opt.*, **17**, 684 (1978).

17. R. M. Brown and R. C. Fry, *Anal. Chem.*, **53**, 532 (1981).

18. R. M. Brown, S. J. Northway, and R. C. Fry, *Anal. Chem.*, **53**, 934 (1981).

19. S. K. Hughes and R. C. Fry, *Anal. Chem.*, **53**, 1111 (1981).

20. S. K. Hughes and R. C. Fry, *Appl. Spectrosc.*, **35**, 493 (1981).

21. S. K. Hughes, R. M. Brown, and R. C. Fry, *Appl. Spectrosc.*, **35**, 396 (1981).

22. A. J. J. Schleisman, W. G. Fateley, and R. C. Fry, *J. Phys. Chem.*, **88**, 398 (1984).

SURFACE ANALYSIS

I. INTRODUCTION

The characterization of molecules adsorbed on the surface of a variety of substrates is vital for the understanding of a large number of processes and reactions. Although many instrumental techniques can provide data on the *elemental* composition of surfaces, vibrational spectrometry is unsurpassed in obtaining detailed information on the nature of adsorbed *molecules*. Before the era of Fourier spectrometry, it was difficult to obtain infrared spectra of species adsorbed at the monolayer or partial monolayer level at high SNR. In the past decade, however, good FT–IR spectra of adsorbed species measured using a variety of different sampling techniques have been reported.

The most popular sampling technique is still a conventional absorption measurement in which a powdered adsorbent such as a metal oxide or transition metal supported on alumina or silica is pressed into a very thin wafer. This technique was first developed over 25 years ago [1]. Since FT–IR spectrometry has decreased the detection limits for adsorbed molecules to far lower levels than could previously be attained, this has allowed other sampling techniques to be used for surface analysis including diffuse reflectance (DR), photoacoustic spectrometry (PAS), reflection–absorption (R–A) measurements, attenuated total reflectance (ATR), and emission spectrometry. It is not within the scope of this chapter to review the rapidly increasing body of *results* obtained using Fourier spectrometers—a separate volume would probably be needed for this task. Rather, the strengths and weaknesses of each sampling technique will be compared, and some representative results will be compared.

II. ABSORPTION SPECTROMETRY

Since the early work of Eischens and Pliskin [1], many workers have prepared thin wafers of powdered adsorbents for characterization of the adsorbate by infrared spectrometry. The books by Hair [2], Little [3],

① IR Temp Controller

Riley

② Recharge Liq. N_2
Dewar on Vac Oven

and Kiselev and Lygin [4] describe many of the early results in this field, whereas the book by Bell and Hair [5] gives a more modern perspective. Disks are usually prepared by pressing the powdered adsorbent into a wafer between 0.1 and 0.25 mm thick. Perhaps the most popular material for this purpose is γ-alumina, often with between 0.5 and 10% (w/w) of a transition metal deposited on the surface. Substantially greater pressure is required to prepare a wafer of a supported metal catalyst than is needed to prepare a wafer of the γ-alumina reference [6]. Even for very thin alumina wafers, the sample is essentially opaque below 1000 cm^{-1} because of the very strong Al–O stretching band. Above 1100 cm^{-1}, the transmittance increases to greater than 50% for very thin samples and then decreases at higher wavenumbers due both to scattering and the broad, intense band absorbing around 3400 cm^{-1} due to surface hydroxyl groups. Haaland [6] reported the transmittance of a 0.2-mm-thick wafer of γ-alumina, which had been treated for several hours *in vacuo* at 350°C and then briefly taken to 500°C to eliminate as many of the hydroxyl groups as possible, can be as low as 0.1% at 3000 cm^{-1}. Therefore, most workers interested in the spectral region between 4000 and 2800 cm^{-1} use thinner (and even more fragile) disks of adsorbent.

Although good data on samples of this kind can be obtained below 2500 cm^{-1} using an FT–IR spectrometer with a DTGS detector, spectra in the region around 3000 cm^{-1} have a poor SNR, especially if the amount of adsorbate is low. Much better sensitivity can be attained with a narrow-range MCT detector. It is also necessary to employ the greatest possible optical throughput; sometimes the optics of commercial spectrometers must be modified to achieve this end. For example, in Chapter 5 it was noted that the standard Nicolet 7199 spectrometer does not allow the full optical throughput allowed by Eq. 1.41 to be met for measurements made at 1 cm^{-1} resolution or poorer. If this instrument is to be used for absorption spectrometry of thin alumina disks, far superior results are obtained if the source and detector optics are replaced by optional optics allowing higher optical throughput [6]. (In this case, if a background spectrum is to be measured with no wafer present in the beam, a screen or reference beam attenuator must be placed in the beam to avoid saturation of the detector and overloading of the ADC.)

Many cells designed for this type of measurement allow the capability for treating the sample away from the infrared beam and subsequently moving it into the beam. A typical cell of this type is shown in Fig. 17.1. The adsorbent wafer can be heated to a temperature of 500°C and evacuated to 10^{-5} torr, treated with the adsorbate, and then moved into the infrared beam. In cells of this type, the sample cannot be maintained at the temperature of the furnace during the infrared measurement. Also,

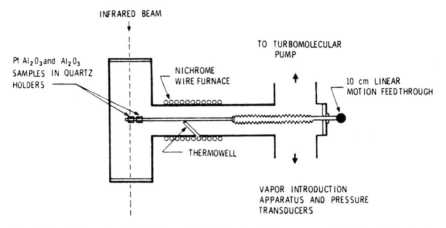

Fig. 17.1. Schematic of cell designed by Haaland to hold compressed pellets of adsorbents for pretreatment at elevated temperature and then, by moving the wafer into the infrared beam, to measure the transmittance spectrum of the disk at ambient temperature. (Reproduced from [6], by permission of North Holland Publishing Company; copyright © 1981.)

the pathlength for gaseous molecules is so great that it would be essentially impossible to study reactions that require the reactants to be present at high pressure while the spectrum is being measured. Using a cell of this design, it can be difficult or impossible to investigate many commercially important heterogeneous catalytic reactions under the exact conditions at which they occur.

A simple cell permitting relatively high pressures (up to 2.4 MPa) and temperatures (up to 200°C) with a very short gaseous absorption path (2.4 mm) has been described by Hicks et al. [7] (see Fig. 17.2), and many more industrially significant reactions could be studied with such a cell. Edwards and Schroder [8] have described a cell that may be taken to slightly higher pressures. A simple flow cell permitting *in situ* measurements of heterogeneous catalytic reactions at temperatures up to 450°C was described by King [9], see Fig. 17.3. The key to this cell is the use of graphite as the gasket material used to seal the windows. It is interesting to note that the NaCl window used in this cell showed no observable erosion even when steam at 450°C was passed through the cell.

Many reports of adsorption and heterogeneous catalysis have appeared in the literature, and several of the results obtained with FT–IR spectrometers have been summarized by Angell [10]. A few representative results will be described here to illustrate the types of systems that can be studied and spectra that can be obtained.

Ngyen and Sheppard [11] have studied the adsorption of allyl benzene on ZnO. Zinc oxide has been proposed to be a base catalyst with the O^{2-} anions leading to the dissociation of C–H bonds of hydrocarbons with pK values of less than 36. They found that allyl benzene may be dissociatively chemisorbed on ZnO, yielding an adsorbed phenylallyl species, $[C_6H_5CHCHCH_2]^-$. When allyl benzene is admitted to a ZnO catalyst

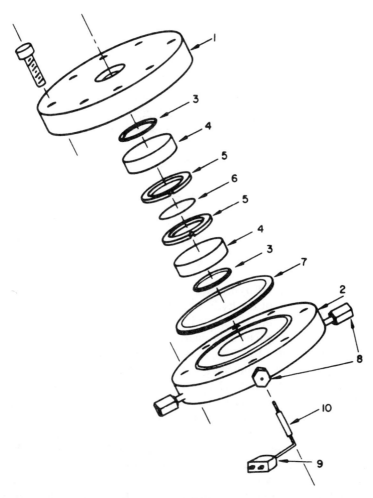

Fig. 17.2. Exploded view of reactor cell designed for *in situ* infrared studies of catalytic reactions by Hicks et al. (1) top flange; (2) bottom flange; (3) Kalrez O-ring; (4) CaF$_2$ window; (5) sample holder; (6) catalyst disk; (7) copper gasket; (8) Swagelok fitting; (9) sheathed thermocouple; (10) sleeve attached to thermocouple sheath. (Reproduced from [7], by permission of Academic Press, Inc.; copyright © 1981.)

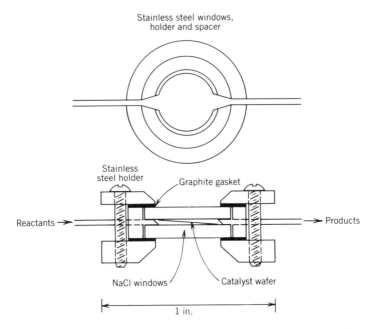

Fig. 17.3. Simple flow reactor designed by King; catalyst wafers are 1 cm in diameter and contain 10–30 mg of catalyst. (Reproduced from [9], by permission of the Society for Applied Spectroscopy; copyright © 1980.)

containing preadsorbed hydrogen, adsorption of the phenylallyl species causes displacement of the hydrogen from the surface. To arrive at these conclusions, Ngyen and Sheppard needed infrared spectra in the C–H stretching region as well as the fingerprint region. Their high-wavenumber spectra are reproduced in Fig. 17.4, illustrating the potential of Fourier spectrometry for measuring spectra at high SNR and adequate resolution (4 cm^{-1} in this case) in this relatively energy starved region.

A more difficult measurement from an instrumental viewpoint was reported by Haaland [12]. Cyclohexane and benzene were each adsorbed on 0.2-mm-thick pressed wafers of 10% platinum on alumina. The spectrum measured immediately after treatment of the catalyst with cyclohexane and subsequent evacuation is shown in Fig. 17.5a. The same sample 18 hr after evacuation is shown in Fig. 17.5b. These data have been interpreted to show that the strong bands in these spectra are assignable to benzene formed by the rapid dehydrogenation of cyclohexane. This conjecture was confirmed by adsorbing benzene on the same adsorbent, see Fig. 17.5c.

An enormous number of spectrometric studies of CO adsorbed on sup-

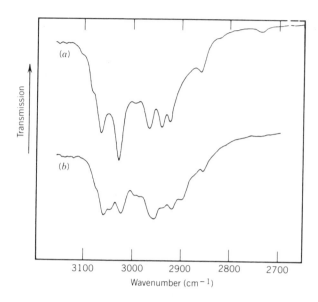

Fig. 17.4. Infrared spectra of adsorbed species of allylbenzene adsorbed on ZnO in the region from 3200 to 2700 cm⁻¹ (a) without hydrogen preadsorption and after evacuation; (b) on ZnO containing preadsorbed hydrogen and before evacuation. These spectra show the high SNR and information content of spectra of adsorbed species in the C–H stretching region of the spectrum, even when measured using a DTGS detector. (Reproduced from [11], by permission of Academic Press; copyright © 1981.)

ported metal catalysts has been reported over the past 30 years, and, of course, many of the more recent studies have been made using Fourier spectrometers. Most of these concern measurements made under equilibrium conditions, but recently one paper appeared that showed that good kinetic data can be obtained for heterogeneous reactions with half-lives of only a few seconds. Cant and Bell [13] have studied the hydrogenation of carbon monoxide over ruthenium using transient response isotopic tracing and *in situ* infrared spectrometry. In an outstanding paper, they show that chemisorbed CO exchanges rapidly with gas phase CO and that under reaction conditions the two species are in equilibrium. A similar conclusion was reached for chemisorbed H atoms and gas phase H_2. The dissociation of molecularly adsorbed CO to form atomic carbon and oxygen was found to require vacant surface sites and to be reversible. It was also found that the rate at which nonoxygenated carbon undergoes hydrogenation is faster than the rate at which adsorbed CO is hydrogenated, supporting the hypothesis that nonoxygenated carbon is an intermediate in CO hydrogenation. Sample spectra are shown in Fig. 17.6. The spectra

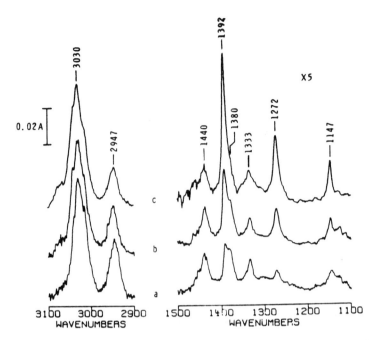

Fig. 17.5. Difference spectra of adsorbates on Pt/Al$_2$O$_3$ after addition and evacuation. (*a*) C$_6$H$_{12}$ immediately after evacuation; (*b*) C$_6$H$_{12}$ 18 hr after evacuation; (*c*) C$_6$H$_6$ immediately after evacuation from catalyst containing carbon residues. Samples were contained in the cell shown in Fig. 17.1. (Reproduced from [12], by permission of North Holland Publishing Company; copyright © 1981.)

of C^{18}O at 323 K are also shown for 8 and 150 sec after the introduction of gaseous CO.

The spectra were measured with a slow interferometer scan speed, but there is really no reason why several spectra per second cannot be measured and stored (e.g., by using GC/FT–IR software) when the transmittance of the catalyst wafer is high (as is usually the case at 2000 cm^{-1}) and the absorbance of adsorbate bands is high (as it was in this work). This paper is one of the first to illustrate the capability of FT–IR spectrometry for studying the kinetics of heterogeneous reactions. Doubtless we can expect many more kinetic studies of rapid reactions to be reported in the future.

These measurements nicely illustrate the usual tradeoff between SNR and measurement time. If the signal is large, spectra can be measured in a short time; however, if the transmittance of the pellet is low in the spectral region of interest and the adsorbate bands are weak, even an optimized FT–IR spectrometer may be hard-pressed to yield good data

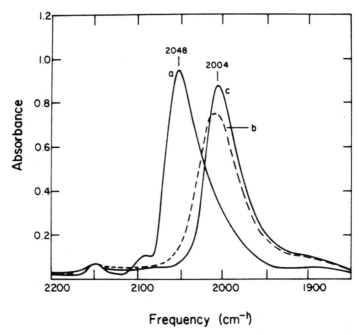

Fig. 17.6. Spectra showing replacement of $C^{16}O$ adsorbed on Ru/SiO$_2$ by $C^{18}O$ at 323 K in the presence of hydrogen. (*a*) $C^{16}O$ alone; (*b*) 8 sec after switching to $C^{18}O$; (*c*) 150 sec after switching to $C^{18}O$. (Reproduced from [13], by permission of Academic Press, Inc.; copyright © 1982.)

from alumina disks in reasonable measurement times (say, 30 min). From a sampling viewpoint, the use of pressed disks of adsorbents also leaves several properties to be desired. For example, the adsorbate may take several seconds to diffuse into the pellet; similarly, any reaction products will take several seconds to effuse out of the disk. Additionally, when the powdered adsorbent is pressed into a pellet under pressures of up to 3000 kg cm^{-2}, the surface area per gram of adsorbent available to adsorbate molecules is almost certainly decreased below the surface area per gram of the loose powder.

In spite of its drawbacks, an enormous quantity of useful information has been obtained by the use of pressed disks, and undoubtedly this technique will continue to be used extensively in the future. If other sampling techniques are to be used as alternative methods to investigate surface chemistry, they must overcome the disadvantages of the pressed pellet technique. The most promising techniques that could achieve these ends are diffuse reflectance and photoacoustic spectrometry, and these techniques will be described in the next two sections.

III. DIFFUSE REFLECTANCE SPECTROMETRY

In theory, diffuse reflectance (DR) spectrometry has several advantages over absorption spectrometry for characterizing species adsorbed on powdered adsorbents. First, loose powders can be sampled without the need for pressing them into pellets. In DR spectrometry, a scattering sample is mandatory. Thus spectra do not show the energy loss at high wavenumbers caused by scattering, which is characteristic of transmittance spectra of pressed pellets of silica or alumina. Second, it has been shown [14] that for bulk samples mixed with a nonabsorbing powder such as KBr, the percentage absorption is greater in DR spectra than if the same amount of sample was pressed into a transparent disk of the same diameter. Furthermore, if the sample is deposited as a thin layer on the surface of the powdered KCl, the band intensity may be up to four times greater than if the components were mixed in bulk. It is probable that the enhancement of band intensity of the absorption bands of surface species is caused by an effective multipassing of the beam due to external and (possibly) internal reflection before it emerges from the sample.

Another significant potential advantage of DR spectrometry for the study of adsorbates on supported metal and pure oxide adsorbents is the fact that the area of contact of the adsorbent with a gas is much greater than if the sample were pressed into a transparent pellet. Diffusion of reactants into, and effusion of products from the sample is very much faster than it would be for pressed pellets. In some cell designs [15–17], the gas can be drawn through the sample so that equilibrium may be established in a matter of seconds throughout the sample.

The disadvantages of DR spectrometry include the fact that, as for absorption spectrometry, the diffuse reflectance may be very nearly zero in spectral regions where the absorptivity of the substrate is high. A warning might also be issued at this point: when the absorptivity of the substrate becomes *very* high, the specular reflectance from the surface of individual particles may become quite large. In these regions, it can appear that the reflectance of the sample is high, that is, that absorption by the adsorbent is low. However, since this radiation does not penetrate the sample, it is very unlikely that absorption bands due to the adsorbate will be observed in these spectral regions.

Another potential disadvantage of DR spectrometry is the lack of reproducibility of band intensities because of variation of the scattering coefficient each time the sample is loaded into the cell. Under these circumstances, complete subtraction of bands due to the adsorbent is impossible. The problem may be alleviated by drawing an inert gas through the adsorbent prior to treatment by the adsorbate [16]. After switching from the inert gas to an equal flow of the adsorbate gas, the background spectrum remains reproducible enough to achieve a fairly good subtraction.

Some of the optics and cell designs for DR spectrometry of adsorbed species were described in Chapter 5, Section III.c. Sample temperatures as high as 600°C have been reached using these cells. Gases can be drawn through the powdered adsorbent, permitting exceptionally efficient activation of supported metal catalysts using hydrogen at high temperature and rapid equilibration between the adsorbate and adsorbent after activation.

Spectra of CO adsorbed on rhodium supported on alumina as a function of the pressure of CO are shown in Fig. 17.7 to illustrate the sensitivity of the technique. Spectra of samples held at 400°C may be measured routinely. It should be noted, however, that when an adsorbent of fairly high emissivity is contained in the cell at high temperature, so much dc flux is incident on the detector that it becomes saturated, and its response drops below the level of the value for ambient temperature samples (see Fig. 17.8).

Although DR spectrometry is a relatively unproven tool for the study of adsorbed species, we believe that it will ultimately become the single

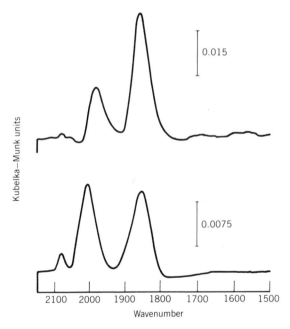

Fig. 17.7. Diffuse reflectance spectra of CO on 1% (lower trace) and 5% (upper trace) Rh–Al$_2$O$_3$ under a pressure of 0.015 torr of CO. (Reproduced from [16], by permission of Academic Press, Inc.; copyright © 1984.)

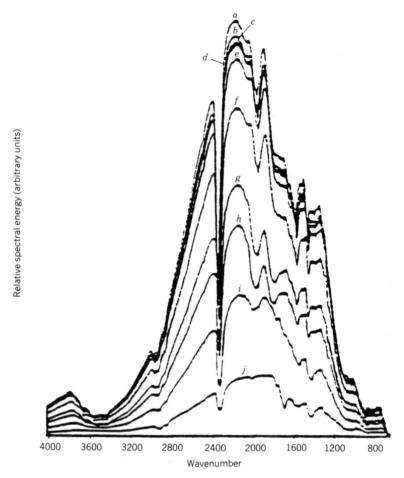

Fig. 17.8. Single-beam DR spectra of 5% Rh–Al₂O₃ measured under vacuum using an MCT detector, with the sample at the following temperatures: (*a*) room temperature; (*b*) 50°C; (*c*) 65°C; (*d*) 75°C; (*e*) 115°C; (*f*) 185°C; (*g*) 250°C; (*h*) 290°C; (*i*) 400°C; (*j*) 600°C. (Reproduced from [16], by permission of Academic Press, Inc.; copyright © 1984.)

most important technique for studying the chemistry of processes taking place on the surface of powdered catalysts of high surface area.

IV. PHOTOTHERMAL SPECTROMETRY

It is sometimes thought that photoacoustic spectrometry is a technique for studying surfaces whereas diffuse reflectance spectrometry is more characteristic of the bulk sample. However, as we saw in Chapter 9, the extent to which this statement is true depends on the relative values of the optical absorption depth μ_β and the thermal diffusion depth μ_s. Since μ_β and μ_s are usually 5 μm or greater, except where the absorption coefficient β is very large, there is apparently little justification for the claim that PAS is a surface-sensitive technique. Even so, there are still several significant differences between PA and DR spectrometry. For example, the intensity of the bands due to the adsorbent are much weaker in PA spectra than in DR spectra of the same sample. This may lead to the possibility of measuring strong adsorbate bands in the region of intense adsorbent bands; however, no measurements to verify this conjecture have yet been reported.

One important disadvantage of PA spectrometry is that the SNR of PA spectra is usually considerably worse than that of absorption or DR spectra measured in the same time using an MCT detector. A second disadvantage that is particularly relevant for the study of surfaces is that the photoacoustic effect is at least two orders of magnitude greater for gases than for solid samples. Thus, if there is an equilibrium between the adsorbate in the gaseous phase and on the surface, the PA signal from the

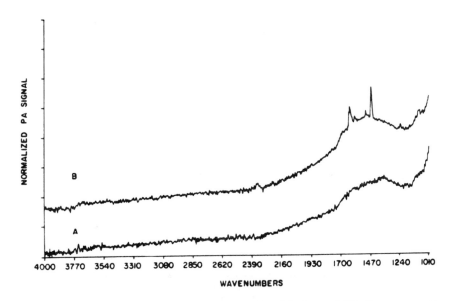

Fig. 17.9. Photoacoustic spectra of calcined γ-Al$_2$O$_3$ (A) before and (B) after exposure to 18 torr of pyridine at 150°C with subsequent purging; absorption features due to pyridine adsorbed on Lewis and Brönsted acid sites can be observed between 1650 and 1400 cm^{-1}. (Reproduced from [18], by permission of the American Chemical Society; copyright © 1982.)

gaseous molecules will dominate the spectrum unless the equilibrium constant is overwhelmingly in favor of adsorption. The adsorbate gas must therefore be flushed from the cell by a nonabsorbing gas such as N_2 or He. Obviously, if the adsorbate is not strongly bound, it will be lost during this process. Photoacoustic cells cannot be evacuated if microphonic detection is needed (which is, of course, usually the case) since no signal could be measured. They also cannot be significantly pressurized because the gas pressure must be approximately equal on both sides of the diaphragm. Finally, the sample temperature cannot be increased much above

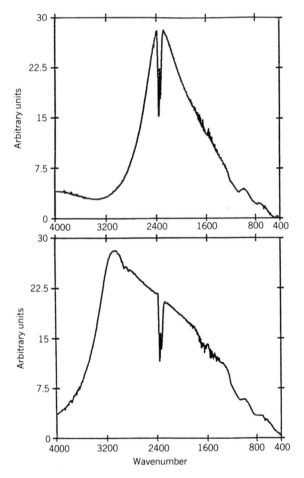

Fig. 17.10. Single-beam PA spectra measured by Yeboah [19] in a resonant cell to show the difference between the response of the cell before modification (lower spectrum) and after modification to permit gases to swept through the powdered sample (upper spectrum).

ambient since the SNR of the spectrum becomes poorer at elevated temperatures.

Riseman et al. [18] have demonstrated that FT–IR photoacoustic spectrometry can be used effectively for studying certain types of chemisorption. Typical spectra measured by these workers for pyridine adsorbed on γ-alumina are shown in Fig. 17.9. Yeboah [19] studied the same system and warned against the possibility of mistaking PA signals from vapor phase species with the signals from the corresponding adsorbed species. Both Riseman et al. and Yeboah suggest that the cell should be flushed with an inert gas for at least half an hour to eliminate any nonadsorbed molecules. Yeboah redesigned the Digilab PA cell to permit gases to be drawn through the adsorbent to ensure that nonadsorbed molecules are swept rapidly from the cell. The response characteristics of the cell changed dramatically after it was redesigned, with the resonant frequency being shifted to a much lower wavenumber, see Fig. 17.10. In spite of the reduced response at high wavenumbers, good spectra below 2200 cm^{-1} could be obtained using this cell.

The principal advantage of PAS for the characterization of surface species arises from the lack of complete saturation in the region of very strong adsorbent bands. This potential advantage is usually more than offset by the rather poor sensitivity of PAS relative to conventional absorption and diffuse reflectance measurements. It is probable that, in practice, PAS will only find an application in surface chemistry to the measurement of relatively high surface coverages of molecules on strongly absorbing adsorbents (i.e., those where μ_β is substantially less than 1 μm).

V. REFLECTION–ABSORPTION SPECTROMETRY

The two possible approaches to reflection–absorption (R–A) spectrometry of species adsorbed on the surface of flat, specularly reflecting metals were discussed in Chapter 5. In one approach, the angle of incidence, α, may be relatively small (45° ± 20°) so that the absorbance of a single molecular layer will be very low but may be enhanced by multiple reflection. The value of the absorbance is not strongly dependent on the nature of the polarization of the incident beam. In the other approach, a rather high incidence angle (preferably greater than 80°) is selected. In this case, only a single reflection is usually achieved. The absorbance of the beam polarized perpendicular to the surface is enhanced over the value calculated solely on geometric grounds whereas the absorbance of a beam polarized parallel to the surface is effectively zero.

Ishitani et al. [20] have published a useful summary of the range of samples that can readily be studied by R–A spectrometry. For organic molecules on the surface of metal substrates, about 10 molecular layers were required to yield a recognizable spectrum in this work, where α was equal to about 70° and a TGS detector was employed.

Lower detection limits were reported by Allara and Swalen [21] for calcium arachidate adsorbed on silver measured using an MCT detector. According to this article, the spot size was 2 mm and the incidence angle was 86°, but these parameters in combination do not appear possible in view of the arguments made in Chapter 5. Identifiable spectra at the monolayer level were measured, see Fig. 17.11. The dashed line is the reflection spectrum of a hypothetical 10-layer sample calculated using optical constants obtained from the transmission spectra of a KBr disk of calcium arachidate. The authors were able to interpret these spectra in terms of the orientation of the alkyl chain and the packing of the layers in multilayer samples. Subsequently, Rabolt et al. [22] published a more

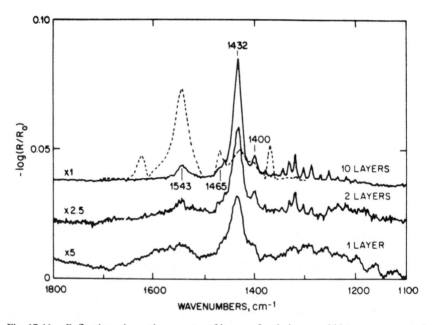

Fig. 17.11. Reflection–absorption spectra of layers of cadmium arachidate on evaporated silver in the region between 1100 and 1800 cm^{-1}. Solid lines indicate measured spectra of 1, 2, and 10 layers of the adsorbate, and the broken line is a spectrum calculated for a 10-monolayer structure obtained by using bulk optical constants corrected for the increased packing density of monolayer films. (Reproduced from [21], by permission of the American Chemical Society; copyright © 1982.)

detailed report on the same molecular system. The data of Allara and Swalen [21] and Rabolt et al. [22] certainly illustrate the potential of R–A spectrometry for surface characterization, but the measurement of the spectra of monolayers cannot yet be claimed to be routine.

For all R–A measurements made using a Fourier spectrometer, the problem of digitization noise can determine the ultimate detection limits

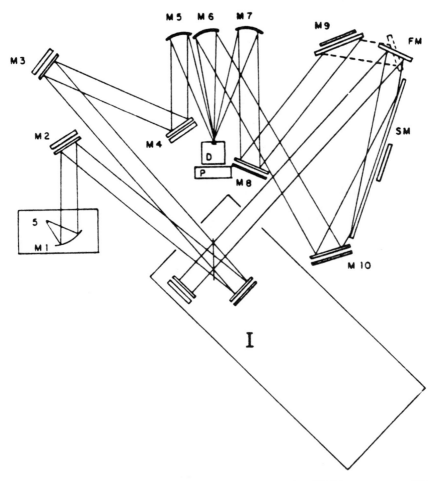

Fig. 17.12. Optical layout of a dual-beam reflection–absorption FT–IR spectrometer. M1, M5, M6, and M7 are off-axis paraboloidal mirrors and M2, M3, M4, M8, M9, FM, SM, and M10 are plane mirrors. Spectra of surface layers on SM are measured by subtracting the background spectrum measured from M7, M8, and M9 from the spectrum measured from M6, M9, M10, and SM. (Reproduced from [23], by permission of the Society for Applied Spectroscopy; copyright © 1981.)

for any sample. In Chapter 8, several ways of reducing the possibility of digitization noise by suppressing the amplitude of the centerburst of the interferogram were discussed, and both optical subtraction and polarization modulation techniques have recently been applied to surface analysis.

The first, reported by Kemeny and Griffiths [23], used the rather cumbersome optical design for dual-beam spectrometry shown in Fig. 17.12.

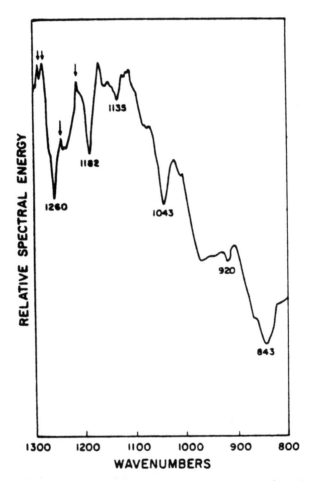

Fig. 17.13. Reflection–absorption spectrum of an approximately 15-Å-thick layer of Epon 828 epoxy resin on an aluminum surface measured using the optical arrangement shown in Fig. 17.12. Wavenumbers of bands due to the surface species are indicated and features caused by uncompensated water vapor lines are marked with arrows. (Reproduced from [23], by permission of the Society for Applied Spectroscopy; copyright © 1981.)

The sample was deposited on mirror S1, which was mounted to provide an incidence angle of about 80°. The transmitted beam from the interferometer could either be reflected from this mirror and passed to the detector or passed along an alternative path by flipping the mirror FM. The reflected beam from the interferometer is passed along a third path, with all three paths being the same length to minimize atmospheric interference. Spectra measured using the two alternate paths for the transmitted beam were subtracted to yield the spectrum of the species on the surface of S1. Fairly good sensitivity was achieved using this dual-beam arrangement, as illustrated in Fig. 17.13.

An alternative and probably better method of eliminating the centerburst utilizes the polarization modulation approach, in which a photoelastic modulator rapidly alternates a beam between parallel and perpendicular polarization. The results of early attempts to characterize surface species by this technique were discussed in Chapter 8. The data of Dowrey and Marcott [24] and Golden et al. [25,26] indicate the potential superiority of this technique over R–A spectrometry with unmodulated parallel polarized radiation with a Fourier spectrometer or polarization modulation with a grating spectrometer. Ultimately, we believe that the use of photoelastic modulators will become routine for R–A spectrometry.

One *caveat* concerning the interpretation of R–A spectra may be given. Distortions in band shapes that depend on the thickness and refractive index of the sample and the angle of incidence of the beam can occur. Allara et al. [27] have found good agreement between calculated and observed spectra of thin polymer films on metal surfaces and warn that care has to be taken when band shifts and splittings in R–A spectra are interpreted in terms of chemical effects.

VI.   ATTENUATED TOTAL REFLECTANCE

Because the depth of penetration of the evanescent wave from an internal reflection element (IRE) is small, it is often relatively simple to measure the attenuated total reflectance (ATR) spectrum of species deposited on the surface of an IRE at a thickness of more than 10 nm. In fact, the use of ATR spectrometry is a rather standard technique for identifying solutes in volatile solvents since the solvent can be readily evaporated, leaving the solute as a thin layer on the surface of the IRE. To a certain extent, ATR can always be regarded as a technique for the study of surface films. However, since the depth of penetration of the evanescent wave into the surface layer is typically approximately equal to one tenth of the wavelength of the radiation, measurement of the ATR spectrum of a few molecular layers can still present a problem, especially if a grating spectrometer is being used. The increased sensitivity of a Fourier spectrometer will alleviate the problem, and a few relevant examples of surface measurements made by FT–IR/ATR will be given below.

Ohnishi et al. [28] have reported ATR spectra of mono- and multilayer films of cadmium arachidate deposited on glass microscope slides. Spectra were measured using an early Digilab FTS-14 equipped with a nichrome wire source and a TGS detector, and it may be noted that an order of magnitude improvement in SNR might be expected if a Globar source and MCT detector had been used. The angle of incidence at the KRS-5/air interface was 45°, and nine reflections occurred within the plate. The spectrum of the bulk sample and spectra of mono- and multi-layer films and the glass substrate are shown in Fig. 17.14. The monolayer spectrum was interpreted to indicate that the carboxylate group interacts strongly with the glass surface. The appearance of a band progression between 1350 and 1200 cm^{-1} in the spectrum of the multilayer films, which is associated with CH_2 wagging and twisting modes, suggested that the hydrocarbon chains are in a planar zig-zag configuration in a way similar to that of the crystalline state, with the molecules oriented along a direction perpendicular to the glass surface.

Gidaly and Kellner [29] have reported the ATR spectra of residual thin (1–10-nm) films of oils and aluminum foil and showed that the coatings could be analyzed quantitatively by ATR spectrometry without the use of calibration curves. Spectra of the same sample were measured using a grating spectrometer (Perkin-Elmer 180) and a Fourier spectrometer (Bruker IFS-113 with a TGS detector), see Figs. 17.15 and 17.16. They show that even though no bands can be observed in the region below 2000 cm^{-1} with the grating spectrometer, a good spectrum is obtained easily with the FT–IR spectrometer.

This paper also addressed the influence of polarization on ATR spectra measured on a Fourier spectrometer. As we saw in Chapter 4, the efficiency of a Ge–KBr beamsplitter is about the same for parallel and perpendicular polarized radiation, but this is far from being the case for grating monochromators. A thin layer of silicone rubber deposited on a KRS-5 IRE should yield different intensities to radiation of each polarization. However, Gidaly and Kellner found that band intensities were insensitive to the polarization of the beam and attributed their findings to the microscopic surface roughness of a KRS-5 IRE. When the same sample was deposited on a germanium IRE, which is much harder and less easily scratched than KRS-5, the band intensities were indeed found to be polarization dependent. It is interesting to note that similar argu-

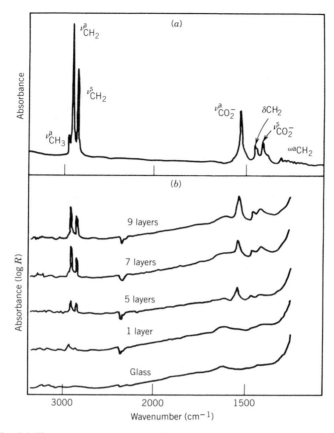

Fig. 17.14. (*a*) Absorption spectrum of bulk cadmium arachidate prepared as a KBr disk. (*b*) ATR spectra of cadmium arachidate for mono- and multilayer films of cadmium arachidate on glass. Note the differences between these spectra and the R-A spectra of cadmium arachidate adsorbed on silver shown in Fig. 17.11. (Reproduced from [28], by permission of the American Chemical Society; copyright © 1978.)

ments have been used to explain the lack of polarization dependence of R–A spectra of monolayers of CO on Pd sheets [30].

The evanescent wave from most IREs generally cannot penetrate thick layers of metals deposited on the surface of the crystal. If the metallic layer is thin enough, however, some penetration to the outer surface can take place. Jakobsen [31] has coated a germanium IRE with a 4-nm layer of iron and measured the ATR spectrum of stearic acid adsorbed on the surface, see Fig. 17.17. Brinda-Konopik et al. [32] have taken this work one step further and used a similar crystal with a thin layer of vapor-

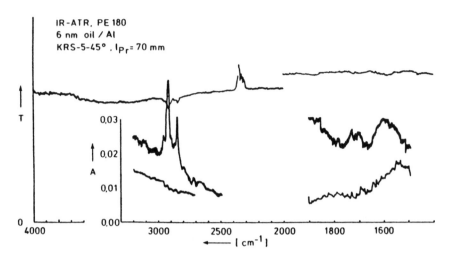

Fig. 17.15. ATR spectra of 6-nm-thick layer of oil on an aluminum substrate measured using a Perkin-Elmer Model 180 grating spectrophotometer. The upper trace is the transmittance spectrum while the lower traces show absorbance spectra of the sample and background. (Reproduced from [29], by permission of Springer Verlag, Vienna; copyright © 1981.)

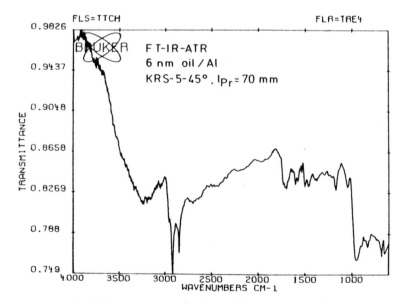

Fig. 17.16. ATR spectrum of the same sample used for the spectrum shown in Fig. 17.15 but measured on a Bruker IFS 113 vacuum FT–IR spectrometer; the increased information content of this spectrum is readily apparent. (Reproduced from [29], by permission of Springer Verlag, Vienna; copyright © 1981.)

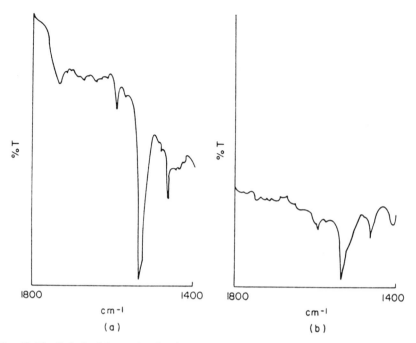

Fig. 17.17. Polarized internal reflection spectra of film adsorbed on an iron-coated Ge crystal from a 1% solution of stearic acid in hexadecane: (*a*) radiation polarized parallel to the face of the IRE; (*b*) radiation polarized perpendicular to the face of the IRE. (Reproduced from [3], by permission of Academic Press; copyright © 1979.)

deposited iron as the working electrode of an electrochemical cell. Spectra were measured at several potentials during the anodic oxidation of the iron in a concentrated solution of alkali, and the existence of different species during the first anodic process was demonstrated. Undoubtedly, this paper will be a forerunner of many future papers describing electrochemical processes studies using FT-IR/ATR spectrometry.

Another extraordinarily elegant example of the use of ATR spectrometry for studying the kinetics of the deposition of monolayers of proteins from whole blood flowing over a germanium IRE coated with a thin polymer film should be mentioned in the context of surface analysis. This experiment is discussed in much greater detail in Chapter 14.

VII. EMISSION SPECTROMETRY

In most measurements of the spectra of adsorbed species, cooled detectors are used to enhance the sensitivity. Any time there is a difference in temperature between the sample and the detector, it is possible to measure the emission spectrum of the sample. For adsorbates on substrates of low emissivity, especially metal surfaces, emission spectrometry might be thought to be a useful sampling technique. Few results have been reported in the literature, however, largely because of the difficulties in measuring emission spectra of high SNR. Problems are also encountered in the interpretation of emission spectra because of the effect of temperature gradients near the surface of the sample [33].

Spectra of oxides of copper [34] and aluminum [35] on the heated metal substrates measured using TGS [34] and MCT [35] detectors have been reported by Kember et al. The thickness of the oxide layer in these samples was not reported but was estimated to be several molecular layers. In order to improve the sensitivity, these authors substituted an MCT detector for the original TGS detector and reduced the noise level in the

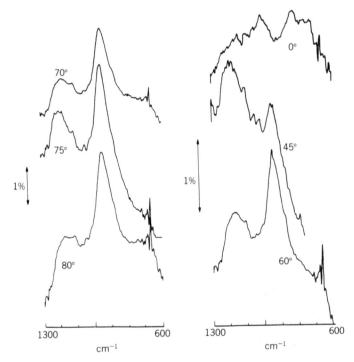

Fig. 17.18. Infrared emission spectra of the oxide coating on commercial aluminum foil as a function of the angle of the normal to the sample with respect to the optical axis of the interferometer. The 1% emittance scales for each set of three spectra are shown. Sloping backgrounds are caused by temperature mismatches between the sample and the blackbody reference. (Reproduced from [35], by permission of Pergamon Press; copyright © 1979.)

spectra by at least an order of magnitude. More significant from a sampling viewpoint was the effect on band intensity of varying the angle of the normal to the sample with respect to the optical axis of the interferometer. The emittance of two bands assignable to aluminum oxide on the surface of aluminum foil increased as this angle was increased (see Fig. 17.18), in agreement with the theoretical predictions of Greenler [36].

Most spectra have been reported for samples held well above ambient temperature to optimize the flux from the surface layer. However, the only requirement for measuring infrared emission spectra is that there is a temperature difference between the sample and the detector. Indeed, Low and Coleman [37] and Chase [38] have shown that it is possible to measure emission spectra of samples whose temperature is below that of the detector, see Fig. 17.19.

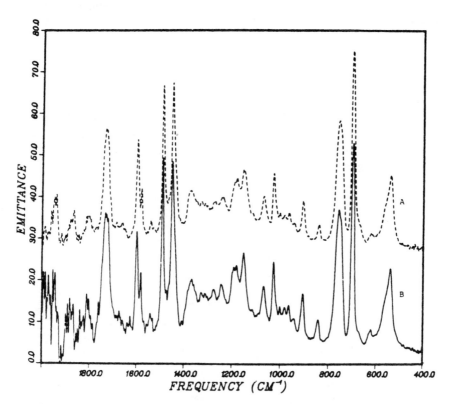

Fig. 17.19. Infrared emittance spectra of a polystyrene film held above and below the temperature of a DTGS detector. Sample temperatures were (*a*) 40°C and (*b*) 10°C. (Reproduced from [38], by permission of the Society for Applied Spectroscopy; copyright © 1981.)

In spite of these measurements, it must still be said that it is extremely difficult to measure the emission spectrum of a monolayer deposited on a metallic surface using a standard commercial FT–IR spectrometer even with a high-sensitivity MCT detector. A liquid-helium-cooled photoconductive copper-doped germanium detector has even higher sensitivity, but only at very low photon flux levels. The flux from interferometers at ambient temperature (or from instruments thermostated above ambient) is sufficient to saturate these detectors. Allara et al. [39] have shown that the combination of a cryogenically cooled interferometer and a Ge–Cu detector permits the measurement of emission spectra of surface species held at ambient temperature. Their equipment and results were discussed in Chapter 15, Section I. It is interesting to note that Allara and Teicher [40] have found that good-quality spectra can be obtained from films deposited on extremely roughened and nonspecularly reflecting surfaces, a result not found for high-angle R–A spectrometry [30].

VIII.   DETECTION SPECTROMETRY

Whereas in emission spectrometry the sample is the infrared source, it is equally possible that the sample can be incorporated as part of the detector. Bailey et al. [41,42] demonstrated remarkable sensitivity for the detection of adsorbed species by measuring the absorption of radiation on a thin sample cooled to liquid helium temperature. A very simple optical arrangement was used in this work, see Fig. 17.20. A thin (100-

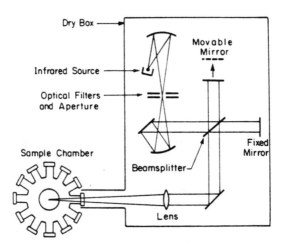

Fig. 17.20.   Optical layout used by Bailey et al. [41,42] for detection spectrometry. (Reproduced from [5], by permission of the American Chemical Society; copyright © 1980.)

nm) layer of a metal was deposited on a doped germanium bolometer, and gases could be adsorbed on the metal. Radiation absorbed by the sample is detected as an increase in the temperature of the bolometer, and absorptances as low as 0.001% could be measured.

Spectra of CO adsorbed on Ni and Cu films are shown in Fig. 17.21. Carbon monoxide adsorption at 77 K resulted in the appearance of only a single band at 2097 cm^{-1} for Ni and at 2109 cm^{-1} for Cu. Further exposure of the sample at 1.5 K resulted in the observation of a second band, due to physisorbed CO, at 2143 cm^{-1} for both metals.

Although this device is of limited application for industrial processes, it should prove very useful for the study of molecules adsorbed on single-crystal surfaces in view of its very high sensitivity.

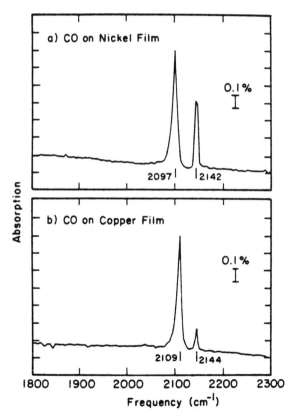

Fig. 17.21. Spectra of CO adsorbed on (*a*) nickel and (*b*) copper films measured by detection spectrometry; note that the sample is at liquid helium temperature. (Reproduced from [5], by permission of the American Chemical Society; copyright © 1980.)

IX. SUMMARY

The characterization of adsorbed species still represents a challenge to infrared spectrometrists, but it is apparent that with so many sampling techniques available, FT–IR spectrometry will be used by surface chemists to a greater extent in the coming years.

REFERENCES

1. R. P. Eischens and W. A. Pliskin, *Adv. Catal.*, **10**, 1 (1958).
2. M. L. Hair, *Infrared Spectroscopy in Surface Chemistry*, Marcel Dekker, New York (1967).
3. L. H. Little, *Infrared Spectra of Adsorbed Species*, Academic Press, New York (1966).
4. A. V. Kiselev and V. I. Lygin, *Infrared Spectra of Surface Compounds*, Wiley, New York (1975).
5. A. T. Bell and M. L. Hair, "Vibrational Spectroscopies for Adsorbed Species," A.C.S. Symp. Ser. 137 (1980).
6. D. M. Haaland, *Surface Sci.*, **102**, 405 (1981).
7. R. F. Hicks, C. S. Kellner, B. J. Savatsky, W. C. Hecker, and A. T. Bell, *J. Catal.*, **71**, 216 (1981).
8. J. F. Edwards and G. L. Schrader, *Appl. Spectrosc.*, **35**, 559 (1981).
9. S. T. King, *Appl. Spectrosc.*, **34**, 632 (1980).
10. C. L. Angell, Chapter 1 in *Fourier Transform Infrared Spectroscopy: Techniques using Fourier Transform Interferometry*, Vol. 3 (J. R. Ferraro and L. J. Basile, eds.), Academic Press, New York (1982).
11. T. T. Ngyen and N. Sheppard, *J. Catal.*, **67**, 402 (1981).
12. D. M. Haaland, *Surface Sci.*, **111**, 555 (1981).
13. N. W. Cant and A. T. Bell, *J. Catal.*, **73**, 257 (1982).
14. M. P. Fuller and P. R. Griffiths, *Appl. Spectrosc.*, **34**, 533 (1980).
15. M. P. Fuller, Ph.D. dissertation, Ohio University, Athens, OH (1980).
16. I. M. Hamadeh, D. King, and P. R. Griffiths, *J. Catal.*, **88**, 264 (1984).
17. Cells with this property are now available from Harrick Scientific, Ossining, New York and Spectra-Tech, Stamford, Connecticut.
18. S. M. Riseman, F. E. Massoth, G. M. Dhar, and E. M. Eyring, *J. Phys. Chem.*, **86**, 1760 (1982).
19. S. A. Yeboah, Ph.D. dissertation, Ohio University, Athens, OH (1982).
20. A. Ishitani, H. Ishida, F. Soeda, and Y. Magasawa, *Anal. Chem.*, **54**, 682 (1982).
21. D. L. Allara and J. D. Swalen, *J. Phys. Chem.*, **86**, 2700 (1982).
22. J. F. Rabolt, F. C. Burns, N. E. Schlotter, and J. D. Swalen, *J. Chem. Phys.*, **78**, 946 (1983).
23. G. J. Kemeny and P. R. Griffiths, *Appl. Spectrosc.*, **35**, 128 (1981).

24. A. E. Dowrey and C. Marcott, *Appl. Spectrosc.*, **36**, 414 (1982).

25. W. G. Golden and D. D. Saperstein, *J. Electron Spec.*, **30**, 43 (1983).

26. W. G. Golden, D. D. Saperstein, M. W. Severson, and J. Overend, *J. Phys. Chem.*, **88**, 572 (1984).

27. D. L. Allara, A. Baca, and C. A. Pryde, *Macromolecules*, **11**, 1215 (1978).

28. T. Ohnishi, A. Ishitani, H. Ishida, N. Yamomoto, and H. Tsubomura, *J. Phys. Chem.*, **82**, 1989 (1978).

29. G. Gidaly and R. Kellner, *Mikrochim. Acta*, **1981I**, 131 (1981).

30. J. B. L. Harkness, Ph.D. dissertation, Massachusetts Institute of Technology, Cambridge, MA (1970).

31. R. J. Jakobsen, Chapter 5 in *Fourier Transform Infrared Spectroscopy: Applications to Chemical Systems*, Vol. 2 (J. R. Ferraro and L. J. Basile, eds.), Academic Press, New York (1979).

32. N. Brinda-Konopik, H. Neugebauer, G. Gidaly, and G. Naner, *Mikrochim. Acta*, Suppl. 9, 329 (1981).

33. P. R. Griffiths, *Appl. Spectrosc.*, **26**, 73 (1972).

34. D. Kember and N. Sheppard, *Appl. Spectrosc.*, **29**, 496 (1975).

35. D. Kember, D. H. Chenery, N. Sheppard, and J. Fell, *Spectrochim. Acta*, **35A**, 455 (1979).

36. R. G. Greenler, *Surface Sci.*, **69**, 647 (1977).

37. M. J. D. Low and I. Coleman, *Appl. Opt.*, **5**, 1453 (1966).

38. D. B. Chase, *Appl. Spectrosc.*, **35**, 77 (1981).

39. D. L. Allara, D. Teicher, and J. F. Durana, *Chem. Phys. Lett.*, **84**, 20 (1981).

40. D. L. Allara and D. Teicher, Paper TA8 at 1981 Int. Conf. on Fourier Spectrosc., Columbia, SC (1981).

41. R. B. Bailey and P. L. Richards, Lawrence Berkeley Laboratory Report LBL-7639, Berkeley, CA (1979).

42. R. B. Bailey, T. Iri, and P. L. Richards, *Surface Sci.*, **100**, 626 (1980).

CHAPTER

18

GC/FT–IR SPECTROMETRY

I. INTRODUCTION

The separation of volatile components of a mixture by gas chromatography (GC) is an important quantitative technique for the analytical chemist. However, the gas chromatograph provides little more qualitative information than the number of components in the mixture, some guide to the polarity and molecular weight of each eluate from the retention time, and possibly the oxidation states or presence of certain atoms in each eluate if the appropriate GC detectors are employed. Hybrid instruments must be developed to identify the chromatographic eluates. One such "hyphenated" technique is gas chromatography/Fourier transform–infrared (GC/FT–IR) spectrometry. Gas chromatographic eluates, or "peaks", are usually small samples ranging from a few tens of micrograms to less than a nanogram and elute in a few seconds depending on the chromatographic technology. Rapid-scanning interferometers have the capability to record the entire spectrum in less than 1 sec, at low resolution. The throughput and multiplex advantages allow the spectrometrist to collect spectra of samples in submicrogram quantities without trapping them. Thus, the FT–IR spectrometer becomes a *detector* for the chromatograph. This detector is nondestructive and can provide *structural* information about GC eluates.

The potential of an FT–IR spectrometer interfaced to a gas chromatograph was realized soon after the commercial development of rapid-scanning interferometers. Low and Freeman [1] used a Block Engineering Model 200 interferometer for the first GC/FT–IR experiment. The GC eluates were passed from a packed column into a small flowthrough cell of about 0.6 mL volume (4 mm diameter × 5 cm long). Synthetic mixtures were prepared and run isothermally into an infrared cell that was heated to 150°C and placed between the source and interferometer. Data were collected and stored on a two-channel analog tape recorder. Interferograms were collected in both a flow-through mode and when the eluates were trapped. Spectra of signal-averaged scans were collected, since signal averaging could be accomplished even with flowing eluates, and the

564

background was removed by spectral subtraction. By signal averaging Low and Freeman were able to record high signal-to-noise ratio (SNR) spectra of strongly absorbing samples, such as methyl acetate. Figure 18.1 shows spectra of this compound in quantities of approximately 1 mg to 5 μg. As the sample size decreased, the number of scans required to achieve an adequate SNR had to be increased. Partially resolved GC peaks were investigated, as shown in Fig. 18.2. In this figure, spectra were collected of fractions on opposite sides of an unresolved GC peak and comparisons to spectra of the pure compounds made. In subsequent work, Low sought to remove the background by optical subtraction rather than by computation [2]. He employed a dual-beam optical layout, Fig. 18.3. This apparatus produced high-SNR spectra of the strong absorber methyl stearate in quantities as low as 10 μg. Although this system pro-

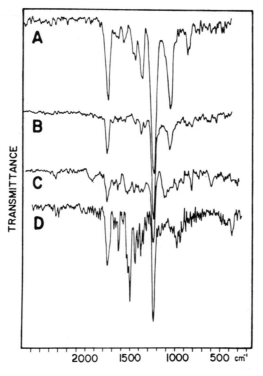

Fig. 18.1. GC/FT–IR spectra of small quantities of methyl acetate. Sample sizes range from approximately 1 mg in A to 5 μg in D. Sample A was collected as the eluate was flowing, with 30 scans coadded. Samples B, C, and D were trapped and 100 scans were coadded for B and 300 scans were coadded for each of C and D. (Reproduced from [1], by permission of the American Chemical Society, copyright © 1967.)

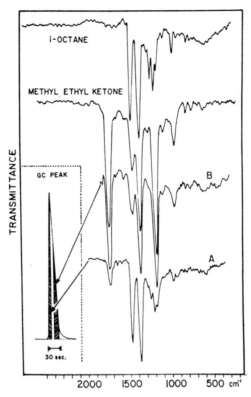

Fig. 18.2. GC/FT–IR spectra from beginning and end of partially resolved GC peak. 100 scans were coadded to produce spectra A and B. Spectra of iso-octane and methyl ethyl ketone are shown for reference. (Reproduced from [1], by permission of the American Chemical Society, copyright © 1967.)

duced relatively high quality spectra, it was found to be rather impractical at the time because of the lack of automation.

In the early 1970s, Digilab produced a commercial GC/FT–IR accessory [3]. This system consisted of a cell that was about 1.5-mL in volume (6 mm diameter × 5 cm) and fit inside the spectrometer sample compartment. Detection limits using this accessory were demonstrated at 4.3 μg of methylene chloride. Mixtures were separated by a packed column gas chromatograph, and the eluates were passed through a thermal conductivity GC detector. When the signal detected at the thermal conductivity detector (TCD) exceeded a specified threshold, signal averaging was initiated after a preset delay to permit the sample to travel from the TCD to the cell. Signal averaging was continued until the GC signal returned below the threshold level. If greater sensitivity was required, the

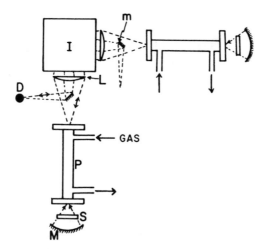

Fig. 18.3. Schematic of dual-beam interferometer for GC/FT–IR. (Reproduced from [2], by permission of Marcel Dekker, Inc.; copyright © 1968.)

cell could be sealed and the eluate trapped so that signal averaging could be performed on the trapped eluate for an extended period of time.

Digilab modified their GC–IR accessory to collect spectra or interferograms for packed-column chromatography in various ways besides the flow-through and trapped modes described above [4,5]. For example, it was possible to stop the carrier gas flow through the column while signal averaging of a trapped eluate was performed. Three different techniques could be employed while the effluent was flowing through the cell (or *light pipe*). Data could be coadded during the time that the GC detector signal exceeded a certain threshold or collected continuously as the effluent passed through the light pipe, with a specified number of scans being coadded into separate files. The on-the-fly method collected a file (or interferogram) for each individual scan. These latter two methods essentially retained all the GC/FT–IR data and were useful for examining peaks carefully, including those that were only partially resolved. Some rather impressive spectra could be collected using the Digilab accessory. Figure 18.4 shows the spectrum of what was reported to be 10 ng of isobutyl methacrylate, an extremely strong infrared absorber. However, this quantity is believed to be the amount of material *in the light pipe* and not the quantity injected into the chromatograph. The eluate was signal averaged for 16 scans and ratioed with a 48-scan reference spectrum.

During this time, rather drastic developments took place in the field of GC/FT–IR spectrometry primarily due to the impetus of Leo V. Azarraga, a research chemist with the U.S. Environmental Protection Agency

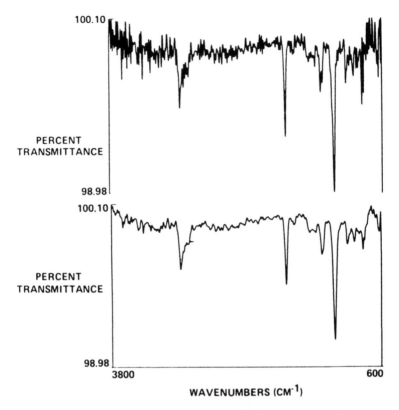

Fig. 18.4. Spectra of 10 ng isobutyl methacrylate on-the-fly, 16 scans coadded versus a 48-scan reference. The bottom spectrum has been spectrally smoothed, whereas the top spectrum is unsmoothed. (Reproduced from [5], by permission of the Society for Applied Spectroscopy; copyright © 1977.)

(EPA). Azarraga and McCall [6,7] were the first to propose and demonstrate the capabilities of flow-through GC/FT–IR spectrometry using capillary column GC. They showed that GC/FT–IR spectrometry could be performed using a 50-ft support-coated open tubular (SCOT) capillary column and a Digilab GC–IR accessory [7]. Spectra were collected either as single scans of the flowing effluent, which has come to be known as *on-the-fly* sampling, or trapped in the cell. Detection limits were found to approach 300 ng of cyclohexanone. The work prompted a more concerted effect in the area of GC/FT–IR spectrometry. The most significant advance was the development by Azarraga of a technique for making a light pipe by gold coating the inside of borosilicate glass tubes (*vide infra*, Section II.b, this chapter). Subsequently, both Digilab and Nicolet, and more recently other vendors, developed light pipes based on the design of Azarraga.

GC/FT–IR spectrometers continued to be developed commercially for packed column chromatography in the mid 1970s, especially by Digilab and Nicolet; however, it was not until the early 1980s that capillary column GC/FT–IR spectrometers became available from the instrument manufacturers. In the interim, between the early work of Azarraga [6,7], Kizer [3], and Mantz [4,5] and the development of commercial capillary GC/FT–IR systems, many technical problems needed to be solved. Of prime importance were sampling procedures, light-pipe design, optics and data collection, and processing. These aspects of GC/FT–IR spectrometry are discussed in more detail below.

II. HARDWARE

The GC/FT–IR system is composed of three main parts, the spectrometer, the gas chromatograph, and the interface. Such a system is represented schematically in Fig. 18.5. Every part of the system has evolved to produce an extremely high sensitivity instrument, and each aspect of the instrumentation will be dealt with in some detail.

a. Sampling

Care must be taken when making GC/FT–IR measurements to ensure that the light pipe has its optimum volume. For example, if the volume

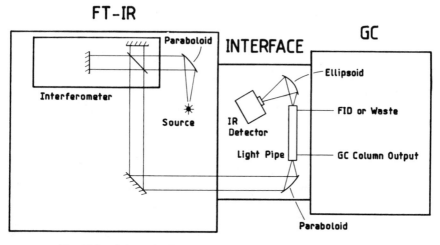

Fig. 18.5. Schematic diagram of a GC/FT–IR spectrometry system.

of the cell is too large, there is an effective dilution of the eluate in the light pipe, and both the infrared signal and the GC resolution may be degraded. If the volume is too small, the quantity of eluate in the cell at any instant is less than the optimum value. Sampling must be concerned with the relative volume of each eluate peak and the volume of the light pipe. The volume of the transfer line to the light pipe from the GC must also be made as small as possible to avoid degrading the resolution of the separated eluates.

Griffiths has established criteria for the sampling of GC peaks by FT–IR spectrometry [8]. These criteria are equally applicable to packed and capillary column gas chromatography. In many of the earlier GC/FT–IR systems, and even in some commercially produced today, data storage could be quite limited. If every interferogram measured during a long chromatographic run were saved, a very large amount of computer storage space would be required. Storage space can be saved if interferograms corresponding to peaks are the only data stored. As mentioned in the introduction, data collection is sometimes triggered by a separate GC detector (either an in-line nondestructive or a destructive detector in conjunction with an outlet splitter) that records a signal above a certain threshold. There are two major flaws with this sampling criterion; insufficient numbers of interferograms may be collected for small peaks, and if the GC baseline drifts, the trigger fails altogether. Griffiths proposed a method whereby the sampling would be based on the peak itself, and coadded interferograms would produce the maximum achievable SNR. His argument was based on the assumption of a Gaussian shape for a GC peak:

$$C = C_{max} \exp\left[\frac{-(V - V_R)^2}{V_e^2}\right] \qquad (18.1)$$

where C is the concentration of the eluate, C_{max} is the maximum sample concentration, V is the volume of carrier gas eluted after injection, V_R is the peak retention volume, and V_e is the volume required to change C_{max} to C_{max}/e. Data were calculated for spectral SNR using Eq. 18.1, assuming the noise is proportional to $t^{-1/2}$ where t is the data acquisition time. The results of these calculations are shown in Fig. 18.6, where x is equal to $(V - V_R)/V_e$. To produce Fig. 18.6, data were calculated for coadding interferograms around V_R. The factor x is directly proportional to t. The maximum SNR occurs when $x = 0.95$, that is, when all the data are coadded between $\pm 0.95(V - V_R)/V_e$. The triggering points are not particularly critical: to maintain a SNR of at least 90% of the optimum value, x may take any value from 0.5 to 1.7. Perhaps the most convenient places to start and finish signal averaging for each peak are the inflection points

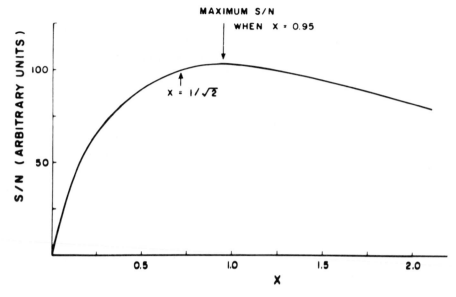

Fig. 18.6. Variation of the SNR of GC/FT–IR spectra where the concentration of the eluates in the cell exhibits a Gaussian profile, when data are collected between $\pm xV/V_e$ of the maximum concentration (V_R). When data are collected between the inflection points of the Gaussian profile ($x = 2^{-1/2}$) the SNR is 97% of the maximum value. (Reproduced from [8], by permission of the Society for Applied Spectroscopy; copyright © 1977.)

of each peak. For peaks with a Gaussian shape, these points occur at $x = \pm 0.707$ and are therefore well within the 90% SNR range. Such a sample trigger is adequate for some partially resolved peaks. Although this method provides a feasible triggering procedure, these authors know of no commercial system that employs this method. Of course, the triggering problem is obviated if storage space is not a problem and data are collected continuously either as single interferograms or as small groups (two to six scans) of coadded interferograms.

Another aspect of sampling is the optimum volume of the light pipe, also discussed by Griffiths in the same paper [8]. The light-pipe volume must be controlled to regulate the concentration in the cell so that the maximum signal can be attained. Various parameters were defined for the calculation of the optimum cell volume:

Q_{total} = total quantity of sample injected
$C_{\max}$ = maximum sample concentration in the GC peak
$V_{1/2}$ = full width at half-height of the GC peak (in milliliters)
F_c = carrier gas flow rate (milliliters per second)

V_{cell} = volume of infrared cell

V_L = volume of the transfer line between the GC detector and the entrance to the cell if a nondestructive in-line detector is used. If an effluent splitter is used, V_L is the difference between the volume from the junction to the entrance to the cell and the volume from the junction to the GC detector.

The simplest treatment arises when no effluent splitter is used and the GC peaks are assumed to be triangular. In the case where $V_{cell} \ll V_{1/2}$, the GC peak concentration profile will be similar to that of a standard GC detector, assuming no broadening occurs in the eluate transfer lines. The cell concentration profile will show a time delay from the GC detector of

$$t'_d = \frac{V_L}{F_c} \tag{18.2}$$

The quantity of eluate in the cell, Q_{cell}, is a fraction of Q_{total}:

$$Q_{cell} = Q_{total} \frac{V_{cell}}{V_{1/2}} \tag{18.3}$$

which implies that as Q_{total} decreases, so does Q_{cell}. Conversely, if $V_{cell} \gg V_{1/2}$, the entire sample, or at least a large portion, can be in the cell at the time, and the concentration is necessarily reduced by

$$C_{cell} = \frac{Q_{total}}{V_{cell}} \tag{18.4}$$

A family of curves can be calculated for Eqs. 18.3 and 18.4. Figure 18.7a illustrates the concentration profiles of eluate peaks as they travel through cells of different volumes with respect to $V_{1/2}$. The concentrations are normalized to C_{max}. In Fig. 18.7b, a similar set of profiles is shown with the quantity of the eluate normalized to Q_{total}. As can be seen, as the cell increases in volume, Q_{cell}/Q_{total} reaches a limiting value of 1.00. The data from the curves in Figs. 18.7a, b can be consolidated into a separate curve, see Fig. 18.7c. This figure shows that the data from the two previous figures intersect when $V_{cell} = V_{1/2}$. It can be shown that this is the best compromise between concentration profile and quantity of eluate in the cell. In fact, most commercial instruments use light-pipe cells that have been empirically optimized for the samples produced by the chromatograph. The major emphasis today is with 0.25- and 0.32-mm-i.d. fused silica wall-coated open tubular (WCOT) capillary columns, and the light pipes have volumes between about 0.1 and 0.4 mL. This range is consistent with the criteria established by Griffiths.

Because the light pipes are rarely of the same internal diameter as the GC column, the light pipe may be considered to have a *dead volume*. When the eluate enters the cell, it must fill the volume and so its effective flow rate through the cell is lower than it is through the column, assuming the internal diameter of the light pipe is greater than that of the column. Griffiths [8] showed that the delay time must be increased by a new

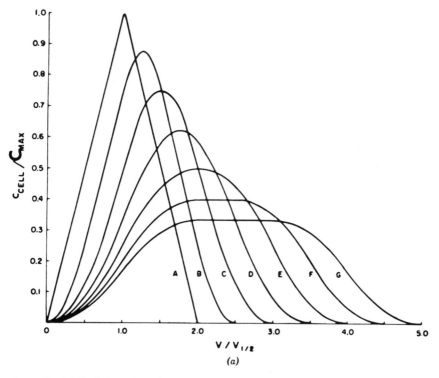

Fig. 18.7. (*a*) Variation of C_{cell}/C_{max}, the ratio of the concentration in the cell to the maximum sample concentration of the GC detector, with $V_{cell}/V_{1/2}$, the ratio of the volume of carrier gas that has passed after the sample was first detected to the FWHH of the GC peak. A triangular GC peak profile is assumed and $t'_d = 0$. The values of $V_{cell}/V_{1/2}$ are: A, 0.00; B, 0.50; C, 1.00; D, 1.50; E, 2.00; F, 2.50; G, 3.00. (*b*) Variation of Q_{cell}/Q_{total} (the fraction of eluate in the cell) versus $V_{cell}/V_{1/2}$, for (*a*) or triangular peak and with $t'_d = 0$. $V_{cell}/V_{1/2}$ has the following values: A, 0.50; B, 1.00; C, 1.50; D, 2.00; E, 2.50. (*c*) Consolidation of data from (*a*) and (*b*), the variation of Q_{cell}/Q_{total} and C_{cell}/C_{max} versus $V_{cell}/V_{1/2}$. The two curves intersect when $V_{cell}/V_{1/2} = 1$. (Reproduced from [8], by the permission of the Society for Applied Spectroscopy; copyright © 1977.)

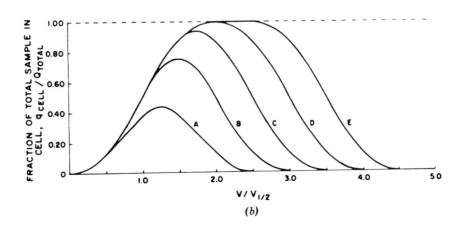

(b)

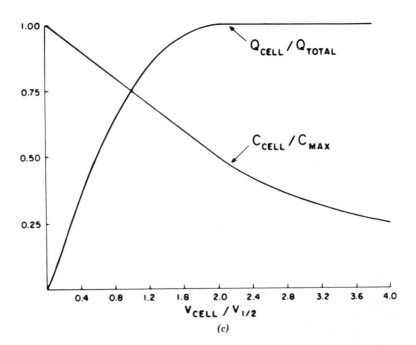

(c)

Fig. 18.7. (*continued*)

duration

$$t_d'' = \frac{V_{cell}}{2F_c} \qquad (18.5)$$

so that the total delay is $t_d' + t_d''$. For baseline-separated peaks that are of equal volume, the overlap of eluates in the light pipe is not significant. Of course, as the chromatographic resolution decreases, the overlap becomes more severe.

An alternative method of increasing the delay time is often employed, that is, supplying a make-up carrier gas directly to the light pipe. This increases the flow of the eluate in the cell, and it can be made to equal or very slightly exceed F_c. Although the mixing problem is largely eliminated, the concentration of the eluates in the light pipe is reduced, thereby increasing the detection limits somewhat. The problem of dead volume between the end of the column and the light pipe has been eliminated in several commercial GC/FT–IR systems by connecting the end of the column to an uncoated narrow fused silica capillary and connecting this directly into the light pipe.

b. Light Pipe

Many of the light pipes originally made for packed column GC/FT–IR spectrometers had volumes of less than 2 mL, whereas a somewhat larger volume is desirable if the optimum detection limits are to be achieved according to the criteria discussed above. This state of affairs was radically changed in 1976 through the work of Azarraga [9].

Azarraga recognized that in order to enhance the sensitivity of GC/ FT–IR systems when the volume of the cell could not exceed the half-width of the narrowest peak, it is necessary to reduce the internal diameter of the cell and increase its length. In order to permit the radiation to be transmitted through a cell of internal diameter of 3 mm or less with high efficiency while retaining chemical inertness, the inside of the cell had to be gold coated. One of Azarraga's main contributions to GC/FT–IR spectrometry was a development of a method for internally coating Pyrex tubes with gold to yield light pipes with a very high transmittance [10]. The dimensions of these first light pipes varied from about 0.7 mm i.d. × 52 cm to about 3 mm i.d. × 20 cm [11]. When these light pipes were interfaced to a capillary column gas chromatograph, Azarraga was able to produce high-quality spectra of 200 ng of methyl salicylate, see Fig. 18.8.

A method for calculating the optimum dimensions for light pipes has

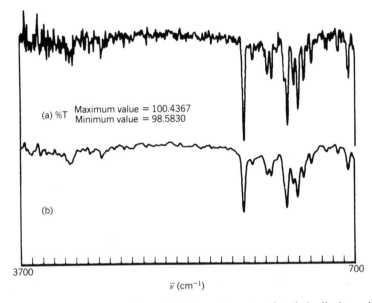

Fig. 18.8. Unsmoothed (*top*) and smoothed (*bottom*) spectra of methyl salicylate collected on the fly. (Reproduced from [9], by permission of the author.)

been proposed by Griffiths [8]. He took several factors into account, including the reflection losses at each window on the ends of a light pipe, the internal reflection losses, and vignetting loss due to the entrance aperture of the light pipe, d_p (in millimeters), being smaller than the diameter of the infrared beam at the focus, d_f. A schematic drawing of a light-pipe effluent cell is shown in Fig. 18.9. Griffiths had at his disposal light pipes of 3 mm i.d. only, and from these he was able to determine that the reflection loss in the cell is a function of the length of the cell and the aperture. It was found that the transmittance of a 50-cm, 3-mm-i.d. light pipe was $32 \pm 2\%$. The transmittance of light pipes was then formulated to be

$$T_p = 0.9 \times \left(\frac{d_p}{d_f}\right)^2 \times 0.32^{(3L/50d_p)} \qquad (18.6)$$

The factor 0.9 is the reflection loss of two alkali–halide windows. The quotient d_p/d_f is the vignetting loss of the light pipe. It was assumed that d_f had a value of 3 mm, so for $d_p \geq 3$ mm, there was no vignetting loss, and the quotient is set to unity. The last term in Eq. 18.6 represents the reflection loss in the cell, which is normalized to the transmission of the

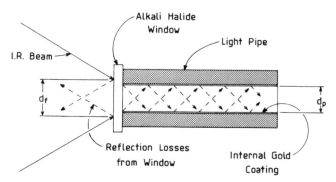

Fig. 18.9. Schematic drawing of a light-pipe effluent cell.

3-mm-i.d. × 50-cm light pipe. This equation was used to calculate a family of curves for the transmission of light pipes versus volume, Fig. 18.10. Lines passing through the origin represent the transmittance versus the length, whereas the intersecting lines represent the transmittance versus aperture. The break in the curves occurs when $d_f = d_p$.

The SNR of a GC/FT–IR spectrum can be related to the length L of the light pipe and to the transmittance. The signal is proportional to L and noise is inversely proportional to the transmittance T_p. Figure 18.11 shows the relationship between LT_p (directly proportional to SNR in the absence of digitization noise) versus light-pipe length L for two specific volumes, 1 and 4 mL.

The larger of the two volume light pipes has a fairly broad curve, so there is a large range of length for which the SNR is high. This volume cell has a maximum transmittance of about 47%. The small light pipe has a pronounced maximum of $d_p = d_f$. The point for both curves where $T_p = 0.070$ is indicated in Fig. 18.11. This is the maximum transmission that can be used with a narrow-band (750–4000-cm^{-1}) MCT detector with a 2-in. aperture interferometer scanning at 3 mm (retardation) per second before the signal at the centerburst becomes digitization noise limited. For the 4-mL light pipe being used with this detector, it is necessary to construct the cell so that it has dimensions of about 2.5 mm i.d. × 80 cm. These dimensions reduce LT_p by a factor of about 3 below the optimum value. The theoretical gain of this MCT detector over a typical TGS detector is about a factor of 14, so the overall advantage to using an MCT detector is approximately a factor of 4.7. When using a smaller-volume light pipe, a shorter narrower cell can be constructed for use with a narrow-range MCT detector.

It is now universally recognized that light pipes must always be optimized in terms of volume of the eluate, transmission, and detector characteristics. Virtually all commercial GC/FT–IR systems use light pipes patterned after the design employed by Azarraga. For example, the current Digilab and pipes are approximately 100 μL in volume. The only exception to this design is manufactured by Accuspec [12,13], who has reverted to the original dimensions used by Low [1]. Because the transmittance of this cell is so high, it cannot be used with a narrow-range

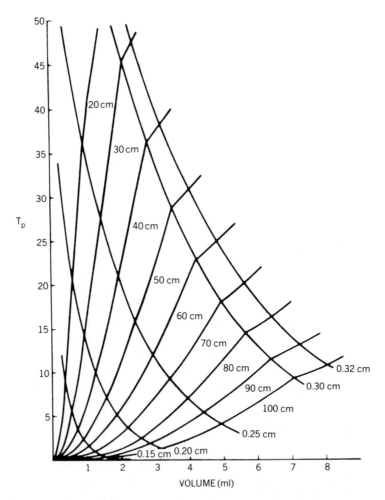

Fig. 18.10. Calculation of the variation in transmittance of cylindrical gold-coated light pipes of various lengths and internal diameters versus the volume of the light pipe. (Reproduced from [8], by permission of the Society for Applied Spectroscopy; copyright © 1977.)

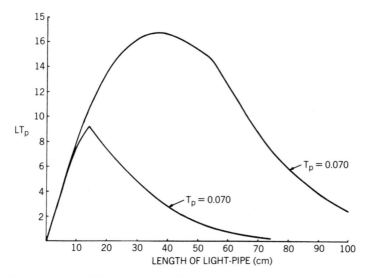

Fig. 18.11. Variation of LT_p (light-pipe length times transmittance) versus length, calculated for a 4-mL light pipe (upper trace) and a 1-mL light pipe. The point at which $T_p = 0.070$ is marked on both curves, the transmittance below which narrow-range mercury cadmium telluride detectors can be used without becoming digitization noise limited. (Reproduced from [8], by permission of the Society for Applied Spectroscopy; copyright © 1977.)

MCT detector and can only be used with a wide-range MCT detector if optics of rather low throughput are employed.

c. Transfer Lines

It is necessary to transport the GC effluent from the chromatographic column to the light pipe or cell. In the earlier packed column systems that used an in-line TCD after the column, simple tubes were connected from the GC outlet to the entrance of the cell. Some systems such as the early Digilab GC–IR accessory [3] had valves that closed the entrance and exit ports to the cell, trapping its contents and venting the effluent to waste until the valves were reopened. As capillary column GC came into prominence, effluent splitters were sometimes employed at the end of the column so that the major portion, or at least half, of the effluent was transported to the infrared cell and the minor portion went to a sensitive destructive detector such as the FID. It is now not unusual for the entire effluent to be passed through the light pipe and then to the FID. This is most often done with narrow-bore WCOT columns where infrared detection limits are very close to the capacity of the column.

There are three important criteria for transfer lines. First, these transfer lines must not have a large dead volume compared to the column, as explained in the section on sampling. If the dead volume is too large, degradation of chromatographic resolution may occur and much of the effort in separating the eluates is wasted. This dead volume includes the entrance to the light pipe or cell. Figure 18.12a schematically represents a transfer line connected to a light pipe. As can be seen, there is a void between the end of the light pipe and the infrared transmitting window used to seal the cell. Voids such as this become more critical as the eluate volume decreases, that is, when narrow-bore capillary columns are used. To avoid this problem, infrared windows may be glued or mechanically fastened directly to the end of the light pipe and a hole bored through the wall of the light pipe itself. This enables the spectrometrist to insert the transfer tube directly into the light pipe. The transfer tube is sometimes

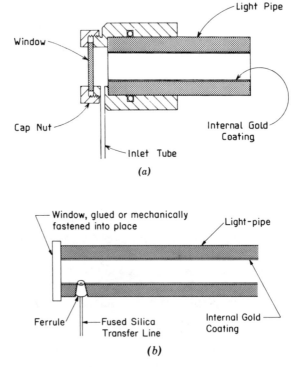

Fig. 18.12. (a) Schematic drawing of a transfer line connected to a light pipe. (b) Schematic drawing of a fused silica transfer line connected directly to the light pipe to avoid dead volumes.

secured with a graphite or vespel ferrule, which seals the entrance hole, see Fig. 18.12b.

The second criterion is that the transfer line (and light pipe) must have no cold spots where any eluate may condense. The majority of eluates studied by this method are vapors rather than gases and consequently are prone to condense. Two basic methods are employed to ensure that no cold spots occur. The transfer lines and light pipe may be external to the gas chromatograph and are thus wrapped in heating tape and then covered with an insulating jacket. This method has been used often, but it suffers from the drawback that it is difficult to heat the entire apparatus uniformly. As a result, cold spots are occasionally found at connections in the effluent path. More recently, light pipes and transfer tubes have been placed inside the chromatograph oven, thus eliminating the cold spot problem. Even though the light pipe is inside the oven, it may have a substantially large thermal mass and hence a lag in temperature behind the column during GC temperature programming. If this is the case, an auxiliary heater must be used to heat the light pipe to the temperature of the column. It is possible, however, to construct a light pipe that resides in the chromatograph oven such that the light pipe does not lag in temperature behind the column [14].

The final important criterion is that the transfer lines must be chemically inert. Because the eluates are at high temperatures (up to 300°C), they can decompose at sites in a metal tube. Some transfer lines are stainless steel, but they are glass lined to prevent chemical decomposition. For those systems that have a light pipe in the gas chromatograph oven, unprotected glass or fused silica transfer lines can be employed. As there is no need to insulate the in-oven systems, the mechanical strength of stainless steel transfer lines is not needed.

There is one application in which a cold spot in the transfer lines may be used to some advantage. Azarraga and Potter demonstrated a thermal trapping system for GC/FT–IR spectrometry [15]. These authors constructed an automated valve system whereby the GC effluent was passed directly into the light pipe and then back to the gas chromatograph. As the effluent entered the valve system, it was passed through a coil immersed in liquid nitrogen that condensed all the eluates. The entire set of eluates could be trapped by this method or it could be restricted to specific peaks. The trapped eluates were then flash vaporized and reintroduced to the gas chromatograph to the same column as before or to another column. A schematic diagram of the GC light-pipe interface is shown in Fig. 18.13a and a diagram of the thermal trap is given in Fig. 18.13b. The reliability of this method was illustrated with an organic test mixture. A mixture of 10 components was injected directly onto a column

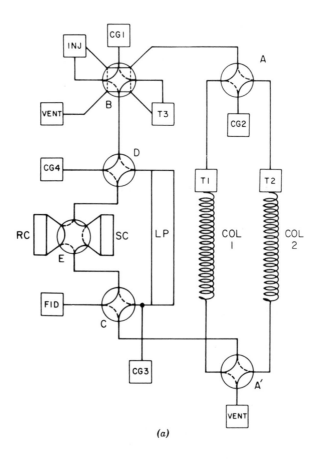

(a)

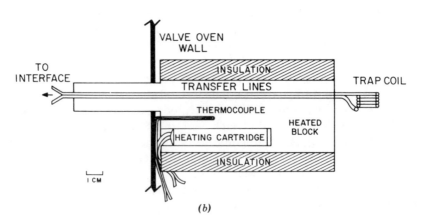

(b)

582

with an OV-101 (nonpolar) stationary phase and then trapped and re-chromatographed through the same column. The results are shown as runs II and III in Fig. 18.14, and examination of these data shows that there is little difference between the two runs.

A practical example of the trapping method can be illustrated with a complex sample from the printing and ink industry. Figure 18.15 shows a chromatogram from the effluent chromatographed on an SP-2250 (intermediate polarity) column. This chromatogram is reconstructed from the interferometric data rather than from an FID (see Section III.b of this chapter). The peaks prior to interferogram 850 were sufficiently well resolved to be identified, but succeeding eluates were not well resolved. The eluates after interferogram 850 were thermally trapped and rechromatographed on an OV-101 column. The resulting chromatogram is shown in Fig. 18.16. Although the separation of these eluates was still not complete, components were more easily identified. Figure 18.17 reproduces some of the single-scan GC/FT–IR spectra from the eluates in Fig. 18.16. The analysis showed that this fraction consisted of aliphatic ketones (412 and 446), siloxane, carbitol homologs, and the last broad peak consisted of alkylethoxyethanols. The first peak, 263, is simply water vapor and was found in all trapped samples.

d. Optics

A general schematic for the optics used in GC/FT–IR spectrometry is presented in Fig. 18.18. Basically, the infrared beam from the interferometer is focused onto the end of the light pipe by an off-axis paraboloid. The focus is made as small as possible so that to avoid vignetting losses, the solid angle should be as large as possible (see Section II.b above). The larger the focus area, the more the radiation is occluded by the entrance aperture of the light pipe. Nevertheless, as the solid angle of the entering beam increases in magnitude, the number of reflections on the inside of the light pipe and the reflection loss also increase. If point sources were used, off-axis high-quality optics could be employed to focus the

---

Fig. 18.13. (a) Schematic drawing of GC/FT–IR light-pipe interface for trapping eluates. Key: A (represented as A and A') and B are eight-port valves; C, D, and E are six-port valves. CG1 is the GC capillary helium carrier gas; CG2 is a regulated source of He to purge the analyte from the sample collection cartridge, SC into the auxiliary cold trap T3. CG3 supplies regulated He as make-up gas to the light pipe, LP, and CG4 is the regulated He tank. T1 and T2 are the on-column cold traps on capillary columns COL1 and COL2. INJ is the capillary column injector part, and RC is a sample recovery cartridge. (b) Schematic drawing of cross section of auxiliary cold trap. (Reproduced from [15], by permission of Huethig Publishing Inc.; copyright © 1981.)

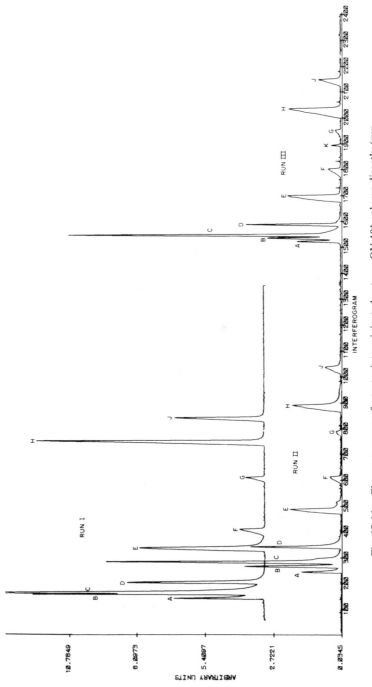

Fig. 18.14. Chromatograms of a test mixture injected onto an OV-101 column directly (run II) and the same column after the eluates were trapped (run III). Component K is a component of unknown origin but is probably residual water that condensed in the cold trap. Run I is the test mixture run on an SP-2250 column. (Reproduced from [15], by permission of Huethig Publishing Inc.; copyright © 1981.)

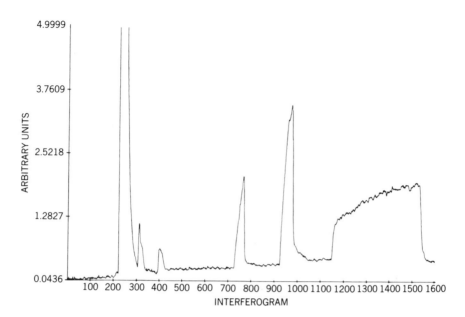

Fig. 18.15. Chromatogram from ink and printing industry, chromatographed on an SP-2250 capillary column. (Reproduced from [15], by permission of Huethig Publishing Inc.; copyright © 1981.)

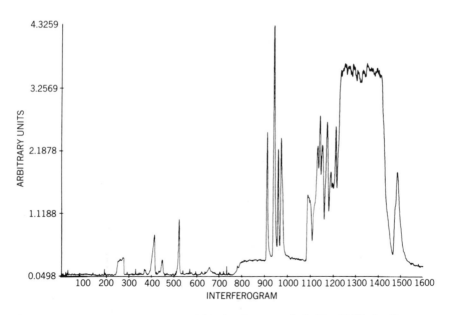

Fig. 18.16. Chromatogram of trapped fraction from sample in Fig. 18.15 after it was rechromatographed on an OV-101 column. The trapped fraction from Fig. 18.15 is interferograms 850–1600. (Reproduced from [15], by permission of Huethig Publishing Inc.; copyright © 1981.)

585

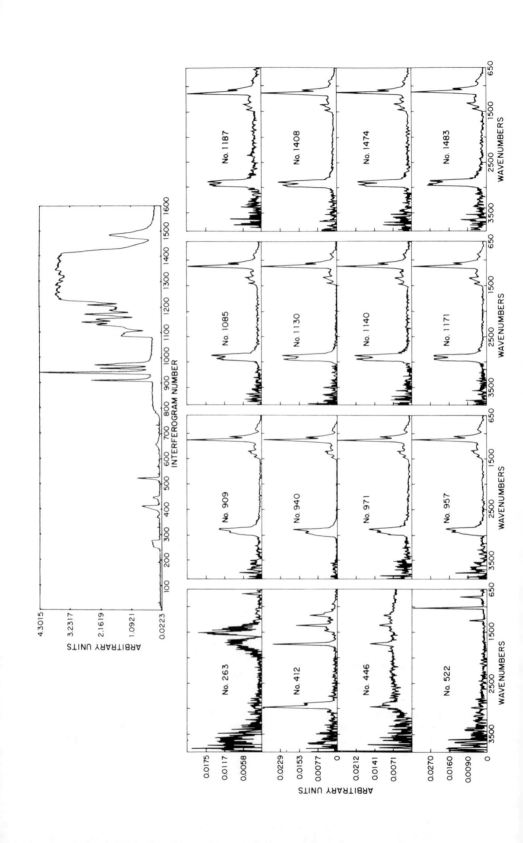

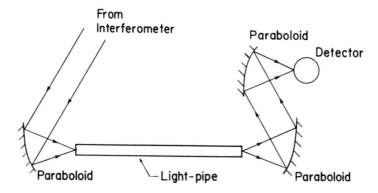

Fig. 18.18. Schematic drawing of GC/FT–IR light-pipe interface optics.

input beam to as small an area as possible. However, point sources are not used, so that even the best off-axis paraboloid cannot focus the beam to its specified blur circle. The collimation of the interferometer beam could be improved by placing an aperture in the source optics (see Chapter 1, Section VII), but this reduces the radiant energy available for transmission through the light pipe. Therefore, some compromise must be made between the focus area of the paraboloid and the radiant energy flux. These criteria will specify the performance of the reflector. These arguments apply, of course, to an off-axis ellipsoid if there is a focus of the infrared beam between the interferometer and light pipe.

The exit optic is either an off-axis ellipsoid that has the exit aperture of the light pipe at one focus and the detector at the other or two off-axis paraboloids. The radiation exiting from the light pipe tends to be somewhat dispersed as the exit aperture is not a point source. As noted above, this aperture may range from 0.7 to 6 mm in diameter. Because the radiation is dispersed by the alkali–halide window and the imperfect edge of the light pipe, it has usually been believed that the collecting mirror should have the greatest possible solid angle of collection. The radiation is focused onto the detector element using the same criteria used when the radiation is focused onto the light-pipe entrance.

Most commercial GC/FT–IR systems use an optical design that is based on the schematic in Fig. 18.18. The "collimated" infrared beam from the interferometer is often passed through an exit port in the spectrometer and directed to the light pipe. In this manner, the GC/FT–IR system does

←

Fig. 18.17. Chromatogram and spectra of constituents from sample in Fig. 18.16. (Reproduced by permission of Huethig Publishing Inc.; copyright © 1981.)

not interfere with the normal sample compartment, and thence the spectrometer may be used for more than one task with a minimum of set-up work. The schematic of the Nicolet 170SX GC/FT–IR system shown in Fig. 18.19 may be considered typical of the commercial optical designs. It should be noted, however, that a few systems have the light pipe in the sample compartment.

Although most commercial spectrometer systems use optics as described above, some workers recently have sought to improve the GC/FT–IR optics. One major problem with GC/FT–IR systems is driving the detector or its preamplifier to a nonlinear state (Chapter 2, Section I), which produces photometric errors. The source of the problem is the radiant flux from the heated light pipe that is added to the modulated signal. If the detector receives too much radiation, some detectors or preamplifiers will not provide the appropriate signal gain, and the gain will vary with wavenumber. To avoid this problem, optics have been designed so that most of the modulated signal from the interferometer and little of the dc emission from the light pipe is passed to the detector. Three different designs have been proposed.

Yang and Griffiths [16] have demonstrated a design where the input mirror has as large a solid angle as possible. The output mirror is a 90° off-axis ellipsoid with a 6:1 focal length ratio and a small blur circle. Using this type of mirror, only the transmitted modulated radiation is collected, and the unmodulated emission from the light-pipe end is eliminated. Yang and Griffiths have demonstrated that the most intense radiation comes directly through the light pipe and that the intensity of off-axis radiation is low [16]. The long focal length of the off-axis ellipsoid collects the radiation from the light pipe, and the short focal length focuses it onto the detector. This optical layout is shown in Fig. 18.20. This work led to the conclusion that a 0.2×0.2-mm detector appears to be optimum for light-pipes that are approximately 1 mm i.d. Commercial GC/FT–IR systems, such as the Digilab FTS–60, are now incorporating designs similar to the one proposed by Yang and Griffiths.

Hirschfeld has modified the above design so that an even smaller detector can be used [17]. This alternate design is shown in Fig. 18.21. The input optics are the same as the design in Fig. 18.20, but the output uses two paraboloids. The first output paraboloid has a long focal length, which is consistent with the design of Yang and Griffiths. The detector window is gold coated so that it becomes a plane mirror. A hole is left in the coating in front of the detector element. The detector window is tilted so that the radiation from the first output paraboloid is directed to an on-axis paraboloid. The focus of the on-axis paraboloid is at the detector element. These optics will allow the radiation to be focused to a small

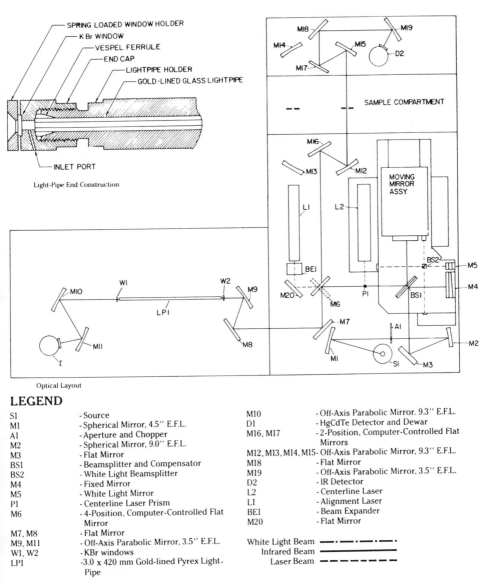

Light-Pipe End Construction

Optical Layout

LEGEND

S1	- Source
M1	- Spherical Mirror, 4.5″ E.F.L.
A1	- Aperture and Chopper
M2	- Spherical Mirror, 9.0″ E.F.L.
M3	- Flat Mirror
BS1	- Beamsplitter and Compensator
BS2	- White Light Beamsplitter
M4	- Fixed Mirror
M5	- White Light Mirror
P1	- Centerline Laser Prism
M6	- 4-Position, Computer-Controlled Flat Mirror
M7, M8	- Flat Mirror
M9, M11	- Off-Axis Parabolic Mirror, 3.5″ E.F.L.
W1, W2	- KBr windows
LP1	- 3.0 x 420 mm Gold-lined Pyrex Light-Pipe

M10	- Off-Axis Parabolic Mirror. 9.3″ E.F.L.
D1	- HgCdTe Detector and Dewar
M16, M17	- 2-Position, Computer-Controlled Flat Mirrors
M12, M13, M14, M15	- Off-Axis Parabolic Mirror, 9.3″ E.F.L.
M18	- Flat Mirror
M19	- Off-Axis Parabolic Mirror, 3.5″ E.F.L.
D2	- IR Detector
L2	- Centerline Laser
L1	- Alignment Laser
BE1	- Beam Expander
M20	- Flat Mirror

White Light Beam — · — · — · —
Infrared Beam ——————
Laser Beam — — — — — —

Fig. 18.19. Optical layout of Nicolet 170SX GC/FT–IR system. (Courtesy of Nicolet Analytical Instruments.)

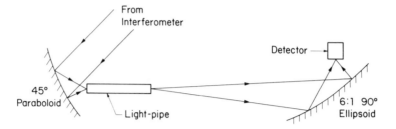

Fig. 18.20.   Optical layout of GC/FT–IR light-pipe interface by Yang and Griffiths [16].

area, so that a 125 × 125-μm detector can be effectively used. Using the appropriate preamplifier and an even smaller detector target, the detection limits are projected to be on the order of 200 pg for strong absorbers (e.g., isobutyl methacrylate) and 1 ng for weak absorbers (e.g., polyaromatic hydrocarbons).

The third optical design is shown in Fig. 18.22 and has been developed by Buijs [18]. This design is for the Bomem spectrometers, which use a beam-centered He–Ne trigger (see Chapter IV). Because the He–Ne is in the center of the infrared beam, the infrared signal is occluded from this area. This necessitates a longer focal length input paraboloid so that a beam concentric to the light pipe is produced. The output optics consist of two ellipsoids rather than paraboloids as found in the preceding design. This is so that an aperture can be placed at the common focus of the two ellipsoids. The aperture limits the unmodulated radiation from the light pipe. Detection limits using this design have not been reported at the time of this writing.

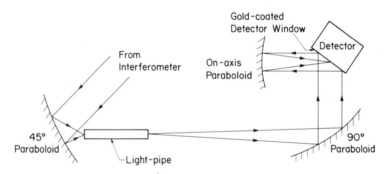

Fig. 18.21.   Schematic representation of GC/FT–IR light-pipe interface optical diagram described by Hirschfeld [17].

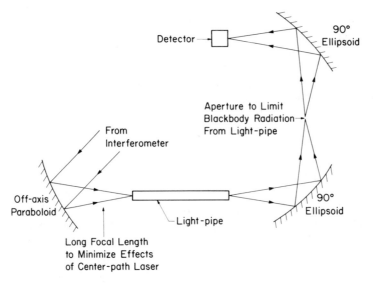

Fig. 18.22. Schematic drawing of optics used for GC/FT–IR light-pipe interface by Buijs, Berube, and Vail [18].

e. Detectors

GC/FT–IR spectra are collected in the mid-infrared region, which extends from about 400 to 4000 cm^{-1}. The most commonly employed detectors in this region are the DTGS pyroelectric detectors and the MCT photoconductive detector (see Chapter 5, Section VI). In general, the MCT detector is anywhere from 4 to 50 times more sensitive than the DTGS detector, depending on its bandwidth. For this reason, the MCT is often the detector of choice for GC/FT–IR spectrometry as the transmission of light pipes may be rather low, often less than 5% of the incident radiation. Care must be taken, however, to prevent the interferogram signal from being digitization noise limited, as explained above in the discussion on sampling. If low transmission light pipes are used, digitization noise is generally not a problem.

Another advantage to the MCT detector over DTGS is the fact that an MCT detector has a relatively short response time, considerably shorter than that of DTGS. In fact, the sensitivity (D^*) of MCT detectors increases with sampling frequency, and most reach the maximum at a data acquisition rate of about 80–100 kHz, as explained in Chapter 5. At these acquisition rates, as many as. twenty 8-cm^{-1}-resolution interferograms can be collected per second. The half-width of most capillary GC peaks is at least 3 sec, so several scans from an interferometer capable of collecting data at this rate can be averaged to improve SNR. The peak width is sufficiently large that the concentration in the light pipe does not change significantly and hence the spectrum is not distorted.

The main disadvantage of using a narrow-band MCT detector is the restricted bandwidth (about 750–4000 cm^{-1}); however, it is possible to purchase wide-range MCT detectors with typical bandwidths of 450–4000 cm^{-1}. The wide-range MCT detectors are less sensitive than the narrow-range detectors by about an order of magnitude in the D^* value but have a less stringent specification on noise-limited operation. It should also be noted that MCT detectors used in GC/FT–IR spectrometry should have a small detector element, usually smaller than 1 × 1 mm (see Section II.d of this chapter). For an interferometer employing as much as possible of the Jacquinot advantage, it is common for MCT detectors to have target elements four times this maximum area (2 × 2 mm). The reason for a small element is that the optical throughput of the beam emerging from the light pipe is much smaller than the throughput of the beam passing through the interferometer. The radiation from this aperture is collected and focused onto the MCT element, but in a correctly designed system, the radiation does not fill an area greater than 1 mm^2. If the detector element is unfilled, those unexposed areas generate noise, which degrades the interferometric signal. Simply defocusing the beam to fill the detector element does not alleviate the problem as the effective flux density is too low.

In FT–IR spectrometry, it is rarely critical whether the sample is placed before or after the interferometer. The obvious exception to this is emission spectrometry, for which the emitting sample must be placed before the interferometer so that the radiation can be modulated. A heated GC/FT–IR light pipe is an infrared emitter; hence the radiation from the light pipe itself would be modulated if the light pipe were placed before the interferometer. For this reason, the light pipe is always placed after the interferometer, so that any emitted radiation that is collected is measured as a dc signal and does not interfere with the interferometric signal.

f.   Gas Chromatographs

There has been a definite trend in GC/FT–IR spectrometry to keep apace with the innovations in gas chromatography. Until the mid-1970s, packed column gas chromatography was the preferred method to separate volatile mixtures, even though open tubular chromatography had been known for about 20 years [19]. The major problem with capillary chromatography was that even rather small quantities of materials would overload the column, so that with the FT–IR technology at the time it was particularly difficult to detect minor peaks. GC/FT–IR detection limits remained on the order of a few micrograms for strongly absorbing compounds in the early 1970s. With the demonstration of long-path light pipes in 1976 by Azarraga [9], detection limits began to fall rather sharply.

It is an unfortunate consequence of gas chromatography that chromatographic resolution is largely dependent on sample size. Due to their length, open tubular columns have a higher number of theoretical plates than packed columns, but the open tubular columns have a smaller sample capacity. That is, the large samples needed to yield an interpretable infrared spectrum cannot be placed on open tubular columns without severe degradation of the chromatographic resolution. Thus, until about 1976, it was generally believed that GC/FT–IR measurements were only applicable for packed column chromatography. When gold-coated light pipes were first demonstrated, a shift toward the use of SCOT columns for GC/FT–IR took place because of the increase in the number of theoretical plates relative to packed columns and the increased sample capacity relative to WCOT columns. The combination of optimally designed light pipes and the MCT detector led to an improvement in the GC/FT–IR detection limits of between one and two orders of magnitude.

Even though the first use of wall-coated open tubular (WCOT) GC columns was reported in 1976 by Azarraga [9], the next report of WCOT columns used for GC/FT–IR spectrometry did not appear until 1980 [20]. Today, most manufacturers have optimized their top-of-the-line GC/FT–IR accessory for separation using fused silica WCOT technology.

An interesting comparison of detection limits was made in a review of GC/FT–IR systems by Erickson in 1979 [21]. He plotted reported GC/FT–IR detection limits versus the year of the report for the period 1968–1978. These data are reproduced in Fig. 18.23. In the 11 years reported, detection limits improved by about three orders of magnitude. Detection limits of conventional single-beam GC/FT–IR measurements improved slowly from 1978 through 1983, but now subnanogram detection limits appear feasible. An alternative approach by which subnanogram detection limits may be reached involves the use of dual-beam, or optical subtraction, FT–IR spectrometry.

g. Dual-Beam GC/FT–IR

Low demonstrated the first dual-beam GC/FT–IR system in 1968, but it was impractical due to the difficulty in balancing the two beams of the interferometer [2]. In addition, two sources and two detectors were used,

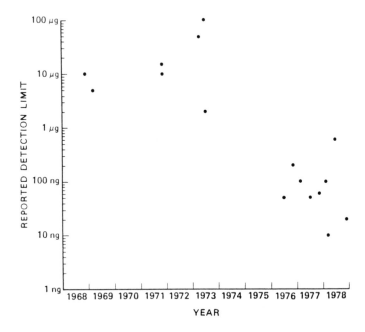

Fig. 18.23. GC/FT–IR detection limits as reported over the period 1968–1978. (Reproduced from [21], by permission of Marcel Dekker, Inc.; copyright © 1979.)

which further complicated the apparatus. A decade later, in 1978, Gomez-Taylor and Griffiths presented a new design for a dual-beam GC/FT–IR system [22]. As can be seen in Fig. 18.24, a single source and single MCT detector are used with a Digilab Model 296 interferometer. With this optical arrangement, the beams from the two arms of the interferometer can be optically subtracted effectively. Like the system devised by Low, the apparatus used by Gomez-Taylor and Griffiths employed two light pipes, one for each beam of the interferometer. Each cell in Low's system had a volume of 22 mL, which undoubtedly contributed to the sensitivity of this system but restricted its use to $\frac{1}{4}$-in.-o.d. packed columns. Gomez-Taylor and Griffiths used a $\frac{1}{8}$-in.-o.d. packed column GC and 4.8-mL light pipes (4 mm $\times$ 4 mm $\times$ 30 cm), which was consistent with the sampling criterion for packed column GC established earlier [8]. Data collection was performed on the fly (one scan per data file) at 8 cm^{-1} resolution. Each interferogram represented optically subtracted data, and theoretically only absorption data should have been present. To ensure that no instrumental data were present in the interferogram and to normalize the data, a 100-scan optically subtracted reference interferogram of the empty cells was digitally subtracted from each sample interferogram. The cor-

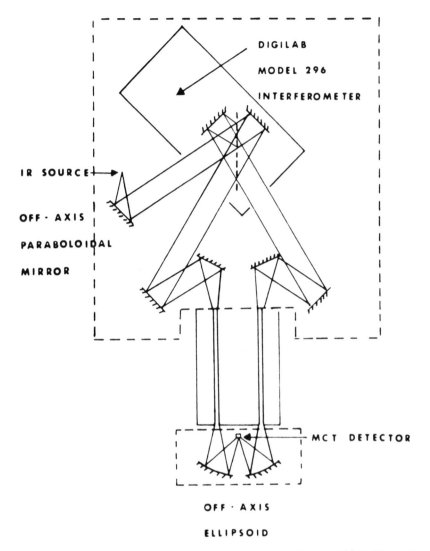

Fig. 18.24. Schematic drawing of the optical layout of a dual-beam GC/FT–IR system. (Reproduced from [22], by permission of the American Chemical Society; copyright © 1978.)

rected optically subtracted interferogram was then transformed to produce the infrared spectrum. A sample spectrum is shown in Fig. 18.25 that is 100 ng of anisole, a strong absorber. This method was found to have considerably improved detection limits over single beam systems, as shown in Table 18.1.

In a later paper, Griffiths and co-workers [23] presented a dual-beam

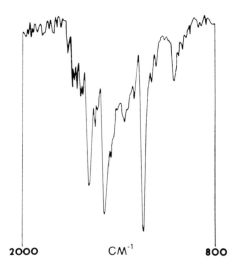

Fig. 18.25. Dual-beam GC/FT–IR spectrum of 100 ng of anisole. (Reproduced from [22], by permission of the American Chemical Society; copyright © 1978.)

Table 18.1. **Experimental Detection Limits for a Dual-Beam System Versus a Single-Beam System Using Both a DTGS and an MCT Detector [22]**

Compound	Single-beam Detection Limits (μg)		Dual-beam Detection Limit (μg)
	DTGS	MCT	MCT
Anisole	0.8	0.20	0.050
Chlorobenzene	1	0.25	0.075
Diethyl malonate	2	0.45	0.100
Acetonitrile	5	1.5	0.400
Benzonitrile	10	2.5	0.750
Aldrin	5	1.5	0.400
Perthane	10	2.5	0.750
p,p'-DDT	6	1.5	0.400
Heptachlor	8	2.0	0.600

GC/FT–IR system that was configured for SCOT column chromatography. The basic design of the system remained the same except the light-pipe volumes were reduced to about 600 μL each (3 mm i.d. × 8 cm). Anisole was again used as a test compound, and Fig. 18.26 shows some single-scan spectra. Figure 18.26a is practically noise free and represents 100 ng of anisole; at 5 ng (Fig. 18.26b), the spectrum is still recognizable. The last spectrum (Fig. 18.26c) is the single-scan optically subtracted spectrum of 1 ng of anisole. Although even lower detection limits may be possible, dual-beam GC/FT–IR spectrometry remains a difficult experiment to perform primarily due to the optical alignment of the two beams of the interferometer. It is not likely that dual-beam GC/FT–IR spectrometry will be a routine method in the near future unless a special-purpose GC/FT–IR instrument were to be introduced.

Although dual-beam GC/FT–IR spectrometry has impressive detection limits, the most sensitive technique appears to be gas chromatography/matrix isolation Fourier transform infrared (GC/MI–FT–IR) spectrome-

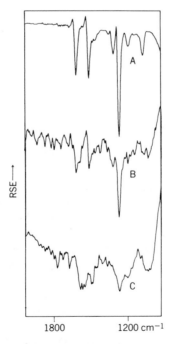

Fig. 18.26. Improved on-the-fly dual-beam GC/FT–IR spectra of (a) 100 ng, (b) 5 ng, and (c) 1 ng of anisole using a SCOT column. (Reproduced from [23], by permission of the Society for Applied Spectroscopy; copyright © 1980.)

ISOBUTYL METHACRYLATE, 10 ng

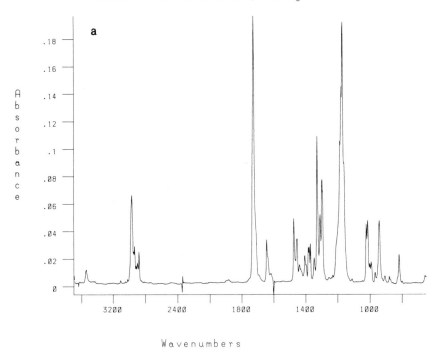

Wavenumbers

Fig. 18.27. GC/MI–FT–IR spectra of (*a*) 10 ng and (*b*) 100 pg of isobutyl methacrylate. Each spectrum was signal averaged for 256 scans. (Courtesy of Mattson Instruments, Inc.)

try, which traps the eluate in a frozen inert matrix so that signal averaging can be accomplished. GC/MI–FT–IR spectrometry is described in detail in Chapter 15, Section II, so the general principles are not reiterated here. Spectra have been obtained of 100 pg of strongly absorbing eluates using this technique. Figure 18.27*a* shows the GC/MI–FT–IR spectrum of 10 ng of isobutyl methacrylate, and Fig. 18.27*b* shows 100 pg of the same compound. Both spectra were signal averaged for 256 scans. GC/MI–FT–IR spectrometry is a useful technique for extremely small quantities of eluates.

h. GC/FT–IR/MS Systems

Infrared spectrometry can provide the structure of many GC eluates when interfaced in a GC/FT–IR system. However, structure identification is not always possible by infrared spectrometry alone; for example, indi-

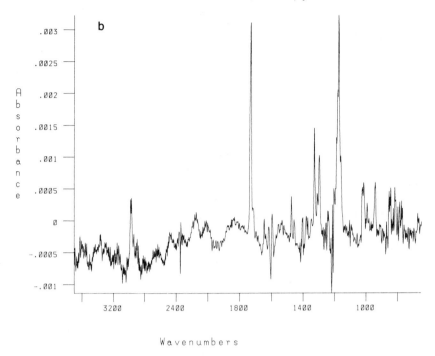

Fig. 18.27. (*continued*)

vidual spectra in a series of low-resolution vapor phase infrared spectra of homologs are difficult to distinguish. Mass spectra can be used to identify each member in the series by its molecular weight. On the other hand, isomers may be very difficult or impossible to distinguish by mass spectrometry, whereas infrared spectra will readily indicate structural differences. At least three groups have been investigating hybrid GC/FT–IR/MS systems [24–26]. All three groups have interfaced both an FT–IR and a mass spectrometer to a single GC. Two methods are available: either the effluent can be split to each spectrometer or the FT–IR light pipe precedes the MS inlet system. Each of the three systems has been able to demonstrate the feasibility of measuring infrared and mass spectra for each eluate. Even more interesting is the development of a GC/FT–IR/FT–MS hybrid system [27], but in this reference the emphasis is on the GC/FT–MS interface rather than on the GC/FT–IR interface. GC/FT–IR/MS has been reviewed elsewhere [28]. It is anticipated that GC/FT–

IR/MS will become a tremendously useful tool in the future. Although the cost of such an instrument will be high (certainly more than $100,000), the reduction in ambiguity of spectral interpretation compared with GC/FT–IR or GC/MS systems individually should lead to their widespread acceptance.

III. SOFTWARE

Software has been an integral part of the development of GC/FT–IR systems. Conceptually, each sample to be analyzed may have a large number of components, and even if only data across each GC peak are collected and baseline data rejected, a large number of interferograms will still have to be stored and transformed and the spectra plotted and identified. The number of interferograms increases drastically if each single scan is stored for an on-the-fly operation. In the case where signal-averaged eluate interferograms are stored, methods must be devised so that eluate peaks are recognized and data collection begins. GC/FT–IR software has evolved to the point where the plethora of data produced by the experiment can be handled rapidly. Let us now consider the development of data collection, transformation, display, and spectral search techniques used in GC/FT–IR spectrometry.

a.   Data Collection

Perhaps the simplest method of collecting GC/FT–IR data is to trap a single GC eluate in a cell and signal-average an appropriate number of interferograms to produce the desired SNR. Such an operation can be done manually and requires virtually nothing more of the spectrometer software than is available for more routine analyses. Clearly, much of the advantage of GC/FT–IR spectrometry is lost with such a method. Spectra of all, or at least the pertinent, eluates is far preferable.

The method that was employed by Low and Freeman [1] was not very sophisticated in terms of computer software, but certainly it was technologically innovative. Low and Freeman devised a system that ran in either a trapped or an on-the-fly mode. Regardless of the operating mode, each single scan of the interferometer was recorded on a 7.5-ips analog magnetic tape. The tape was rewound and read to convert each analog interferogram to computer-compatible (digital) data. When interferograms were recorded in the trapped mode, as many interferograms as necessary were written on the tape to produce a spectrum of the desired SNR. In this mode, interferograms were individually recorded, and it was possible to signal-average as many or as few as was desired.

Over the past 15 years, methods have been developed that transpose some of these ideas to systems with dedicated minicomputers. Clearly, the major problem in GC/FT–IR spectrometry is the collection of all data so that nothing significant is lost. Many GC runs, especially those using capillary columns, exceed 20 min. It takes a very large disk to store more than 1200 interferograms, even at 8 cm^{-1} resolution. In 1975, the U.S. EPA contracted with Digilab to produce a specialized GC/FT–IR data collection package [29]. Azarraga (who was the contract officer) was experimenting with the light pipe cell design and determined that it was necessary to save each single-scan interferogram from GC runs. Rather than store the data on disk, Azarraga chose a computer magnetic tape drive. In this system, interferograms at 8 cm^{-1} resolution were collected at the rate of approximately one per second and written immediately to tape. This was conceptually the same procedure as used by Low and Freeman [1]; however, Azarraga used a digitally encoded tape that removed the need for analog-to-digital conversion upon reading the tape. A standard magnetic tape can store more than 2000 interferograms. Although magnetic tape does not have the random access capability of a disk drive, the added storage capacity and relative low cost of the storage medium more than compensate for the inconvenience.

The shortcoming of Azarraga's system was that it was tedious to identify those interferograms that pertained to eluate peaks. The earliest version of the spectrometer Azarraga used, a Digilab FTS-14, did not even have an oscilloscope to display spectra. Hence, on this system, it was necessary to transform every interferogram that was suspected of being collected while an eluate was in the light pipe and then plot the spectrum on a digital hard copy plotter. It was reputed that such a procedure could take as long as two man-weeks to analyze one 20-min GC run [29]. When a display oscilloscope was added to the system, the visual inspection of the data became considerably faster. Nonetheless, a method was lacking whereby eluate interferograms could be located rapidly and accurately.

Once methods had been developed whereby eluate interferograms could be located directly from the entire set of stored interferograms, more sophisticated GC/FT–IR systems were developed. At least four of the major instrument manufacturers offered extensive GC/FT–IR packages in 1985. Digilab, Nicolet, Mattson and IBM have systems that collect data to magnetic tape or hard disk whereas Analect and Perkin-Elmer market systems that collect data on floppy disks. Present-day disk systems are available that have capacities of 300 Mbytes or more, which far exceed the storage capacity of a reel of magnetic tape. The more sophisticated systems provide data collection methods whereby the gas chromatograms can be constructed from the interferogram data, rapid transformation of the data, comprehensive plot packages, and search systems.

b. Gas Chromatogram Construction

The major drawback to collecting every single-scan interferogram in a GC/FT–IR run is the identification of the eluate interferograms. Of course, it is possible to put a GC detector in the system and know when each component is eluted, but this detector gives a signal that is displaced from the interferometric data and is subject to some error. In practice, it is far preferable to measure the infrared response directly. In 1977, two techniques were reported that constructed chromatograms directly from the interferometric data [30,31]. The first of the two methods was devised by Coffey and Mattson and relied on the Fourier transform algorithm [30].

Coffey and Mattson collected single-scan interferograms and constructed functional group specific chromatograms. Each single-scan interferogram was collected at a resolution of 8 cm^{-1}, that is, 2048 data points per interferogram. A fast Fourier transform was performed on 512 data points around the centerburst to yield a 32-cm^{-1}-resolution spectrum. This FFT took less than 1 sec of computer time. The spectrum was divided in five spectral regions, or "windows." Integrated infrared absorbance was calculated across each of the spectral regions to produce five separate chromatograms. The infrared absorbance within each window corresponds to the presence of a specific peak and hence a specific functional group. Coffey, Mattson, and Wright presented a detailed description of the functional group construction in a later paper [32]. The authors showed the reconstructed functional group chromatogram for the synthetic mixture listed in Table 18.2. Five spectral regions were selected where the infrared absorbance was integrated for each single-scan interferogram collected. The five regions had the spectral ranges of 3600–3570 cm^{-1}, the O–H stretch region; 3080–3020 cm^{-1}, the aromatic and olefinic C–H stretch region; 1760–1680 cm^{-1}, the C=O stretch region; 1560–1520 cm^{-1}, aromatic C–C stretching frequencies; and 1170–1155 cm^{-1}, the C–O–C stretch. Figure 18.28 shows the constructed chromatograms for the five spectral regions. These data were used to identify which interferograms were recorded when eluates were in the light pipe. A distinct advantage of this type of construction is that eluates can be identified, to a certain extent at least, by chemical class, which is helpful if specific eluates are sought. The selected regions are left to the discretion of the operator so this can be made a custom system.

Table 18.2. Mixture Composition for Chromatograms in Figure 18.28[a]

Compound	Concentration, ppm (v/v)
Guaiacol	10000
Toluene	5000
o-Nitrotoluene	1000
Cyclohexanone	500
Isobutyl methacrylate	100

[a] Methyl chloride was the solvent and the total injection volume was 6 μL.

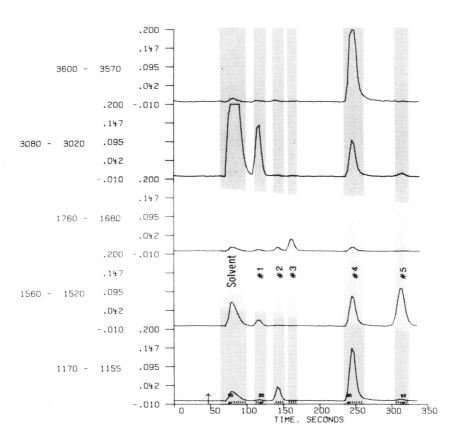

Fig. 18.28. Constructed chromatogram of a five-component mixture using the transform method. (Reprinted from *American Laboratory*, **10**(5), 126, 1978. Copyright © 1978 by International Scientific Communications, Inc.)

603

The authors who developed this technique designed this construction specifically for a Nicolet FT–IR spectrometer. The early GC/FT–IR spectrometers manufactured by this vendor did not have magnetic tapes, so all the data was stored on disk. Disk space could be rapidly exhausted so the constructed chromatogram was used to identify which eluate interferograms should be stored. All baseline data were discarded. Nicolet spectrometers later had a magnetic tape option so that an entire GC/FT–IR run could be stored. Nicolet dubbed this functional group specific constructed chromatogram the Chemigram®, but other vendors rapidly produced similar packages such as the Infragram® and Absorptiogram® by Digilab. The Chemigram and Absorptiogram are ostensibly identical and an Infragram is a similar process that uses a single datum to represent the absorbance instead of several data integrated within a range. With the increased computational speed of the Nicolet 1280 minicomputer over the speed of the 1180, for which the Chemigram software was originally written, it is now possible to transform 2-k points per scan instead of the original 512 points.

The other method whereby gas chromatograms were constructed from interferometric data was developed by de Haseth and Isenhour [31,33]. This method does not involve any Fourier transforms but relies on the multiplex nature of the interferogram, so that each datum in the interferogram contains information about every datum in the spectrum (see Chapter 7). Hence, a small portion of the interferogram should contain ample information to determine if an eluate is present in the light pipe or not. The method by which interferograms may be distinguished is based on a technique in vector algebra called Gram–Schmidt vector orthogonalization.

For the purposes of this study, a vector is a series of data points, such as the consecutive values of a discrete interferogram. These discrete values represent coordinates in free space, and if more than three coordinates (dimensions) are present, the free space is referred to as hyperspace. In this case, an interferogram I with n data points may be represented as an n-dimensional vector in hyperspace as

$$\mathbf{I} = (i_1, i_2, i_3, \ldots, i_n) \qquad (18.7)$$

where $\mathbf{I}$ is the vector notation for the interferogram I and i_1, i_2, i_3, . . . , i_n are the n consecutive values of the interferogram, or more appropriately, the coordinates.

Gram–Schmidt vector orthogonalization is in reality a normalized subtraction routine. The procedure takes two vectors and removes from one any common components found in the other. Thus, if two interferograms are recorded when the light pipe is empty, theoretically there is no difference between them, and the orthogonalization would show that there was no uncommon element. The resultant is zero. If one interferogram has eluate information and the other is simply background, the orthogonalization would yield a nonzero difference. The magnitude of the difference depends on the concentration of the eluate in the light pipe.

Gram–Schmidt vector orthogonalization is general and is performed as follows. The first step is to form a unit vector U_1 from an interferogram vector I_1. Vector I_1 is an interferogram that was recorded when no eluate was present in the light pipe.

$$U_1 = \frac{I_1}{\sqrt{I_1^T I_1}} \tag{18.8}$$

where I_1^T is the transpose of I_1 and is simply a mathematical formality to denote the dot product of two vectors. For two vectors J and K, the dot product is a scalar equal to $\sum_{m=1}^{n} j_m k_m$. Thus, the term $\sqrt{I_1^T I_1}$ is the scalar length of I_1.

If I_1 were noise free, U_1 would be a good representation of an empty light-pipe interferogram. Unfortunately, these data are rather noisy and more than one interferogram is necessary to characterize the empty light pipe. A second empty light-pipe (or background) interferogram is chosen, I_2, and all components common to the first interferogram are removed:

$$V_2 = I_2 - c_1 U_1 \tag{18.9}$$

where V_2 is the vector that remains after the common components are removed and c_1 is a scalar that causes V_2 to be orthogonal to U_1. If two vectors are orthogonal, their dot product is zero; hence,

$$U_1^T V_2 = U_1^T (I_2 - c_1 U_1) = 0 \tag{18.10}$$

But $U_1^T U_1 = 1$ as they are unit vectors, so that

$$c_1 = U_1^T I_2 \tag{18.11}$$

and

$$V_2 = I_2 - (U_1^T I_2) U_1 \tag{18.12}$$

Both U_1 and V_2 are background interferograms, but the unit vector form is more useful; thus,

$$U_2 = \frac{V_2}{\sqrt{V_2^T V_2}} \qquad (18.13)$$

The combination of U_1 and U_2 describes the background signal in an interferogram more accurately than U_1 alone. This combination is called a *basis*. In some instances, two interferograms may be insufficient to characterize the background signal and more vectors must be added to the basis set. This is a sequential operation (one added at a time) and can be expressed as

$$V_k = I_k - \sum_{j=1}^{k-1} (U_j^T I_k) U_j \qquad (18.14)$$

and the appropriate unit vector is

$$U_k = \frac{V_k}{\sqrt{V_k^T V_k}} \qquad (18.15)$$

Once the background has been characterized by the basis, interferograms can be compared to the basis by an operation similar to that described in Eq. 18.14; that is,

$$V_p = I_p - \sum_{j=1}^{k} (U_j^T I_p) U_j \qquad (18.16)$$

where I_p is not in the set 1–k. The magnitude of the vector $V_p(\sqrt{V_p^T V_p})$ is the deviation of the interferogram under study from the basis, or background characterization. Each single-scan interferogram in a GC/FT–IR run can be compared to a basis, and the resulting scalar from each interferogram indicates the quantity of eluate in the light pipe.

It was stated above that the orthogonalization procedure applies because of the multiplex nature of the interferogram. This means that it is not necessary to use the entire interferogram to construct the chromatogram. It was found that an optimum signal was produced if 100 data points are selected from the interferogram in the region starting ~50 data points from the centerburst. (This assumes that a He–Ne laser is being used to sample a mid-infrared interferogram with one point sampled per wavelength.) Most constructed chromatograms take 10 or fewer vectors, and an example of a constructed chromatogram calculated using the Gram–Schmidt vector orthogonalization method is shown in Fig. 18.29. This algorithm is very rapid and can be done simultaneously to GC/FT–IR data collection. Several studies have been completed investigating this method of data analysis for GC/FT–IR in more detail [34–37].

The success of these reconstruction techniques led to further computer data processing for GC/FT–IR, including search systems.

c. GC/FT–IR Search Systems

Infrared search systems have been studied for more than 30 years. A detailed history of the progress of infrared searches is beyond the scope of this text; however, it should be noted that virtually all searches were performed remote to the spectrometer prior to 1978. This was due, in large part, to the fact that most infrared spectrometers were not interfaced to a computer. A major incentive to bring on-line searching to infrared spectrometry was the need associated with the large number of spectra produced by GC/FT–IR systems.

In 1975, the Coblentz Society formed a subcommittee to specify the format of vapor phase spectrometric data with the intention of producing a universally acceptable format for the collection and exchange of these data [38]. Using these specifications, the U.S. EPA contracted for the collection of 3300 vapor phase spectra, which has come to be known as

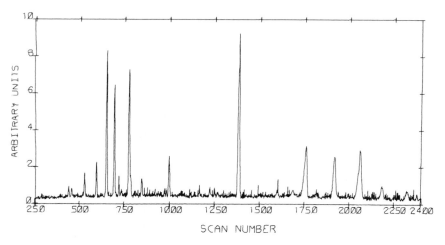

Fig. 18.29. Constructed chromatogram using the Gram–Schmidt vector orthogonalization method.

EPALIB. These data formed the library for a search system devised by Azarraga and Hanna as part of a software package for a Digilab spectrometer they named GIFTS (Gas Chromatography/Infrared Fourier Transform Software) [39]. This package is a FORTRAN and assembler language based program for the collection, chromatogram construction, and searching of GC/FT–IR data.

The search system itself consists of taking each library spectrum and condensing the data. This is done by restricting the bandwidth of the absorbance spectrum to two regions, 700–2100 cm^{-1} and 2700–3300 cm^{-1}. The interval between data points was increased from 4 to 8 cm^{-1}, and each absorbance value was restricted to 1 part in 200. Thus, the spectra were "deresolved" and the precision of the absorbance reduced. The advantage to such a library storage format is that each spectrum is stored in less than one hundred and twenty-eight 16-bit words of disk storage. When an unknown spectrum is collected, it too is reduced to the same format as the library and compared to each entry in the library. Comparison is made by the dot product operation or one of its closely allied operations, such as Euclidean distance. The closest matches in the library are listed, and these results may be used to aid in the identification of unknown spectra. Positive identification cannot be made unless the unknown spectra is visually compared and matched with a library spectrum. Most of the commercial GC/FT–IR systems incorporate similar search systems, and several modifications have been made, including the possibility of performing the search with interferograms or the inverse Fourier transforms of absorbance spectra [40–46].

IV. CONCLUSIONS

GC/FT–IR spectrometry has progressed considerably since the first demonstration in 1967. At the present time, detection limits for single-beam FT–IR systems remain at the low-nanogram or high-picogram level. It is doubtful that these limits will be decreased by more than an order of magnitude. Optical subtraction GC/FT–IR appears to be more sensitive than the single-beam method, but by less than one order of magnitude. GC/MI–FT–IR now appears to be even more sensitive but is very expensive to implement and does not yield spectra in real time. Nevertheless, GC/FT–IR spectrometry remains a powerful tool for the structural analysis of complex volatile mixtures and is an excellent complement to GC/MS. Combined GC/FT–IR/MS systems will almost certainly become commercially available in the very near future, so that qualitative and quantitative analysis of complex mixtures of volatile components will become increasingly routine.

REFERENCES

1. M. J. D. Low and S. K. Freeman, *Anal. Chem.,* **39,** 194 (1967).
2. M. J. D. Low, *Anal. Lett.,* **1,** 819 (1968).
3. K. L. Kizer, *Am. Lab.,* **5**(6), 40 (1973).
4. A. W. Mantz, *Ind. Res.,* **19**(2), 90 (1977).
5. D. L. Wall and A. W. Mantz, *Appl. Spectrosc.,* **31,** 552 (1977).
6. L. V. Azarraga and A. C. McCall, Infrared Fourier Transform Spectrometry of Gas Chromatography Effluents, EPA-660/2-73-034, 61 pp (1974).
7. L. V. Azarraga and A. C. McCall, 25th Annual Pittsburgh Conference on Analytical Chemistry and Applied Spectroscopy, Cleveland, Ohio, March 1974, paper no. 328.
8. P. R. Griffiths, *Appl. Spectrosc.,* **31,** 284 (1977).
9. L. V. Azarraga, 27th Annual Pittsburgh Conference on Analytical Chemistry and Applied Spectroscopy, Cleveland, Ohio, March 1976, paper no. 334.
10. L. V. Azarraga, *Appl. Spectrosc.,* **34,** 224 (1980).
11. L. V. Azarraga, U.S. Environmental Protection Agency, Athens, Georgia, personal communication.
12. V. Rossiter, *Am. Lab.,* **14**(2), 144 (1982).
13. V. Rossiter, *Am. Lab.,* **14**(6), 71 (1982).
14. J. A. de Haseth, unpublished results.
15. L. V. Azarraga and C. A. Potter, *HRC & CC,* **4,** 60 (1981).
16. P. W. Yang and P. R. Griffiths, *Appl. Spectrosc.,* **38,** 816 (1984).
17. T. Hirschfeld, P. Griffiths, and P. W. Yang, 35th Annual Pittsburgh Conference on Analytical Chemistry and Applied Spectroscopy, Atlantic City, New Jersey, March 1984, paper no. 269.
18. H. Buijs, J. N. Berube, and G. Vail, 35th Annual Pittsburgh Conference on Analytical Chemistry and Applied Spectroscopy, Atlantic City, New Jersey, March 1984, paper no. 273.
19. M. J. E. Golay, in *Gas Chromatography* (V. J. Coates, H. J. Noebels, and I. S. Fagerson, eds.), Academic Press, New York, pp. 1–13 (1958).
20. K. H. Shafer, A. Bjørseth, J. Tabor, and R. J. Jakobsen, *HRC & CC,* **3,** 87 (1980).
21. M. D. Erickson, *Appl. Spectrosc. Rev.,* **15,** 261 (1979).
22. M. M. Gomez-Taylor and P. R. Griffiths, *Anal. Chem.,* **50,** 422 (1978).
23. D. Kuehl, G. J. Kemeny, and P. R. Griffiths, *Appl. Spectrosc.,* **34,** 222 (1980).
24. K. H. Shafer, T. L. Hayes, and J. E. Tabor, *Proc. Soc. Photo-Opt. Instrum. Eng.,* **289,** 160 (1981).

25. C. L. Wilkins, G. N. Giss, G. M. Brissey, and S. Steiner, *Anal. Chem.*, **53**, 113 (1981).

26. R. W. Crawford, T. Hirschfeld, R. H. Sanborn, and C. M. Wong, *Anal. Chem.*, **54**, 817 (1982).

27. C. L. Wilkins, G. N. Giss, R. L. White, G. M. Brissey, and E. C. Onyiriuka, *Anal. Chem.*, **54**, 2260 (1982).

28. P. R. Griffiths, J. A. de Haseth, and L. V. Azarraga, *Anal. Chem.*, **55**, 1361A (1983).

29. L. V. Azarraga, U.S. Environmental Protection Agency, Athens, Georgia, personal communication.

30. P. Coffey and D. R. Mattson, International Conference on Fourier Transform Infrared Spectroscopy, Columbia, South Carolina, June 1977, paper no. MB6.

31. J. A. de Haseth and T. L. Isenhour, International Conference on Fourier Transform Infrared Spectroscopy, Columbia, South Carolina, June 1977, paper no. MB7.

32. P. Coffey, D. R. Mattson, and J. C. Wright, *Am. Lab.*, **10**(5), 126 (1978).

33. J. A. de Haseth and T. L. Isenhour, *Anal. Chem.*, **49**, 1977 (1977).

34. R. C. Wieboldt, B. A. Hohne, and T. L. Isenhour, *Appl. Spectrosc.*, **34**, 7 (1980).

35. B. A. Hohne, G. Hangac, G. W. Small, and T. L. Isenhour, *J. Chromatogr. Sci.*, **19**, 283 (1981).

36. R. L. White, G. N. Giss, G. M. Brissey, and C. L. Wilkins, *Anal. Chem.*, **53**, 1778 (1981).

37. G. M. Brissey, D. E. Henry, G. N. Giss, P. W. Yang, P. R. Griffiths, and C. L. Wilkins, *Anal. Chem.*, **56**, 2002 (1984).

38. P. R. Griffiths, L. V. Azarraga, J. A. de Haseth, R. W. Hannah, R. J. Jakobsen, and M. M. Ennis, *Appl. Spectrosc.*, **33**, 543 (1979).

39. L. V. Azarraga and D. A. Hanna, GIFTS, Athens ERL GC/FT–IR Software and Users' Guide (U.S. EPA/ERL, Athens, GA, 1979).

40. D. A. Hanna, J. C. Marshall, and T. L. Isenhour, *J. Chromatogr. Sci.*, **17**, 434 (1979).

41. G. W. Small, G. T. Rasmussen, and T. L. Isenhour, *Appl. Spectrosc.*, **33**, 444 (1979).

42. K. Krishnan, R. H. Brown, S. L. Hill, S. C. Simonoff, M. L. Olsen, and D. Kuehl, *Am. Lab.*, **13**(3), 122 (1981).

43. S. R. Lowry and D. A. Huppler, *Anal. Chem.*, **53**, 889 (1981).

44. G. Hangac, R. C. Wieboldt, R. B. Lam, and T. L. Isenhour, *Appl. Spectrosc.*, **36**, 40 (1982).

45. L. V. Azarraga, R. R. Williams, and J. A. de Haseth, *Appl. Spectrosc.*, **35**, 466 (1981).

46. J. A. de Haseth and L. V. Azarraga, *Anal. Chem.*, **53**, 2292 (1981).

CHAPTER

19

THE HPLC/FT–IR INTERFACE

I. FLOW CELL TECHNIQUES

Unlike GC/FT–IR, the interface between a high-performance liquid chromatograph and a Fourier transform infrared spectrometer (HPLC/FT–IR) cannot be said to be at an advanced state of development. To understand the reasons why, it is helpful to compare gas chromatography and liquid chromatography, especially from the viewpoint of the nature of the mobile phase and the widths of the peaks.

In gas chromatography, the mobile phase is usually helium or nitrogen, neither of which absorbs infrared radiation. The nature of the mobile phase therefore has no effect on the design or dimensions of the light pipe. As we have seen in Chapter 18, the volume of the light pipe is optimally set equal to or slightly less than the full width at half-height (FWHH) of the narrowest peak in the chromatogram [1]. The values of the length and area of the light pipe are set to yield the optimum combination of maximum absorption due to each peak and minimum noise on the baseline of the spectrum. If the light pipe is too wide and too short, digitization noise may limit the sensitivity, whereas if it is too long, the sensitivity will be limited by reflection losses. Nevertheless the transmittance of the GC mobile phase is always 100%. It should also be noted that reflection losses occurring in the light pipe attenuate the signal by approximately the same factor across the entire spectrum.

In the simplest HPLC/FT–IR interfaces, the effluent from the column is also passed directly through a flow-through cell; several prototype systems were described in the 1970s [2–5]. The actual dimensions of these flow cells are strongly dependent on the size of the chromatographic column and the composition of the mobile phase, so that the calculation of the optimum cell dimensions cannot be made in an analogous fashion to GC/FT–IR systems.

For peaks eluting from a conventional (4.6-mm-i.d.) HPLC column, the typical FWHH of a fairly narrow peak is 250 μL. It is unnecessary to design a cell with a greater diameter than 4 mm, since the beam from an FT–IR spectrometer is easily condensed below this size. The largest cross-sectional area for the cell is therefore 12.5 mm^2. If the mobile phase were transparent to infrared radiation, the optimum pathlength for an HPLC/FT–IR flow cell would be 20 mm (i.e., 250/12.5) if the same criteria developed earlier for GC/FT–IR systems [1] were applied. Because of absorption by the solvent, however, substantially shorter pathlengths are required for HPLC/FT–IR flow cells.

There is no single pathlength suitable for all solvents used in liquid chromatography. The optimum SNR for spectra of solutes present in dilute solution is obtained when the transmittance is e^{-1}, or 36.8% [6], but organic solvents do not, of course, absorb uniformly across the spectrum. It is therefore common to work with pathlengths at which the transmittance of the stronger absorption bands of the solvent is much less than 36.8%. In practice, the pathlength is chosen so that one or two bands "black out" in order that the SNR of solute bands in solvent window regions is high enough to permit them to be observed. If the mobile phase is a weak infrared absorber, such as CCl$_4$, it is possible to work with pathlengths as long as 1 mm, but this is still over an order of magnitude below the optimum value of 20 mm derived above. For solvents with very strong absorption bands, such as water, it is necessary to reduce the pathlength below even 0.1 mm. It is convenient to categorize solvents as being weak, intermediate, or strong infrared absorbers depending on the optimum pathlength for HPLC/FT–IR measurements.

Liquid chromatographic separations may also be divided into three categories according to the *mechanism* of the separation. These mechanisms are as follows:

1. Adsorption, or normal-phase, chromatography: here a silica column is usually employed, although less frequently silicas with polar groups bonded onto the surface are used. The less polar the solvent, the longer the retention time. Paraffins such as hexane are the most common solvents, and an increasing concentration of a more polar solvent, such as ethyl acetate or 2-propanol, may sometimes be programmed in to speed the elution of the more strongly retained components.

2. Reverse-phase chromatography: here the column is packed with silica to which a nonpolar group (typically C$_8$ or C$_{18}$ chains) has been bonded. For columns of this type, the more polar the solute, the less strongly it is retained. Water is the weakest solvent and another less polar solvent, usually methanol or acetonitrile, is often programmed in to speed up the elution of more strongly retained components.

3. Size-exclusion, or gel permeation chromatography (GPC): in GPC, porous packings separate molecules according to their size; the smaller the solute molecule, the more pores in the packing it is able to penetrate

and the later it elutes. The nature of the mobile phase is less critical for GPC than for the other two mechanisms, although it is obvious that the analyte must at least be soluble in the chosen solvent.

In terms of their infrared absorption, the solvents used for normal-phase chromatography have fairly strong absorption bands at several places across the spectrum. A cell with a pathlength of between 0.1 and 0.2 mm usually gives the best solute spectra. This thickness is about two orders of magnitude less than the optimum value of 20 mm derived above. The transmittance of water and H_2O–CH_3OH or H_2O–CH_3CN mixtures used in reverse-phase chromatography is so low below 1700 cm^{-1} that a pathlength no greater than 25 μm is required if the entire spectrum is to be observed. The infrared absorption characteristics of the solvents used in GPC cannot be categorized so easily. Sometimes the sample is only soluble in water, in which case very thin cells would have to be used. More frequently, solvents like toluene or tetrahydrofuran (THF) are required, in which case a 0.2-mm cell would probably be needed. In the best case for GPC/FT–IR measurements, the sample is soluble in dichloromethane, chloroform, or (ideally) carbon tetrachloride. For these solvents, a pathlength of 1 mm can usually be tolerated.

It was noted above that the solvent transmittance required to give the optimum SNR in HPLC/FT–IR measurements is $1/e$, or 36.8% [6]. In an excellent review article, Vidrine [7] has reported the spectra of several solvents plotted on a log absorbance scale and calibrated in units of optimum cell thickness. Several of these are reproduced in Figs. 19.1–19.3. In Fig. 19.1, the spectra of CCl_4, $CHCl_3$, and CH_2Cl_2 are shown; these solvents are ideal for many GPC/FT–IR measurements. The strong band due to the C–Cl stretch is opaque at all useful cell thicknesses, but it can be seen that much of the rest of the spectrum is optimally observed at a thickness of approximately 1 mm, with relatively short regions being lost in the spectra of $CHCl_3$ and CH_2Cl_2. In Fig. 19.2, the spectra of ethyl acetate, THF, and hexane are shown. Except for the region of the aliphatic C–H stretching bands and the short regions around 1450 and 1380 cm^{-1}, a 0.5-mm cell is probably optimal for hexane. However, a much shorter cell should be used for ethyl acetate and THF. Therefore, if ethyl acetate is mixed with hexane to accelerate the elution of more polar solutes being separated by normal-phase HPLC, a thinner cell is required. For THF, not only is the region from 3000 to 3800 cm^{-1} lost but several other spectrometrically important regions are also opaque. Nevertheless, a surprisingly large fraction of the spectrum can be observed even when a cell as thick as 0.2 mm is used, and it is interesting to compare the spectra shown in Fig. 19.5 with the spectrum of THF shown here in order to get

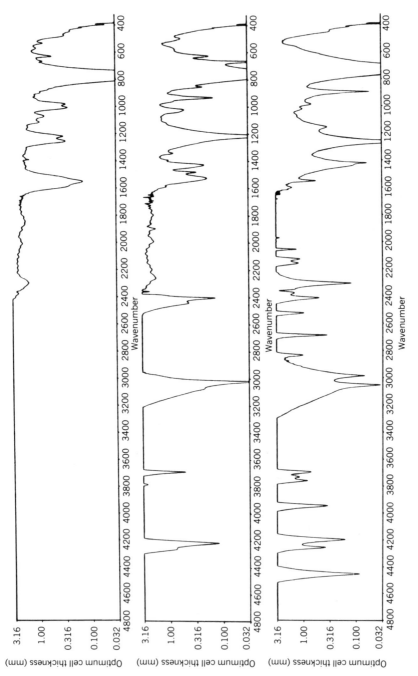

Fig. 19.1. Spectra of (*above*) carbon tetrachloride, (*middle*) chloroform, and (*below*) methylene chloride, plotted with the pathlength required to give a transmittance of $1/e$ as the ordinate. These solvents are representative of the optimum solvents for HPLC/FT–IR using flow cells. (Reproduced from [7], by permission of Academic Press, copyright © 1979.)

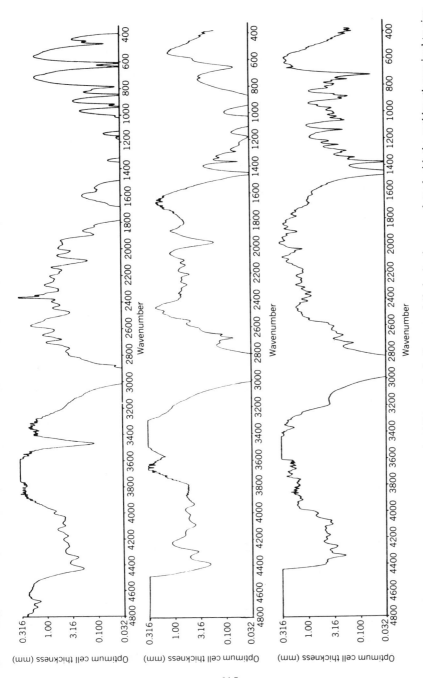

Fig. 19.2. Spectra of (*above*) ethyl acetate, (*middle*), tetrahydrofuran, and (*below*) *n*-hexane, plotted with the pathlength required to give a transmittance of $1/e$ as the ordinate. These solvents are typical of the most commonly used mobile phases for normal-phase HPLC, but which have several opaque regions for HPLC/FT–IR using flow cells. (Reproduced from [7], by permission of Academic Press; copyright © 1979.)

615

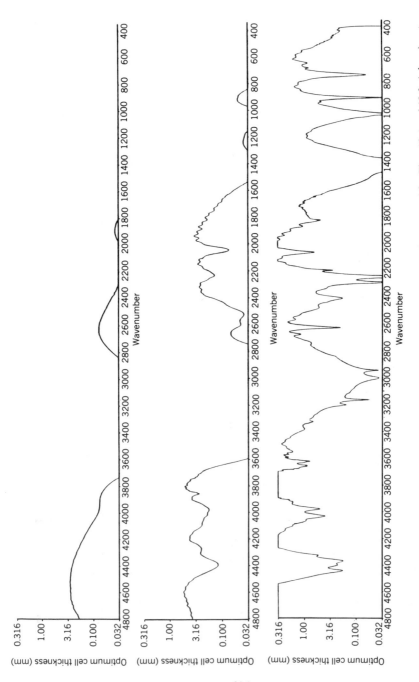

Fig. 19.3. Spectra of typical solvents used for reverse-phase HPLC plotted in the same format used for Figs. 19.1 and 19.2: (*above*) water; (*middle*) methanol; (*below*) acetonitrile. Note that water and methanol are essentially opaque in the fingerprint region of the spectrum, below 1800 cm^{-1} for reasonable pathlengths. (Reproduced from [7], by permission of Academic Press; copyright © 1979.)

a better idea of the potential of HPLC/FT–IR measurements with solvents of intermediate absorptivity such as the ones shown in Fig. 19.2.

The spectra of the three most commonly used solvents for reverse-phase HPLC are shown in Fig. 19.3. Although the spectrum of acetonitrile is a little less strongly absorbing than that of methanol, it is apparent that the presence even of 10 or 20% of water is going to cause reverse-phase HPLC/FT–IR measurements to be extremely difficult, and very narrow cells would be needed. As a result, few studies that fall into this class have been reported.

Assuming that the diameter of the flow cell used need be no greater than 4 mm, the maximum volumes of cells with 1-mm, 200-μm, and 25-μm pathlengths are 12, 2.5, and 0.3 μL, respectively. The FWHH of typical HPLC peaks is rarely less than 250 μL unless microbore columns are used. Therefore, for most HPLC/FT–IR measurements, it is rare that even 1% of any peak is present in the cell at any instant during the measurement. This may be compared with a GC/FT–IR measurement where as much as 50% of each peak may be present in the light pipe at the moment of maximum concentration.

It should also be noted that there is a difference between the effect of changing the pathlength of the cell on GC/FT–IR and HPLC/FT–IR measurements. If the length of a GC/FT–IR light pipe is doubled, its transmittance is reduced fairly uniformly across the spectrum. Provided the SNR is not limited by digitization noise, the signal loss may often be compensated by using a more sensitive detector, so that the sensitivity of the measurement is increased. On the other hand, if the pathlength of an HPLC/FT–IR flow cell is doubled, the transmittance in the region of strong bands may be reduced by as much as a factor of 10 whereas the solute absorbance is only increased by a factor of 2. Choosing a cell that is too thick can therefore be much worse than selecting one that is too thin. It should also be realized that because changing the pathlength of the cell does not cause the signal to change uniformly across the spectrum, the reduction in signal cannot be recovered simply by changing to a detector of increased sensitivity. Most of the energy reaching the detector is usually from the region between 2800 and 1800 cm^{-1} (see, e.g., the spectra in Fig. 19.2) so that an increase in pathlength hardly reduces the energy in this region.

One other area where HPLC/FT–IR spectrometry compares adversely with GC/FT–IR spectrometry concerns gradient elution. Temperature programming is commonly used in gas chromatography to decrease the measurement time and increase the sensitivity by reducing the widths of the more strongly retained components. In liquid chromatography, solvent programming (gradient elution) serves the same purpose, but gradient elution presents a severe problem for HPLC/FT–IR measurements because of the need to use different reference spectra throughout the chromatogram to subtract the solvent bands. Flow programming has been suggested as an alternative to solvent programming [7]; however, flow programming is not as effective as gradient elution in sharpening broad peaks. In fact, flow programming actually increases the width of HPLC peaks if the FWHH is measured in units of volume rather than of time.

In light of the above discussion, we will summarize the present state-of-the-art HPLC/FT–IR systems using flow cells separately for each chromatographic mechanism.

a. Gel Permeation Chromatography

Provided the sample is soluble in a chlorinated solvent, GPC/FT–IR spectrometry can be an extremely powerful technique and most HPLC/FT–IR results reported to date have involved separations based on molecular size. One of the more impressive reports of a GPC/FT–IR measurement involved the use of THF as the mobile phase with a 200-μm-thick flow cell [7]. In this work, 500 μL of a solution containing 0.2% poly(butyl acrylate) and 0.8% polystyrene (i.e., 1 mg of butyl acrylate polymer and 4 mg of polystyrene) was injected on a 100-Å μ-Styragel column using a 1 mL min^{-1} flow rate of THF. The reconstructed chromatogram in four spectral regions is shown in Fig. 19.4 and the spectra of the polymer peaks are reproduced in Fig. 19.5. Because of the poor transmittance characteristics of THF, various regions of the spectrum cannot be observed. These regions are not plotted since all solute features would be obscured by noise. This separation involved milligram quantities of the analyte. Although samples this large can be injected onto wide-bore GPC columns without severely degrading the resolution, most other types of HPLC columns do not have this large a capacity.

From spectra such as those shown in Fig. 19.5, it is obvious that GPC/FT–IR spectrometry will become an important tool for the study of polymers and other large molecules. However, similar techniques are less likely to be productive when applied to normal- and reverse-phase HPLC.

b. Normal-Phase HPLC

The transmittance of most solvents used in normal-phase HPLC is equal to or greater than that of THF at the same pathlength. In spite of this, HPLC/FT–IR spectra as good as those shown in Fig. 19.5 cannot be obtained for normal-phase separations because milligram quantities of solutes cannot be injected onto conventional (4.6-mm-i.d.) silica columns

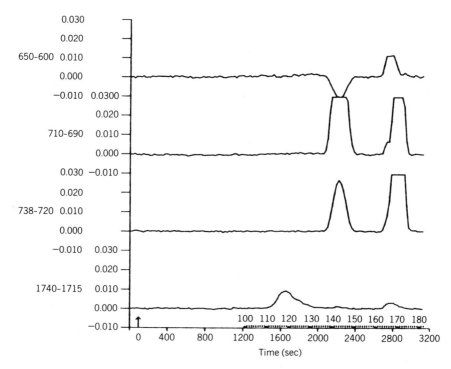

Fig. 19.4. Reconstructed chromatograms (Chemigrams) for the spectral regions 650–600, 710–690, 738–720, and 1740–1715 cm^{-1} for a gel permeation separation of a mixture a 0.2% poly(butyl acrylate) and 0.8% polystyrene in 0.5 mL of THF. The stationary phase was a 100-Å pore size μ-Styragel and the mobile phase was THF at a flow rate of 1.0mL min^{-1}. The cell pathlength was 200 μm. Note that the FWHH of each peak in the chromatogram is several milliliters whereas the cell volume is only 1.5 μL. (Reproduced from [7], by permission of Academic Press; copyright © 1979.)

without greatly exceeding their capacity. The capacity of these columns is typically between 2 and 50 μg. The maximum concentration of most solutes as they emerge from a normal-phase HPLC column is about 20 ppm. Thus, it is extremely difficult to detect an eluate eluting at its maximum allowed concentration, let alone as a minor peak in the chromatogram. There have therefore been few successful normal-phase HPLC/FT–IR measurements with conventional solvents that have been performed using flow cells.

A practical example where normal-phase HPLC/FT–IR spectrometry has been used very successfully was described by Brown et al. [8,9]. This group has developed several spectroscopic interfaces for liquid chromatographic columns for characterizing the very complex mixtures of

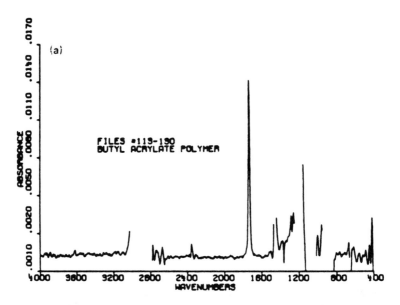

Fig. 19.5. (*a*) Spectrum of poly(butyl acrylate) obtained from files 113–130 in the reconstructed chromatogram shown in Fig. 19.4. (*b*) Spectrum of polystyrene obtained from files 137–149 in the reconstructed chromatogram. Note the good quality of the spectrum in the window regions of the solvent spectrum but the lack of any information in the regions where the solvent is opaque. (Reproduced from [7], by permission of Academic Press; copyright © 1979.)

compounds present in solvent refined coal (SRC). Infrared spectrometry gives information on organic species containing heteroatoms, such as ethers, phenols, ketones, acids, esters, lactones, which are believed to be present in SRC. Initially a single-wavelength infrared detector involving a circularly variable filter (a Foxboro-Wilks Miran 1A) was used for this work [9]. Although this instrument gave a few useful results, the data were still somewhat limited, both because of the low resolution of the filter and its single-wavelength detection. However, the HPLC/FT–IR technique applied subsequently remedied both these disadvantages [8].

In one example cited by Brown et al [8], fractions separated from an SRC using sequential elution selected solvent chromatography (SESC) were separated using a Polar Amino-Cyano (PAC) column. A 50-μL aliquot of one particular fraction (SESC 7) was injected onto the PAC column and eluted with $CDCl_3$ at a flow rate of 1 mL min^{-1} through an FT–IR flow cell. The cell had a pathlength of 1 mm and a diameter of 3.1 mm (i.e., a volume of 23 μL). Reconstructed chromatograms from this fraction of two SRCs are shown in Fig. 19.6, and the spectra at four different retention times for one of these samples are shown in Fig. 19.7. Full details

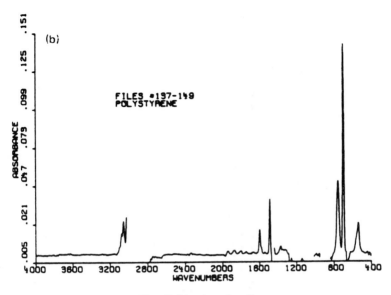

Fig. 19.5.b (*continued*)

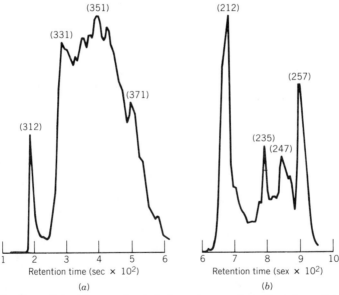

Fig. 19.6. Gram–Schmidt reconstructed chromatograms from the intermediate polarity fractions of two solvent refined coals: (*a*) a hydrogenated process solvent (HPS) and (*b*) a nonhydrogenated process solvent (PS). Each was separated on a 25-cm-long, 4.6-mm-i.d. PAC column using CDCl$_3$ as the mobile phase at a flow rate of 1 mL min^{-1}. Fifty microliters of the sample was injected for each run. (Reproduced from [8], by permission of the American Chemical Society; copyright © 1983.)

621

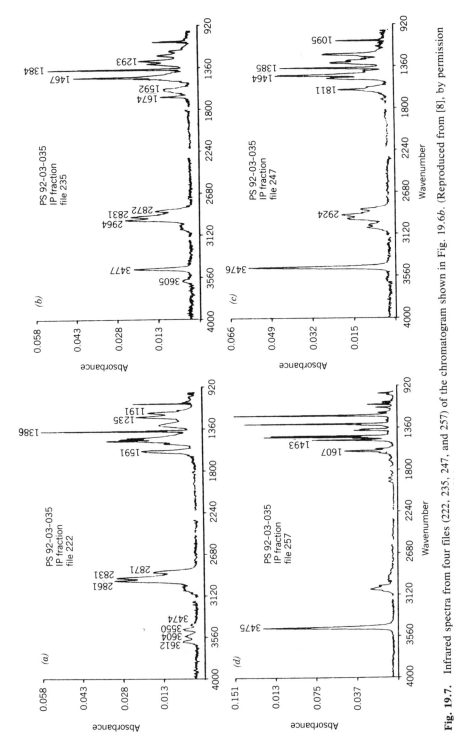

Fig. 19.7. Infrared spectra from four files (222, 235, 247, and 257) of the chromatogram shown in Fig. 19.6b. (Reproduced from [8], by permission of the American Chemical Society; copyright © 1983.)

of the interpretation of these spectra are given in the paper, but it is apparent even from a quick glance at the spectra that their information content is quite high. The particular fraction under study (SESC 7) was of the correct polarity to be separated using $CDCl_3$; however, fractions of higher or lower polarity would require a different solvent. Whereas $CDCl_3$ has good spectral properties for HPLC/FT–IR, most other solvents do not. This work should not therefore be considered as representative of the type of spectral quality and interpretability that can usually be achieved by normal-phase HPLC/FT–IR spectrometry.

c.   Reverse-Phase HPLC

It has been estimated that at least 80% of all HPLC is currently performed using reverse-phase columns [10], so there is a much greater need for identifying unknowns eluting from reverse-phase columns than from any other type of HPLC column. However, since these separations almost invariably involve aqueous solvents, the problems discussed above for normal-phase HPLC/FT–IR measurements are increased by an order of magnitude for reverse-phase HPLC/FT–IR separations. In the opinion of these authors, conventional reverse-phase HPLC/FT–IR measurements using flow cells do not appear to have a promising future.

II.   SOLVENT ELIMINATION TECHNIQUES

a.   Normal-Phase HPLC

From the above discussion, it is easy to see that most of the problems for normal- and reverse-phase HPLC/FT–IR measurements are caused by the presence of the solvent in the flow cell. Several workers have attempted to develop techniques for the automated on-line elimination of the mobile phase. Whereas none of these can presently be claimed to be at a state where commercial production is on the immediate horizon, a discussion of the approaches currently being taken might shed some light on the future development of a high-sensitivity HPLC/FT–IR interface.

The first device described for this purpose [11,12] involved a two-step solvent elimination process, with each solute being deposited on powdered potassium chloride prior to the measurement of its diffuse reflectance spectrum. The effluent from the column was first concentrated by about a factor of 10 by spraying it into a short heated tube, see Fig. 19.8. A signal generated from the ultraviolet detector of the HPLC indicated when a peak was eluting from the column, and a microcomputer calculated

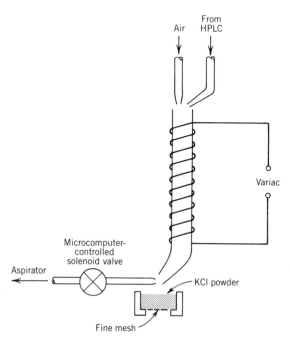

Fig. 19.8. Concentrator for the solvent elimination HPLC/DR–FT–IR device constructed by Kuehl and Griffiths. (Reproduced from [11], by permission of Preston Publishing Co.; copyright © 1979).

the time delay required for each peak to travel from the ultraviolet detector to the end of the concentrator. Only when a solute was present would the effluent from the concentrator be permitted to drop onto the KCl. At all other times the solvent was drawn off by an aspirator and not studied spectrometrically.

Several cups containing powdered KCl were mounted on a carousel. When the solvent containing the first "peak" reached the end of the concentrator tube, a solenoid valve was actuated to permit the liquid to drop onto one of the cups. The base of each cup was fritted to allow air or nitrogen to be drawn through the cup so that any solvent remaining would evaporate. After the first peak had been deposited, the wheel was rotated and the next peak deposited in a different cup. The process was repeated for each peak detected by the ultraviolet detector until the first cup reached the sample position of the spectrometer. In the prototype design of Kuehl and Griffiths [11,12], this occurred when the fifth peak was being deposited. At this time, the diffuse reflectance (DR) spectrum of the solute deposited on the KCl powder was measured. The process

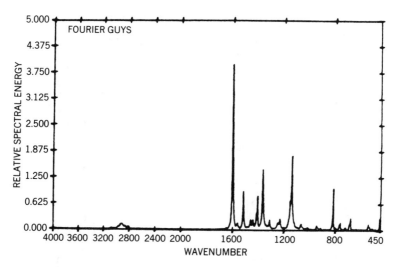

Fig. 19.9. Diffuse reflectance spectrum of 1 μg (injected) of Butter Yellow dye separated from other dyes using a 25-cm-long, 4.6-mm-i.d. column packed with 10-μm silica after elimination of the solvent; the mobile phase was 2% methanol in *n*-hexane. (Reproduced from [11], by permission of Preston Publishing Co.; copyright © 1979.)

was continued automatically until the spectrum of each peak had been obtained.

Submicrogram detection limits were obtained for several compounds separated by normal-phase HPLC. Perhaps the most impressive spectrum measured using this device was from 1 μg of Butter Yellow dye, see Fig. 19.9. This dye had been injected onto a 25-cm-long column packed with 10 μm silica along with several other dyes contained in Stahl's test dye solution. The mobile phase was hexane containing 1% 2-propanol. Chromatographic resolution was retained surprisingly well in view of all the places it could have been degraded [12].

Undoubtedly, the principal advantage of this device is its sensitivity with respect to the flow cell technique. Brown et al. [13] have compared spectra measured using the flow cell technique and the solvent elimination–diffuse reflectance approach for samples of chlorinated pesticides eluting from a silica column. The mobile phase was a mixture of ethyl acetate and isooctane in the ratio 5:1. Figure 19.10 shows the spectra of Endrin measured using a flow cell and the solvent elimination–diffuse reflectance technique. For the flow cell measurement, 400 μg of Endrin was injected, whereas only 10 μg was needed for the diffuse reflectance spectrum. A comparison of the SNR of these spectra indicates a difference of at least two orders of magnitude in the detection limits of these two

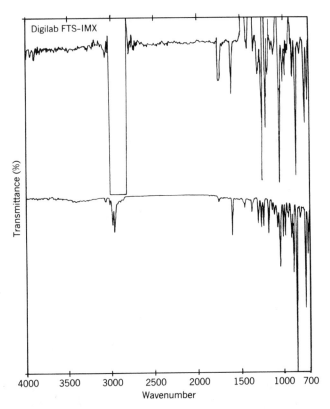

Fig. 19.10. Spectra of Endrin after elution from a 25-cm-long, 4.6-mm-i.d. silica column using a a mobile phase of ethyl acetate (80%) and octane (20%). (*Above*) Solvent-compensated transmittance spectrum for 400 μg injected, measured using a flow cell. (*Below*) Diffuse reflectance spectrum for 10 μg injected after solvent elimination. Note both the improved SNR of the DR spectrum despite the fact that 40 times less sample was injected and the absence of total absorption by the solvent in any spectral region. (Reproduced from [13], by permission of Digilab Division of Bio-Rad Laboratories.)

techniques in spectral regions where the transmittance of the solvent in the flow cell is high. With flow cells, all spectral information is lost in regions where the solvent is totally absorbing.

Brown et al. [13] also noted that when samples are injected in a different solvent than the mobile phase, elutes emerging near the void volume of the column cannot be identified using the flow cell technique because of interference by the solvent. This problem is not encountered in the solvent elimination technique, where all volatile materials are evaporated.

The solvent elimination device described above does, of course, have several disadvantages. It does not give good results with aqueous solvents since they cannot be concentrated quickly enough, because of both the high surface tension and the high latent heat of vaporization of water. Even if concentration could be effected, the solvent would dissolve the KCl in the sample cup on contact. Even for normal-phase separations, where solvent effects are minimized, every peak must have a chromophore absorbing at the wavelength of the ultraviolet detector, otherwise some peaks will be completely missed. Also, the contents of each cup have to be replaced after every run, which can be a very time consuming operation.

There are certain advantages derived from the solvent elimination technique. For example, since each separated solute is collected on KCl, spectra can be remeasured at the end of the chromatogram by averaging many scans if the SNR of the on-line spectrum was not high enough to permit the sample to be identified. Conversely, if the SNR of the spectrum was good enough but the compound still could not be unequivocally identified, the sample can be washed from the KCl for off-line characterization by other spectrometric techniques such as NMR or mass spectrometry.

b. Reverse-Phase HPLC

The most obvious disadvantage with a solvent elimination technique of the type described by Kuehl and Griffiths relates to its unsuitability for HPLC separations effect using aqueous solvents. Two methods of removing water to permit reverse-phase HPLC/FT–IR measurements to be made have been described. In the first, reported initially by Duff et al. [14] and later in more detail by Conroy et al. [15], solutes emerging from the HPLC column are continuously extracted into an immiscible solvent ·(usually dichloromethane). The organic phase is subsequently separated from the aqueous phase and then removed by evaporation with the solute deposited on KCl cups for diffuse reflectance spectrometry in an analogous fashion to the method of Kuehl and Griffiths.

The system described by Conroy et al. [15] is shown schematically in Fig. 19.11. The two phases separate into segments and each solute is distributed between the two phases. Passing this two-phase liquid down a coiled tube speeds the rate at which equilibrium is attained. The liquid is then passed into a PTFE-lined separation tee, where the segments are broken up and converted to an upper (aqueous) and lower (CH_2Cl_2) layer, see Fig. 19.12. The upper layer is aspirated or pumped to waste whereas the lower layer is fed to a concentrator. The solution emerging from the concentrator is dripped onto cups containing powdered KCl, at which point the remaining CH_2Cl_2 is rapidly evaporated. The DR spectrum of each solute can then be measured. One of the disadvantages of this tech-

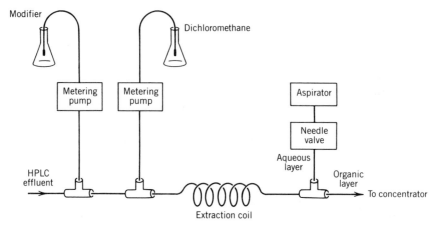

Fig. 19.11. Schematic representation of the HPLC/FT–IR interface in which each component eluting from a reverse-phase column is continuously extracted into dichloromethane and passed to a concentrator before being deposited on KCl powder. The modifier is added to increase the distribution ratio of each component between dichloromethane and the aqueous phase. (Reproduced from [15], by permission of the American Chemical Society; copyright © 1984.)

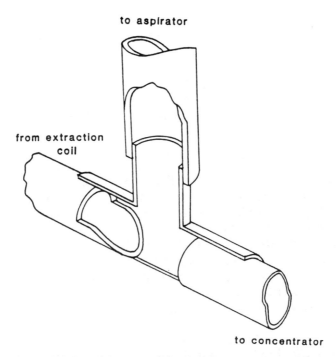

Fig. 19.12. Expanded view of the separation tee used in the continuous extraction HPLC/FT–IR interface shown in Fig. 19.11. (Reproduced from [17].)

nique is that the aqueous and organic phases can be miscible if the concentration of CH_3OH or CH_3CN in the aqueous phase is too high. In this case, it is necessary to add water to the eluent from the column before the addition of dichloromethane.

Submicrogram detection limits were easily achieved provided that the distribution ratio D was favorable. Even in unfavorable cases, it is often possible to increase D by adding water, a dilute strong acid or base, or a buffer solution through the first mixing tee. Using this technique, even mixtures separated by ion-pairing or ion-suppression chromatography have been characterized since the ionic modifier is not extracted into dichloromethane [16,17], see Fig. 19.13.

In the second method, developed by Kalasinsky et al. [18,19], water present in the mobile phase is continuously removed by reaction with 2,2'-dimethoxypropane (DMP):

$$H_2O + CH_3-C(OCH_3)_2-CH_3 \xrightarrow{H^+} CH_3COCH_3 + 2CH_3OH$$

The acetone and methanol formed in this reaction may be removed together with the excess DMP by evaporation, and the DR spectrum of each

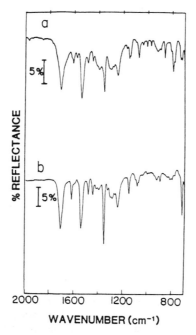

Fig. 19.13. Diffuse reflectance spectra of 1 μg (injected) each of (a) *ortho*-nitrobenzoic acid and (b) *meta*-nitrobenzoic acid separated by ion-suppression HPLC after elution from the continuous extraction interface shown in Fig. 19.11. (Reproduced from [17].)

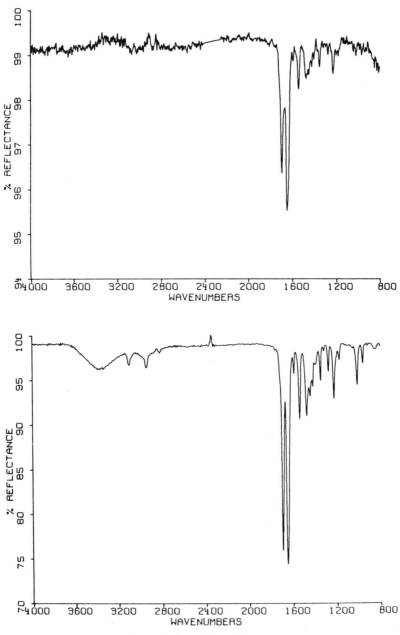

Fig. 19.14. (*Above*) DR spectrum of caffeine after elution from C_{18} column with a 30:70 methanol–water mobile phase. Five micrograms of caffeine was injected, and approximately 500 ng was in the beam for this spectrum. (*Below*) Authentic reference spectrum of 2 μg of caffeine. (Reproduced by permission of Kathryn S. Kalasinsky, Mississippi State Chemical Laboratory.)

eluate is again measured. Although this system is simpler than the extraction technique described above, it is also less versatile. For example, if nonvolatile ionic modifiers have been added to the mobile phase (e.g., for ion-pairing chromatography), they cannot be separated from the analyte, and their absorption features dominate the resulting spectrum. Nevertheless, when the mobile phase is a simple H_2O–CH_3OH or H_2O–CH_3CN mixture, this technique appears to be quite promising, as may be seen from the spectra reproduced in Fig. 19.14.

Recently, a third approach for reverse-phase HPLC/FT–IR measurements has been described. This technique involves aspirating the column effluent directly onto industrial-grade diamond powder. The first report of this approach was by Azarraga [20], who used a simple nebulizer to effect rapid evaporation of the mobile phase while trapping the solutes on the diamond powder. Subsequently, Conroy and Griffiths [21] showed that a thermospray, which had been developed earlier for the HPLC/MS interface [22,23], could also be used efficiently for this purpose. Typical data from the latter group are shown in Fig. 19.15.

Like Kalasinsky's method, the direct aspiration technique is inappropriate when a nonvolatile modifier has been added to the mobile phase.

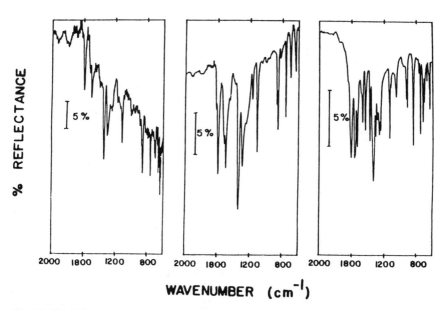

Fig. 19.15. DR spectra of 10 μg (injected) each of (*left*) p-methoxyphenol, (*middle*) p-nitrophenol, and (*right*) 2,4-dinitrophenol deposited on diamond powder using a Thermospray, after separation using a 4.6-mm-i.d., 25-cm-long C_{18} column with a mobile phase of 2% methanol in water. (Reproduced from [17].)

Nonetheless, its simplicity makes it a quite attractive technique for re-verse-phase HPLC/FT–IR measurements effected with H_2O–CH_3OH or H_2O–CH_3CN mobile phases.

III.   MICROBORE HPLC/FT–IR

a.   Flow Cells

It has been suggested in several articles that the use of HPLC columns with a low internal diameter should be beneficial for HPLC/FT–IR measurements [25–27]. The concentration of minor peaks is considerably greater for microbore HPLC (μHPLC) than if the same quantity of material were injected onto a conventional (4.6-mm-i.d.) column. The low flow rates involved with μHPLC columns mean that the use of expensive solvents (especially deuterated materials) is no longer prohibitively expensive, since the total solvent consumption may be less than 1 mL. The spectra of perdeuterated solvents often have "window" regions that co-incide with the characteristic functional group frequencies of nondeuter-ated analytes [27,28]. Freon-113 has also been used as a mobile phase for μHPLC/FT–IR measurements for this reason [29]. Finally it is possible to attain high chromatographic efficiency by coupling several microbore columns together [30].

Brown and Taylor [27] have suggested that the combination of all these beneficial properties make microbore columns much more suitable for HPLC/FT–IR measurements using simple flow cells than conventional columns. They have reported μHPLC/FT–IR measurements of mixtures of phenols and amines separated on a 1-m-long × 1-mm-i.d. polar amino cyano (PAC) column using $CDCl_3$ as the mobile phase. Not only was good chromatographic resolution obtained but detection limits for several of the stronger bands in the solute spectra were less than 1 μg. The recon-structed chromatogram for a synthetic test mixture and the spectrum of 6 μg (injected) of indole are shown in Figs. 19.16 and 19.17, respectively.

Despite the promise of these data, μHPLC/FT–IR measurements using flow cells do have several drawbacks. First, chloroform is not commonly used as a mobile phase, even with PAC columns, so that the chromato-graphic conditions employed by Brown and Taylor are somewhat atypical. Like HPLC/FT–IR measurements with 4.6-mm-i.d. columns, μHPLC/FT–IR measurements using flow cells cannot be made successfully if gradient elution is required for the separation of all components in a given mixture or for reverse-phase chromatography using an aqueous mobile phase. It may also be noted that the efficiency of microbore columns

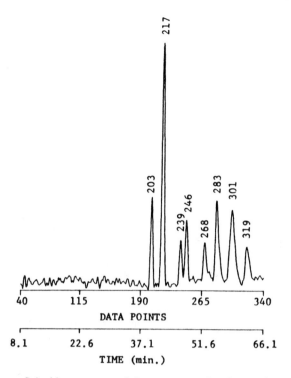

Fig. 19.16. Gram–Schmidt reconstructed chromatogram of a mixture of model amines separated on a 1-m-long, 1-mm-i.d. PAC column using $CDCl_3$ as the mobile phase at a flow rate of 20 μL min^{-1}. Three micrograms of the first two components and 6 μg of the remaining components were injected. (Reproduced from [27], by permission of the American Chemical Society; copyright © 1983.)

packed with a certain stationary phase is approximately the same as that of wide-bore columns of the same length packed with the same stationary phase. However, the capacity of the microbore column is decreased in proportion to the ratio of the cross-sectional areas of the two columns.

The implication of these statements may be seen through the following example. The capacity of a 1-mm-i.d. column of a certain length is 20 [i.e., $(4.6)^2$] times less than that of a 4.6-mm-i.d. column of the same length. It is therefore possible to use a 10-μL sampling loop on the injection valve for conventional HPLC measurements whereas a 0.5-μL loop would have to be used for the microbore column. If the flow rate of the mobile phase through the 4.6-mm-i.d. column is 1 mL min^{-1}, the flow rate for the 1-mm-i.d. column to achieve the same linear velocity would be 50 μL min^{-1}. Under these conditions, the concentrations of the peaks

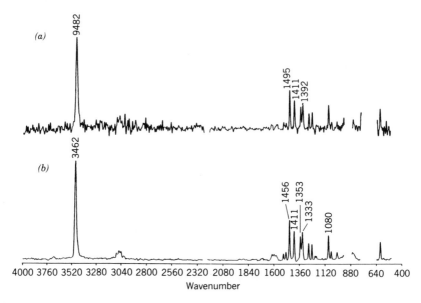

Fig. 19.17. (*a*) Spectrum of the final peak in the chromatogram shown in Fig. 19.16 due to 6 μg of indole. (*b*) Indole reference spectrum also measured using CDCl₃ as the solvent. (Reproduced from [27], by permission of the American Chemical Society; copyright © 1983.)

eluting from each column are the same so that, assuming the pathlengths of the flow cells are identical, the measured HPLC/FT–IR and μHPLC/ FT–IR spectra would have exactly the same SNR. Provided that at least 10 μL of analyte is available, the only advantage to the microbore column is the 20-fold reduction in solvent consumption. The use of deuterated solvents (such as CDCl₃) is therefore more feasible with microbore columns. On the other hand, greater care has to be taken for μHPLC to minimize the dead volume of the tubing, connections, and cell so that the chromatographic resolution is not degraded.

b. Solvent Elimination Techniques

The first successful μHPLC/FT–IR interface in which the solvent was eliminated prior to measuring the spectrum of the solute was the "buffer memory" technique developed by Jinno et al. [24–26,31]. In this device, the outlet from a microbore column and detector is held just above the surface of a KBr plate. The plate is translated at a slow rate, and warm nitrogen is blown across the outlet of the column so that the solvent is rapidly evaporated and each separated eluate is deposited over a small area of the plate. A schematic diagram of the interface is shown in Fig. 19.18. A "memorized" record of the entire chromatogram remains on the plate, which can then be transferred to a spectrometer so that the absorption spectrum of each region may be recorded.

The columns employed by Jinno's group are quite short (typically 10–20 cm in length) and between 0.5 and 0.25 mm in diameter. The efficiency of these columns is good, but because of their small diameter, their capacity is generally less than 100 ng per component for peaks eluting early in the chromatogram. Detection limits for absorption spectra appear to be on the order of a few hundred nanograms (see Fig. 19.19) so that this technique still exhibits the limited dynamic range of measurements involving the use of flow cells.

Conroy et al. [32] have modified the solvent elimination–diffuse reflectance technique of Kuehl and Griffiths [11,12] for use with microbore

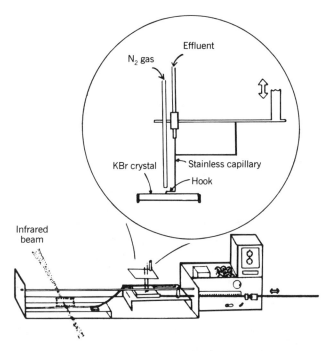

Fig. 19.18. "Buffer memory" μHPLC/FT–IR interface constructed by Jinno et al. The deposition step is shown in greater detail in the inset portion. After deposition, the plate is translated through the infrared beam, and transmission spectra are measured using similar software to that used for GC/FT–IR and flow cell HPLC/FT–IR. (Reproduced from [25], by permission of the Society for Applied Spectroscopy; copyright © 1982.)

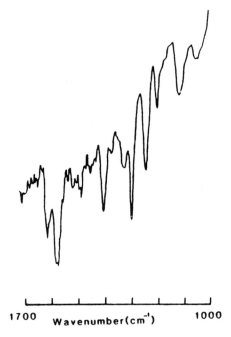

1700   Wavenumber(cm⁻¹)   1000

Fig. 19.19. Spectrum of approximately 3 µg of the di-*n*-amyl ketone adduct of dinitrophenylhydrazine separated by normal-phase chromatography and measured using the buffer memory technique. (Reproduced from [25], by permission of the Society for Applied Spectroscopy; copyright © 1982.)

columns. By reducing the diameter of each sample cup to 2 mm, acceptable spectra were measured from as little as 10 ng of a strong absorber injected into the chromatograph, see Fig. 19.20. For reverse-phase µHPLC/FT–IR detection limits of about 100 ng may be achieved if water is eliminated by reaction with DMP, and even lower limits still appear to be feasible using a thermospray. For separations effected by ion-pairing or ion-suppression chromatography, µHPLC/FT–IR measurements have not been reported, since extraction of each eluate into an immiscible organic solvent does not appear to be readily feasible because of the peak broadening introduced by the separation tee.

In summary, the benefit of reduced detection limits for µHPLC/FT–IR measurements is offset by the decreased capacity of the columns. The largest practical advantage of µHPLC/FT–IR spectrometry is the drastically reduced solvent consumption and the concomitant ease by which the solvent can be eliminated in an environmentally safe manner.

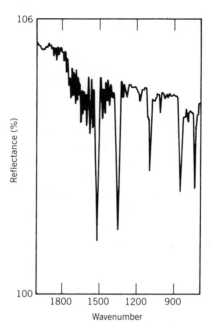

Fig. 19.20. DR spectrum of 10 ng of 4-chloronitrobenzene after elution from a 1-mm-i.d., 25-cm-long silica column using a mobile phase of 2% methanol in hexane at a flow rate of 35 μL min^{-1}. (Reproduced from [17]).

IV. SUPERCRITICAL FLUID CHROMATOGRAPHY

Supercritical fluid chromatography (SFC) bridges the gap between GC and HPLC in many respects. The mobile phase is a fluid that is held above its critical temperature and pressure. The viscosity of supercritical fluids is intermediate between that of gases and liquids, so that the diffusion coefficients of solutes being separated by SFC also fall between the corresponding values for GC and HPLC. Both wall-coated open tubular (WCOT) GC columns and packed HPLC columns have been used for SFC. WCOT columns of less than 100 μm i.d. afford the highest resolution, but HPLC columns usually have considerably higher capacity.

From an applications standpoint, SFC should be most useful for the separation of relatively high molecular weight materials. The components may be neither volatile nor thermally stable enough to be studied by GC, and the resolution of conventional packed HPLC columns is often insufficient to effect the separation of chemically similar materials of high molecular weight. Whereas SFC cannot be claimed to have been universally accepted by chromatographers at the time of this writing, it is already certain that it will ultimately assume an important role in the separation scientist's arsenal. Methods for the on-line identification of species separated by SFC are therefore important, and already a few SFC/FT–IR measurements have been reported.

The first of these was by Shafer and Griffiths [33], who separated a simple three-component mixture of substituted aromatics on a fused silica WCOT column using supercritical CO_2 as the mobile phase and measured the spectrum of each component using a 1-cm-pathlength flow cell. Carbon dioxide is remarkably transparent to infrared radiation near the critical point, but as the pressure is increased, certain symmetry-forbidden bands become more allowed, see Fig. 19.21. Apparently, the CO_2 molecules become slightly bent when subjected to high pressures. This result is significant because pressure programming is often used for SFC in a similar manner to temperature programming in GC and gradient elution (solvent programming) in HPLC, namely, to reduce the retention times of strongly retained peaks. It can therefore be seen that the scaled addition of CO_2 reference spectra measured at different pressures may be necessary if absorption bands due to the mobile phase are to be accurately compensated throughout a pressure-programmed chromatogram.

Shafer and Griffiths used a mobile phase of pure CO_2 at constant pressure in their work and were able to realize detection limits of approximately 1 μg (injected) for each component, see Fig. 19.22. The resolution of the 300-μm-i.d. WCOT column used in this work was poor even though its capacity was quite high. To obtain high chromatographic resolution in SFC, it has been shown that the internal diameter of fused silica columns should be less than 100 μm. However, the capacity of such narrow columns is too low to permit good SFC/FT–IR measurements to be achieved using a flow cell without overloading the column.

Recently, Olesik et al. [34] have reported SFC/FT–IR measurements for which a flow cell of considerably improved design was employed. An exploded view of this cell is shown in Fig. 19.23. These workers separated a test mixture of five aromatic compounds using a 200-μm-i.d. WCOT column with a stationary phase of 1 μm thickness. They noted that the maximum sample loading for GC/FT–IR measurements for this column is about 100 ng per component, so that the dynamic range problem noted above for HPLC/FT–IR measurements should still represent a problem for SFC/FT–IR systems. Nevertheless, they also mention that since the solute diffusivities in the mobile and stationary phases are closer, the capacity of WCOT columns used for SFC should increase relative to GC.

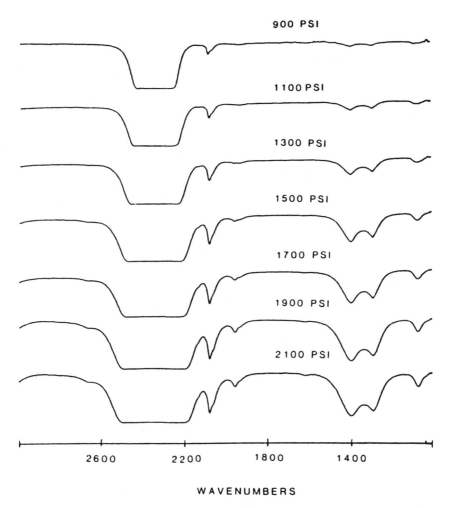

Fig. 19.21. Transmittance spectra of supercritical CO_2 in a 1-cm-pathlength flow cell at 40°C and pressures between 900 and 2100 psi. The intensification of several bands in the spectrum, especially those around 1350 cm^{-1}, should be noticed. (Reproduced from [33], by permission of the American Chemical Society; copyright © 1983.)

Preliminary data using this technique were interesting. The Gram–Schmidt real-time chromatogram of the test mixture separated using pure CO_2 under a pressure program showed a large tilt (see Fig. 19.24), as would be expected because of the increase in absorbance of the forbidden bands of CO_2 with time. Nevertheless, the detection limits of the stronger bands in the spectra of several of the analytes were less than 100 ng. A representative spectrum of 2 μg (injected) of benzaldehyde, a strong infrared absorber, is shown in Fig. 19.25.

All early SFC/FT–IR work involved the use of pure CO_2 as the mobile phase, so that very long pathlength flow cells could be used. It is often found, however, that it is necessary to add polar modifiers such as methanol to the CO_2 to reduce retention times for strongly retained components

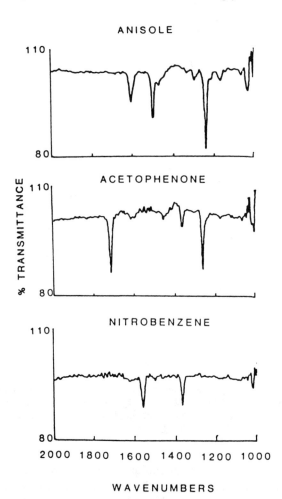

Fig. 19.22. Transmittance spectra of anisole, acetophenone, and nitrobenzene after separation on a 300-μm-i.d. WCOT column using supercritical CO_2 as the mobile phase; 3 μg of each component was injected. (Reproduced from [33], by permission of the American Chemical Society; copyright © 1983.)

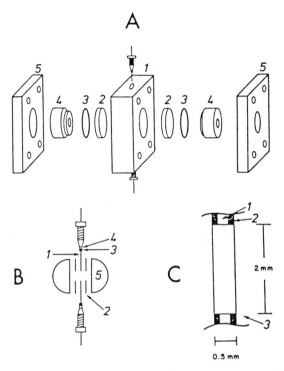

Fig. 19.23. (*A*) Schematic diagram of high-pressure flow cell constructed by Olesik et al. for SFC using WCOT columns: 1, cell body; 2, window; 3, O-ring; 4, support ring; 5, face plates. (*B*) Exploded view of components which define cell volume: 1, capillary column; 2, indium gasketing; 3, indium washers; 4, PTFE tubing; 5, lead spacers. (*C*) Cross-sectional view of detection region: 1, capillary column; 2, indium washers; 3, lead spacer (Reproduced from [34], by permission of Pergamon Press; copyright © 1984.)

when HPLC-type columns are used for the separation. Even if only 1% CH_3OH is added to CO_2, very strong bands are observed when the mixture is passed into a 1-cm-pathlength cell. For strongly retained components, it may even be necessary to use a polar compound as the major component of the mobile phase (e.g., one of the volatile Freons or pure ammonia). In this case, if a simple flow-cell were to be employed for SFC/FT–IR measurements, a very narrow pathlength cell would be needed to prevent the total obscuration of much of the infrared spectrum. When any polar material is included in the mobile phase, it is not difficult to recognize that the sensitivity of SFC/FT–IR measurements made using flow-through cells is limited by the same factors that control HPLC/FT–IR measurements using flow cells; these factors were discussed in Section I of this chapter.

To avoid problems of this sort, Shafer et al. [35,36] have investigated the possibility of eliminating the mobile phase in an analogous fashion to the methods described in Section II of this chapter. In practice, since many of the materials used as mobile phases for SFC are quite volatile, they may be eliminated rather easily. Preliminary results are promising. For example, it is possible to deposit a solute eluting from a packed microbore silica column (0.5-mm i.d.) using 2% CH_3OH in CO_2 as the mobile phase in a spot of less than 0.5 mm in diameter. By depositing each eluate over such a small area, the concentration is increased and hence the band intensities from a given quantity of sample are also increased.

One of the other advantages of SFC/FT–IR over HPLC/FT–IR measurements apparent from this work is that in SFC/FT–IR spectrometry the eluates can be continuously deposited onto a very narrow (0.5-mm-wide) strip of KCl powder with minimal spreading, and hence without seriously degrading the chromatographic resolution. In solvent elimination HPLC/FT–IR measurements, each drop emerging from the concentrator (or directly from the column for μHPLC) has to be deposited on

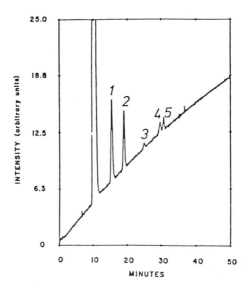

Fig. 19.24. Gram–Schmidt reconstructed chromatogram of test mixture separated on a 200-μm-i.d. WCOT column with 1-μm film thickness using supercritical CO_2 under a pressure program as the mobile phase. The baseline slope is caused by the change in the absorption spectrum of supercritical CO_2 with pressure. Two micrograms each of the following components were injected: 1, benzaldehyde; 2, o-chlorobenzaldehyde; 3, 2,6-di-t-butylphenol; 4, 2-naphthol; 5, benzophenone. (Reproduced from [34], by permission of Pergamon Press; copyright © 1984.)

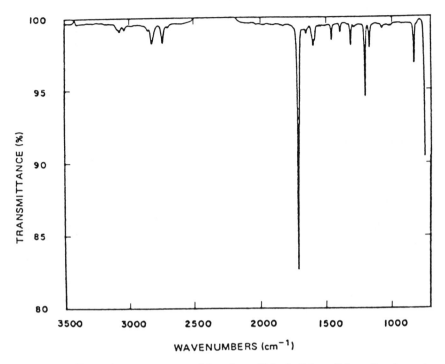

Fig. 19.25.   Transmittance spectrum of 2 μg benzaldehyde (injected) from the chromatogram shown in Fig. 19.24. (Reproduced from [34] by permission of Pergamon Press; copyright © 1984.)

a separate sample cup to prevent the solute from spreading over a wide area because of capillarity. In their second paper on SFC/FT-IR, Shafer et al. [36] showed that it is possible to move a strip of KCl powder, prepared in a fashion somewhat analogous to a TLC plate, continuously under the restrictor of a supercritical fluid chromatograph and thence to the focus of a diffuse reflectance accessory.

To illustrate the potential of this technique, a mixture of five quinones was separated on a 1-mm-i.d., 25-cm-long column packed with unmodified silica using a mobile phase of 5% CH_3OH in CO_2. Identifiable spectra of sample quantities less than 200 ng (injected) could be obtained on-line, and lower quantities yielded recognizable spectra if the strip was held stationary and interferograms were signal-averaged. Figure 19.26 shows the spectrum of 50 ng of acenaphthenequinone measured in this way. The key factor in this work was the fact that the mobile phase evaporated so rapidly that the effects of capillarity were minimal and a continuous record of the chromatogram could be obtained. The chromatogram measured by

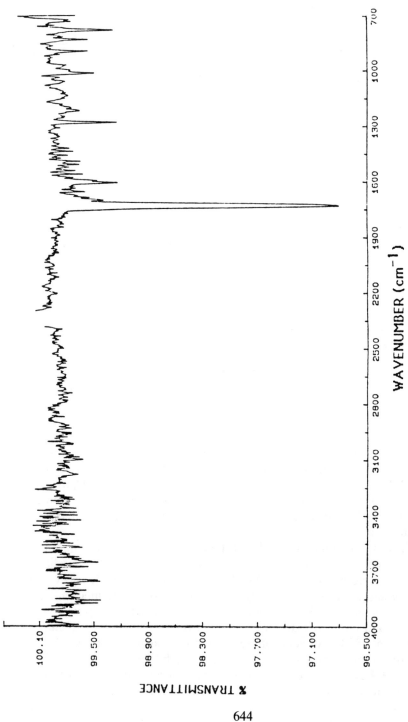

Fig. 19.26. DR spectrum of 50 ng of acenaphthenequinone after separation from a mixture of other quinones by SFC using a 1-mm-i.d., 25-cm-long packed silica column with 5% CH_3OH in CO_2 as the mobile phase. The sample was deposited on a moving KCl strip and moved back into the IR beam at the end of the chromatogram. 65 scans were signal-averaged. (Reproduced from [36] by permission of the American Chemical Society; copyright © 1986.)

an ultraviolet detector (UVD) is shown in Fig. 19.27 together with the chromatogram reconstructed from the infrared absorption spectrum between 1400 and 1200 cm^{-1}. Only a slight degradation of the chromatographic resolution may be observed and the artifacts in the UVD trace are completely eliminated.

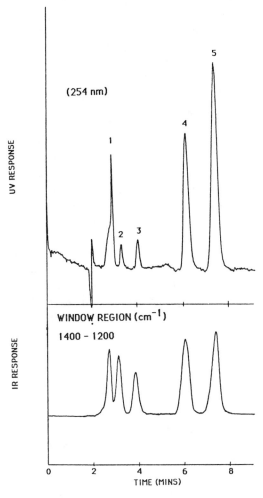

Fig. 19.27. (A) Chromatogram of a mixture of quinones separated by SFC under the same conditions described in the caption for Fig. 19.26, measured using an ultraviolet detector. (B) Chromatogram reconstructed from the measured DR infrared spectra of the deposited quinones. The maximum injected quantity of any of these components was 2 μg. (Reproduced from [36] by permission of the American Chemical Society; copyright © 1986.)

The sensitivity demonstrated by these preliminary experiments is sufficiently high that the technique may be expected to be applicable to SFC separations of nonvolatile compounds. In view of the low capacity of WCOT columns, however, on-line identification techniques must have detection limits of less than 100 ng of each injected component if they are to be useful. Thus the techniques described above appear to have the potential to form the basis for an excellent method for SFC/FT–IR measurements. Indeed, if a means of rapidly evaporating liquid mobile phases can be developed, it might even be possible to construct an interface that is equally applicable to SFC/FT–IR and HPLC/FT–IR spectrometry.

V. CONCLUSIONS

Like many other results shown throughout this book, the data reported in this chapter cannot be considered to be the "last word." Indeed, all SFC/FT–IR studies should be considered as very preliminary and undoubtedly much more exciting work will be described in the future.

The recent improvements in HPLC/FT–IR and SFC/FT–IR spectrometry typify the history of Fourier transform infrared spectrometry over the past decade. Between 1974 and 1984, the number of major manufacturers of FT–IR spectrometers in North America has increased from 3 to 10. Significant competition is appearing from Japanese and European manufacturers. The number of papers describing work in which an FT–IR spectrometer was used is rising at an almost exponential rate. We hope that this book not only has given a good indication of the status of many areas of analytical chemistry where FT–IR spectrometry can be used profitably, but also provides a clue as to the future directions of this exciting field.

REFERENCES

1. P. R. Griffiths, *Appl. Spectrosc.,* **31,** 284 (1977).
2. K. L. Kizer, A. W. Mantz, and L. C. Bonar, *Amer. Lab.,* 7(5), 85 (1975).
3. D. W. Vidrine and D. R. Mattson, *Appl. Spectrosc.,* **32,** 502 (1978).
4. D. W. Vidrine, *J. Chromatogr. Sci.,* **17,** 477 (1979).
5. K. H. Shafer, S. V. Lucas, and R. J. Jakobsen, *J. Chromatogr. Sci.,* **17,** 464 (1979).
6. See, for example, G. D. Christian, *Analytical Chemistry,* 3rd ed., p. 400, Wiley, New York (1980).
7. D. W. Vidrine, Chapter 4 in *Fourier Transform Infrared Spectroscopy: Applications to Chemical Systems,* Vol. 2 (J. R. Ferraro and L. J. Basile, eds.), Academic Press, New York (1979).

8. R. S. Brown and L. T. Taylor, *Anal. Chem.*, **55**, 723 (1983).

9. R. S. Brown, D. W. Hausler, L. T. Taylor, and R. C. Carter, *Anal. Chem.*, **53**, 197 (1981).

10. J. J. Kirkland, E. I. du Pont de Nemours & Co., personal communication (1981).

11. D. Kuehl and P. R. Griffiths, *J. Chromatogr. Sci.*, **17**, 471 (1979).

12. D. T. Kuehl and P. R. Griffiths, *Anal. Chem.*, **52**, 1394 (1980).

13. R. H. Brown, J. Knecht, and H. Witek, *Proc. Soc. Photo-Opt. Instrum. Eng.*, **289**, 51 (1981).

14. P. J. Duff, C. M. Conroy, P. R. Griffiths, B. L. Karger, P. Vouros, and D. P. Kirby, *Proc. Soc. Photo-Opt. Instrum. Eng.*, **289**, 53 (1981).

15. C. M. Conroy, P. J. Duff, P. R. Griffiths, and L. V. Azarraga, *Anal. Chem.*, **56**, 2636 (1984).

16. P. J. Duff, Ph.D. dissertation, Ohio University, Athens, OH (1984).

17. C. M. Conroy, Ph.D. dissertation, University of California, Riverside (1984).

18. K. S. Kalasinsky, J. T. McDonald, and V. F. Kalasinsky, Paper No. 357, Pittsburgh Conf. on Anal. Chem. and Appl. Spectrosc., Atlantic City, NJ (1983).

19. K. S. Kalasinsky, J. A. Smooter Smith, and V. F. Kalasinsky, Paper No. 659, Pittsburgh Conf. on Anal. Chem. and App. Spectrosc., Atlantic City, NJ (1984).

20. L. V. Azarraga, paper presented at Eastern Analytical Symposium, New York, NY (1983).

21. P. R. Griffiths and C. M. Conroy, *Adv. Chromatogr.*, in press (1986).

22. C. R. Blakley, M. I. McAdams, and M. L. Vestal, *J. Chromatogr.*, **158**, 261 (1978).

23. C. R. Blakley and M. L. Vestal, *Anal. Chem.*, **55**, 750 (1983).

24. K. Jinno and C. Fujimoto, *J. High Res. Chromatogr. Chromatogr. Commun.*, **4**, 532 (1981).

25. K. Jinno, C. Fujimoto, and Y. Hirata, *Appl. Spectrosc.*, **36**, 67 (1982).

26. K. Jinno, C. Fujimoto, and D. Ishii, *J. Chromatogr.*, **239**, 625 (1982).

27. R. S. Brown and L. T. Taylor, *Anal. Chem.*, **55**, 1492 (1983).

28. K. Jinno, C. Fujimoto, and G. Vematsu, *Amer. Lab.*, **16**(2), 39 (1984).

29. C. C. Johnson and L. T. Taylor, *Anal. Chem.*, **55**, 436 (1983).

30. R. P. W. Scott and P. Kucera, *J. Chromatogr.*, **169**, 51 (1979).

31. K. Jinno, *Spectrosc. Lett.*, **14**, 659 (1981).

32. C. M. Conroy, P. R. Griffiths, and K. Jinno, *Anal. Chem.*, **57**, 822 (1985).

33. K. H. Shafer and P. R. Griffiths, *Anal. Chem.*, **55**, 1939 (1983).

34. S. V. Olesik, S. B. French, and M. Novotny, *Chromatographia*, **18**, 489 (1984).

35. K. H. Shafer, S. L. Pentoney, and P. R. Griffiths, *J. High Res. Chromatogr. Chromatogr. Commun.*, **7**, 707 (1984).

36. K. H. Shafer, S. L. Pentoney and P. R. Griffiths, *Anal. Chem.*, in press (1986).

INDEX